Lecture Notes in Computer Science   10359

Commenced Publication in 1973
Founding and Former Series Editors:
Gerhard Goos, Juris Hartmanis, and Jan van Leeuwen

Editorial Board

Giovanni Livraga · Sencun Zhu (Eds.)

Data and Applications Security and Privacy XXXI

31st Annual IFIP WG 11.3 Conference, DBSec 2017
Philadelphia, PA, USA, July 19–21, 2017
Proceedings

 Springer

Editors
Giovanni Livraga (iD)
Università degli Studi di Milano
Crema
Italy

Sencun Zhu
Pennsylvania State University
Philadelphia, PA
USA

ISSN 0302-9743   ISSN 1611-3349   (electronic)
Lecture Notes in Computer Science
ISBN 978-3-319-61175-4   ISBN 978-3-319-61176-1   (eBook)
DOI 10.1007/978-3-319-61176-1

Library of Congress Control Number: 2017943855

LNCS Sublibrary: SL3 – Information Systems and Applications, incl. Internet/Web, and HCI

Printed on acid-free paper

This Springer imprint is published by Springer Nature
The registered company is Springer International Publishing AG
The registered company address is: Gewerbestrasse 11, 6330 Cham, Switzerland

Preface

This volume contains the papers selected for presentation at the 31st Annual IFIP WG 11.3 Conference on Data and Applications Security and Privacy (DBSec 2017), held in Philadelphia, PA, USA, on July 19–21, 2017.

In response to the call for papers of this edition, 59 submissions were received, and all submissions were evaluated on the basis of their significance, novelty, and technical quality. The Program Committee, comprising 45 members, performed an excellent task and with the help of additional reviewers all submissions went through a careful anonymous review process (three or more reviews per submission). The Program Committee's work was carried out electronically, yielding intensive discussions. Of the submitted papers, 21 full papers and nine short papers were selected for presentation at the conference.

The success of DBSec 2017 depends on the volunteering effort of many individuals, and there is a long list of people who deserve special thanks. We would like to thank all the members of the Program Committee and all the external reviewers, for all their hard work in evaluating the papers and for their active participation in the discussion and selection process. We are very grateful to all people who gave their assistance and ensured a smooth organization process, in particular Krishna Kant and Peng Liu for their efforts as DBSec 2017 General Chairs; Sabrina De Capitani di Vimercati (IFIP WG11.3 Chair) for her guidance and support; and Fengjun Li (Publicity Chair) for helping with publicity. A special thanks goes to the keynote speaker, who accepted our invitation to deliver a keynote talk at the conference.

Last but certainly not least, thanks to all the authors who submitted papers and all the conference attendees. We hope you find the proceedings of DBSec 2017 interesting, stimulating, and inspiring for your future research.

July 2017

Giovanni Livraga
Sencun Zhu

Organization

IFIP WG 11.3 Chair

Sabrina De Capitani   Università degli Studi di Milano, Italy
  di Vimercati

General Chairs

Krishna Kant   Temple University, USA
Peng Liu   Pennsylvania State University, USA

Program Chairs

Giovanni Livraga   Università degli Studi di Milano, Italy
Sencun Zhu   Pennsylvania State University, USA

Publicity Chair

Fengjun Li   The University of Kansas, USA

Program Committee

Alessandro Armando	FBK and Università di Genova, Italy
Vijay Atluri	Rutgers University, USA
Marina Blanton	SUNY Buffalo, USA
Soon Ae Chun	CUNY, USA
Frédéric Cuppens	Telecom Bretagne, France
Nora Cuppens-Boulahia	Telecom Bretagne, France
Sabrina De Capitani di Vimercati	Università degli Studi di Milano, Italy
Josep Domingo-Ferrer	Universitat Rovira i Virgili, Spain
Carmen Fernandez-Gago	University of Málaga, Spain
Simon Foley	Telecom Bretagne, France
Sara Foresti	Università degli Studi di Milano, Italy
Joaquin Garcia-Alfaro	Telecom SudParis, France
William Garrison	University of Pittsburgh, USA
Stefanos Gritzalis	University of the Aegean, Greece
Ehud Gudes	Ben-Gurion University, Israel
Yuan Hong	University at Albany, USA
Sushil Jajodia	George Mason University, USA
Sokratis Katsikas	Giøvik University College, Norway
Florian Kerschbaum	University of Waterloo, Canada

Yingjiu Li	Singapore Management University, Singapore
Javier Lopez	University of Málaga, Spain
Fabio Martinelli	IIT-CNR, Italy
Catherine Meadows	NRL, USA
Aziz Mohaisen	SUNY Buffalo, USA
Martin Olivier	University of Pretoria, South Africa
Stefano Paraboschi	Università degli Studi di Bergamo, Italy
Günther Pernul	Universität Regensburg, Germany
Silvio Ranise	FBK Security and Trust Unit, Italy
Indrajit Ray	Colorado State University, USA
Indrakshi Ray	Colorado State University, USA
Pierangela Samarati	Università degli Studi di Milano, Italy
Ravi Sandhu	University of Texas at San Antonio, USA
Andreas Schaad	WIBU-SYSTEMS AG, Germany
Scott Stoller	Stony Brook University, USA
Tamir Tassa	The Open University of Israel, Israel
Mahesh Tripunitara	University of Waterloo, Canada
Jaideep Vaidya	Rutgers University, USA
Cong Wang	City University of Hong Kong, Hong Kong, SAR China
Lingyu Wang	Concordia University, Canada
Edgar Weippl	Vienna University of Technology, Austria
Yi Yang	Fontbonne University, USA
Meng Yu	University of Texas at San Antonio, USA
ShengZhi Zhang	Florida Institute of Technology, USA
Yuqing Zhang	Chinese Academy of Sciences, China
Quanyan Zhu	New York University, USA

Additional Reviewers

Isaac Agudo	Rudolf Mayer
Hafiz Asif	Alessio Merlo
Anis Bkakria	Georg Merzdovnik
Daniel Borbor	Meisam Mohamady
Juntao Chen	David Nuñez
Luis Del Vasto	Javier Parra
Philip Derbeko	Alexander Puchta
Sebastian Groll	Stefan Rass
Panagiotis Kintis	Johannes Sänger
Michael Kunz	Ankit Shah
Giovanni Lagorio	Jordi Soria-Comas
Costas Lambrinoudakis	Tanay Talukdar
Suryadipta Majumdar	Iman Vakilinia
Sergio Martínez	Andrea Valenza

Contents

Cloud Security

Secure Storage in the Cloud

Secure Systems

Security in Networks and Web

Access Control

Cryptographically Enforced Role-Based Access Control for NoSQL Distributed Databases

Yossif Shalabi and Ehud Gudes[✉]

Ben-Gurion University, 84105 Beer-Sheva, Israel
shalabiyossif@gmail.com, ehud@cs.bgu.ac.il

Abstract. The support for Role-Based Access Control (RBAC) using cryptography for NOSQL distributed databases is investigated. Cassandra is a NoSQL DBMS that efficiently supports very large databases, but provides rather simple security measures (an agent having physical access to a Cassandra cluster is usually assumed to have access to all data therein). Support for RBAC had been added almost as an afterthought, with the Node Coordinator having to mediate all requests to read and write data, in order to ensure that only the requests allowed by the Access Control Policy (ACP) are allowed through.

In this paper, we propose a model and protocols for cryptographic enforcement of an ACP in a cassandra like system, which would ease the load on the Node Coordinator, thereby taking the bottleneck out of the existing security implementation. We allow any client to read the data from any storage node(s) – provided that only the clients whom the ACP grants access to a datum, would hold the encryption keys that enable these clients to decrypt the data.

1 Introduction

Security has been a notable weakness in almost every NoSQL database, a fact that was highlighted in a 2012 InformationWeek special report entitled *"Why NoSQL Equals No Security."* [2] Since the inception of NoSQL databases, their whole point was seen as guaranteeing rapid, unfettered access to big data. Thus, naturally, enforcing access control was seen by the Big Data community mainly as a hindrance in the way of fast data access: At the same time, NoSQL databases are now used by big financial institutions, healthcare companies, government services, and even by millitary intelligence – so, these systems must be able to handle sensitive data, providing the necessary security guarantees. *"And yet, the NoSQL ecosystem is woefully behind in incorporating even basic security,"* [2] This shortcoming affects all aspects of data security: users authentication, access control, transport-level security of inter-node communication, etc., so, an ever-growing deployment of NoSQL systems can subject them to many attacks which are likely to catch these systems entirely unprepared.

Recognizing this situation, a major focus on security had been put in Cassandra since version 3.0, even though the security subsystem is not enabled by

© IFIP International Federation for Information Processing 2017
Published by Springer International Publishing AG 2017. All Rights Reserved
G. Livraga and S. Zhu (Eds.): DBSec 2017, LNCS 10359, pp. 3–19, 2017.
DOI: 10.1007/978-3-319-61176-1_1

default. Cassandra's built-in authorization module does not use encryption, and instead enforces the Access Control Policy (ACP) by relying on *security monitors*, i.e. privileged components that handle a client's requests for access [1,14]. At the same time, Cassandra 3.0 introduced data encryption, applicable to whole tables. The encryption keys are stored on the server, and are not directly related to an ACP; the same keys may be used for all encrypted tables, irrespective of the access permissions for these tables. Data encryption in Cassandra is designed to protect the data from an attacker bypassing the system's security monitor, e.g. by getting a root access to one of the nodes, or by physically stealing the disks from one of the nodes; the assumption is that the server would only decrypt data as part of a granted access. None of the existing modules for Cassandra, however, implement *cryptographic access control*, which would allow all read requests unconditionally, and use the centralized security monitors to mediate only the write requests. The benefit of such access control system for a distributed database is to avoid the bottleneck of the security monitor when most requests are to read data – which is indeed the case for most NoSQL deployments.

In this paper we present a model and protocols for enforcing cryptographically based RBAC using Cassandra. Cassandra was chosen as the platform for our proof-of-concept implementation owing to the following reasons:

- It is a popular NoSQL database. It is used by many companies, including Facebook and Twitter, and recently Cassandra is also being used by Cisco and Platform64 for personalized television streaming.
- It is open-source and well-documented; – although, as a mature industrial-quality DBMS, its code is very sophisticated and not easy to modify;
- "Out of the box", it includes at least the basic support for RBAC, whereas less mature open-source NoSQL DBMSs, such as Druid or Voldemort, which are considerably simpler and therefore easier to modify, don't include any support for any sort of access control, so that implementing RBAC in these DBMSs would require a major redesign of their core.

In this paper we suggest to enforce cryptography based RBAC using Predicate encryption [9] and adapt it especially to a distributed architecture like that of NoSQL Cassandra. The main contribution of the paper lies in the novelty of the protocol and the detailed description of its proposed implementation.

The rest of this paper is organized as follows. Section 2 provides a more detailed background on cryptographic access control and the various access control systems suggested for cloud storage, including a survey of the related work. Section 3 discusses the proposed protocol and its implementation. Section 4 concludes the paper and outlines future research.

2   Background and Related Work

Whereas the classic schemes for private-key and public-key encryption were concerned with privacy of *transmission*, with a predefined recipient or set of recipients that must hold the relevant encryption key to be able to decrypt the message, – *attribute-based encryption* (ABE) [6] extends it to privacy of *data storage*, where different users may have access to different subsets of the stored data. It would be inefficient to store multiple copies of the data, one copy for each user that must have access to the data, each copy encrypted with its user's personal key; furthermore, it would be inconvenient for each user to keep a separate key for each file that he's authorized to access. Instead, ABE operates on a set of *attributes* held by the users; each datum would be encrypted only once, with the encryption key derived from the ACP, expressed as a conjunction of attribute equalities or attribute ranges that a user must satisfy to be able to decrypt the data. *Predicate encryption* (PE) [9] generalizes ABE by allowing policy expressions consisting of conjunctions, disjunctions, and more complicated equations on the users' attributes. A user can decrypt data only if the data's *access predicate,* evaluated on the user's attribute(s), is logically true.

An alternative approach [7] based on *symmetric-key encryption* (e.g. DES), is to store each user's set of keys, known as the user's *key-chain,* in the file system, encrypted with the user's *master key.* To access a file, the user will have to first read his key-chain file, and to decrypt the relevant data key with his master key. Such a scheme is, in fact, used in most modern web browsers and personal operating systems. Another option for a symmetric-key-based system is to store the keys corresponding to each file together with the file, in a *keys record;* once again, each data key is encrypted with the corresponding user's master key. The advantage of this option over the first one is that when deleting a file, all corresponding keys are deleted together with the file.

The conventional cloud systems impose certain trade-offs [5]: the users can either get advanced functionality, but will have to trust the *cloud service provider* (CSP), giving it unrestricted access to the data; or, conversely, they can use advanced security systems, withholding the encryption keys from the CSP, – but will then get limited functionality and/or performance, as the CSP cannot perform any local data processing. To help strike a balance between privacy and performance, one may use ABE, and split the ACP into two layers: the *inner encryption layer* (IEL), with keys unknown to the CSP, ensures that the data remain protected from the CSP; and then the CSP itself applies the *outer encryption layer* (OEL) on top. If an ACP update conforms to the access restrictions set by the IEL, then the affected data may be re-encrypted in the cloud, eliminating the need for the data owner to download and to re-upload the data.

The benefits of such *two-layer encryption* (TLE) [11] depend entirely on the decomposition of the ACPs into sub-ACPs for the IEL and the OEL: the more of the future ACP updates are restricted to the OEL sub-ACPs, the better. This dynamic aspect of the two layers policy, i.e. setting it up in anticipation of future ACP updates, is more fully addressed in a TLE scheme when the ACP is an *access control matrix,* [15] instead of a set of predicates. Our own proposed scheme is based on such a policy, named Delta_SEL, in which the initial ACP is translated into the IEL, and the OEL is initially empty. An example of an ACP update in a Delta_SEL system is as follows: suppose user u_1 has access permissions for files f_1 and f_2, so that the IEL encrypts both with a key $k_{1,2}$, known to u_1, and another user u_2 is granted access to f_1. This is handled by granting u_2 the IEL key $k_{1,2}$, and at the same time, encrypting f_2 in the OEL with a new key k'_2, issued only to u_1. Revocation is handled in a similar way: if u_1 is now revoked access to f_1, a new OEL key k''_1 is issued to u_2, and f_1 is re-encrypted with k''_1. A major advantage of this TLE scheme is that the IEL keys never change, and therefore ACP updates never require the expensive retransmission and re-encryption by the data owner.

Another relevant paper presents a formal model for analyzing cRBAC (cryptographically enforced RBAC) systems [3]. A subsequent paper [4] describes an automated tool for verification of role reachability in RBAC systems by converting them into formally-verifiable imperative programs, and then applying a combination of abstract approximation and precise simulation of ACP updates, both of them operating heuristically and non-deterministically. The authors provide a formal analysis of their approach, however, they do not analyze any specific cRBAC protocols, especially implemented in the context of a distributed system like Cassandra. We present such detailed protocols in the next section, and the longer version of this paper includes a semi-formal analysis. Another recent paper by Garrison et al. [8] discusses the use of ABE/IBE cryptography for enforcing cloud based policies like RBAC, but does not address a distributed architecture like that of Cassandra which we focus on.

The main goal of our own work is implementing cRBAC in a distributed cloud environment using Predicate encryption PE. The scheme we're using for PE [9] allows attributes from $\mathbb{Z}_N^n$, for some $N = p \cdot q \cdot r$, where p, q, r are three distinct primes; and equality predicates on inner (dot) products of the attribute vectors. Alternatively, the scheme allows scalar attributes from $\mathbb{Z}_N$, and equality predicates on polynomials over $\mathbb{Z}_N$. A "dual" of the latter construction allows the attributes to be polynomials, and the predicates to correspond to evaluation at a fixed point. Access predicates expressed as DNF or CNF formulae are easily convertible into the polynomial form. A detailed discussion of predicate encryption is out of scope for this paper. We just depict the main steps involved:

Setup: The setup algorithm takes a security parameter and outputs a public key PK and a secret master key MSK

Key Derivation: Key derivation takes as input the master secret key and a vector of predicate values (X1,X2,...Xn) and outputs a secret key SK associated with this vector

Encryption: Encrypt takes as input the public key PK, a set of attribute values (Y1,Y2,...Yn) and the message M and generates an encrypted message M'

Decryption: Decrypt takes the encrypted message M', the secret key SK and the list of attributes values Y. The decryption succeeds only if the scalar product of the two vectors X and Y is zero

This encryption scheme for polynomial predicates can be extended to boolean formulae, as follows: $(x = a_1) \vee (x = a_2)$ is converted into the polynomial predicate $(x - a_1)(x - a_2) = 0$; $(x_1 = b_1) \wedge (x_2 = b_2)$ is converted into the polynomial predicate $r \cdot (x_1 - b_1) + (x_2 - b_2)$, for a random $r \in \mathbb{Z}_N$. These ideas extend to more complex combinations of disjunctions and conjunctions, meaning that the predicate encryption scheme can handle arbitrary CNF or DNF formulae.

Cassandra is a distributed storage system for managing very large amounts of structured data spread out across many commodity servers, while providing highly available service with no single point of failure [10]. Cassandra aims to run on top of an infrastructure of hundreds of nodes. In Cassandra, columns are grouped together into sets called column families, which may be nested. The rows are dynamically partitioned over a set of storage nodes in the cluster, using an order preserving hash function on the row ID. The output range of the hash function is treated as a ring, and each storage node is assigned a random position on the ring. The hash based distribution is done by the node coordinator while each storage node is responsible for its share of the range of records. The location of a row (record) can be computed easily by the node coordinator. and the read or write of a record is directed to one of the storage nodes containing it. The support for RBAC was added in recent releases of Cassandra [14]. In Cassandra syntax, users are represented as roles that have a permission to log in, i.e. a user is a special kind of role. Cassandra roles are first-class database objects, and so they have permissions defined on themselves, too: a role may be assigned permissions (ALTER, DROP, GRANT, REVOKE) on other roles, or equally on itself. As was mentioned, there is no connection between Cassandra's support for RBAC and Cassandra's support for encryption, which is a major motivation to this paper.

3   Proposed Design

In our design, we apply the ideas from cryptographically based RBAC [3], Predicate encryption [9], the two layers encrypton of [5], and Cassandra specific distributed architecture [10]. The main principles of the design is that the ACP is implemented by encryption at the storage node level. Thus once the location of the record is determined, the information requested is sent directly to the client by the node storage manager, and the client can decrypt it only if it has the correct decryption key. The new cRBAC component supply the built-in data encryption module with the relevant encryption keys, based on the user identity. Nodes send the requested data directly to the client, and do not require the coordinator to relay the data; this avoids a possible bottleneck in data throughput. The role of the coordinator in handling read requests is limited to finding a node which has a copy of the data, and referring the client to that node. This is performed in the same way as normally, by maintaining a $(key \rightarrow nodes)$ mapping; our changes to the access control do not affect this in any way. Another principle employed by our scheme is the ability to verify written records by any client. This will be detailed later. In summary the main principles of our model are:

1. Read can be performed by any client, but the content is meaningful only for a client authorized by the ACP

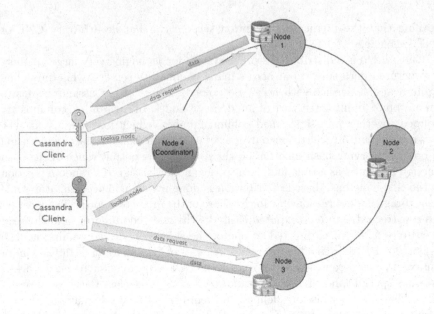

Fig. 1. Handling read requests

2. Write can be performed by any client, but only ACP authorized users will have a valid signature on the written record.
3. Any client can verify the validity of a written record even if it cannot decrypt its content

3.1   Read/Write Access

Handling read requests is illustrated in Fig. 1. We use the cRBAC predicate encryption scheme described earlier. We employ a combination of a *master key*, from which multiple *encryption keys* and *decryption keys* are derived according to the ACP. The encryption keys (one per data file) are used to encrypt the data initially, and the same encryption keys are distributed to the clients who have write access to the corresponding files. The decryption keys (one per role) are distributed to all clients according to their role: each client's decryption key allows him to decrypt the data that he has read access to, according to the ACP.

Handling of write requests, in a way that does not require the coordinator to mediate each request, is slightly more complicated. Our proposal relies on the fact that most cloud storage systems (including Cassandra) only support appending to existing data, but not deleting or overwriting them. In our cRBAC-based system, any user may append new records, by sending them directly to the storage node, and only involving the coordinator for looking up the relevant node for his key value(s). Each written record must include the writer's encrypted signature; and a subsequent reader, upon receiving the data from a storage node, needs to perform a little extra work in order to verify the signature of the record, and discard any invalid records. The signature is simply an encrypted hash of the record data; all that matters is that any client can easily tell whether it's valid or not. The key derivation for encrypting these signatures is done similarly as for the read access keys, but distributed in a different way: the decryption keys (one per data file) are known to all readers (who can use them for verification), and the encryption keys (one per role) are private to each writer.

Example. In this example, there are three roles A, B, C and three files X, Y, Z. The access control policy is shown below in matrix form:

	X	Y	Z
A	rw		
B		rw	
C	r	rw	rw

Initially, the server generates a master key mk and random numbers a, b, c. Then, the server derives encryption keys

$$k_X = derive_e(mk, (a \cdot c, -a - c, 1)),$$
$$k_Y = derive_e(mk, (b \cdot c, -b - c, 1)),$$
$$k_Z = derive_e(mk, (c, -1, 0))$$

for the three files, and decryption keys

$$k_A = derive_d(mk, (1, a, a^2)),$$
$$k_B = derive_d(mk, (1, b, b^2)),$$
$$k_C = derive_d(mk, (1, c, c^2))$$

for the three roles. Each client gets a decryption key according to his role. Predicate encryption [9] guarantees that

$$decrypt(k_A, encrypt(k_X, X)) = decrypt(k_C, encrypt(k_X, X)) = X,$$

and so on for the other two files; it also guarantees that $decrypt(k_B, encrypt(k_X, X))$ is a random bit string that does not reveal any information about the contents of X; and so are $decrypt(k_A, encrypt(k_Y, Y))$, $decrypt(k_A, encrypt(k_Z, Z))$, and $decrypt(k_B, encrypt(k_Z, Z))$. This is because

$$(1, a, a^2) \cdot (a \cdot c, -a - c, 1) = (1, c, c^2) \cdot (a \cdot c, -a - c, 1) = 0, while$$
$$(1, a, a^2) \cdot (b \cdot c, -b - c, 1) \neq 0,$$
$$(1, b, b^2) \cdot (a \cdot c, -a - c, 1) \neq 0,$$
$$(1, a, a^2) \cdot (c, -1, 0) \neq 0,$$
$$(1, b, b^2) \cdot (c, -1, 0) \neq 0.$$

For the purpose of write access control, the server generates a separate master key wk, and derives encryption keys

$$w_A = derive_e(mk, (1, a, a^2)),$$
$$w_B = derive_e(mk, (1, b, b^2)),$$
$$w_C = derive_e(mk, (1, c, c^2))$$

for the three roles, and decryption keys

$$w_X = derive_d(wk, (a, -1, 0)),$$
$$w_Y = derive_d(wk, (b \cdot c, -b - c, 1)),$$
$$w_Z = derive_d(wk, (c, -1, 0))$$

for the three files. Each client gets an encryption key according to his role, as well as a complete set of decryption keys (for write validation). Predicate encryption guarantees that $decrypt(w_X, encrypt(w_A, hash(X))) = hash(X)$, and therefore any client can, by using the publicly known w_X, validate that X had indeed been written and signed by A. Furthermore, predicate encryption guarantees that $decrypt(w_X, encrypt(w_C, hash(X))) \neq hash(X)$, and therefore an attempt by C to write and sign X will be detected by a subsequent reader, and ignored. These guarantees can be demonstrated in the same way as before, by computing the dot-products of derivation vectors for the keys.

The data are stored (possibly by different nodes) in the following manner, with the encrypted data of each file preceded by the writer's signature:

$encrypt(w_A, hash(X))$	$encrypt(k_X, X)$
$encrypt(w_?, hash(Y))$	$encrypt(k_Y, Y)$
$encrypt(w_C, hash(Z))$	$encrypt(k_Z, Z)$

A has $k_A, w_A, k_X, w_X, w_Y, w_Z$;
B has $k_B, w_B, k_Y, \dot{w}_X, w_Y, w_Z$;
C has $k_C, w_C, k_Y, k_Z, w_X, w_Y, w_Z$.

Additional Notes:

– In the illustration above, $w_?$ could be either w_B or w_C: when several roles have write access to a file, the signature does not disclose which one of them performed the write – only that the write was valid. There is a privacy problem here in case there is only one writer to a file. This will be dealt with in future work.

– A client who does not have read access to a file cannot validate whether it had been written legally or not, since the validation requires knowing the hash of the file contents. Every client needs validation keys for all files that he can read; validation keys for files that he cannot read are useless to him, but there's no harm and no waste in distributing all validation keys to all users, and this is simpler to manage.

– A signature cannot be "reused" for a file with different contents, or for a different file; however, it can be "reused", by a client without write access to a file, to revert the file to its earlier valid version. If this is undesirable, then a version number or a timestamp must be stored inside the file contents.

– The computational overhead for the cryptographic operations does not directly depend on the number of files or roles (each file has either one or two layers of encryption [11,15]), and this overhead is expected to remain far below the data transmission latency. When there are many files or many roles, the task of traversing a user's key-chain and looking up the correct key for a file may appear computationally demanding; but in fact, storing these in a sufficiently big hash table, indexed either by file ID, user ID or any functional equivalent will make looking up a key a very efficient operation, with effectively a constant time complexity.

3.2   Write Access Issues

In the proposed system, there is no central authority to decide, for each write request, whether to allow or to deny it; instead, the clients themselves decide, for each written record, whether it had been written legally. This is similar to how the Blockchain [12] operates – anybody can issue a write, but any invalid writes will be ignored by the readers. Blockchain is a "distributed ledger" which keeps a verifiably immutable (but appendable) record of all transactions since the system's initialization, with the integrity preserved via a "proof of work", via signatures by trusted parties, or by other means. Originally implemented in 2008 for the Bitcoin cryptocurrency, the blockchain is public and permissionless, allowing any user to submit new blocks.

There are, however, some difficulties associated with cryptographic access control on write. One such difficulty is the coherence of access policy updates: as there is no central authority to handle all write requests and policy updates, it's impossible to tell which "happened" earlier, a revocation of a write permission, or a write request relying on the same permission. In the simplest (and most secure) case, the readers verify the signatures using their present signature validation keys (SVKs) – discarding all records whose writers don't currently have the write access, whether or not they had write access at the time the records were written. A more complicated implementation would have to keep track of historic SVKs, and correlate each signature with an SVK which could be in effect at the time of the writing.

A further difficulty with cryptographic access control on write is a possibility for a DoS attack, where a user with no write access appends excessive amounts of invalid records, running the system out of storage capacity. However, the traditional security model for NoSQL databases [2] had always assumed that the cluster operates deep in the back-end of a high-load system, and is not exposed to any external agents which could be malicious; and that some of the security can be sacrificed in order to improve the cluster performance. This is also the reason why the NoSQL databases have adopted the append-only data model, where an excessive amount of data updates can run the system out of storage capacity. Even though readers may report the occurrence of invalid writes to the nodes coordinator, this is of no help against a DoS attack, since the written records (whether valid or invalid) are anonymous, and therefore, it's impossible to track down and block a client which writes excessive amounts of records (whether valid or invalid). Once again, this is not a defect in our proposed system, but a deliberate design decision upon which the NoSQL DBMSs are built.

3.3   ACP Updates

ACP updates require the data to be re-encrypted with new keys. We want to minimize the coordinator's involvement in this; therefore, the re-encryption should be performed locally to the data, by the nodes. This requires the new cRBAC component to plug into the node daemon as well, which seems to be rather uncommon for an authorizer implementation, and may require some adaptation of the core Cassandra node daemon to accept the plug-in. The biggest part of the cRBAC component, however, would run at the coordinator server to implement the keys management, i.e. to handle the changes to the ACP by generating new encryption keys and issuing them to the relevant users, as well as instructing the nodes to re-encrypt the affected data with the new keys. It is important to note that reencryption is done mostly at the **second level of encryption**, and the content itself is not re-encrypted.

Examples. Suppose that in the example above, C is revoked access to Y:

	X	Y	Z
A	rw		
B		rw	
C	r	~~rw~~	rw

For this, the server derives a new key $k'_Y = derive_e(mk, (b, -1, 0))$, and orders over-encryption of Y with the new key:

$encrypt(w_A, hash(X))$	$encrypt(k_X, X)$
$encrypt(w_?, hash(Y))$	$encrypt(k'_Y, encrypt(k_Y, Y))$
$encrypt(w_C, hash(Z))$	$encrypt(k_Z, Z)$

$(1, c, c^2) \cdot (b, -1, 0) \neq 0$, and therefore C can no longer read Y:
Since $decrypt(k_C, encrypt(k'_Y, encrypt(k_Y, Y)))$ results with a random bit string

that does not reveal any information about the contents of $encrypt(k_Y, Y)$ – which, in turn, C would still be able to decrypt.

To make sure C's signature on Y will no longer be accepted, the server derives a new decryption key $w'_Y = derived_d(wk, (b, -1, 0))$, and distributes it to all clients, so that:

$$A \text{ has } k_A, w_A, k_X, w_X, (w_Y), w'_Y, w_Z;$$
$$B \text{ has } k_B, w_B, (k_Y), k'_Y, w_X, (w_Y), w'_Y, w_Z;$$
$$C \text{ has } k_C, w_C, (k_Y), k_Z, w_X, (w_Y), w'_Y, w_Z.$$

(the keys which are no longer of any use to the holding client are parenthesized).

The existing signatures remain valid: e.g., if Y had been written and signed by B, then

$$decrypt(w'_Y, encrypt(w_B, hash(Y))) = decrypt(w_Y, encrypt(w_B, hash(Y))) = hash(Y)$$

even though the decryption key had been updated; this is because

$$(b, -1, 0) \cdot (1, b, b^2) = 0.$$

However, a signature made by C will now be detected as invalid:

$$decrypt(w'_Y, encrypt(w_C, hash(Y)) \neq hash(Y)$$

Now suppose that A is granted read/write access to Y:

	X	Y	Z
A	rw	rw	
B		rw	
C	r		rw

In this case, no over-encryption is necessary: since access of A is strictly wider than the access of B, the server simply hands over B's keys to A, so that:

$$A \text{ has } k_A, k_B, w_A, w_B, k_X, k'_Y, w_X, (w_Y), w'_Y, w_Z;$$
$$B \text{ has } k_B, w_B, (k_Y), k'_Y, w_X, (w_Y), w'_Y, w_Z;$$
$$C \text{ has } k_C, w_C, (k_Y), k_Z, w_X, (w_Y), w'_Y, w_Z.$$

The effect is that A acquires all of the permissions that B had. However, if B is subsequently granted new permissions that A does not have, then B will need to be issued a new set of private keys (k'_B, w'_B). For example, suppose that B is granted read/write access to Z:

	X	Y	Z
A	rw	rw	
B		rw	rw
C	r		rw

The server now has to generate new random numbers b' and c', and derive the new keys for B and C:

$$k'_B = derive_d(mk, (1, b', b'^2)),$$
$$k'_C = derive_d(mk, (1, c', c'^2)),$$
$$w'_B = derive_e(mk, (1, b', b'^2)),$$
$$w'_C = derive_e(mk, (1, c', c'^2)).$$

Then the server derives new decryption keys

$$w''_Y = derive_d(wk, (a \cdot b', -a - b', 1)),$$
$$w'_Z = derive_d(wk, (b' \cdot c', -b' - c', 1)),$$

– and distributes them to the clients; next, the server derives a new encryption key for X:

$$k'_X = derive_e(mk, (a \cdot c', -a - c', 1)),$$

– and orders over-encryption of X with the new key:[1]

$encrypt(w_A, hash(X))$	$encrypt(k'_X, encrypt(k_X, X))$
$encrypt(w_?, hash(Y))$	$encrypt(k'_Y, encrypt(k_Y, Y))$
$encrypt(w_C, hash(Z))$	$encrypt(k_Z, Z)$

$(1, b', b'^2) \cdot (b' \cdot c, -b' - c, 1) = 0$, but $(1, b', b'^2) \cdot (c, -1, 0) \neq 0$; therefore, B can read $encrypt(k_Y, Y)$ with his new key k'_B, but cannot yet read Y. To complete the grant of read access to Y, the server hands over k_C to B; this is secure because the only files that k_C enables decrypting are X and Z, and X has been over-encrypted with k'_X to protect it from B. Finally, the server sends k'_X to A because it has the write access. After the grant is complete:

$$A \text{ has } k_A, k_B, w_A, (w_B, k_X), k'_X, k'_Y, w_X, (w_Y, w'_Y), w''_Y, (w_Z), w'_Z;$$

---

[1] It may appear wasteful that granting B read access for Z requires a re-encryption of X, and possibly of many other files that C has read access to. Local re-encryption of Z by the storage node (decryption with k_C and encryption with a new key) would be the most efficient alternative, but it would have compromised the privacy of Z by disclosing its plain-text to the storage node; therefore, this is not an option for our proposed system. Therefore, B has to be given k_C, since the encryption with k_Z can only be decrypted using k_C; and if X is not re-encrypted, then B would be able to read X using k_C. This overhead (switching from one-layer encryption to TLE) may only happen once for each file; and the possible alternatives (either downloading Z, decrypting it, encrypting it with a new key, and uploading it again; or double-encrypting all files at system initialization time, paying all of the possible overhead upfront) are in fact much less efficient.

B has $k_B, k'_B, k_C, (w_B), w'_B, (k_Y), k'_Y, w_X, (w_Y, w'_Y), w''_Y, (w_Z), w'_Z$;
C has $k_C, k'_C, (w_C), w'_C, (k_Y, k_Z), w_X, (w_Y, w'_Y), w''_Y, (w_Z), w'_Z$.

Note that A no longer needs w_B, as the updated w''_Y allows him to sign his writes to Y using his own encryption key w_A. Also note that k'_X had to be derived using the new value of c' in order to protect X from B, who had been given k_C as part of granting read access for Z.

Management of Keys. As shown in the examples above, each user has to keep a large set of keys:

- To *read* a file, a user needs his own key to decrypt the data, and also one key per file he has read access to, to validate the writers' signatures;
- To *write* a file, a user needs one key per file he has write access to, to encrypt the data; and also his own key to create a valid signature.

Managing so many keys may be inconvenient to the user. To facilitate the keys management, we can use any of the techniques from [7], such as key-chain files, keys records attached to the files, or hierarchical keys management. For our implementation, we chose the key-chain technique, since it involves less overhead or knowledge on behalf of the ACP manager.

3.4   Formal Description of the Protocol

To simplify the description, we're presenting here a variant of our system where all files are double-encrypted at system initialization time (Full_SEL), instead of incrementally as part of the ACP updates (Delta_SEL) [15].

- Involved parties:
    1. data owner/ACP manager MGR
    2. user belonging to ROLE (the client)
    3. nodes coordinator CTR
    4. data storage node NODE
- Note that key management and distribution is done by the MGR, while second layer encryption is done by the CTR. The CTR never has to relay any data transmission.

1. **Setup**:
    (a) MGR derives (from the ACP) the initial encryption keys
    (b) MGR encrypts the initial data
    (c) MGR issues an "initialize" request to CTR, and receives the ID(s) of NODE(s) where to upload the initial data
    (d) MGR uploads the (encrypted) initial data to NODE(s)
    (e) MGR distributes the encryption keys to ROLEs, according to the ACP
2. **Read**:
    (a) ROLE issues a "lookup" request to CTR, and receives the ID of the NODE where the data is stored

   (b) ROLE downloads the encrypted data from NODE

   (c) For each record, starting from the latest:

      i. ROLE validates the signature item on the record

      ii. If the record's signature is valid, ROLE proceeds to (d)

      iii. Otherwise, ROLE proceeds to the next record read

   (d) ROLE decrypts the data

3. **Write**:

   (a) ROLE encrypts the data with his *read* key

   (b) ROLE signs the data with his *write* key

   (c) ROLE issues a "lookup" request to CTR, and receives the ID of NODE where the data is stored

   (d) ROLE uploads the encrypted data to NODE

4. **Grant read access for X to ROLE U**

   (a) If U's role key k_U had been shared with another ROLE V, who does not have read access to X, then

      i. MGR derives a new role key k_U' and sends it to U

      ii. MGR derives a new encryption key k_X', forming the derivation vector as described earlier, based on the identities of all ROLEs (including U's new identity) which are to have read access to X

      iii. MGR proceeds to step (d) as described below

   (b) Otherwise, if another ROLE W already has read access to X, and U already has read access to all data that W has read access to, then

      – no re-encryption is necessary: instead, MGR hands over W's role key k_W to U

   (c) Otherwise, [i.e. if k_U had not been shared with any ROLE V which does not have read access to X, and there's no such ROLE W which already has read access to X, and for which U already has read access to all data that W has read access to], MGR derives a new encryption key k_X', forming the derivation vector as described earlier, based on the identities of all ROLEs (including U) which are to have read access to X

   (d) MGR distributes k_X' to all ROLEs which have write access to X

   (e) If U's role key k_U does not allow decrypting the 1st layer encryption on X (i.e. if U didn't have read access to X at setup time, and had not been given a role key of a ROLE which had read access to X at setup time), then

      i. MGR chooses any role key $k_{U'}$ which allows decrypting the 1st layer encryption on X (i.e. U' had read access to X at setup time),

      ii. MGR hands over $k_{U'}$ to U

   (f) MGR issues an "update ACP" request to CTR, passing k_X' as part of the request

   (g) CTR handles the request by relaying k_X' to NODE(s) where X is stored

   (h) NODE(s) CTR re-encrypt the data locally, using k_X' for the 2nd layer encryption

5. **Revoke read access from X**

   (a) MGR derives a new encryption key k_X', forming the derivation vector as described earlier, based on the identities of all ROLEs which still have read access to X

    (b) MGR distributes k'_X to all ROLEs which have write access to X

    (c) MGR issues an "update ACP" request to CTR, passing k'_X as part of the request

    (d) CTR handles the request by relaying k'_X to NODE(s) where X is stored

    (e) NODE(s) CTR re-encrypt the data locally, using k'_X for the 2nd layer encryption

6. **Grant write access for X to ROLE U**

    (a) MGR sends the encryption key k_X to U

    (b) If U's role key w_U had been shared with another ROLE V, who does not have write access to X, then

       i. MGR derives a new encryption key w'_U and sends it to U

      ii. MGR derives a new decryption key w'_X, forming the derivation vector as described earlier, based on the identities of all ROLEs which are to have write access to X

     iii. MGR distributes w'_X to all ROLEs which have read access to X

    (c) Otherwise, if another ROLE W already has write access to X, and U already has write access to all data that W has write access to, then

     – there's no need to derive new keys: instead, MGR hands over W's encryption key w_W to U

    (d) Otherwise, MGR derives a new decryption key w'_X, forming the derivation vector as described earlier, based on the identities of all ROLEs (including U) which are to have write access to X

    (e) MGR distributes w'_X to all ROLEs which have read access to X

    (f) No re-encryption is necessary

7. **Revoke write access from X**

    (a) MGR derives a new decryption key w'_X, forming the derivation vector as described earlier, based on the identities of all ROLEs which still have write access to X

    (b) MGR distributes w'_X to all ROLEs which have read access to X

    (c) No re-encryption is necessary

8 **Granting ROLE membership**

    (a) MGR sends the new ROLE member all keys held by other ROLE members

9. **Revoking ROLE membership**

    (a) MGR re-issues keys for all remaining ROLE members

    (b) MGR orders re-encryption of all data accessible by the remaining ROLE members

    (c) MGR updates its local representation of the ACP, which for each user lists the effective access permissions, with all role memberships expanded

    (d) MGR handles the changes in effective access permissions, as detailed above

To prove correctness of the protocols above, one has to show formally that every ACP change results only with the intended Read or Write permissions. This will be included in the longer version of this paper.

4 Conclusions

This paper proposes a scheme for cryptographic enforcement of RBAC, using Cassandra for the proof-of-concept implementation. It combines several pre-existing techniques and algorithms, such as: Predicate encryption, Second level encryption and Cassandra's distributed architecture, for providing a flexible and efficient scheme to apply cRBAC for ACP enforcement in NoSQL databases. It presents a formal description of the resulting protocol, and presents examples of its operation. Currently we are implementing the protocols as part of a real-life Cassandra database. Results of this experimental evaluation will be reported in the future. We also plan to investigate further some of the issues mentioned, such as Privacy or DDOS attacks.

References

1. DataStax: Securing Cassandra (2015). https://docs.datastax.com/en/cassandra/3.0/cassandra/configuration/secureIntro.html
2. Davis, M.A.: Why NoSQL equals NoSecurity. InformationWeek (2012)
3. Ferrara, A.L., Fuchsbauer, G., Warinschi, B.: Cryptographically enforced RBAC. In: 2013 IEEE 26th Computer Security Foundations Symposium (CSF), pp. 115–129. IEEE (2013)
4. Ferrara, A.L., Madhusudan, P., Nguyen, T.L., Parlato, G.: VAC - verifier of administrative role-based access control policies. In: Biere, A., Bloem, R. (eds.) CAV 2014. LNCS, vol. 8559, pp. 184–191. Springer, Cham (2014). doi:10.1007/978-3-319-08867-9_12
5. Foresti, S.: Data security and privacy in the cloud. In: 29th Annual IFIP WG 11.3 Working Conference on Data and Applications Security and Privacy (2015)
6. Goyal, V., Pandey, O., Sahai, A., Waters, B.: Attribute-based encryption for fine-grained access control of encrypted data. In: Proceedings of the 13th ACM Conference on Computer and Communications Security, CCS 2006, Alexandria, VA, USA, 30 October–3 November 2006, pp. 89–98 (2006)
7. Gudes, E.: The design of a cryptography based secure file system. IEEE Trans. Softw. Eng. **5**, 411–420 (1980)
8. Iii, W.C.G., Shull, A., Myers, S., Lee, A.J.: On the practicality of cryptographically enforcing dynamic access control policies in the cloud. In: IEEE Symposium on Security and Privacy, SP 2016, San Jose, CA, USA, 22–26 May 2016, pp. 819–838 (2016)
9. Katz, J., Sahai, A., Waters, B.: Predicate encryption supporting disjunctions, polynomial equations, and inner products. In: Smart, N. (ed.) EUROCRYPT 2008. LNCS, vol. 4965, pp. 146–162. Springer, Heidelberg (2008). doi:10.1007/978-3-540-78967-3_9
10. Lakshman, A., Malik, P.: Cassandra: a decentralized structured storage system. Oper. Syst. Rev. **44**(2), 35–40 (2010)
11. Nabeel, M., Bertino, E.: Privacy preserving delegated access control in public clouds. IEEE Trans. Knowl. Data Eng. **26**(9), 2268–2280 (2014)
12. Pilkington, M.: Blockchain technology: principles and applications. In: Research Handbook on Digital Transformations (2015)

13. MIT Csail Computer Systems Security Group: Crypto tutorial (2010). http://css. csail.mit.edu/security-seminar/cryptoslides.ppt
14. Tunnicliffe, S.: Role based access control in Cassandra (2015). http://www. datastax.com/dev/blog/role-based-access-control-in-cassandra
15. Vimercati, S.D.C.D., Foresti, S., Jajodia, S., Paraboschi, S., Samarati, P.: Encryption policies for regulating access to outsourced data. ACM Trans. Database Syst. (TODS) **35**(2), 12 (2010)

Resilient Reference Monitor for Distributed Access Control via Moving Target Defense

Dieudonne Mulamba and Indrajit Ray[⊠]

Department of Computer Science,
Colorado State University, Fort Collins, CO 80523, USA
{indrajit,mulamba}@cs.colostate.edu

Abstract. Effective access control is dependent not only on the existence of strong policies but also on ensuring that the access control enforcement subsystem is adequately protected. Protecting this subsystem has not been adequately addressed in the literature. In general, it is assumed to be implemented as a reference monitor in a trusted computing base (TCB) that is tamper-proof. However, in distributed access control, ensuring TCB security kernel to be tamper proof is not always feasible. It needs to be implemented in software and on platforms that can potentially have vulnerabilities. We posit that allowing a very limited opportunity to the attacker to enumerate exploitable vulnerabilities in the access control subsystem can considerably facilitate its protection. Towards this end we propose a moving target defense framework for access control in a distributed environment. In this framework, access control is provided by cooperation of several distributed modules that materialize randomly, announce their services, enforce access control and then disappear to be replaced by another module randomly. As a result, the attacker does not know which process can be targeted to compromise the access control system.

1 Introduction

Many emerging distributed applications rely on on-demad network-enabled access to a shared pool of computing resources. Examples of such applications are IoT applications, sensor networks or enterprise level distributed workflow systems. This distributed computing model brings with it some unique challenges to access control that require re-visiting the traditional TCB approach [3]. In traditional systems, access control is implemented by the cooperation of four functional modules that are part of the trusted computing base:

1. *Policy Administration Point* (PAP): The PAP is a repository for the authorization policies that are expressed in terms of the actions that subjects (human users, devices, processes, organizations etc.) can take on various objects in the system. The authorization policies are essentially an instantiation of the access control model tailored towards the organization. It is the main component for the authorization portion of access control.

© IFIP International Federation for Information Processing 2017
Published by Springer International Publishing AG 2017. All Rights Reserved
G. Livraga and S. Zhu (Eds.): DBSec 2017, LNCS 10359, pp. 20–40, 2017.
DOI: 10.1007/978-3-319-61176-1_2

2. *Policy Information Point* (PIP): The PIP is the component that gathers together all the attribute information that are needed to evaluate an authorization policy.
3. *Policy Decision Point* (PDP): The PDP gets relevant information from the PIP and consults the PAP to arrive at a decision whether to grant or deny an access request.
4. *Policy Enforcement Point* (PEP): The PEP receives access requests from subjects in the external world, hands them to the PDP for evaluation, and after receiving the grant or deny response from the PDP, ensures the appropriate action is taken.

One of the requirements of a TCB is that it implements the concept of reference monitor [3]; that is, the TCB mediates all access to objects by subjects, is tamper-proof and cannot be bypassed, and is small enough to be thoroughly tested and analyzed. In the newer distributing computing environments however, making the TCB small and tamper-proof is very difficult. This is because access control in such environment needs to be achieved via the cooperation of both local as well as remote access control engines. To remain within the confines of the TCB paradigm, not only all of these separate components need to be made tamperproof, but also all communications and coordinations among the components. As a result, the trusted computing base needs to be enlarged in scope and functionality, which violates the principles of reference monitor. Moreover, the sheer size of these distributed systems, the degree of heterogeneity among the different devices (potentially virtual machines), and the dynamicity of the whole system, compound the problem many fold. It appears, therefore, that it is next to impossible to rely on a single TCB to provide access control in these environments. The access control subsystem should try to satisfy as many properties of a TCB as possible but should also incorporate certain self-defending strategies to make it secure.

In this work, we treat access control in such a distributed environment as a service that needs to be proactively protected. From a functional perspective, this service is achieved by the four functional units – PAP, PIP, PDP and PEP. We assume that like any other service the access control service can be attacked by an attacker and hence needs to be protected. An attacker intent on damaging the access control service will launch reconnaissance efforts seeking exploitable vulnerabilities for this subsystem. We propose being proactive and allow only limited opportunity to the attacker to enumerate exploitable vulnerabilities, thus reducing the attack surface of the access control subsystem. Towards this end, we propose employing a Moving Target Defense (MTD) paradigm for the access control subsystem. The four functional modules are effectuated by randomly materializing processes that announce their services, enforce access control and then disappear to be replaced by another module randomly. As a result, the attacker does not know which processes can be targeted to compromise the system. Moreover, the window of opportunity for targeting processes is varied to further reduce opportunities of attack. We describe an implementation of this system to handle RBAC policies using COTS components.

The rest of the paper is organized as follows: Sect. 2 reviews previous works on access controls, leader election protocols, moving target as well as on service discovery protocols. In Sect. 3 we first present the reference architecture for access control in distributed environments. We then give an overview of our moving target defense approach for protecting the access control subsystem. Section 4 presents the moving target defense architecture. We present our implementation in Sect. 5 as well as an analysis of the security of the proposed approach. Finally, we conclude this paper in Sect. 6 and give some directions for future works.

2   Background and Related Works

2.1   Protection of Access Control Subsystems

One of the most important aspect of security is ensuring that users access only resources to which they are authorized. Research on designing and deploying access control in computers and networks can be traced back several decades [43]. Early standards of access control included discretionary and mandatory access control [13,17,34,42]. However, Role based Access Control (RBAC) represented a major leap forward in term of flexibility. RBAC is built on the principle that users do not have discretionary access to enterprise objects. In a RBAC model, roles are created and users belong administratively to these roles, while permissions are administratively assigned to the roles. This arrangement provides more flexibility and simplicity to the management of authorization [13]. RBAC has been traditionally implemented for centralized systems. In recent years, several works have been done to provide the capabilities of this access control model to distributed systems and the cloud. For instance, in [21] authors present an access control tailored for distributed control systems. [41] explains how one can provide access control to anonymous users while verifying their authorization in a decentralized manner.

In several works, including recently in [14], researchers have worried that a malicious program may tamper with the operation of an access control system. The notion of a trusted computing base implementing the reference monitor concept was proposed by Anderson [3], in order to address this problem. Security kernels such as Scomp [15] and GEMSOS [45], included a reference validation mechanism to satisfy the reference monitor concept of the TCB. Other operating systems such as Trusted Solaris [36], the Linux Security Modules (LSM) framework [52], TrustedBSD [49], Mac OS X, and the Xen hypervisor provide' various degree of support for reference validation so as to enable some shade of reference monitor. However, the major problem with these systems is that the tamperproof property that needs to be ensured for provably implementing a reference monitor, is hard to achieve. Tamperproofing requires the TCB to have a very small footprint. It can be shown that a general algorithm to prove that an arbitrary program behaves correctly reduces to solving the Halting problem. While current algorithms can prove correctness properties of specific programs, the variety of reference validation code and the complexity of correctness properties preclude verification for all but the smallest, most specialized systems.

Unfortunately, for most of these systems, the TCB is too large to determine whether tampering is prevented. Moreover, for practicality and functionality, many systems allow user-level processes to modify the kernel. However, none of these user-level processes are immune to tampering, thus becoming one of the weakest links. In this work, we are interested in the protection of the access control subsystem where ensuring the tamperproof property of a reference monitor is challenging.

2.2 Moving Target Defense

Several works have been done on Moving Target Defense (MTD). In order to improve the understanding of MTD, [54] presents key concepts and their basic properties. Other works on MTD are mainly focused on network-based MTD [2,4,10]. In addition, in [12], the effectiveness of MTD using low-level techniques to defend a computing system has been studied. Those low-level techniques include Address Space Randomization, Instruction Set Randomization, and Data Randomization. In [19] two measures are designed that allow a defender to quantify its gain in security while deploying a MTD system. In order to chose a particular MTD technique, one needs to know its effectiveness. For that purpose, [53] proposes a comparison of different MTD techniques based on their effectiveness. However, none of these techniques are applicable for the problem we are addressing.

2.3 Leader Election

In a distributed system, leader election is a fundamental problem that requires that a unique leader be elected from among a set of given nodes. The goal of a leader election algorithm is to elect a good processor as a leader in a setting where there are n processors among which a certain number $m < n$ are bad, while ensuring that no bad processor get elected as a leader [27].

A distributed computing system often requires that active nodes continue performing their task after a failure has occurred. This reorganization or reconfiguration necessitate that a coordinator be elected in the first place [16]. This is the reason of the wide interest the leader election problem [31] has received. Several works have been done on Leader Election [1,6,16,44,47].

2.4 Consensus Algorithms

One approach for building fault-tolerant applications is the Lamport's approach. The core of this approach involves two primitives: consensus and atomic broadcast [28]. Leader election protocols are generally used to solve the consensus problem.

Bully algorithm [5] and Ring algorithm [48] are among the most used algorithms for solving the consensus problem. A hugely popular algorithm is the

Paxos algorithm proposed by Lamport [30]. Another popular algorithm, considered even superior to Paxos due to its simplicity, is the Raft consensus algorithm [20]. Raft provides the capabilities for Leader election and log replication.

ZooKeeper [22], an open-source replicated service for coordinating web applications, and Chubby [7] are some practical systems exploiting these algorithms. Recently, GIRAFFE [46] has been proposed to provide a coordination service in scalable distributed system. Another practical system is the Apache Kafka [51] which allows the building of a replicated logging system. However, these algorithms and protocols do not take into account Byzantine failures. In addition, their approach for electing a new leader does not prevent a malicious host from being elected as a leader.

2.5 Byzantine Fault Tolerance

A computer system can be affected by a type of failure that can cause it to behave in an arbitrary way. After being affected, the computer system can be led either to process requests incorrectly, to corrupt their local state, and/or to produce incorrect or inconsistent outputs. This type of failure is called Byzantine failures. The problem for coping with this type of failure is known as the Byzantine Generals Problem [29]. The goal of Byzantine fault tolerance is to allow computer system to be immune against Byzantine failures.

Several works have been proposed to reach a consensus in the face of Byzantine failures. Building on Paxos, the authors in [9] have proposed an improvement that allows Paxos to support Byzantine fault tolerance with a modest latency. Castro and Liskovs proposed the Practical Byzantine Fault-tolerance protocol [8] that reduces the number of messages exchanged to only four messages. [33] looked at improving the number of communications in the Byzantine Paxos protocols. In [11], authors took the task of improving Raft to support Byzantine fault-tolerance. They reach their goal by combining the ideas from the original Raft algorithm and from Practical Byzantine Fault-tolerance protocol [9].

2.6 Service Location Protocol

A zero configuration approach is a self-management networking approach that allows network devices to be automatically configured, discover services automatically, and to access service without the involvement of a network specialist [Intelligent Self-Management Home Multimedia Service System]. Three major self-management technologies have been proposed. The Internet Engineering Task Force (IETF) promoted SLP [18,38] as an intranet standard for automatic network resource discovery. Intel and Microsoft, on their part, proposed the Universal Plug and Play (UPnP) [39] as a standard for automatic communication between network devices using XML messages. Apple Inc. proposed a protocol called Bonjour [23] as its zero configuration networking standard. Recently, z2z [32] has been proposed for the discovery of network services beyond the local network.

In this work, our contribution includes the development of a system that leverages these existing concepts into a single framework so as to address the problem of protecting a reference monitor.

3   Architecture Overview

We assume that the access control model is Role-Based Access Control. We start with a high level operational architecture of access control (AC) in the distributed setup. The resources we are concerned about are the shared resources. The AC architecture is composed of four functional entities. Each entity has specific functions and participates in the communication as a client, as a server or both.

3.1   Access Control Architecture Components

The four entities are:

1. *A users client*: It is an endpoint entity whose main objective is to access protected resources. It is responsible to initiate or terminate a session with the Resource Access Server. It resides on the individual devices running the applications that require access to shared resources.
2. *Resource Access Service (RAS)*: It is an entity that manages the various distributed resources and controls the access to it. It acts as both client and server when receiving or replying to access requests from the client. The decision to grant or deny the access to protected resources is received from the Authorization Control Server. The Resource Access Server is responsible for reinforcing that decision.
3. *Authorization Control Service (ACS)*: It is an endpoint entity that governs the access to each protected resource. It hosts the Access Control engine. The access control engine is based on RBAC model (other models are also possible) and is designed to prevent unauthorized access to protected resources. The policies defining the access to protected resources are also managed by the Authorization Control Service. This service receives any client request and replies with the decision to grant or deny the access to the needed resources.
4. *Discovery Service*: Since the protection of the Authorization Control Service requires this later to be moved to a different location in a non predictable manner, the clients need to be able to discover the new location of the Authorization Control Service. The Discovery Service provides such capability.

Using these entities, the distributed access enforcement proceeds as follows. When a client needs to access a protected resource, it sends a request to the Authorization Control Service (ACS). The ACS verifies the policy governing the access to the needed resources, and replies with a decision to grant or deny access to the requested resource. If the decision was to grant access, the request is forwarded to the Resource Access Service along with a limited life-time token given to the client allowing it to access that particular resource. The occurrence of an election triggers the Authorization Control Service to be Switched to a different location. In this case, the ACS will register its new location with the Discovery Service. In addition, the client will need to consult the Discovery Service in order to find the new location of the ACS.

3.2 Threat Model

In our architecture, we consider access control (AC) as a service and we assume that the Access Control Service can be attacked and compromised. To mitigate the vulnerability of the open network in which an attacker passively listens to various communications, we make all access related communications go over secure channels. This makes the communication secure, but does not protect the endpoints, particularly the Resource Access Server and the Authorization Control Server where the AC engine components reside. We assume that an attacker can masquerade as one of these servers. Alternately, some numbers of these servers can themselves be compromised by malware and behave in a Byzantine manner. An attacker masquerading as a valid server or corrupting a server are treated similarly.

To protect these two entities, we propose the Moving Target Defense strategy. Our motivation for this approach follows from the observation that an attacker needs a reconnaissance window to explore the vulnerabilities in a system before attacking it. The moving target defense strategy reduces this window of opportunity. It requires both the Resource Access Server and the Authorization Control Server to be replicated. At each instance there is only one Resource Access Server and one Authorization Control Server that is responsible to handle access requests. These are called respectively, RAS leader and ACS leader servers. In addition, both leader servers are periodically replaced by a pair of leader servers randomly chosen by following a Byzantine consensus process executed by the existing candidate RAS and ACS servers. The replacement is achieved through a migration process that relies on a secure service discovery process.

Using moving target defense in this manner to protect the Access Control Engine raises several challenges. Those challenges are as follows.

1. How does one avoid migrating a leader server to a malicious server during the migration process?
2. Which access control component is going to be migrated? And,
3. After the migration process, how does one discover which server is currently providing the services?

In the following sections we are going to address these challenges (Figs. 1 and 4).

4   Distributed Access Control Architecture

We present an architecture that provides access control services. In order to protect the access control service against certain attacks, we have proposed to implement a Moving Target Defense strategy on the access control service architecture. In this section, we present the different components that constitute our Moving Target Defense architecture.

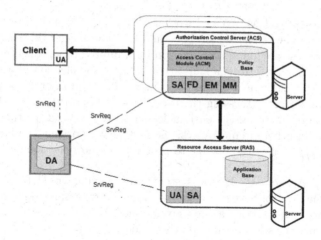

Fig. 1. Moving Target Defense architecture

4.1 The Client

It is an endpoint entity whose main objective is to access protected resources. It is responsible to initiate or terminate a session with the Resource Access Service. It resides on the individual devices running the applications that require access to shared resources.

4.2 The Authorization Control Service

In this section we present the different components that allow the Authorization Control Service to provide access control service, to be elected as a leader, and to announce its services once elected as a leader.

Access Control Engine. The access control engine is based on RBAC model (other models are also possible) and is designed to prevent unauthorized access to protected resources. It comprises the Policy Enforcement Point (PEP), the Policy Decision Point (PDP), the Policy Administration Point (PAP), and the Policy Information Point (PIP).

Fault Detector Module (FD). The Fault Detector Module is designed to detect the byzantine faults occurring in the server providing the Access Control service. The failure and the compromise of the current leader server are reasons to trigger the Moving Target Defense. The Fault Detector is able to detect any kind of byzantine faults that are local to the leader server. The unavailability of the leader server is detected by the Fault Detector of any other server that probes the aliveness of the leader server. The occurrence of either failure, compromise, or unavailability is used to trigger the election of a new leader server that will be responsible to provide access control services.

Election Module (EM). The Election Module is responsible for processing the election of a new leader server. Several causes can trigger this election. Since the leader server is assuming this function for a limited time, called here a term, the expiration of this term is a cause that triggers the election of a new leader server. We add randomness to the duration of this term in order to prevent an attacker from correctly guessing the occurrence of the next leader election. At the expiration of his term, the current leader server proceeds to the election of a new leader. In other circumstances, the first server noticing the failure of the current leader server, is responsible to proceed to a new election.

Leader Election Protocol. In our system, after having opted to replicate the Authorization Control Service among several servers, a single server is responsible to provide this service at any given time. We call this server the leader. At any instance, this leader may be the subject of attacks or of failures. Thus to realize the moving target defense, the leader is required to be periodically changed. This change can occur at the expiration of the current leader's lifetime or when the current leader fails. Moreover, the next server is not pre-determined but is elected by existing servers, each of which is a candidate. All this is done in an environment where we assume that some servers may be malicious attackers. Thus, we need a leader election algorithm that can ensure that a malicious server cannot be elected as leader. Once a server is elected, it sets a random lifetime for itself.

For the sake of maintaining our distributed system in a good functioning state, it becomes crucial to prevent faulty nodes from becoming leader. We adapt the algorithm from [26] in order to realize a leader election. The election process proceeds in the form of a distributed protocol as follows.

The election algorithm proceeds in rounds and in each round there is a node that is coordinating the consensus, called the coordinator. For each round r there is a coordinator c known a priori by each participating node by computing $c \equiv (r \bmod n) + 1$ with n being the total number of nodes. At each node, there are several local variables that are maintained, among which there is the estimate value which is an input value selected by the node, its current election round, its current coordinator c_p, and a timestamp ts_p. The consensus algorithm runs by exchanging messages between nodes participating in the distributed system. These messages include the types ESTIMATE, SELECT, CONFIRM, READY/NREADY, and SUSPECT. The algorithm runs in a sequence of five tasks that are concurrently executed.

The consensus algorithm, (see Algorithm 1), works as follows. The algorithm starts with each node p picking its estimate of the input value, and sending an ESTIMATE message to all nodes. The coordinator, after receiving $n - k$ ESTIMATE messages that it was waiting for, selects a value es based on all the estimate values received. It then sends a SELECT message carrying the es value to all nodes. Each node p, upon receiving SELECT message from the coordinator, sends a CONFIRM message carrying the es value to all other nodes. The es value should be the same for a given round r. After receiving a CONFIRM message from $\lfloor (n + k)/2 \rfloor + 1$ distinct nodes, each node p updates its local variables.

It then sends a READY message or a NREADY message depending on whether it had received the same es value or not from $\lfloor (n+k)/2 \rfloor +1$ CONFIRM messages. It should be noted that if the CONFIRM messages received by a node p did not contain the same es value, p will assume that the coordinator had deviated from the algorithm. The node p will therefore add the coordinator to its list of Suspect, and will send a SUSPECT message containing the id of the suspected coordinator to all. After a node p received the same es value as content of READY message from $\lfloor (n + k)/2 \rfloor + 1$ distinct nodes, it will decide on that value. A node q is confirmed to be malicious by a node p and added to $Output(D)_p$ if and only if node q have been reported malicious by at least $k + 1$ nodes. $Output(D)_p$ is the final list of malicious nodes. Any round in which the coordinator has not been reported with malicious nodes will end with a consensus on the input value and the coordinator being confirmed as the new leader. Otherwise, a new round will start with a new coordinator.

Migration Module (MM). The Migration Module is responsible for executing the migration protocol. The Migration Module receives a notification from the Election Module that a new leader has been elected. This notification contains the identity of the new leader server. Upon receiving the notification from the Election Module, this module executes the migration protocol, which transfers the Access Control service to the new elected leader server.

Service Migration Protocol. In our architecture, replacing the current leader server by a new one necessitates the migration of the Authorization Control Service provided by the current leader server to the new one. For this purpose, we need to put in place a service migration protocol that can handle this task.

Process migration is the movement of a running process from one host to another. A process migration protocol can have several components like the transfer policy, the selection policy, and the location policy. Since we are interested with the migration of an access control service, we define these policies in terms of the requirements of the access control service.

1. *The Transfer Policy:* This policy determines when a host needs to send a process to another host. In our case, it determines when the current leader server needs to send the access control services to a newly elected leader server. In our architecture, this decision is triggered by the successful completion of the leader election protocol.
2. *The Location Policy:* This policy determines the destination host to which to transfer the process to be migrated. In our architecture, this information is provided by the leader election protocol that communicates the identity of the new elected leader server to all the hosts. This new Authorization Control Service is the destination host for the migration protocol.
3. *The Selection Policy:* Determines which resource to transfer. It is question here to determine which component of the access control service needs to be migrated. We have designed the Authorization Control Server to be fully replicated. We assume the existence of a replicated protocol. Therefore, the

Algorithm 1. Leader Election

```
1:  /* Each node p executes the following */
2:  /* Initialization */
3:  e_p = V_P  {Chosen value}
4:  r_p = 0  {Initial round}
5:  ts_p = 0  {Initial timestamps}
6:  Estimates_p = ∅  {list of estimate msg}
7:  Confirms_p = ∅  {list of comfirm msg}
8:  Suspected_p = ∅  {list of suspected nodes}
9:  Output_p = ∅  {blacklisted nodes}
10: for r in listnode do
11:     Suspecting_p[r] = ∅
12: end for
13: COBEGIN  {Concurrent tasks}
14: {Task 1:}
15: while true do
16:     r_p ← r_p + 1
17:     {Select a Coordinator c_p}
18:     c_p ← (r_p mod n) + 1
19:     {Task 1, Phase 1 : each node creates esti-
        mate msg}
20:     estimate_p = (ESTIMATE, p, r_p, e_p, ts_p)
21:     Send estimate_p to all
22:     {Task 1, Phase 2: Coordinator counts
        received estimate msg}
23:     if [p = c_p] then
24:         [Wait until received (n − k) distinct
            estimate_q messages from q nodes]
25:         Estimates_p ← estimate_q
26:         ts = largest ts_q : estimate_q ∈
            Estimates_p
27:         if [ts = 0 and (at least (k + 1) distinct
            estimate_q ∈ Estimates_p have common
            value e)] then
28:             es ← e
29:         else
30:             es ← e_p
31:         end if
32:         {Coordinator creates select msg}
33:         select_p = ((SELECT, p, r_p, e_s)_p)
34:         Send select_p to all
35:     end if
36:     {Task 1, Phase 3 : receiving confirm msg}
37:     [Wait until received [(n+k)/2]+1 distinct
        confirm_q messages or c_p ∈ Output_p]
38:     confirm_p = ((SELECT, p, r_p, e_s)_p)
39:     if [[(n + k)/2] + 1 distinct confirm_q ∈
        Confirms_p have common value e] then
40:     ts_p ← r_p
41:     e_p ← e
42:     Confirm_p ← (CONFIRM, q, r_p, e)_q
43:     ready_p = (READY, p, r_p, e)_p)
44:     Send ready_p to all
45: else
46:     nready_p = (NREADY, p, r_p, e)_p
47:     Send nready_p to all
48: end if
49: end while
50: {Task 2: }
51: while true do
52:     if [p received select_p msg from c =
        (r_p mod n) + 1 and p has not sent
        confirm_q msg] then
53:         Selects_p ← ((SELECT, c, r, e, ts)_c)
54:         confirm_p = (CONFIRM, p, r, e)_p
55:         send confirm_p to all
56:     end if
57: end while
58: {Task 3:}
59: while true do
60:     [Wait until received [(n + k)/2] + 1 dis-
        tinct ready_q messages from q nodes with
        common r,e]
61:     decide(e)
62: end while
63: {Task 4:}
64: while true do
65:     {Send list of suspected nodes}
66:     Suspected_p ← D_1
67:     suspect_p = (SUSPECT, p, Suspected_p)_p
68:     Send suspect_p to all
69: end while
70: {Task 5:}
71: for r in S do
72:     When p receives Suspect_q from q
73:     if r in Suspected_q then
74:         Suspecting_p[r] ← Suspecting_p[r] ∪ (q)
75:     else
76:         Suspecting_p[r] ← Suspecting_p[r] − (q)
77:     end if
78:     if |Suspecting_p[r]| ⩾ k + 1 then
79:         Output_p = Output_p ∪ (r)
80:     else
81:         Output_p = Output_p − (r)
82:     end if
83: end for
```

discussion about the replication protocol is beyond the scope of this paper. Being fully replicated, each Authorization Control Service has the same PDP, PAP, and PEP. There is no need to migrate those entities. However, only the current leader server has the information about the granted access requests in addition to having the most up to date policy database. Granted access requests information is stored in the session history of the current Authorization Control Service. Therefore, the session history and the policy database need to be migrated to the new Authorization Control Service to allow users with granted access a continuous use of the allowed resources.

4.3   The Discovery Service

Implementing a Moving Target Defense for an Access Control service requires switching frequently but randomly the server that is responsible for offering the Access Control service. Once the Authorization Control service has been switched to a newly elected leader server, users need a way to rediscover the new server offering the service. In this section, we adapt the Service Location Protocol or SLP in order to enable users to discover the new location of the services they need.

SLP Module. The SLP Module is responsible for running the service discovery protocol. This service is provided to clients that need to consult the SLP Module in order to discover the new Authorization Control Service location.

SLP Overview. Service Location Protocol (SLP) is a protocol designed by the Internet Engineering Task Force to eliminate the need of a manual configuration from users of communication networks in order to discover services, applications, and devices available in those networks. Since users, mainly mobile users, increasingly experience changing environments and the fact that the Internet has became more service oriented, service location is becoming more helpful in today's complex networks [18,50].

SLP Architecture. The SLP framework includes three main components called "Agents" to process SLP information. These agents are: User Agents (UA), Service Agents (SA), and Directory Agents (DA).

- *User Agents (UA)*: They are responsible for requesting services on behalf of the users or applications.
- *Service Agents (SA)*: They are entities that advertise the location and description of services on behalf of services. To advertise services, the SAs embed the service information into an URL. These information include the IP address, the port number, the service type and the path. Each service type is characterized by specific attributes along with their default values. All of these are specified by the Service Templates.
- *Directory Agents (DA)*: They are central repositories that aggregate SLP information. Since service information are embedded in a URL, these URL are stored by the DA which provides them to any UA that have issued a request which matched some attributes in the URL.

  At the beginning of the protocol, any service provider needs to advertise its services. For that purpose, its SA registers the service with the DA. This step is known as the Service Registration. The DA acknowledges the registration by issuing a service ACK message to the SA. A user that needs to use

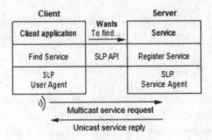

Fig. 2. Service discovery flow

the given service needs to have his UA issuing a query with the appropriate attributes to the DA. This is known as the Service Request step. The DA may reply back to the UA with the address and characteristics of the desired service (Service Replay). This is the general approach of the SLP protocol as illustrated on Fig. 2.

There is a major issue with the general approach of the SLP protocol as explained above. No one from both the UA and the SA knows the address of the DA. Before registering a service with the DA, the SA needs to discover the existence of the DA. The same thing applies to the UA.

Three different methods are used to discover the location of the DA: static, active, and passive. When using the static discovery method, both the UA and the SA learn the address of the DA from a DHCP server. In case of an active discovery, SLP agents contact the DA by sending service requests to the SLP multicast address on which the DA is configured to listen to for incoming communications. Upon receiving a service request, DA responds directly to the requesting agent via the agent's unicast address. The passive discovery method involves DAs periodically advertising their existence through the SLP multicast address. The other SLP agents discover the DAs location after listening the multicast advertisements. They can then contact the DAs directly through their unicast addresses for other operations [37].

Besides the basic SLP architecture involving SAs, DAs, and UA, it is possible to set a SLP architecture without DAs. In this case, UAs and SAs need to communicate directly to each other. In order to discover available services, UAs repeatedly send out their service requests to the SLP multicast address. On the other hand, SAs are listening for incoming requests on the SLP multicast address. Upon receiving a request corresponding to a service they are advertising, SAs reply through unicast address to UAs.

4.4   The Resource Access Service

It is a server that manages the various protected resources. It acts as both client and server when receiving or replying to access requests from clients. The decision to grant the access to those resources is received from the Authorization Control Service.

5   Implementation

In this section, we introduce the proof-of-concept implementation of our proposed architecture. We have designed a test case to exemplify the functioning of the protocol.

5.1   Clients and Resource Access Service

We assume that we have a set of users represented by client applications. Those users are grouped into a set of roles (Undergraduate, Graduate, and Faculty). We also have a set of resources, in this case files stored on a file server. This file server constitutes our Resource Access Service. We have also defined a set of actions to be performed on those files by users. We have chosen basic Linux actions: Read, Write, and Execute. The different permissions given to users over those files are represented on Fig. 3.

Role/Resource	syllabus.txt	research.txt	grades.txt	assignment.txt	certificate.txt
Undergrad	r		r	rwx	r
Graduate	rwx	rwx	r	rx	
Faculty	rwx	rwx	rwx	r	rwx

Fig. 3. Roles-Permissions assignment

5.2   Authorization Control Server

Using socket programming, we have implemented five Authorization Control Services named $ACS1$, $ACS2$, $ACS3$, $ACS4$, and $ACS5$. Those have been implemented as Java client-servers. Each of those Authorization Control Services has an Access Control Module that is responsible for verifying user's authorization to resources that are stored on the Resource server. At each time, only one Authorization Control Service is responsible for providing the access control service.

Access Control. We start our process with the Authorization service being provided by, let us say, the Authorization Control Server ACS1. We consider that user Tom submits a request to read a file named certificate.txt. This request is intercepted by ACS1 which run the Access Control Module. The Access Control Module has been implemented using the Balana [24] open source implementation of XACML. Tom's query will be a tuple user-id, action, resource-name where in this particular case user-id is Tom, action is read, and resource-name is the file certificate.txt requested by Tom.

Policy Enforcement Point. Tom's query is intercepted by the Policy Enforcement Point (PEP). In fact, the PEP intercepts all queries sent to the Resource Access Service [40]. We have developed a wrapper that converts the original query into a XACML request. The request is then sent to the PDP for verification. Figure 5 illustrates the form of the XACML request.

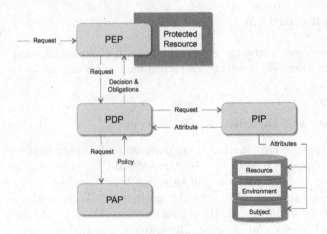

Fig. 4. XACML architecture

```
-<Request CombinedDecision="false" ReturnPolicyIdList="true">
 -<Attributes Category="urn:oasis:names:tc:xacml:3.0:attribute-category:resource">
  -<Attribute AttributeId="urn:oasis:names:tc:xacml:1.0:resource:resource-id" IncludeInResult="false">
     <AttributeValue DataType="http://www.w3.org/2001/XMLSchema#string">resource</AttributeValue>
   </Attribute>
  </Attributes>
 -<Attributes Category="urn:oasis:names:tc:xacml:1.0:subject-category:access-subject">
  -<Attribute AttributeId="http://wso2.org/claims/role" IncludeInResult="false">
     <AttributeValue DataType="http://www.w3.org/2001/XMLSchema#string">Role</AttributeValue>
   </Attribute>
  </Attributes>
 -<Attributes Category="urn:oasis:names:tc:xacml:3.0:attribute-category:action">
  -<Attribute AttributeId="urn:oasis:names:tc:xacml:1.0:action:action-id" IncludeInResult="false">
     <AttributeValue DataType="http://www.w3.org/2001/XMLSchema#string">operation</AttributeValue>
   </Attribute>
  </Attributes>
 </Request>
```

Fig. 5. User request sample

Policy Decision Point. The Policy Decision Point (PDP) receives Tom's request coming from the PEP. It needs to analyse if Tom fulfills the required conditions to read the file certificate.txt. The PDP will consult the Policy file to determine what actions Tom is allowed to perform on the file certificate.txt. Balana [24] provides us with an API call that allows us to create a PDP.

Policy Administration Point. To write policies, we have made use of the Simple Policy Editor. This policy editor is part of WSO2 Identity Server [25]. Simple Policy Editor allows anyone to create XACML 3.0 policies without an extensive knowledge of XACML language. However, an understanding of access control rules is required. Figure 6 is a sample of our policy file.

In addition to the Policy file, the PDP also consults the user-role assignment table. After determining Tom's role, which is undergraduate, and consulting the Policy file, the PDP reaches the conclusion to authorize Tom to read the desired file. The PDP passes that decision back to the PEP. That response is represented as a XACML file. A sample of the response XACML file is exhibited on Fig. 7.

The PEP then replies to Tom with a response granting him access to the file. Tom can now access the file server.

```
-<Rule Effect="Permit" RuleId="Rule-5">
 -<Target>
  -<AnyOf>
   -<AllOf>
    -<Match MatchId="urn:oasis:names:tc:xacml:1.0:function:string-equal">
       <AttributeValue DataType="http://www.w3.org/2001/XMLSchema#string">certificate.txt</AttributeValue>
       <AttributeDesignator AttributeId="urn:oasis:names:tc:xacml:1.0:resource:resource-id" Category="urn:oasis
    </Match>
   </AllOf>
  </AnyOf>
 </Target>
 -<Condition>
  -<Apply FunctionId="urn:oasis:names:tc:xacml:1.0:function:string-at-least-one-member-of">
   -<Apply FunctionId="urn:oasis:names:tc:xacml:1.0:function:string-bag">
      <AttributeValue DataType="http://www.w3.org/2001/XMLSchema#string">read</AttributeValue>
      <AttributeValue DataType="http://www.w3.org/2001/XMLSchema#string">write</AttributeValue>
      <AttributeValue DataType="http://www.w3.org/2001/XMLSchema#string">execute</AttributeValue>
   </Apply>
      <AttributeDesignator AttributeId="urn:oasis:names:tc:xacml:1.0:action:action-id" Category="urn:oasis:names:tc
  </Apply>
 </Condition>
</Rule>
```

Fig. 6. Policy file sample

```
-<Response>
 -<Result>
   <Decision>Deny</Decision>
  -<Status>
     <StatusCode Value="urn:oasis:names:tc:xacml:1.0:status:ok"/>
  </Status>
  -<PolicyIdentifierList>
     <PolicyIdReference>univ_undergrad_policy</PolicyIdReference>
  </PolicyIdentifierList>
 </Result>
</Response>
```

Fig. 7. Response sample

Leader Election. Our objective is to protect the Access Control Module by regularly switching the Authorization Control Service providing the Access Control service at any given time. For the sake of demonstrating, we have chosen to switch the Authorization Control Service after every 10 min plus a random number of seconds. The random time is added to cancel the predictability of the time when the election takes place. An attacker knowing when the new leader is elected can schedule his attack accordingly.

An election is called after the end of term of the current Leader. In this instance, that term is set to ten minutes and some random seconds. The current leader being ACS1, it is the one responsible to call for an election. Using Java, the Leader Election is implemented according to the adaptation of the protocol presented in Sect. 4.2. At the end of the protocol a new leader is elected. This leader is different from ACS1, for instance ACS3 has been selected a the new leader. This is the server that is going to be responsible of providing the Authorization Control Service until next election. We made all Authorization Control Services probe the leader after every ninety seconds by sending a IsAlive message to it. This is done in order to detect the failure of the current leader.

Tom want to request another file stored on the file server, but the Authorization Control Service has been moved from ACS1 to ACS3. Any attacker who was in the middle of preparing an attack against ACS1 will be attacking the wrong Authorization Control Service, which is the intended goal of our architecture. However, Tom will also be sending his authorization request to the wrong server.

Migration. We have implemented the Migration Module as a mechanism to simply transfer the session history file and the policy file from the previous leader ACS1 to the elected leader ACS3. As stated in Sect. 4.2, the other access control modules are the same across all Authorization Control Services. The reason for transferring ACS1 Policy file to ACS3 is that while ACS1 was providing the Authorization service, policies, resources and users may have been updated. To avoid disruption in the access control service, ACS3 needs to have the most recent policy file.

Other alternatives to this migration can be envisioned. One option is to store the policies in a Policy Database, and replicate the database across all the Authorization Control Services accordingly. Another option would be to migrate the database from the current leader to the new leader at the end of an election. An additional option would be to use a single Policy database that would be shared by all the Authorization Control Services. This last option can create a potential issue by making that single policy database a single point of failure attractive for would be attacker.

5.3 Discovery Service

We need to let Tom know that the Authorization Control Server ACS3 has been elected as the new leader, and therefore he should send his request to ACS3. The Discovery Service allows us to achieve that goal through the adaptation of SLP protocol.

We have implemented the Discovery Service using a tool called OpenSLP [35] which is an open source implementation of the Service Location Protocol. OpenSLP can be used either in a three components mode or in a two components mode. In the first mode, we can have a User Agent (UA), a Service Agent (SA) and Directory Agent (DA). The User Agent is the Agent requesting services. The Service Agent is the Agent providing the services, while the Directory Agent is the repository of services. In a two component mode, we can have only the User Agent and the Service Agent. In this case, the Service Agent plays also the role of a Directory Agent (DA). For the sake of this demonstration, we have implemented the later option. We have installed OpenSLP and made sure that *slpd*, which is the OpenSLP daemon, is running.

Service Agent. Since our setting do not use a Directory Agent, the new leader will have to register its services with *slpd* upon being elected. The old Authorization Control Service, previously registered, is unregistered to avoid confusing users. The new leader registers its access control service by issuing a query in the form of a ServiceURL. The ServiceURL has the following form: *service:ServiceName://IPAddress:PortNo*, where ServiceName is the Authorization Service, IPAddress is the IP address of the new leader, and PortNo is the port where the Authorization Service is running.

User Agent. In our setting, the User Agent is Tom who needs to find the location of the new leader which is providing the access control service. In order to discover the Authorization Service, Tom's client sends a multicast packet with a ServiceURL. The Service Agent, which is the new leader, will verifies if Tom's query matches the registered service. In case of a match, the Service Agent replies to Tom with the ServiceURL informing him how to access the access control service.

6 Conclusion and Future Work

We have sketched a Moving Target Defense architecture aiming at defending an Access Control Reference Monitor. The design allows a master Resource Access Server and a master Authorization Control Server to be periodically and randomly switched to other ones. This mechanism allows the disruption of any ongoing attack on the Access Control Reference Monitor. This work opens a new direction in research on Moving Target Defense of an Access Control Reference Monitor. This architecture can benefit from some improvements. For instance, we do not believe that the election algorithm is an optimal one in term of computation and the number of messages exchanged during the election process.

Acknowledgement. This work was partially supported by funding from CableLabs, the US National Science Foundation under grant number 1650573, and the US Department of Energy under contract DE-NE0008571. Any opinions, findings, and conclusions or recommendations expressed in this material are those of the authors and do not necessarily reflect the views of CableLabs, the National Science Foundation, or the Department of Energy.

References

1. Abu-Amara, H., Lokre, J.: Election in asynchronous complete networks with intermittent link failures. IEEE Trans. Comput. **43**(7), 778–788 (1994)
2. Al-Shaer, E.: Toward network configuration randomization for moving target defense. In: Jajodia, S., Ghosh, A.K., Swarup, V., Wang, C., Wang, X.S. (eds.) Moving Target Defense, vol. 54, pp. 153–159. Springer, New York (2011)
3. Anderson, J.: Computer Security Technology Planning Study. Technical report ESD-TR-73-51, Electronic Systems Division, Hanscom Air Force Base, Hanscom, MA (1974)
4. Antonatos, S., Akritidis, P., Markatos, E.P., Anagnostakis, K.G.: Defending against hitlist worms using network address space randomization. Comput. Netw. **51**(12), 3471–3490 (2007)
5. Arghavani, A., Ahmadi, E., Haghighat, A.: Improved bully election algorithm in distributed systems. In: 2011 International Conference on Information Technology and Multimedia (ICIM), pp. 1–6. IEEE (2011)
6. Brunekreef, J., Katoen, J.P., Koymans, R., Mauw, S.: Design and analysis of dynamic leader election protocols in broadcast networks. Distrib. Comput. **9**(4), 157 (1996)
7. Burrows, M.: The chubby lock service for loosely-coupled distributed systems. In: Proceedings of the 7th Symposium on Operating systems design and implementation, pp. 335–350. USENIX Association (2006)

8. Castro, M., Liskov, B.: Practical byzantine fault tolerance and proactive recovery. ACM Trans. Comput. Syst. (TOCS) **20**(4), 398–461 (2002)
9. Castro, M., Liskov, B., et al.: Practical byzantine fault tolerance. In: OSDI 1999, pp. 173–186 (1999)
10. Compton, M.D.: Improving the quality of service and security of military networks with a network tasking order process (2010)
11. Copeland, C., Zhong, H.: Tangaroa: a byzantine fault tolerant raft
12. Evans, D., Nguyen-Tuong, A., Knight, J.: Effectiveness of moving target defenses. In: Jajodia, S., Ghosh, A.K., Swarup, V., Wang, C., Wang, X.S. (eds.) Moving Target Defense, vol. 54, pp. 29–48. Springer, New York (2011)
13. Ferraiolo, D., Cugini, J., Kuhn, D.R.: Role-based access control (RBAC): features and motivations. In: Proceedings of 11th Annual Computer Security Application Conference, pp. 241–48 (1995)
14. Ferreira, A., Chadwick, D., Farinha, P., Correia, R., Zao, G., Chilro, R., Antunes, L.: How to securely break into RBAC: the BTG-RBAC model. In: Annual Computer Security Applications Conference, ACSAC 2009, pp. 23–31. IEEE (2009)
15. Fraim, L.J.: Scomp: a solution to the multilevel security problem. IEEE Comput. **16**(7), 26–34 (1983)
16. Garcia-Molina, H.: Elections in a distributed computing system. IEEE Trans. Comput. **31**(1), 48–59 (1982)
17. Gilbert, M.D.M.: An examination of federal and commercial access control policy needs. In: National Computer Security Conference, 1993 (16th) Proceedings: Information Systems Security: User Choices, p. 107. DIANE Publishing (1995)
18. Guttman, E.: Service location protocol: Automatic discovery of IP network services. IEEE Internet Comput. **3**(4), 71–80 (1999)
19. Han, Y., Lu, W., Xu, S.: Characterizing the power of moving target defense via cyber epidemic dynamics. In: Proceedings of the 2014 Symposium and Bootcamp on the Science of Security, p. 10. ACM (2014)
20. Howard, H., Schwarzkopf, M., Madhavapeddy, A., Crowcroft, J.: Raft refloated: do we have consensus? ACM SIGOPS Oper. Syst. Rev. **49**(1), 12–21 (2015)
21. Huh, J.H., Bobba, R.B., Markham, T., Nicol, D.M., Hull, J., Chernoguzov, A., Khurana, H., Staggs, K., Huang, J.: Next-generation access control for distributed control systems. IEEE Internet Comput. **20**(5), 28–37 (2016)
22. Hunt, P., Konar, M., Junqueira, F.P., Reed, B.: Zookeeper: wait-free coordination for internet-scale systems. In: USENIX Annual Technical Conference, vol. 8, p. 9 (2010)
23. Inc, A.: Bonjour. https://support.apple.com/bonjour. Accessed: 26 Feb 2017
24. Info, X.: Balana. http://xacmlinfo.org/2012/12/18/getting-start-with-balana. Accessed: 26 Feb 2017
25. Info, X.: Wso2 identity server. http://xacmlinfo.org/category/wso2is/. Accessed: 26 Feb 2017
26. King, V., Saia, J., Sanwalani, V., Vee, E.: Scalable leader election. In: Proceedings of the 17th Annual ACM-SIAM Symposium on Discrete Algorithm, Miami, FL, USA (2006)
27. King, V., Saia, J., Sanwalani, V., Vee, E.: Scalable leader election. In: Proceedings of the Seventeenth Annual ACM-SIAM Symposium on Discrete Algorithm, pp. 990–999. Society for Industrial and Applied Mathematics (2006)
28. Lamport, L.: The part-time parliament. ACM Trans. Comput. Syst. (TOCS) **16**(2), 133–169 (1998)
29. Lamport, L., Shostak, R., Pease, M.: The byzantine generals problem. ACM Trans. Program. Lang. Syst. (TOPLAS) **4**(3), 382–401 (1982)

30. Lamport, L., et al.: Paxos made simple. ACM Sigact News **32**(4), 18–25 (2001)
31. Le Lann, G.: Distributed systems-towards a formal approach. In: IFIP Congress, Toronto, vol. 7, pp. 155–160 (1977)
32. Lee, J.W., Schulzrinne, H., Kellerer, W., Despotovic, Z.: z2z: discovering zeroconf services beyond local link. In: 2007 IEEE Globecom Workshops, pp. 1–7. IEEE (2007)
33. Martin, J.P., Alvisi, L.: Fast byzantine consensus. IEEE Trans. Dependable Secure Comput. **3**(3), 202–215 (2006)
34. Mohammed, I., Dilts, D.M.: Design for dynamic user-role-based security. Comput. Secur. **13**(8), 661–671 (1994)
35. OpenSLP: Service location protocol. http://www.openslp.org/. Accessed: 26 Feb 2017
36. ORACLE: Trusted Solaris Operating System. http://www.oracle.com/technetwork/server-storage/solaris/overview/index-136311.html
37. Perkins, C., Kaplan, S.: Service location protocol. In: ACTS Mobile Networking Summit/MMITS Software Radio Workshop (1999)
38. Perkins, C.E., et al.: Dhcp options for service location protocol (1999)
39. Presser, A., Farrell, L., Kemp, D., Lupton, W.: UPnP device architecture 1.1. In: UPnP Forum, vol. 22 (2008)
40. Rissanen, E., et al.: Extensible access control markup language (xacml) version 3.0 (2013)
41. Ruj, S., Stojmenovic, M., Nayak, A.: Decentralized access control with anonymous authentication of data stored in clouds. IEEE Trans. Parallel Distrib. Syst. **25**(2), 384–394 (2014)
42. Sandhu, R.S., Coyne, E.J., Feinstein, H.L., Youman, C.E.: Role-based access control models. Computer **29**(2), 38–47 (1996)
43. Sandhu, R.S., Samarati, P.: Access control: principle and practice. IEEE Commun. Mag. **32**(9), 40–48 (1994)
44. Sayeed, H.M., Abu-Amara, M., Abu-Amara, H.: Optimal asynchronous agreement and leader election algorithm for complete networks with byzantine faulty links. Distrib. Comput. **9**(3), 147–156 (1995)
45. Schell, R., Tao, T., Heckman, M.: Designing the GEMSOS security kernel for security and performance. In: Proceedings of the 8th National Computer Security Conference, Gaithersburg, MD (1985)
46. Shi, X., Lin, H., Jin, H., Zhou, B.B., Yin, Z., Di, S., Wu, S.: Giraffe: a scalable distributed coordination service for large-scale systems. In: 2014 IEEE International Conference on Cluster Computing (CLUSTER), pp. 38–47. IEEE (2014)
47. Singh, G.: Leader election in the presence of link failures. IEEE Trans. Parallel Distrib. Syst. **7**(3), 231–236 (1996)
48. Soundarabai, P.B., Thriveni, J., Manjunatha, H., Venugopal, K., Patnaik, L.: Message efficient ring leader election in distributed systems. In: Chaki, N., Meghanathan, N., Nagamalai, D. (eds.) Computer Networks & Communications (NetCom), pp. 835–843. Springer, New York (2013)
49. The TrustedBSD Project: Trustedbsd. http://www.trustedbsd.org
50. Veizades, J., Perkins, C.E.: Service location protocol (1997)
51. Wang, G., Koshy, J., Subramanian, S., Paramasivam, K., Zadeh, M., Narkhede, N., Rao, J., Kreps, J., Stein, J.: Building a replicated logging system with apache kafka. Proc. VLDB Endowment **8**(12), 1654–1655 (2015)
52. Wright, C., Cowan, C., Smalley, S., Morris, J., Kroah-Hartman, G.: Linux security modules: general security support for the linux kernel. In: Proceedings of the 11th USENIX Security Symposium, San Francisco, CA (2002)

53. Xu, J., Guo, P., Zhao, M., Erbacher, R.F., Zhu, M., Liu, P.: Comparing different moving target defense techniques. In: Proceedings of the First ACM Workshop on Moving Target Defense, pp. 97–107. ACM (2014)
54. Zhuang, R., DeLoach, S.A., Ou, X.: Towards a theory of moving target defense. In: Proceedings of the First ACM Workshop on Moving Target Defense, pp. 31–40. ACM (2014)

Preventing Unauthorized Data Flows

Emre Uzun[1]([⊠]), Gennaro Parlato[2], Vijayalakshmi Atluri[3], Anna Lisa Ferrara[2],
Jaideep Vaidya[3], Shamik Sural[4], and David Lorenzi[3]

[1] Bilkent University, Ankara, Turkey
emreu@bilkent.edu.tr
[2] University of Southampton, Southampton, UK
gennaro@ecs.soton.ac.uk, al.ferrara@soton.ac.uk
[3] MSIS Department, Rutgers Business School, Newark, USA
{atluri,jsvaidya,dlorenzi}@cimic.rutgers.edu
[4] Department of Computer Science and Engineering,
IIT Kharagpur, Kharagpur, India
shamik@cse.iitkgp.ernet.in

Abstract. Trojan Horse attacks can lead to unauthorized data flows
and can cause either a confidentiality violation or an integrity violation.
Existing solutions to address this problem employ analysis techniques
that keep track of all subject accesses to objects, and hence can be expensive. In this paper we show that for an unauthorized flow to exist in an
access control matrix, a flow of length one must exist. Thus, to eliminate
unauthorized flows, it is sufficient to remove all one-step flows, thereby
avoiding the need for expensive transitive closure computations. This
new insight allows us to develop an efficient methodology to identify and
prevent all unauthorized flows leading to confidentiality and integrity
violations. We develop separate solutions for two different environments
that occur in real life, and experimentally validate the efficiency and
restrictiveness of the proposed approaches using real data sets.

1 Introduction

It is well known that access control models such as Discretionary Access Control
(DAC) and Role Based Access Control (RBAC) suffer from a fundamental weakness – their inability to prevent leakage of data to unauthorized users through
malware, or malicious or complacent user actions. This problem, also known
as a Trojan Horse attack, may lead to an unauthorized data flow that may
cause either a confidentiality or an integrity violation. More specifically, (i) a
confidentiality violating flow is the potential flow of sensitive information from
trusted users to untrusted users that occurs via an illegal read operation, and

---

The work of Parlato and Ferrara is partially supported by EPSRC grant no.
EP/P022413/1.

© IFIP International Federation for Information Processing 2017
Published by Springer International Publishing AG 2017. All Rights Reserved
G. Livraga and S. Zhu (Eds.): DBSec 2017, LNCS 10359, pp. 41–62, 2017.
DOI: 10.1007/978-3-319-61176-1_3

(ii) *integrity violating flow* is the potential contamination of a sensitive object that occurs via an illegal write operation by an untrusted user. We now give an example to illustrate these two cases.

Table 1. Access control matrix

Subject	o_1	o_2	o_3	o_4	o_5	o_6	o_7
s_1	r	r	w	w	w		
s_2	r	r	w	w	w		
s_3			r	r	r	w	w
s_4			r	r	r	w	w
s_5						r	

Example 1. Consider a DAC policy represented as an access control matrix given in Table 1 (r represents *read*, and w represents *write*).

Confidentiality Violating Flow: Suppose s_3 wants to access data in o_1. s_3 can simply accomplish this (without altering the access control rights) by exploiting s_1's read access to o_1. s_3 prepares a malicious program disguised in an application (i.e., a *Trojan Horse*) to accomplish this. When run by s_1, and hence using her credentials, the program will read contents of o_1 and write them to o_3, which s_3 can read. All this is done without the knowledge of s_1. This unauthorized data flow allows s_3 to *read* the contents of o_1, without explicitly accessing o_1.

Integrity Violating Flow: Suppose s_1 wants to contaminate the contents of o_6, but she does not have an explicit write access to it. She prepares a malicious program. When this is run by s_3, it will read from o_3, that s_1 has write access to and s_3 has read access, and write to o_6 using s_3's credentials, causing o_6 to be contaminated by whatever s_1 writes to o_3. This unauthorized flow allows s_1 to *write* to o_6 without explicitly accessing o_6.

Such illegal flows can occur in the many of the most common systems that we use today because they employ DAC policies instead of a more restrictive MAC policy [2]. For example, in UNIX, the key system files are only readable by root, however, the access control rights of the other files are determined solely by the users. If a Trojan horse program is run with the root user's privileges, the data in the system files, such as the user account name and password hashes could be leaked to some untrusted users. As another example, a similar flow might occur in Social Networks as well. For instance, Facebook offers a very extensive and fine-grained privacy policy to protect the data posted on user profiles. However, this policy is under the user's control. A Trojan horse attack is likely when the users grant access to third party Facebook applications, that usually request access to user profile data. An untrusted application could violate the user's privacy settings and access confidential information.

The first step for eliminating occurrences like the ones depicted in the example above is to perform a security analysis. To date, existing solutions to address such problems give the impression that such unauthorized flows could only be efficiently prevented in a dynamic setting (i.e., only by examining the actual operations), while preventing them in a static setting (i.e., by examining the authorization specifications) would require the computation of the transitive closure and therefore be very expensive. However, in this paper, we show that a transitive closure is not needed for the static case and less expensive analyses can be used to solve this problem. More precisely, we have discovered that merely identifying and then restricting a single step data flow, as opposed to the entire path, is sufficient to prevent the unauthorized flow. This new insight has significantly changed the dimensions of the problem and allows us to offer a variety of strategies that fit different situational needs.

Consider the following situations which have differing solution requirements. For example, in embedded system environments complex monitors cannot be deployed due to their computation or power requirements and therefore existing dynamic preventive strategies are not applicable. Similarly, there are solutions for cryptographic access control [8,13], where accesses are not mediated by a centralized monitor and therefore easily offer distributed trust. In such cases, the access control policy needs to be "data leakage free" by design. In other situations, when there are no special computational or power constraints, a monitor can be used, and therefore can be utilized to prevent data leakages. However, there may also be situations where access needs to be granted even if a data leakage may occur and then audited after the fact. This would happen in emergencies, which is why break-glass models exist [5,23,25].

Therefore, in this paper, we develop different solutions to address both the confidentiality and integrity violations. Specifically, we propose a data leakage free by design approach that analyzes the access control matrix to identify "potential" unauthorized flows and eliminates them by revoking necessary read and write permissions. Since this eliminates all potential unauthorized flows, regardless of whether they actually occur or not, this could be considered too restrictive. However, it is perfectly secure in the sense that no data leakages can ever occur, and of course this is the only choice when monitoring is not feasible. Although it may seem very restrictive in the first place, we apply this only to the *untrusted* sections of the access control system. It is important to note that in all potential unauthorized flows one can only be sure of a violation by performing a content analysis of the objects. This is outside the scope of the paper. We also develop a monitor based approach, in which object accesses are tracked dynamically at each read and write operation. Thus, any suspicious activity that could lead to an unauthorized data flow can be identified and prevented at the point of time that it occurs. Thus, this approach only restricts access if there is a *signal* for an unauthorized data flow.

The fact that it is adequate to identify and eliminate one-step flows allows us to identify a limited set of accesses that are both necessary and sufficient to prevent all confidentiality and integrity violations. On the other hand, earlier approaches proposed in the literature [17,21,32] keep track of all the actions and maintain information relevant to these to eliminate unauthorized flows, and therefore are more expensive than our proposed approach. Moreover, while Mao et al. [21] and Zimmerman et al. [32] address the issue of integrity violation, Jaume et al. [17] address the issue of confidentiality violation, however, none of them tackle both of these problems.

This paper is organized as follows. In Sect. 2, we present preliminary background for our analysis, and in Sects. 3 and 4 we present the details of the two strategies. In Sect. 5, we present the results of our empirical evaluation. In Sect. 6, we review the related work. In Sect. 7, we give our concluding remarks and provide an insight into our future work on this problem. Some of the proofs of the theorems and lemmas are presented in the Appendix.

2   Preliminaries

Access Control Systems. An *access control system* (ACS for short) $\mathcal{C}$ is a tuple $(S, O, \rightarrow_r, \rightarrow_w)$, where S is a finite set of *subjects*, O is a finite set of *objects*, $\rightarrow_r \subseteq O \times S$, and $\rightarrow_w \subseteq S \times O$. We always assume that O and S are disjoint. A pair $(o, s) \in \rightarrow_r$, also denoted $o \rightarrow_r s$, is a permission representing that subject s can *read* object o. Similarly, a pair $(s, o) \in \rightarrow_w$, denoted $s \rightarrow_w o$, is a permission representing that subject s can *write* into object o. For the sake of simplicity, we consider only read and write permissions as any other operation can be rewritten as a sequence of read and write operations.

Graph Representation of ACS. An ACS can be naturally represented with a bipartite directed graph [30]. The *graph of an ACS*, $\mathcal{C} = (S, O, \rightarrow_r, \rightarrow_w)$, denoted $G_{\mathcal{C}}$, is the bipartite graph $(S, O, \rightarrow)$ whose partition has the parts S and O with edges $\rightarrow = (\rightarrow_r \cup \rightarrow_w)$. Figure 1 shows the graph representation of the ACS shown in Table 1.

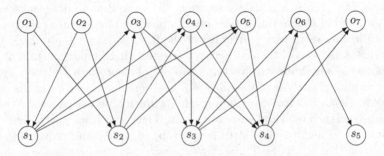

Fig. 1. Graph representation of the ACS given in Table 1

Vulnerability Paths. In an access control system $\mathcal{C}$, a *flow path* from object o to object o', denoted $o \rightsquigarrow o'$, is a path in $G_{\mathcal{C}}$ from o to o', which points out the possibility of copying the content of o into o'. The *length* of a flow path corresponds to the number of subjects along the path. For example, $o_1 \rightarrow_r s_1 \rightarrow_w o_3$ (denoted as $o_1 \rightsquigarrow o_3$) is a flow path of length 1, while $o_1 \rightarrow_r s_1 \rightarrow_w o_3 \rightarrow_r s_3 \rightarrow_w o_6$ (denoted as $o_1 \rightsquigarrow o_6$) is a flow path of length 2 of the ACS shown in Fig. 1. In all, there are 12 flow paths of length 1, while there are 4 flow paths of length 2 in the ACS shown in Fig. 1.

Confidentiality Vulnerability: An ACS $\mathcal{C}$ has a *confidentiality vulnerability*, if there are two objects o and o', and a subject s such that $o \rightsquigarrow o' \rightarrow_r s$ (*confidentiality vulnerability path* or simply *vulnerability path*), and $o \not\rightarrow_r s$. A confidentiality vulnerability, shows that subject s (the *violator*) can *potentially* read the content of object o through o', though s is not allowed to read directly from o. We represent confidentiality vulnerabilities using triples of the form (o, o', s). For example, the ACS depicted in Fig. 1 has the confidentiality vulnerability (o_1, o_3, s_3) since $o_1 \rightsquigarrow o_3$ and $o_3 \rightarrow_r s_3$ but $o_1 \not\rightarrow_r s_3$. Similarly, (o_2, o_6, s_5) is another confidentiality vulnerability since $o_2 \rightsquigarrow o_6$ and $o_6 \rightarrow_r s_5$ but $o_2 \not\rightarrow_r s_5$. In total, there are 15 confidentiality vulnerabilities:

(o_1, o_3, s_3), (o_1, o_3, s_4), (o_1, o_4, s_3), (o_1, o_4, s_4), (o_1, o_5, s_3), (o_1, o_5, s_4), (o_2, o_3, s_3), (o_2, o_3, s_4), (o_2, o_4, s_3), (o_2, o_4, s_4), (o_2, o_5, s_3), (o_2, o_5, s_4), (o_5, o_6, s_5), (o_1, o_6, s_5), (o_2, o_6, s_5).

Integrity Vulnerability: An ACS $\mathcal{C}$ has an *integrity vulnerability*, if there exist a subject s, and two objects o and o' such that $s \rightarrow_w o$, $o \rightsquigarrow o'$ (*integrity vulnerability path* or simply *vulnerability path*) and $s \not\rightarrow_w o'$. An integrity vulnerability, shows that subject s (the *violator*) can indirectly write into o' using the path flow from o to o', though s is not allowed to write directly into o'. We represent integrity vulnerabilities using triples of the form (s, o, o'). For example, the ACS depicted in Fig. 1 has the integrity vulnerability (s_1, o_3, o_6) since $o_3 \rightsquigarrow o_6$ and $s_1 \rightarrow_w o_3$ but $s_1 \not\rightarrow_w o_6$. In total, there are 12 integrity vulnerabilities:

(s_1, o_3, o_6), (s_1, o_3, o_7), (s_1, o_4, o_6), (s_1, o_4, o_7), (s_1, o_5, o_6), (s_1, o_5, o_7), (s_2, o_3, o_6), (s_2, o_3, o_7), (s_2, o_4, o_6), (s_2, o_4, o_7), (s_2, o_5, o_6), (s_2, o_5, o_7).

When an ACS has either a confidentiality or an integrity vulnerability, we simply say that $\mathcal{C}$ has a *vulnerability*, whose *length* is that of its underlying vulnerability path. Thus, for the ACS depicted in Fig. 1, there are $15 + 12 = 27$ vulnerabilities.

Data Leakages. A vulnerability in an access control system does not necessarily imply that a data leakage (confidentiality or integrity violation) occurs. Rather, a leakage can potentially happen unless it is detected and blocked beforehand, using for example a monitor. Before we define this notion formally, we first develop the necessary formalism.

A *run* of an ACS $\mathcal{C}$ is any finite sequence $\pi = (s_1, op_1, o_1) \ldots (s_n, op_n, o_n)$ of triples (or *actions*) from the set $S \times \{read, write\} \times O$ such that for every $i \in [1, n]$ one of the following two cases holds:

(Read) $op_i = read$, and $o_i \rightarrow_r s_i$;
(Write) $op_i = write$, and $s_i \rightarrow_w o_i$.

A run π represents a sequence of allowed read and write operations executed by subjects on objects. More specifically, at step $i \in [n]$ subject s_i accomplishes the operation op_i on object o_i. Furthermore, s_i has the right to access o_i in the op_i mode. A run π has a *flow* from an object $\hat{o}_1$ to a subject $\hat{s}_k$ provided there is a flow path $\hat{o}_1 \rightarrow_r \hat{s}_1 \rightarrow_w \hat{o}_2 \ldots \hat{o}_k$ and $\hat{o}_k \rightarrow_r \hat{s}_k$ such that $(\hat{s}_1, read, \hat{o}_1)(\hat{s}_1, write, \hat{o}_2) \ldots (\hat{s}_k, read, \hat{o}_k)$ is a sub-sequence of π. Similarly, we can define flows from subjects to objects, objects to objects, and subjects to subjects.

Confidentiality Violation: A run π of an ACS $\mathcal{C}$ has a *confidentiality violation*, provided there is a confidentiality vulnerability path from an object o to a subject s and π has a flow from o to s. An ACS $\mathcal{C}$ has a *confidentiality violation* if there is a run of $\mathcal{C}$ with a confidentiality violation.

Thus, for example, in the ACS depicted in Fig. 1, a *confidentiality violation* would occur if there was a sequence $(s_1, read, o_1)(s_1, write, o_3)(s_3, read, o_3)$ which was a sub-sequence of π.

Integrity Violation: A run π of an ACS $\mathcal{C}$ has an *integrity violation*, provided there is an integrity vulnerability path from a subject s to an object o and π has a flow from s to o. An ACS $\mathcal{C}$ has an *integrity violation* if there is a run of $\mathcal{C}$ with an integrity violation.

As above, in the ACS depicted in Fig. 1, a *integrity violation* would occur, for example, if there was a sequence $(s_2, write, o_4)(s_3, read, o_4)(s_3, write, o_7)$ which was a sub-sequence of π.

An ACS has a *data leakage* if it has either a confidentiality or an integrity violation. From the definitions above it is straightforward to see that the following property holds.

Proposition 1. *An access control system is data leakage free if and only if it is vulnerability free.*

The direct consequence of the proposition above suggests that a vulnerability free access control system is data leakage free by design, hence it does not require a monitor to prevent data leakages.

Fundamental Theorem. We now prove a simple and fundamental property of ACS that constitutes one of the building blocks for our approaches for checking and eliminating vulnerabilities/data leakages as shown later in the paper.

Theorem 1. *Let $\mathcal{C}$ be an access control system. $\mathcal{C}$ has a vulnerability only if $\mathcal{C}$ has a vulnerability of length one. In particular, let $\rho = o_0 \rightarrow_r s_0 \rightarrow_w o_1 \ldots s_{n-1} \rightarrow_w o_n$ be vulnerability path of minimal length. Then, if ρ is a confidentiality* *(resp., integrity) vulnerability then $o_0 \to_r s_0 \to_w o_1$ (resp., $o_0 \to_r s_0 \to_w o_n$) is a confidentiality (resp., integrity) vulnerability of length one.*

Proof. The proof is by contradiction. Assume that n is greater than one by hypothesis. We first consider the case of confidentiality vulnerability. Let s be the violator. Since ρ is of minimal length, all objects along ρ except o_0 can be directly read by s (i.e., $o_i \to_r s$ for every $i \in [1, n]$), otherwise there is an confidentiality vulnerability of smaller length. Thus, $o_0 \to_r s_0 \to_w o_1$ is a confidentiality vulnerability of length one, as s can read from o_1 but cannot read from o_0. A contradiction.

We give a similar proof for integrity vulnerabilities. Again, since ρ is of minimal length, all objects along ρ, except o_0, can be directly written by s_0, i.e., $s_0 \to_w o_i$ for every $i \in \{1, \ldots, n\}$. But, this entails that $o_0 \to_r s_0 \to_w o_n$ is an integrity vulnerability of length one (as s can write into o_0 but cannot directly write into o_n). Again, a contradiction.

We now present two alternative strategies for preventing data flows, which fit different environments.

3 Access Control Systems Data Leakage Free by Design

When a monitor is not possible or even doable the only solution to get an access control that is free of data leakages is that of having the ACS free of vulnerabilities (see Proposition 1). In this section, we propose an automatic approach that turns any ACS into one free of vulnerabilities by revoking certain rights.

This can be naively achieved by removing all read and write permissions. However, this would make the whole approach useless. Instead, it is desirable to minimize the changes to the original access control matrix so as not to disturb the users' ability to perform their job functions, unless it is absolutely needed. Furthermore, the removal of these permissions should take into account the fact that some of them may belong to *trusted users* (i.e. subjects), such as system administrators, and therefore we want to prevent the removal of these permissions.

We show that this problem is NP-complete (see Sect. 3.1). Therefore, an efficient solution is unlikely to exist (unless P = NP). To circumvent this computational difficulty, we propose compact encodings of this optimization problem into integer linear programming (ILP) by exploiting Theorem 1 (see Sects. 3.2 and 3.3). The main goal is that of leveraging efficient solvers for ILP, which nowadays exist. We show that this approach is promising in practice in Sect. 5.

Maximal Data Flow Problem (MDFP). Let $\mathcal{C} = (S, O, \to_r, \to_w)$ be an access control system, and $T = (\to_r^t, \to_w^t)$ be the sets of *trusted permissions* where $\to_r^t \subseteq \to_r$ and $\to_w^t \subseteq \to_w$. A pair $Sol = (\to_r^{sol}, \to_w^{sol})$ is a *feasible solution* of $\mathcal{C}$ and T, if $\to_r^t \subseteq \to_r^{sol} \subseteq \to_r$, $\to_w^t \subseteq \to_w^{sol} \subseteq \to_w$ and $\mathcal{C}' = (S, O, \to_r^{sol}, \to_w^{sol})$ does not have any threat. The size of a feasible solution Sol, denoted $size(Sol)$, is the value $| \to_r^{sol} | + | \to_w^{sol} |$. The *MDFP* is to maximize $size(Sol)$.

3.1 MDFP is NP-complete

Here we show that the decision problem associated to MDFP is NP-complete. Given an instance $I = (\mathcal{C}, T)$ of MDFP and a positive integer K, the decision problem associated to MDFP, called D-MDFP, asks if there is a feasible solution of I of size greater or equal to K.

Theorem 2. *D-MDFP is NP-complete.*

See Appendix 7.2 for the proof.

3.2 ILP Formulation

Here we define a reduction from MDFP to integer linear programming (ILP). In the rest of this section, we denote by $I = (\mathcal{C}, T)$ to be an instance of MDFP, where $\mathcal{C} = (S, O, \rightarrow_r, \rightarrow_w)$ and $T = (\rightarrow_r^t, \rightarrow_w^t)$.

The set of variables $\mathcal{V}$ of the ILP formulation is:

$$\mathcal{V} = \{r_{o,s} \mid o \in O \wedge s \in S \wedge o \rightarrow_r s\} \cup \{w_{s,o} \mid s \in S \wedge o \in O \wedge o \rightarrow_r s\}$$

The domain of the variables in $\mathcal{V}$ is $\{0, 1\}$, and the intended meaning of these variables is the following. Let $\eta_I : \mathcal{V} \rightarrow \{0, 1\}$ be an assignment of the variables in $\mathcal{V}$ corresponding to an optimal solution of the ILP formulation. Then, a solution for I is obtained by removing all permissions corresponding to the variables assigned to 0 by η_I. Formally, $Sol_{\eta_I} = (\rightarrow_r^{sol}, \rightarrow_w^{sol})$ is a solution for I, where

$$\rightarrow_r^{sol} = \{(o, s) \mid o \in O \wedge s \in S \wedge o \rightarrow_r s \wedge \eta_I(r_{o,s}) = 1\}$$

$$\rightarrow_w^{sol} = \{(s, o) \mid s \in S \wedge o \in O \wedge s \rightarrow_w o \wedge \eta_I(w_{s,o}) = 1\}.$$

The main idea on how we define the ILP encoding, hence its correctness, derives straightforwardly from Theorem 1: we impose that every flow path of length one, say $o \rightarrow_r \hat{s} \rightarrow_w o'$, if these permissions remain in the resulting access control system $\mathcal{C}' = (S, O, \rightarrow_r^{sol}, \rightarrow_w^{sol})$, then it must be the case that for every subject $s \in S$ if s can read from o' in $\mathcal{C}'$, s must also be able to read from o in $\mathcal{C}'$ (CONFIDENTIALITY), and if s that can write into o in $\mathcal{C}'$, s must be also able to write into o' in $\mathcal{C}'$ (INTEGRITY). Formally, the linear equations of our ILP formulation is the minimal set containing the following.

Confidentiality Constraints: For every sequence of the form $o \rightarrow_r \hat{s} \rightarrow_w \hat{o} \rightarrow_r s$, we add the constraint: $r_{o,s} + w_{\hat{s},\hat{o}} + r_{\hat{o},s} - G \leq 2$ where G is $r_{o,s}$ in case $o \rightarrow_r s$, otherwise $G = 0$. For example, for the sequence $o_1 \rightarrow_r s_1 \rightarrow_w o_3 \rightarrow_r s_2$, in the ACS depicted in Fig. 1(a), we have $r_{o_1,s_1} + w_{s_1,o_3} + r_{o_3,s_2} - 0 \leq 2$.

Integrity Constraints: For every sequence of the form $s \rightarrow_w o \rightarrow_r \hat{s} \rightarrow_w \hat{o}$, we add the constraint: $w_{s,o} + r_{o,\hat{s}} + w_{\hat{s},\hat{o}} - G \leq 2$ where G is $w_{s,\hat{o}}$ in case $s \rightarrow_w \hat{o}$, otherwise $G = 0$. As above, for the sequence $s_2 \rightarrow_w o_4 \rightarrow_r s_3 \rightarrow_w o_7$, in the ACS depicted in Fig. 1(a), we add the constraint $w_{s_2,o_4} + r_{o_4,s_3} + w_{s_3,o_7} - 0 \leq 2$.

Trusted Read Constraints: For every $o \rightarrow_r^t s$, we have the constraint: $r_{o,s} = 1$.

Trusted Write Constraints: For every $s \rightarrow_w^t o$, we have the constraint: $w_{s,o} = 1$.

It is easy to see that any variable assignment η that obeys all linear constraints defined above leads to a feasible solution of I.

Objective Function: Now, to maximize the number of remaining permissions (or equivalently, minimize the number of removed permissions) we define the objective function of the ILP formulation as the sum of all variables in $\mathcal{V}$. Compactly, our ILP-FORMULATION$(\mathcal{C}, T)$ is as shown in Fig. 2.

$$max \quad \sum_{v \in \mathcal{V}} v$$

$$subject\ to$$

$$r_{o,\hat{s}} + w_{\hat{s},\hat{o}} + r_{\hat{o},s} - r_{o,s} \leq 2, \forall\ o \rightarrow_r \hat{s} \rightarrow_w \hat{o} \rightarrow_r s, o \rightarrow_r s$$

$$r_{o,\hat{s}} + w_{\hat{s},\hat{o}} + r_{\hat{o},s} \qquad \leq 2, \forall\ o \rightarrow_r \hat{s} \rightarrow_w \hat{o} \rightarrow_r s, o \not\rightarrow_r s$$

$$w_{s,\hat{o}} + r_{\hat{o},\hat{s}} + w_{\hat{s},o} - w_{s,o} \leq 2, \forall\ s \rightarrow_w \hat{o} \rightarrow_r \hat{s} \rightarrow_w o, s \rightarrow_w o$$

$$w_{s,\hat{o}} + r_{\hat{o},\hat{s}} + w_{\hat{s},o} \qquad \leq 2, \forall\ s \rightarrow_w \hat{o} \rightarrow_r \hat{s} \rightarrow_w o, s \not\rightarrow_w o$$

$$r_{o,s} = 1, \forall\ o \rightarrow_r^t s; \quad w_{s,o} = 1, \forall\ s \rightarrow_w^t o; \quad v \in \{0,1\}, \forall v \in \mathcal{V}$$

Fig. 2. ILP formulation of MDFP.

We now formally state the correctness of our ILP approach, which is entailed from the fact that we remove the minimal number of permissions from $\mathcal{C}$ resulting in a new ACS that does not have any threat of length one, hence from Theorem 1 does not have any threat at all.

Theorem 3. *For any instance I of MDFP, if η_I is an optimal solution of* ILP - FORMULATION(I) *then Sol_{η_I} is an optimal solution of I.*

We note that while the ILP formulation gives the optimal solution, solving two subproblems (one for confidentiality followed by the one for integrity each with only the relevant constraints) does not give an optimal solution.

For example, for the ACS depicted in Fig. 1(a), if we only eliminate the 15 confidentiality vulnerabilities, the optimal solution is to revoke 5 permissions $(o_1 \rightarrow_r s_1, o_2 \rightarrow_r s_1, o_1 \rightarrow_r s_2, o_2 \rightarrow_r s_2,$ and $o_6 \rightarrow_r s_5)$. This eliminates all of the confidentiality, while all of the original integrity vulnerabilities still exist. No new vulnerabilities are added. Now, if the integrity vulnerabilities are to be eliminated, the optimal solution is to revoke 4 permissions $(s_3 \rightarrow_w o_6,$ $s_3 \rightarrow_w o_7, s_4 \rightarrow_w o_6, s_4 \rightarrow_w o_7)$. Thus, the total number of permissions revoked is 9. However, if both confidentiality and integrity vulnerabilities are eliminated together (using the composite ILP in Fig. 2), the optimal solution is to simply revoke 6 permissions $(o_3 \rightarrow_r s_3, o_4 \rightarrow_r s_3, o_5 \rightarrow_r s_3, o_3 \rightarrow_r s_4, o_4 \rightarrow_r s_4,$ $o_5 \rightarrow_r s_4)$, which is clearly lower than 9.

3.3 Compact ILP Formulation

We now present an improved encoding that extends the ILP formulation described in Sect. 3.2 by merging subjects and objects that have the same permissions. This allows us to get a much reduced encoding, in terms of variables, with better performances in practice (see Sect. 5).

Equivalent Subjects: For an instance $I = (\mathcal{C}, T)$ of MDFP with $\mathcal{C} = (S, O, \rightarrow_r, \rightarrow_w)$ and $T = (\rightarrow_r^t, \rightarrow_w^t)$, two subjects are *equivalent* if they have the same permissions. Formally, for a subject $s \in S$, let $read_I(s)$ (respectively, $read_I^t(s)$) denote the set of all objects that can be read (respectively, trust read) by s in $\mathcal{C}$, i.e., $read_I(s) = \{o \in O \mid o \rightarrow_r s\}$ (respectively, $read_I^t(s) = \{o \in O \mid o \rightarrow_r^t s\}$). Similarly, we define $write_I(s) = \{o \in O \mid s \rightarrow_w o\}$ and $write_I^t(s) = \{o \in O \mid s \rightarrow_w^t o\}$. Then, two subjects s_1 and s_2 are *equivalent*, denoted $s_1 \approx s_2$, if $read_I(s_1) read_I(s_2)$, $read_I^t(s_1) = read_I^t(s_2)$, $write_I(s_1) = write_I(s_2)$, and $write_I^t(s_1) = write_I^t(s_2)$.

For every $s \in S$, $[s]$ is the equivalence class of s w.r.t. $\approx$. Moreover, $S^{\approx}$ denotes the quotient set of S by $\approx$. Similarly, we can define the same notion of equivalent objects, with $[o]$ denoting the the equivalence class of $o \in O$, and $O^{\approx}$ denoting the quotient set of O by $\approx$.

Given a read relation $\rightarrow_r \subseteq O \times S$ and two subjects $s_1, s_2 \in S$, $\rightarrow_r [s_1/s_2]$ is a new read relation obtained from $\rightarrow_r$ by assigning to s_2 the same permissions that s_1 has in $\rightarrow_r$: $\rightarrow_r [s_1/s_2] = (\rightarrow_r \setminus (O \times \{s_2\})) \cup \{(o, s_2) \mid o \in O \wedge o \rightarrow_r s_1\}$.

Similarly, $\rightarrow_w [s_1/s_2] = (\rightarrow_w \setminus (\{s_2\} \times O)) \cup \{(s_2, o) \mid o \in O \wedge s_1 \rightarrow_w o\}$. A similar substitution can be defined for objects.

The following lemma states that for any given optimal solution of I it is always possible to derive a new optimal solution in which two equivalent subjects have the same permissions.

Lemma 1. *Let $I = (\mathcal{C}, T)$ be an instance of the MDFP problem, s_1 and s_2 be two equivalent subjects of I, and $Sol' = (\rightarrow_r^{sol}, \rightarrow_w^{sol})$ be a optimal solution of I. Then, $Sol'' = (\rightarrow_r^{sol} [s_1/s_2], \rightarrow_w^{sol} [s_1/s_2])$ is also an optimal solution of I.*

See Appendix 7.1 for the proof.

The following property is a direct consequence of Lemma 1.

Corollary 1. *Let $I = (\mathcal{C}, T)$ with $\mathcal{C} = (S, O, \rightarrow_r, \rightarrow_w)$ be an instance of the MDFP problem that admits a solution. Then, there exists a solution $Sol = (\rightarrow_r^{sol}, \rightarrow_w^{sol})$ of I such that for every pair of equivalent subjects $s_1, s_2 \in S$, s_1 and s_2 have the same permissions in $\mathcal{C} = (S, O, \rightarrow_r^{sol}, \rightarrow_w^{sol})$.*

Lemma 1 and Corollary 1 also hold for equivalent objects. Proofs are similar to those provided above and hence we omit them here.

Compact ILP formulation. Corollary 1 suggests a more compact encoding of the MDFP into ILP. From $\mathcal{C}$, we define a new ACS $\mathcal{C}^{\approx}$ by collapsing all subjects and objects into their equivalence classes defined by $\approx$, and by merging permissions consequently (edges of $G_{\mathcal{C}}$). Formally, $\mathcal{C}^{\approx}$ has $S^{\approx}$ as set of subjects and $O^{\approx}$ as set

$$max \sum_{[o] \rightarrow_r^{\approx} [s]} \left(|\,[o]\,| \cdot |\,[s]\,| \cdot r_{[o],[s]} \right) \quad + \quad \sum_{[s] \rightarrow_w^{\approx} [o]} \left(|\,[s]\,| \cdot |\,[o]\,| \cdot w_{[s],[o]} \right)$$

$$subject\ to$$

$$r_{[o],[\hat{s}]} + w_{[\hat{s}],[\hat{o}]} + r_{[\hat{o}],[s]} - r_{[o],[s]} \leq 2,\ \forall [o] \rightarrow_r^{\approx} [\hat{s}] \rightarrow_w^{\approx} [\hat{o}] \rightarrow_r^{\approx} [s] \wedge [o] \rightarrow_r [s]$$

$$r_{[o],[\hat{s}]} + w_{[\hat{s}],[\hat{o}]} + r_{[\hat{o}],[s]} \leq 2,\ \forall [o] \rightarrow_r^{\approx} [\hat{s}] \rightarrow_w^{\approx} [\hat{o}] \rightarrow_r^{\approx} [s] \wedge [o] \not\rightarrow_r [s]$$

$$w_{[s],[\hat{o}]} + r_{[\hat{o}],[\hat{s}]} + w_{[\hat{s}],[o]} - w_{[s],[o]} \leq 2,\ \forall [s] \rightarrow_w^{\approx} [\hat{o}] \rightarrow_r^{\approx} [\hat{s}] \rightarrow_w^{\approx} [o] \wedge [s] \rightarrow_w [o]$$

$$w_{[s],[\hat{o}]} + r_{[\hat{o}],[\hat{s}]} + w_{[\hat{s}],[o]} \leq 2,\ \forall [s] \rightarrow_w^{\approx} [\hat{o}] \rightarrow_r^{\approx} [\hat{s}] \rightarrow_w^{\approx} [o] \wedge [s] \not\rightarrow_w [o]$$

$$r_{[o],[s]} = 1, \forall [o] \rightarrow_r^t [s];\quad w_{[s],[o]} = 1, \forall [s] \rightarrow_w^t [o];\quad v \in \{0,1\},\ \forall v \in \mathcal{V}^{\approx}$$

Fig. 3. ILP formulation of MDFP based on equivalence classes.

of objects, where the read and write permission sets are defined as follows: $\rightarrow_r^{\approx} =$ $\{ \,([o],[s]) \mid o \in O \wedge s \in S \wedge o \rightarrow_r s \,\}, \rightarrow_w^{\approx} = \{ \,([o],[s]) \mid s \in S \wedge o \in O \wedge$ $s \rightarrow_w o \,\}$. Similarly, we define the trusted permissions of $\mathcal{C}^{\approx}$ as $T^{\approx} = (\rightarrow_r^{t\ \approx}, \rightarrow_w^{t\ \approx})$ where $\rightarrow_r^{t\ \approx} = \{ \,([o],[s]) \mid o \in O \wedge s \in S \wedge o \rightarrow_r^t s \,\}, \rightarrow_w^{t\ \approx} = \{ \,([o],[s]) \mid$ $s \in S \wedge o \in O \wedge s \rightarrow_w^t o \,\}$.

We now define a new ILP encoding, COMPACT-ILP-FORMULATION(I), for MFDP on the instance $(\mathcal{C}^{\approx}, T^{\approx})$, which is similar to that of Fig. 2 with the difference that now edges may have a weight greater than one; reflecting the number of edges of $\mathcal{C}$ it represents in $\mathcal{C}^{\approx}$. More specifically, each edge from a node x_1 to x_2 in $G_{\mathcal{C}^{\approx}}$ represents all edges from all nodes in $[x_1]$ to all nodes in $[x_2]$, i.e., its weight is $|[x_1]| \cdot |[x_2]|$. Figure 1(b) shows the compact representation of Fig. 1(a), where the edges have the appropriate weights.

Figure 3 shows COMPACT-ILP-FORMULATION(I) over the set of variables $\mathcal{V}^{\approx}$. The set of linear constraints is the same as those in Fig. 2 with the difference that now they are defined over $\mathcal{C}^{\approx}$ rather than $\mathcal{C}$. Instead, the objective function is similar to that of Fig. 2, but now captures the new weighting attributed to edges in $G_{\mathcal{C}^{\approx}}$.

Let $\eta_I^{\approx} : \mathcal{V} \rightarrow \{0,1\}$ be a solution to the ILP instance of Fig. 3. Define $\widehat{Sol}_{\eta_I^{\approx}} = (\widehat{\rightarrow}_r^{sol}, \widehat{\rightarrow}_w^{sol})$ where $\widehat{\rightarrow}_r^{sol} = \{ \,(o,s) \in O \times S \mid o \rightarrow_r s \wedge \eta_I^{\approx}(r_{[o],[s]}) \geq 1 \,\}$ and $\widehat{\rightarrow}_w^{sol} = \{ \,(s,o) \in S \times O \mid s \rightarrow_w o \wedge \eta_I^{\approx}(w_{[s],[o]}) \geq 1 \,\}$.

We now prove that $Sol_{\eta_I^{\approx}}$ is an *optimal* solution of I.

Theorem 4. *For any instance I of MDFP, if $\eta_I^{\approx}$ is an optimal solution of* COMPACT-ILP-FORMULATION(I) *then $\widehat{Sol}_{\eta_I^{\approx}}$ is an optimal solution of I. Furthermore, if I admits a solution then $\eta_I^{\approx}$ also exists.*

See Appendix 7.3 for the proof.

4 Preventing Data Leakages with Monitors

A *data-leakage monitor* or simply *monitor* of an access control system $\mathcal{C}$ is a computing system that by observing the behaviors on $\mathcal{C}$ (i.e., the sequence of read and write operations) detects and prevents data leakages (both confidentiality and integrity violations) by blocking subjects' operations. In this section, we present a monitor based on a tainting approach. We first define monitors as language acceptors of runs of $\mathcal{C}$ that are data leakage free. We then present a monitor based on tainting and then conclude with an optimized version of this monitor that uses only 2-step tainting, leading to better empirical performances.

Monitors. Let $\mathcal{C} = (S, O, \rightarrow_r, \rightarrow_w)$ be an ACS, $\Sigma = S \times \{read, write\} \times O$ be the set of all possible *actions* on $\mathcal{C}$, and $R = \{accept, reject\}$. A *monitor* $\mathcal{M}$ of $\mathcal{C}$ is a triple (Q, q_{st}, δ) where Q is a set of *states*, $q_{st} \in Q$ is the *start state*, and $\delta : (Q \times R \times \Sigma) \rightarrow (Q \times R)$ is a (deterministic) *transition function*.

A *configuration* of $\mathcal{M}$ is a pair (q, h) where $q \in Q$ and $h \in R$. For a word $w = \sigma_1 \ldots \sigma_m \in \Sigma^*$ with actions $\sigma_i \in \Sigma$ for $i \in [1, m]$, a *run* of $\mathcal{M}$ on w is a sequence of $m + 1$ configurations $(q_0, h_0), \ldots (q_m, h_m)$ where q_0 is the start state q_{st}, $h_0 = accept$, and for every $i \in [1, m]$ the following holds: $h_{i-1} = accept$ and $(q_i, h_i) = \delta(q_{i-1}, h_{i-1}, \sigma_i)$, or $h_{i-1} = h_i = reject$ and $q_i = q_{i-1}$.

A word w (run of $\mathcal{C}$) is *accepted* by $\mathcal{M}$ if $h_m = accept$. The *language* of $\mathcal{M}$, denoted $L(\mathcal{M})$, is the set of all words $w \in \Sigma^*$ that are accepted by $\mathcal{M}$.

A monitor $\mathcal{M}$ is *maximal data leakage preserving* (MDLP, for short) if $L(\mathcal{M})$ is the set of all words in Σ^* that are confidentiality and integrity free. For any given ACS $\mathcal{C}$, it is easy to show that an MDLP monitor can be built. This can be proved by showing that $L(\mathcal{M})$ is a regular language: we can easily express the properties of the words in $L(\mathcal{M})$ with a formula φ of monadic second order logic (MSO) on words and then use an automatic procedure to convert φ into a finite state automaton [14]. Although, this is a convenient way of building monitors for regular properties, it can lead to automata of exponential size in the number of objects and subjects. Hence, it is not practical for real access control systems.

Building Maximal Data-Leakage Preserving Monitors. A monitor based on tainting can be seen as a dynamic information flow tracking system that is used to detect data flows (see for example [17, 21, 22]).

An MDLP monitor $\mathcal{M}_{taint}$ based on tainting associates each subject and object with a subset of subjects and objects (*tainting sets*). $\mathcal{M}_{taint}$ starts in a state where each subject and object is tainted with itself. Then, $\mathcal{M}_{taint}$ progressively scans the sequence of actions on $\mathcal{C}$. For each action, say from an element x_1 to an element x_2, $\mathcal{M}_{taint}$ updates its state by propagating the tainting from x_1 to x_2. These tainting sets can be seen as a way to represent the endpoints of all flows: if x_2 is tainted by x_1, then there is a flow from x_1 to x_2. Thus, by using these flows and the definitions of confidentiality and integrity violations, $\mathcal{M}_{taint}$ detects data leakages.

More formally, an $\mathcal{M}_{taint}$ state is a map $taint : (S \cup O) \rightarrow 2^{(S \cup O)}$. A state $taint$ is a start state if $taint(x) = \{x\}$, for every $x \in (S \cup O)$. The transition relation δ of $\mathcal{M}_{taint}$ is defined as follows. For any two states $taint, taint', h, h' \in R$ and $\sigma = (s, op, o) \in \Sigma$, $\delta(taint, h, \sigma) = (taint', h')$ if either $h = h' = reject$ and $taint' = taint$, or $h = accept$ and the following holds:

(Data Leakage) $h' = reject$ iff either (Confidentiality Violation) $op = read$ and $\exists \hat{o} \in taint(o)$ such that $\hat{o} \not\rightarrow_r s$, or (Integrity Violation) $op = write$ and $\exists \hat{s} \in taint(s)$ such that $\hat{s} \not\rightarrow_w o$.

(Taint Propagation) either (Read Propagation) $op = read$, $taint'(s) = (taint(s) \cup taint(o))$, and for every $x \in (S \cup O) \setminus \{s\}$, $taint'(s) = taint(s)$; or (Write Propagation) $op = write$, $taint'(o) = (taint(o) \cup taint(s))$, and for every $x \in (S \uplus O) \setminus \{o\}$, $taint'(x) = taint(s)$.

Theorem 5. $\mathcal{M}_{taint}$ *is an MDLP monitor.*

MDLP Monitor Based on 2-Step Tainting: The tainting sets of $\mathcal{M}_{taint}$ progressively grow as more flows are discovered. In the limit each tainting set potentially includes all subjects and objects of $\mathcal{C}$. Since for each action the time for checking confidentiality and integrity violations is proportional to the size of the tainting sets of the object and subject involved in that action, it is desirable to reduce the sizes of these sets to get better performances. We achieve this, by defining a new tainting monitor $\mathcal{M}_{taint}^2$ that keeps track only of the flows that across at most two adjacent edges in $G_\mathcal{C}$. The correctness of our construction is justified by the correctness of $\mathcal{M}_{taint}$ and Theorem 1.

The 2-step tainting monitor $\mathcal{M}_{taint}^2$ is defined as follows. A state of $\mathcal{M}_{taint}^2$ is (as for $\mathcal{M}_{taint}$) a map $taint : (S \cup O) \rightarrow 2^{(S \cup O)}$. Now, a state $taint$ is a start state if $taint(x) = \emptyset$, for every $x \in (S \cup O)$.

The transition relation δ^2 of $\mathcal{M}_{taint}^2$ is defined to guarantee that after reading a violation free run π of $\mathcal{C}$:

- for every $s \in S$, $x \in taint(s)$ iff either (1) $x \in O$, (o, s) is an edge of $G_\mathcal{C}$, and there is a direct flow from x to s in π, or (2) $x \in S$, for some subject $\hat{o} \in O$, $(x, \hat{o}, s)$ is a path in $G_\mathcal{C}$, and there is a 2-step flow from x to s in π;
- for every $o \in O$, $x \in taint(o)$ iff either (1) $x \in S$, (s, o) is an edge of $G_\mathcal{C}$, and there is a direct flow from x to o in π, or (2) $x \in O$, for some subject $\hat{s} \in S$, $(x, \hat{s}, o)$ is a path in $G_\mathcal{C}$, and there is a 2-step flow from x to o in π.

Formally, for any two states $taint, taint'$, $h, h' \in R$ and $\sigma = (s, op, o) \in \Sigma$, $\delta^2(taint, h, \sigma) = (taint', h')$ if either $h = h' = reject$ and $taint' = taint$, or $h = accept$ and the following holds:

(Data Leakage) same as for $\mathcal{M}_{taint}$;

(Taint Propagation) either (Read Propagation) $op = read$, $taint'(s) = taint(s) \cup \{o\} \cup (taint(o) \cap S)$, and for every $x \in (S \cup O) \setminus \{s\}$, $taint'(s) = taint(s)$; or (Write Propagation) $op = write$, $taint'(o) = taint(o) \cup \{s\} \cup (taint(s) \cap O)$, and for every $x \in (S \cup O) \setminus \{o\}$, $taint'(x) = taint(s)$.

From the definition of $\mathcal{M}_{taint}^2$ it is simple to show (by induction) that the following property holds.

Theorem 6. $\mathcal{M}_{taint}^2$ *is an MDLP monitor. Furthermore, for every $\mathcal{C}$ run $\pi \in \Sigma^*$, if $(taint_0, h_0), \ldots (taint_m, h_m)$ and $(taint'_0, h'_0), \ldots (taint'_m, h'_m)$ are, respectively, the run of $\mathcal{M}_{taint}$ and $\mathcal{M}_{taint}^2$ on π, then $taint'_i(x) \subseteq taint_i(x)$, for every $i \in [1, m]$ and $x \in (S \cup O)$.*

Therefore, in practice we expect that for large access control systems $\mathcal{M}^2_{taint}$ is faster than $\mathcal{M}_{taint}$ as each tainting sets of $\mathcal{M}^2_{taint}$ will be local and hence much smaller in size than those of $\mathcal{M}_{taint}$. To show the behavior of the monitor the based approach, consider again the access control system shown in Table 1, along with the potential sequence of operations shown in Table 2. Table 2 shows the taints and monitor's action for each operation in the sequence. Note that the monitor blocks a total of six permissions (2 each on operations (2), (3), and (5)).

Table 2. Sample sequence of actions and monitor's behavior

	User's operation	Actions taken
1	s_1, r, o_1	$taint(s_1) = \{o_1\}$
2	s_1, w, o_3	$taint(o_3) = \{s_1, o_1\}$ Monitor will block $o_3 \rightarrow_r s_3$ $o_3 \rightarrow_r s_4$ to remove the confidentiality vulnerabilities
3	s_1, w, o_4	$taint(o_4) = \{s_1, o_1\}$ Monitor will block $o_4 \rightarrow_r s_3$ $o_4 \rightarrow_r s_4$ to remove the confidentiality vulnerabilities
4	s_2, w, o_4	$taint(o_4) = \{s_1, o_1, s_2\}$
5	s_4, r, o_4	$taint(s_4) = \{s_1, s_2, o_4\}$ Monitor will block $s_4 \rightarrow_w o_6$ and $s_4 \rightarrow_w o_7$ to remove the integrity vulnerability
6	s_3, r, o_3	Access denied
7	s_4, w, o_7	Access denied

5   Experimental Evaluation

We now present the experimental evaluation which demonstrates the performance and restrictiveness of the two proposed approaches. We utilize four real life access control data sets with users and permissions – namely, (1) fire1, (2) fire2, (3) domino, (4) hc [12]. Note that these data sets encode a simple access control matrix denoting the ability of a subject to access an object (in any access mode). Thus, these data sets do not have the information regarding which particular permission on the object is granted to the subject. Therefore, we assume for all of the datasets that each assignment represents both a read and a write permission on a distinct object.

For the data leakage free by design approach, we use the reduced access control matrices obtained by collapsing equivalent subjects and objects, as discussed in Sect. 3. The number of subjects and objects in the original and reduced matrices are given in Table 3. Note that collapsing subjects and objects significantly reduces the sizes of the datasets (on average the dataset is reduced by 93.99%). Here, by size, we mean the product of the number of subjects and objects. Since the number of constraints is linearly proportional to the number of permissions which depends on the number of subjects and objects, a reduction in their size leads to a smaller ILP problem.

Table 3. Dataset details

Dataset	Name	Original size		Reduced size		Percentage
		Subjects	Objects	Subjects	Objects	Reduction
1	fire1	365	709	90	87	96.97 %
2	fire2	325	590	11	11	99.94 %
3	domino	79	231	23	38	95.21 %
4	hc	46	46	18	19	83.84 %

We implement the solution approaches described above. For the data leakage free by design approach (Sect. 3), we create the appropriate ILP model as per Fig. 3. The ILP model is then executed using IBM CPLEX (v 12.5.1) running through callable libraries within the code. For the monitor based approach, the $\mathcal{M}_{taint}^2$ monitor is implemented. The algorithms are implemented in C and run on a Windows machine with 16 GB of RAM and Core i7 2.93 GHz processor.

Table 4 presents the experimental results for the Data Leakage Free by Design approach. The column "Orig. CPLEX Time", shows the time required to run the ILP formulation given in Fig. 2, while the column "Red. CPLEX Time" gives the time required to run the compact ILP formulation given in Fig. 3. As can be seen, the effect of collapsing the subjects and objects is enormous. fire1 and fire2 could not be run (CPLEX gave an out of memory error) for the original access control matrix, while the time required for hc and domino was several orders of magnitude more. Since we use the reduced datasets, as discussed above, the column "Threats" reflects the number of threats in the reduced datasets to be eliminated. The next three columns depict the amount of permission revocation to achieve a data leakage free access matrix. Note that, here we list the number of permissions revoked in the original access control matrix. On average, 25.28% of the permissions need to be revoked to get an access control system without any data leakages.

When we have a monitor, as discussed in Sect. 4, revocations can occur on the fly. Therefore, to test the relative performance of the monitor based approach, we have randomly generated a set of read/write operations that occur in the order they are generated. The monitor based approach is run and the number of

Table 4. Results for data leakage free access matrix

Dataset	Orig. CPLEX Time (s)	Red. CPLEX Time (s)	Threats	# Perm. Init. Assn	# Perm. Revoked	% Revoked
1	-	2582	34240	63902	14586	22.83 %
2	-	0.225	514	72856	12014	16.49 %
3	8608.15	6.01	3292	1460	421	28.84 %
4	1262.82	0.27	1770	2972	980	32.97 %

Table 5. Results for monitor based approach

Dataset	# Perm. Init. Assn.	Number permissions blocked						% Finally blocked
		10%	50%	100%	1000%	5000%	10000%	
1	63902	0	140	532	14221	24031	26378	41.28 %
2	72856	0	13	26	3912	8129	9025	12.39 %
3	1460	0	36	41	130	283	364	24.93 %
4	2972	0	0	0	557	1123	1259	42.36 %

permissions revoked is counted. Since the number of flows can increase as more operations occur, and therefore lead to more revocations, we actually count the revocations for a varying number of operations. Specifically, for each dataset, we generate on average 100 operations for every subject (i.e., we generate $100 * |S|$ number of random operations). Thus, for hc, since there are 46 subjects, we generate 4600 random operations, where as for fire1 which has 365 subjects, we generate 36500 random operations. Now, we count the number of permissions revoked if only $10\% * |S|$ operations are carried out (and similarly for $50\% * |S|$, $100\% * |S|$, $1000\% * |S|$, $5000\% * |S|$, and finally $10000\% * |S|$). Table 5 gives the results. Again, we list the number of permissions revoked in the original access control matrix. As we can see, the number of permissions revoked is steadily increasing, and in the case of fire1 and hc the final number of permissions revoked is already larger than the permissions revoked in the data leakage free method. Also, note that in the current set of experiments, we have set a window size of 1000 – this means that if the gap between a subject reading an object and then writing to another object is more than 1000 operations, then we do not consider a data flow to have occurred (typically a malicious software would read and then write in a short duration of time) – clearly, the choice of 1000 is arbitrary, and in fact, could be entirely removed, to ensure no data leakages. In this case, the number of permission revocations would be even larger than what is reported, thus demonstrating the benefit of the data leakage free approach when a large number of operations are likely to be carried out.

6   Related Work

The importance of preventing inappropriate leakage of data, often called the confinement problem in computer systems, first identified by Lampson in early 70's [20], is defined as the problem of assuring the ability to limit the amount of damage that can be done by malicious or malfunctioning software. The need for a confinement mechanism first became apparent when researchers noted an important inherent limitation of DAC – the Trojan Horse Attack, and with the introduction of the Bell and LaPadula model and the MAC policy. Although MAC compliant systems prevent inappropriate leakage of data, these systems are limited to multi-level security.

While MAC is not susceptible to Trojan Horse attacks, many solutions proposed to prevent any such data leakage exploit employing labels or type based access control. Boebert et al. [3], Badger et al. [1] and Boebert and Kain [4] are some of the studies that address confidentiality violating data flows. Mao et al. [21] propose a label based MAC over a DAC system. The basic idea of their approach is to associate read and write labels to objects and subjects. These object labels are updated dynamically to include the subject's label when the subject reads or writes to that object. Moreover, the object label is a monotonically increasing set of items, with the cardinality in the order of the number of users read (wrote) the object. Their approach detects integrity violating data flows. Zimmerman et al. [32] propose a rule based approach that prevents any integrity violating data flow. Jaume et al. [17] propose a dynamic label updating procedure that detects if there is any confidentiality violating data flow.

Information Flow Control (IFC) models [10,18] are closely related to our problem. IFC model is a fine-grained information flow model which is also based on tainting and utilizes labels for each piece of data that is required to be protected using the lattice model for information flow security by [9]. The models can be at software or OS level depending on the granularity of the control and centralized or decentralized depending on the authority to modify labels [24]. However, these models do not consider the permission assignments, which makes them different than our model.

Dynamic taint analysis is also related to our problem. Haldar et al. [16] propose a taint based approach for programs in Java, and Lam et al. [19] propose a dynamic taint based analysis on C. Enck et al. [11] provide a taint based approach to track third party Android applications. Cheng et al. [6], Clause et al. [7] and Zhu et al. [31] propose software level dynamic tainting.

Sze et al. [26] study the problem of self-revocation, where a revocation in the permission assignments of any subject on an object while editing it might cause confidentiality and integrity issues. They also study the problem of integrity violation by investigating the source code and data origin of suspected malware and prevent any process that is influenced from modifying important system resources [27]. Finally, the work by Gong and Qian [15] focuses on detecting the cases where confidentiality and integrity flows occur due to interoperation of distinct access control systems. They study the complexity to detect such violations.

7   Conclusions and Future Work

In this paper, we have proposed a methodology for identifying and eliminating unauthorized data flows in DAC, that occur due to Trojan Horse attacks. Our key contribution is to show that a transitive closure is not required to eliminate such flows. We then propose two alternative solutions that fit different situational needs. We have validated the performance and restrictiveness of the proposed approaches with real data sets. In the future, we plan to propose an auditing based approach which eliminates unauthorized flows only if the flows get realized. This might be useful to identify the data leakage channels that are actually utilized. We also plan to extend our approach to identify and prevent the unauthorized flows in RBAC, which is also prone to Trojan Horse attacks. Analysis on RBAC is more challenging since there is an additional layer of complexity (roles) that must be taken into account. The preventive action decisions must overcome the dilemma of whether to revoke the role from the user or revoke the permission from the role.

Appendix

7.1   Proof of Lemma 1

Proof. Assume that S and O are the set of subjects and objects of $\mathcal{C}$, respectively. Let $\mathcal{C}' = (S, O, \to_r^{sol}, \to_w^{sol})$ and $\mathcal{C}'' = (S, O, \to_r^{sol}[s_1/s_2], \to_w^{sol}[s_1/s_2])$.

We first prove (by contradiction) that Sol'' is a **feasible** solution of I. Assume that $\mathcal{C}''$ has a threat. This threat is witnessed by a flow path, say ρ, that must contain s_2. If ρ does not involve s_2 then ρ would also be a threat in $\mathcal{C}'$, which cannot be true as Sol' is a feasible solution of I. Now, observe that s_2 can always be replaced by s_1 along any flow path of $\mathcal{C}''$, as s_2 and s_1 have the same neighbor in $G_{\mathcal{C}''}$. Thus, the flow path obtained by replacing s_2 with s_1 along ρ, also witnesses a threat in $\mathcal{C}'$. Again a contradiction. Therefore, Sol'' is a feasible solution of I.

We now prove that Sol'' is also **optimal** (that is, $size(Sol') = size(Sol'')$) by showing that s_1 and s_2 have the same number of incident edges in $G_{\mathcal{C}'}$. Let n_1 (respectively, n_2) be the number of incident nodes of s_1 (respectively, s_2) in $G_{\mathcal{C}'}$. By contradiction, and w.l.o.g., assume that $n_1 > n_2$. Since $\mathcal{C}''$ is obtained from $\mathcal{C}'$ by removing first the permissions of s_2 and then adding to s_2 the same permissions of s_1, it must be the case that $size(Sol'') > size(Sol')$. This would entail that Sol' is not an optimal solution, which is a contradiction.

7.2   Proof of Theorem 2

NP-membership. Let $Sol = (\to_r', \to_w')$ such that $\to_r', \to_w' \subseteq S \times O$. To check whether Sol is a feasible solution of I, we need to check that (1) $\to_r^t \subseteq \to_r' \subseteq \to_r$, (2) $\to_w^t \subseteq \to_w' \subseteq \to_w$, (3) $\mid \to_r' \mid + \mid \to_w' \mid \geq K$, and more importantly, (4) that $(S, O, \to_r', \to_w')$ is an ACS that does not contain any threat. The first three properties are easy to realize in polynomial time. Concerning the last property, we exploit Theorem 1. To check that there is no confidentiality threat, we build all sequences of the form $o_0 \to_r' s_0 \to_w' o_1 \to_r' s_1$ and then verify the existence of the read permission $o_0 \to_r' s_1$. Similarly, for integrity threat we build all sequences such that $s_0 \to_w' o_0 \to_r' s_0 \to_w' o_1$ and then check the existence of the write permission $s_0 \to_w' o_1$. Note that, all these sequences can be built in $O(O^2 \cdot S^2)$ and these checks can all be accomplished in polynomial time. This shows that D-MDFP belongs to NP.

NP-hardness. For the NP-hardness proof, we provide a polynomial time reduction from *the edge deletion transitive digraph* problem (ED-TD) to D-MDFP.

The ED-TD asks to remove the minimal number of edges from a given directed graph such that the resulting graph corresponds to its transitive closure. ED-TD problem is known to be NP-complete (see [28] Theorem 15, and [29]).

The reduction is as follows. Let $G = (V, E)$ be a directed graph with set of nodes $V = \{1, 2, \ldots n\}$ and set of edges $E \subseteq (V \times V)$. We assume that nodes of G do not have self-loops. We now define the instance $I_G = (\mathcal{C}_G, T_G)$ of D-MDFP to which G is reduced to. Let $\mathcal{C}_G = (S, O, \rightarrow_r, \rightarrow_w)$ and $T_G = (\rightarrow_r^t, \rightarrow_w^t)$. $\mathcal{C}_G$ has a subject s_i and an object o_i, for each node $i \in V$. Moreover, there is a read permission from o_i to s_i, and a write permission from s_i to o_i, for every node $i \in V$. These permissions are also trusted, i.e., belonging to $\rightarrow_r^t$ and $\rightarrow_w^t$, respectively; and no further permissions are trusted. Furthermore, for every edge $(i, j) \in E$, there is a read permission from o_i to s_j, and a write permission from s_i to o_j. Formally, $S = \{s_i \mid i \in V\}$ and $O = \{o_i \mid i \in V\}$; $\rightarrow_r^t = \{(o_i, s_i) \mid i \in V\}$; $\rightarrow_w^t = \{(s_i, o_i) \mid i \in V\}$; $\rightarrow_r = \rightarrow_r^t \cup \{(o_i, s_j) \mid (i, j) \in E\}$; $\rightarrow_w = \rightarrow_w^t \cup \{(s_i, o_j) \mid (i, j) \in E\}$.

Lemma 2. *Let G be a directed graph with nodes $V = \{1, 2, \ldots, n\}$, and $Sol = (\rightarrow_r', \rightarrow_w')$ be a feasible solution of I_G. For any $i, j \in V$ with $i \neq j$, $o_i \rightarrow_r' s_j$ if and only if $s_i \rightarrow_w' o_j$.*

Proof. The proof is by contradiction. Consider first the case when $o_i \rightarrow_r' s_j$ and $s_i \not\rightarrow_w' o_j$. Observe that, $s_i \rightarrow_w' o_i$ and $s_j \rightarrow_w o_j$ exist as both of them are trusted permissions of I_G. Thus, $s_i \rightarrow_w' o_i \rightarrow_r' s_j \rightarrow_w o_j$ is an integrity threat, leading to a contradiction. The case when $o_i \not\rightarrow_r' s_j$ and $s_i \rightarrow_w' o_j$ is symmetric, and we omit it here.

We now show that the transformation defined above from G to I_G is indeed a polynomial reduction from ED-TD to D-MDFP. The NP-hardness directly follows from the following lemma.

Lemma 3. *Let G be a directed graph with n nodes. G contains a subgraph G' with K edges whose transitive closure is G' itself if and only I_G admits a feasible solution Sol of size $2 \cdot (n + K)$.*

Proof. Let $G = (V, E)$ with $V = \{1, 2, \ldots, n\}$, $G' = (V, E')$, $I_G = (\mathcal{C}_G, T_G)$ where $\mathcal{C}_G = (S, O, \rightarrow_r, \rightarrow_w)$ and $T_G = (\rightarrow_r^t, \rightarrow_w^t)$, and $Sol = (\rightarrow_r', \rightarrow_w')$.

"only if" direction. Assume that G' is the transitive closure of itself and $|E'| = K$. We define Sol as follows: $\rightarrow_r' = \rightarrow_r^t \cup \{o_i \rightarrow_r s_j \mid (i, j) \in E'\}$ and $\rightarrow_w' = \rightarrow_w^t \cup \{s_i \rightarrow_w o_j \mid (i, j) \in E'\}$. From the definition of I_G, it is straightforward to see that $size(Sol) = 2 \cdot (n + K)$. To conclude the proof we only need to show that Sol is a feasible solution of I_G. Since $\rightarrow_r^t \subseteq \rightarrow_r'$ and $\rightarrow_w^t \subseteq \rightarrow_w'$ we are guaranteed that Sol contains all trusted permissions of T_G. We now show that $\mathcal{C}' = (S, O, \rightarrow_r', \rightarrow_w')$ does not contain any threat. Assume that there is a threat in $\mathcal{C}'$. By Theorem 1, there must be a threat of length one. If it is a confidentiality threat, then $o_i \rightarrow_r' s_k \rightarrow_w' o_z \rightarrow_r' s_j$ and $o_i \not\rightarrow_r' s_j$, for some $i, k, z, j \in V$ with $i \neq j$. From the definition of I_G, it must be the case that there is a path from node i to node j in G'

and $(i', j) \notin E$ which leads to a contradiction. The case of integrity vulnerabilities is symmetric.

"if" direction. Assume that Sol is a feasible solution of I_G of size $2 \cdot (n + K)$. We define $E' = \{(i, j) \mid i \neq j \wedge o_i \rightarrow'_r s_j\}$. Note that, in the definition of E' using permission $s_i \rightarrow'_w o_j$ rather than $o_i \rightarrow'_r s_j$ would lead to the same set of edges E' (see Lemma 2). By the definition of I_G and Lemma 2, it is direct to see the G' is a subgraph of G and $|E'| = K$. We now show that the transitive closure of G' is again G'. By contradiction, assume that there is a path from node i to node j in G' and there is no direct edge from i to j. But this implies that in the access control system $(S, O, \rightarrow'_r, \rightarrow'_w)$ there is a sequence of alternating read and write operations from object o_i to subject s_j and $o_i \not\rightarrow'_r s_j$, which witnesses a confidentiality threat. This is a contradiction as Sol is a feasible solution of I_G.

7.3   Proof of Theorem 4

Proof. Let $I = (\mathcal{C}, T)$, $\widehat{Sol}_{\eta_I^{\approx}} = (\widehat{\rightarrow_r}^{sol}, \widehat{\rightarrow_w}^{sol})$, $\mathcal{C}' = (S, O, \widehat{\rightarrow_r}^{sol}, \widehat{\rightarrow_w}^{sol})$, and $\mathcal{C}^{\approx} = (S^{\approx}, O^{\approx}, \rightarrow_r^{\approx}, \rightarrow_w^{\approx})$. We first show that $\widehat{Sol}_{\eta_I^{\approx}}$ is a **feasible** solution of I. Assume by contradiction that $\mathcal{C}'$ has a one-step confidentiality threat, say $o \rightarrow_r^{sol} \hat{s} \rightarrow_w^{sol} o' \rightarrow_r^{sol} s \wedge o \not\rightarrow_r^{sol} s$. It is easy to see that $[o] \rightarrow_r^{\approx} [\hat{s}] \rightarrow_w^{\approx} [o'] \rightarrow_r^{\approx} [s] \wedge [o] \not\rightarrow_r^{\approx} [s]$ holds, but this is not possible since COMPACT-ILP-FORMULATION(I) contains a constraint that prevents that these relations hold conjunctly. A similar proof exists for integrity vulnerabilities. Therefore, $\widehat{Sol}_{\eta_I^{\approx}}$ is a feasible solution of I.

Now, we show that $\widehat{Sol}_{\eta_I^{\approx}}$ is also *optimal*. Assume by contradiction that $\widehat{Sol}_{\eta_I^{\approx}}$ is not optimal, and $Sol = (\rightarrow_r^{sol}, \rightarrow_w^{sol})$ is an optimal solution of I where all equivalent subjects/objects have the same permissions. The existence of Sol is guaranteed by Corollary 1. Now, we reach a contradiction showing that η_I is not optimal for COMPACT-ILP-FORMULATION(I). For every $s \in S, o \in O$, $\eta(r_{[o],[s]}) = 1$ (respectively, $\eta(w_{[s],[o]}) = 1$) if and only if $o \rightarrow_r^{sol} s$ (respectively, $s \rightarrow_w^{sol} o$) holds. Notice that η is well defined because all subjects/objects in the same equivalent class have the same permissions in Sol. It is straightforward to prove that η allows to satisfy all linear constraints of COMPACT-ILP-FORMULATION(I), and more importantly leads to a greater value of the objective function. Note that, for the variable assignment η the objective function has a value $n_\eta = size(Sol)$ whereas has value $n_{\eta_I} = size(\widehat{Sol}_{\eta_I^{\approx}})$ for the assignment $\eta_I^{\approx}$. Now, $n_\eta > n_{\eta_I}$, and it cannot be true because $\eta_I^{\approx}$ is an optimal assignment. The definition of η and the fact that it satisfies all linear constraints shows that if I admits a solution then it shows that COMPACT-ILP-FORMULATION(I) admits a solution. Therefore, $\eta_I^{\approx}$ also exists.

References

1. Badger, L., Sterne, D.F., Sherman, D.L., Walker, K.M., Haghighat, S.A.: Practical domain and type enforcement for UNIX. In: IEEE S&P, pp. 66–77 (1995)
2. Bell, D.E., LaPadula, L.J.: Secure computer systems: mathematical foundations. Technical report, DTIC Document (1973)
3. Boebert, W., Young, W., Kaln, R., Hansohn, S.: Secure ADA target: issues, system design, and verification. In: IEEE S&P (1985)
4. Boebert, W.E., Kain, R.Y.: A further note on the confinement problem. In: Proceedings Security Technology, pp. 198–202. IEEE (1996)
5. Brucker, A.D., Petritsch, H.: Extending access control models with break-glass. In: SACMAT, pp. 197–206. ACM (2009)
6. Cheng, W., Zhao, Q., Yu, B., Hiroshige, S.: Tainttrace: efficient flow tracing with dynamic binary rewriting. In: ISCC, pp. 749–754. IEEE (2006)
7. Clause, J., Li, W., Orso, A.: Dytan: a generic dynamic taint analysis framework. In: ISSTA, pp. 196–206. ACM (2007)
8. Crampton, J.: Cryptographic enforcement of role-based access control. In: Degano, P., Etalle, S., Guttman, J. (eds.) FAST 2010. LNCS, vol. 6561, pp. 191–205. Springer, Heidelberg (2011)
9. Denning, D.E.: A lattice model of secure information flow. Commun. ACM **19**(5), 236–243 (1976)
10. Efstathopoulos, P., Krohn, M., VanDeBogart, S., Frey, C., Ziegler, D., Kohler, E., Mazieres, D., Kaashoek, F., Morris, R.: Labels and event processes in the asbestos operating system. In: SOSP, vol. 5, pp. 17–30 (2005)
11. Enck, W., Gilbert, P., Chun, B.G., Cox, L.P., Jung, J., McDaniel, P., Sheth, A.: Taintdroid: An information-flow tracking system for realtime privacy monitoring on smartphones. In: OSDI, vol. 10, pp. 255–270 (2010)
12. Ene, A., Horne, W., Milosavljevic, N., Rao, P., Schreiber, R., Tarjan, R.E.: Fast exact and heuristic methods for role minimization problems. In: SACMAT, pp. 1–10 (2008)
13. Ferrara, A., Fuchsbauer, G., Warinschi, B.: Cryptographically enforced RBAC. In: CSF, pp. 115–129, June 2013
14. Flum, J., Grohe, M.: Parameterized Complexity Theory. Texts in Theoretical Computer Science. An EATCS Series. Springer, New York (2006)
15. Gong, L., Qian, X.: The complexity and composability of secure interoperation. In: 1994 IEEE Computer Society Symposium on Research in Security and Privacy 1994, Proceedings, pp. 190–200. IEEE (1994)
16. Haldar, V., Chandra, D., Franz, M.: Dynamic taint propagation for Java. In: ACSAC, pp. 303–311 (2005)
17. Jaume, M., Tong, V.V.T., Mé, L.: Flow based interpretation of access control: detection of illegal information flows. In: ICISS, pp. 72–86 (2011)
18. Krohn, M., Yip, A., Brodsky, M., Cliffer, N., Kaashoek, M.F., Kohler, E., Morris, R.: Information flow control for standard OS abstractions. In: ACM SIGOPS Operating Systems Review, vol. 41, pp. 321–334. ACM (2007)
19. Lam, L.C., Chiueh, T.: A general dynamic information flow tracking framework for security applications. In: ACSAC, pp. 463–472 (2006)
20. Lampson, B.W.: A note on the confinement problem. Commun. ACM **16**(10), 613–615 (1973)
21. Mao, Z., Li, N., Chen, H., Jiang, X.: Trojan horse resistant discretionary access control. In: SACMAT, pp. 237–246. ACM (2009)

22. Mao, Z., Li, N., Chen, H., Jiang, X.: Combining discretionary policy with mandatory information flow in operating systems. ACM TISSEC **14**(3), 24:1–24:27 (2011)
23. Marinovic, S., Craven, R., Ma, J., Dulay, N.: Rumpole: a flexible break-glass access control model. In: SACMAT, pp. 73–82. ACM (2011)
24. Myers, A.C., Liskov, B.: A decentralized model for information flow control. In: SIGOPS Operating Systems Review, vol. 31, pp. 129–142. ACM (1997)
25. Petritsch, H.: Break-Glass: Handling Exceptional Situations in Access Control. Springer, Heidelberg (2014)
26. Sze, W.K., Mital, B., Sekar, R.: Towards more usable information flow policies for contemporary operating systems. In: SACMAT (2014)
27. Sze, W.K., Sekar, R.: Provenance-based integrity protection for windows. In: ACSAC 2015, New York, NY, USA, pp. 211–220. ACM, New York (2015)
28. Yannakakis, M.: Node-and edge-deletion NP-complete problems. In: Lipton, R.J., Burkhard, W.A., Savitch, W.J., Friedman, E.P., Aho, A.V. (eds.) STOC, pp. 253–264. ACM (1978)
29. Yannakakis, M.: Edge-deletion problems. SIAM J. Comput. **10**(2), 297–309 (1981)
30. Zhang, D., Ramamohanrao, K., Ebringer, T.: Role engineering using graph optimisation. In: SACMAT, pp. 139–144 (2007)
31. Zhu, Y., Jung, J., Song, D., Kohno, T., Wetherall, D.: Privacy scope: a precise information flow tracking system for finding application leaks. Ph.D. thesis, UC, Berkeley (2009)
32. Zimmermann, J., Mé, L., Bidan, C.: An improved reference flow control model for policy-based intrusion detection. In: Snekkenes, E., Gollmann, D. (eds.) ESORICS 2003. LNCS, vol. 2808, pp. 291–308. Springer, Heidelberg (2003). doi:10.1007/978-3-540-39650-5_17

Object-Tagged RBAC Model
for the Hadoop Ecosystem

Maanak Gupta[✉], Farhan Patwa, and Ravi Sandhu

Department of Computer Science, Institute for Cyber Security,
University of Texas at San Antonio, One UTSA Circle, San Antonio, TX 78249, USA
gmaanakg@yahoo.com, {farhan.patwa,ravi.sandhu}@utsa.edu

Abstract. Hadoop ecosystem provides a highly scalable, fault-tolerant and cost-effective platform for storing and analyzing variety of data formats. Apache Ranger and Apache Sentry are two predominant frameworks used to provide authorization capabilities in Hadoop ecosystem. In this paper we present a formal multi-layer access control model (called HeAC) for Hadoop ecosystem, as an academic-style abstraction of Ranger, Sentry and native Apache Hadoop access-control capabilities. We further extend HeAC base model to provide a cohesive object-tagged role-based access control (OT-RBAC) model, consistent with generally accepted academic concepts of RBAC. Besides inheriting advantages of RBAC, OT-RBAC offers a novel method for combining RBAC with attributes (beyond NIST proposed strategies). Additionally, a proposed implementation approach for OT-RBAC in Apache Ranger, is presented. We further outline attribute-based extensions to OT-RBAC.

Keywords: Access control · Hadoop ecosystem · Big Data · Data lake · Role based · Attributes · Groups hierarchy · Object Tags

1 Introduction

Over the last few years, enterprises have started harvesting data from 'anything' to discover business and customer needs. It is estimated that 163 zettabytes of data will be generated annually by year 2025 as quoted by IDC [5]. Such massive and varied collections of data, referred to as Big Data, are considered 21st century gold for data miners. Enterprises gain useful insights from analysis to offer targeted marketing, fraud detection, accident forecasting, traffic patterns and even strong love matching. With volume, variety and velocity of data burgeoning, massive storage and compute clusters are required for analysis.

Apache Hadoop [1] has established itself as an important open-source framework for cost-efficient, distributed storage and computing of data in timely fashion. The platform offers resilient infrastructure for sophisticated analytical and pattern recognition techniques for multi-structured data. Hadoop ecosystem includes several open-source and commercial tools (Apache Hive, Apache

© IFIP International Federation for Information Processing 2017
Published by Springer International Publishing AG 2017. All Rights Reserved
G. Livraga and S. Zhu (Eds.): DBSec 2017, LNCS 10359, pp. 63–81, 2017.
DOI: 10.1007/978-3-319-61176-1_4

Storm, Apache HBase, Apache Ambari etc.) built to leverage the full capabilities of Hadoop framework. These tools along with Apache Hadoop 2.x core modules (Hadoop Common, Hadoop Distributed File System (HDFS), YARN and MapReduce) empower users to harness the potential of data assets.

As Hadoop is widely used in government and private sector, its security has been a major concern and widely studied. Multi-tenant Data Lake offered by Hadoop, stores and processes sensitive information from several critical sources, such as banking and intelligence agencies, which should only be accessed by legitimate users and applications. Threats—including denial of resources, malicious user killing YARN applications, masquerading Hadoop services like NameNode, DataNode etc.—can have serious ramifications on confidentiality and integrity of data and ecosystem resources. The distributed nature and platform scale makes it more difficult to protect the infrastructure assets.

Apache Ranger [3] and Apache Sentry [4] are two important software systems used to provide fine-grained access across several Hadoop ecosystem components. In this paper we present the multi-layer access control model for Hadoop ecosystem (referred as HeAC), formalizing the authorization model in Apache Ranger (release 0.6) and Sentry (release 1.7.0) in addition to access controls capabilities in core Hadoop 2.x. We further propose an Object-Tagged Role Based Access Control (OT-RBAC) model which leverages the merits of RBAC and provides a novel approach of adding object attribute values (called object tags) in RBAC model. We also outline extensions to OT-RBAC to incorporate NIST proposed strategies [27] for adding attributes in RBAC. To our knowledge this is the first work to consider formal authorization models specific to Hadoop ecosystem.

The remainder of this paper is as follows. Section 2 discusses current authorization capabilities in Hadoop ecosystem. In Sect. 3, we present a formal Hadoop ecosystem access control model called HeAC. We introduce the Object-Tagged Role Based Access Control (OT-RBAC) model in Sect. 4, followed by proposed implementation in Sect. 5. In Sect. 6, we present attributes-based authorization extensions to OT-RBAC. Section 7 reviews previous related work, followed by Sect. 8 which gives our conclusions.

2   Multi-layer Authorization in Hadoop Ecosystem

The most critical assets required to be secured in Hadoop ecosystem involve services, data and service objects, applications and cluster infrastructure resources. In this section we discuss the multi-layer authorization capabilities provided in Hadoop ecosystem in line with Apache Hadoop 2.x, along with access control features offered by Apache Ranger, Apache Sentry and Apache Knox.

Service Access: The first layer of defense is provided by service level authorization which checks if a user or application is allowed to access the Hadoop ecosystem services and Hadoop core daemons. This check is done before data and service objects permissions are evaluated, thereby preventing unauthorized access early in the access request lifecycle. ACLs (Access Control Lists) are specified with users and groups to restrict access to services. For example, ACL `security.job.client.protocol.acl` is checked to allow a user to communicate with YARN ResourceManager for job submission or application status inquiry. This layer also restricts cross-service communication to prevent malicious processes interaction with Hadoop daemon services (NameNode, ResourceManager etc.). Another ACL `security.datanode.protocol.acl` is checked for interaction between DataNodes and NameNode for heartbeat or task updates. A user making API requests to individual ecosystem services like Apache Hive, HDFS, Apache Storm etc., is restricted by implementing single gateway (e.g. Apache Knox [2]) access point—which enforces policies to allow or deny users to access ecosystem services before operating on underlying objects.

Data and Service Objects Access: Hadoop Distributed File System (HDFS) enforces POSIX style model and ACLs for setting permissions on files and directories holding data. Multiple other ecosystem services require different objects to be secured. For example, Apache Hive requires table and columns, whereas Apache Kafka secures topic objects from unauthorized operations by users. Some services like Apache Hive or Apache HBase also have native access control capabilities to secure different data objects. Security frameworks like Apache Ranger or Sentry provide plugins for individual ecosystem services, where centralized policies are set for different data and service objects. In Apache Ranger, authorization policies can also be formulated on Tags, which are attribute values associated with objects. For example, a tag PII can be associated with table SSN and a policy is created for tag PII. Such tag-based policy will then control access to table SSN. Tags allow controlling access to resources along several services without need to create separate policies for individual services. It should be noted that data access allowed at one service may be restricted by permissions at underlying HDFS, thereby requiring user to have multiple object permissions at different services.

Application and Cluster Resources Access: Multi-tenant Hadoop cluster requires sharing of finite resources among several users, controlled by Apache YARN capacity (or fair) scheduler queues in Hadoop 2.x. Queue level authorization enables designated users to submit or administer applications in different queues. This restricts user from submitting applications in cluster and prevents rogue users from deleting or modifying other user applications. Further, cluster resources are not consumed by certain applications requiring more resources as queues have limited resources allocated. It should be noted that application owner and queue administrator can always kill or modify jobs in queue. These queues support hierarchical structure where permissions to parent queues descend to all child queues. Hadoop implements these authorization configurations using ACL's. Configuration file can also be associated with applications to specify users who can modify or kill an application. Access to cluster nodes can be restricted by assigning node labels. Each queue can be associated with node labels to restrict nodes where applications submitted to queues can run.

Figure 1 reflects multi-layer authorization architecture provided in Hadoop ecosystem. An authenticated user passes through several access control mechanisms to operate on objects and services in Hadoop cluster. Gateway (such as

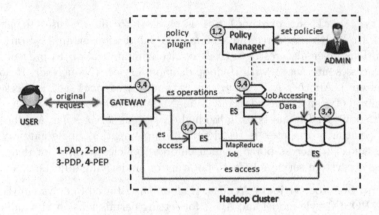

Fig. 1. Example Hadoop Ecosystem Authorization Architecture

Apache Knox) offers single access point to all REST APIs and provides first layer of access control to check if services inside the cluster are allowed access by outside users. Once user is approved through the gateway policy plugin, ecosystem services apply policies cached from central policy manager to validate requests of user. User trying to access objects (files, tables etc.) in ecosystem services like HDFS or Apache Hive (shown as ES in Fig. 1) is checked by policy plugins attached to the services to enforce access decisions. A user wanting to submit an application or to get submitted application status should be allowed through gateway policies to communicate with YARN ResourceManager. Apache YARN queue permissions are then checked and enforced by plugin to know if a user is allowed to submit or administer application in queues. Cross-services access (between Hadoop daemons) for information passing or task status update is mainly enforced using core Hadoop service ACLs.

As shown in Fig. 1, security frameworks like Apache Ranger provides centralized Policy adminstration (1-PAP) and information point (2-PIP). Enforcement and decision points (3-PEP, 4-PDP) are plugins attached to each service which cache policies periodically from central server and enforce access decisions.

3   Hadoop Ecosystem Access Control Model

In this section we present the formal multi-layer access control model (HeAC) for Hadoop ecosystem based on Apache Hadoop 2.x. The model also covers access capabilities provided by two predominant Apache projects, Ranger (release 0.6) and Sentry (release 1.7.0). Apache Ranger supports permissions through users and groups, while Sentry assigns permissions to roles which are assigned to groups and via groups to member users. We will now discuss the formal definitions of HeAC model as specified in Table 1 and shown in Fig. 2.

The basic components of HeAC include: Users (U), Groups (G), Roles (R), Subjects (S), Hadoop Services (HS), Operations (OP_{HS}) on Hadoop Services,

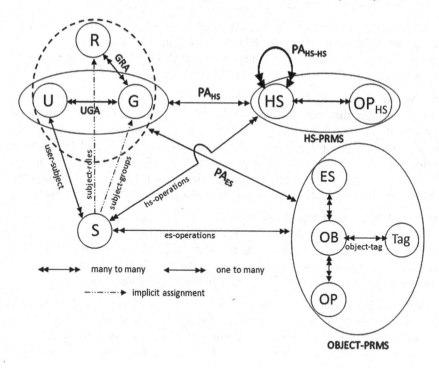

Fig. 2. A Conceptual Model of HeAC

Ecosystem Services (ES), Data and Service Objects (OB) belonging to Ecosystem Services, Operations (OP) on objects, and Object Tags (Tag).

Users, Groups, Roles and Subjects: A user is a human who interacts with computer to access services and objects inside the Hadoop ecosystem. A group is a collection of users in the system with similar organizational requirements. A role is a collection of permissions which can be assigned to different entities in the system. Permissions are assigned to users, groups or roles. In the current model roles can only be assigned to groups, thereby giving permissions to member users of groups indirectly. A subject is an application running on behalf of the user to perform operations in the Hadoop ecosystem. In HeAC model subjects always run with full permissions of the creator user.

Hadoop Services: These services are background daemon processes, like HDFS NameNode, DataNode, YARN ResourceManager, ApplicationMaster etc., which provide core functionalities in Hadoop 2.x framework. User access these services to submit applications, data block recovery or application status updates. Besides interaction with end user, these daemon services also communicate with each other for resource monitoring or task updates. It should be noted that these services do not have objects associated with them.

Operations on Hadoop Services: These are actions allowed on Hadoop services. In most cases, the general action allowed is to access a service.

For example, ACL `security.client.protocol.acl` is used to determine which user is allowed to access HDFS NameNode service. These ACLs are part of Hadoop native access control capabilities (referred as service level authorization).

Ecosystem Services: Data and objects inside the Hadoop ecosystem are accessed through different platforms which we consider as Ecosystem Services. Example of such services include Apache HDFS, Apache Hive, Apache HBase, Apache Storm, Apache Kafka etc. These services can either have data objects (tables, columns) or other type of resources (queues, topics) which they support. Access to the ecosystem services is first required before operation on supported objects. We consider Data Services as one instance of Ecosystem Services.

Data and Service Objects: Ecosystem services manage different types of resources (objects) inside the cluster. For example, Apache HDFS supports files and directories, while Apache HBase has data objects like column-family, cells etc. YARN manages queue objects and Apache Solr has collections. These are resources which are protected from unauthorized operations from users.

Operations on objects: Multiple data and service objects support different operations to perform actions on them. Apache Hive has select, create, drop, alter for tables and columns while Apache HBase data objects (column family, column) support read, write, create etc. YARN queues have operations to submit applications or administer the queue.

Object Tags: Objects inside ecosystem can be assigned attributes based on sensitivity, content or expiration date. Such classification is done using attribute values called Tags. An object can have multiple tags associated with it and vice versa. For example, PII tag can be attached to sensitive data table SSN.

As shown in Table 1, a user can be assigned to multiple groups defined by directUG function. Groups are also assigned to multiple roles as reflected by function directGR. Relation object-tag denotes a many-to-many relation between objects and associated attribute values called tags. Hadoop ecosystem has two different sets of permissions to perform actions on services and objects. OBJECT-PRMS is the set of data and service object permissions which is power set of the cross product of ecosystem services (ES), objects (OB) or object tags (Tag), and operations (OP). Here permissions can be set either on object or object tags, and policies can allow or deny operations on the object based on its associated tags or the object itself. OBJECT-PRMS also include ecosystem service as part of permission thereby taking into account the requirement of service access before object operations. Another set of permissions called Hadoop service permissions (HS-PRMS) is the power set of the cross product of HS and OP_{HS}. These are required for application submission or other non-data or object operations. Depending on the type of operations to be performed, a user may require either one or both type of permissions.

A many-to-many relation PA_{HS} specifies the assignment of HS-PRMS to users or groups. In this way a user can be assigned HS-PRMS directly or through group membership. OBJECT-PRMS can be assigned to users, groups or roles

Table 1. Hadoop Ecosystem Access Control (HeAC) Model Definitions

---

Basic Sets and Functions
- U, G, R, S (finite set of users, groups, roles and subjects respectively)
- HS, OP_{HS} (finite set of Hadoop services and operations respectively)
- ES, OB (finite set of ecosystem services and objects respectively)
- OP, Tag (finite set of object operations and object tags respectively)
- directUG : $U \rightarrow 2^G$, mapping each user to a set of groups, equivalently UGA $\subseteq$ U $\times$ G
- directGR : $G \rightarrow 2^R$, mapping each group to a set of roles, equivalently GRA $\subseteq$ G $\times$ R
- object-tag $\subseteq$ OB$\times$Tag, relation between object and object tags
- OBJECT-PRMS $= 2^{ES \times (OB\ \cup\ Tag) \times OP}$, set of data and service object permissions
- HS-PRMS $= 2^{HS \times OP_{HS}}$, set of Hadoop services permissions

Permission Assignments
- $PA_{HS} \subseteq$ (U $\cup$ G)$\times$HS-PRMS, mapping entities to Hadoop service permissions. Alternatively, hs_{prms} : (x) $\rightarrow 2^{HS-PRMS}$, defined as hs_{prms}(x) = {p | (x,p) $\in PA_{HS}$, x $\in$ (U $\cup$ G)}
- $PA_{ES} \subseteq$ (U $\cup$ G $\cup$ R)$\times$OBJECT-PRMS, mapping entities to object permissions. Alternatively, es_{prms} : (x) $\rightarrow 2^{OBJECT-PRMS}$, defined as es_{prms}(x) = {p | (x,p) $\in PA_{ES}$, x $\in$ (U $\cup$ G $\cup$ R)}

Hadoop Cross Services Access
- $PA_{HS\text{-}HS} \subseteq$ HS $\times$ HS-PRMS, mapping Hadoop service to Hadoop service access. Alternatively, $hs\text{-}hs_{prms}$: (hs:HS) $\rightarrow 2^{HS-PRMS}$, defined as $hs\text{-}hs_{prms}$(hs) = {p | (hs,p) $\in PA_{HS\text{-}HS}$ }

Effective Roles of Users (Derived Functions)
- effectiveR : $U \rightarrow 2^R$, defined as effectiveR(u) = $\bigcup\limits_{\forall g\ \in\ directUG(u)}$ directGR(g)

Effective Permissions of User
- effectiveHS$_{prms}$: $U \rightarrow 2^{HS-PRMS}$, defined as effectiveHS$_{prms}$(u) = hs_{prms}(u) $\cup \bigcup\limits_{\forall g\ \in\ \{directUG(u)\}}$ hs_{prms}(g)
- effectiveES$_{prms}$: $U \rightarrow 2^{OBJECT-PRMS}$, defined as effectiveES$_{prms}$(u) = es_{prms}(u) $\cup \bigcup\limits_{\forall x\ \in\ \{directUG(u)\ \cup\ effectiveR(u)\}}$ es_{prms}(x)

User Subject
- userSub : $S \rightarrow U$, mapping each subject to its creator user, where the subject gets all the permissions of the creator user.

Ecosystem Service Object Operation Decision
A subject s $\in$ S is allowed to perform an operation op $\in$ OP on an object ob $\in$ OB in ecosystem service es $\in$ ES if the effective object permissions of userSub(s) include permissions for object ob or for tag t $\in$ Tag associated with object ob. Formally,
(es,ob,op) $\in$ effectiveES$_{prms}$ (userSub(s)) $\vee$
($\exists$ t) [(ob,t) $\in$ object-tag $\wedge$ (es,t,op) $\in$ effectiveES$_{prms}$ (userSub(s))]

(shown by PA_{ES}). A group can get the object permissions directly or through roles, which will then enable it to the member users. It should be noted that a user may need multiple data object permissions across several data services to operate on a data object. For example, in case of Apache Hive table, besides having permission on the table, a user may be required to have permissions on the underlying data file in HDFS. $PA_{HS\text{-}HS}$ encapsulates the access requirement between several Hadoop services inside the cluster for task updates or resource monitoring (e.g. communication between DataNodes and NameNode). The effective roles of user are covered by effectiveR which is union of roles assigned to all member groups. The effective permissions on Hadoop services attained by user (reflected by effectiveHS$_{prms}$) is the direct permissions on HS and permissions inherited through group membership. The final set of ES object permissions for a user is union of direct permission and permissions assigned through group membership and effective roles as shown in effectiveES$_{prms}$.

A subject is created by a user as expressed by userSub. It inherits all the permissions assumed by the user to perform actions. In last section of Table 1, a subject is allowed to perform operations on objects in ES service depending on either direct permission on objects or permission on tags associated with objects.

It should be noted that Apache Ranger provides context enrichers, which are used to add contextual information to user request based on location, IP address or other attribute. We treat such information as environment attributes and include these in attribute-based model in Sect. 6. It should also be mentioned that data ingestion into Hadoop cluster is beyond the scope of this paper and for the access control points discussed, we assume data already present inside the cluster. Further, we ignore deny access request, for the sake of simplicity.

4   Object-Tagged RBAC for Hadoop Ecosystem

In this section we propose Object-Tagged Role-Based Access Control model for the Hadoop Ecosystem, which we denote as OT-RBAC. With respect to HeAC model, this model assigns both objects and Hadoop service permissions to users only through roles, consistent with the basic principle of RBAC. The model presents a novel approach for combining attributes and RBAC [33] besides NIST proposed approaches (i.e., Dynamic Roles, Attribute-Centric and Role Centric) [27]. Hence the convenient administrative benefits of RBAC, along with a finer-grained attributes authorization, are incorporated in this model.

The conceptual model for OT-RBAC is shown in Fig. 3 followed by formal definitions in Table 2. The remainder of this section discusses the new and modified components introduced in OT-RBAC model (marked** and †† respectively) with respect to HeAC model. In OT-RBAC model users are directly assigned to multiple roles specified by function directUR. Group hierarchy (GH) is introduced into the system, defined by a partial order relation on G and written as $\succeq_g$. The inheritance of roles is from low to high, i.e., $g_1 \succeq_g g_2$ means g_1 inherits roles from g_2. In such cases, we say g_1 is the senior group and g_2 is the junior group. The HS-PRMS and OBJECT-PRMS permissions are assigned to roles only, specified by many-to-many relations PA$_{HS}$ and PA$_{ES}$ respectively. This is modified with respect to the original HeAC, where HS-PRMS were assigned to users or groups and OBJECT-PRMS to user, groups or roles also. This reflects the advantage of RBAC model where permissions are allotted or removed from users by granting or revoking their roles. Both OBJECT-PRMS and HS-PRMS can be assigned to same role in the Hadoop ecosystem. With group hierarchy (GH), the effective roles of a group (expressed by effectiveGR) is the union of

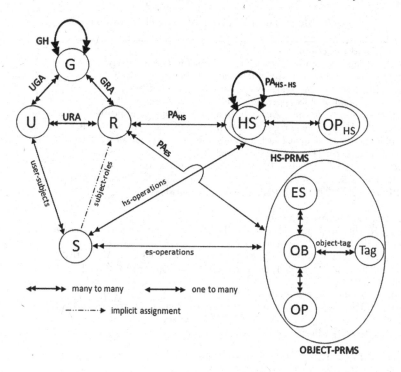

Fig. 3. Conceptual OT-RBAC Model for Hadoop Ecosystem

direct roles assigned to group and effective roles of all its junior groups. It should be noted that this definition is recursive where the junior-most groups have same direct and effective roles. The effective roles of the user (defined by effectiveR) is then the union of direct user roles and effective roles of the groups to which the user is directly assigned. For example, assuming group Grader is assigned roles Student and Graduate and a senior group TA is assigned to role Doctoral. Then the effective roles of group TA would be Student, Graduate and Doctoral. A user u_1 can be directly assigned to role Staff. If u_1 also becomes a member of group TA, u_1 has the effective roles of Student, Graduate, Doctoral and Staff. The important advantage of user group membership is convenient assignment and removal of multiple roles from users with single administrative operation.

A subject S (similar to sessions in RBAC [17]) created by the user can have some or all of the effective roles of the creator user. The effective permissions available to a subject (expressed by effectiveES$_{\text{prms}}$ and effectiveHS$_{\text{prms}}$) will then be the object and Hadoop service permissions assigned to all the effective roles activated by the subject. A subject might need to have multiple permissions to access different services or objects inside Hadoop ecosystem which may result in requiring multiple roles. The prime advantage of OT-RBAC model over HeAC model is the assignment of permissions only to roles instead of assigning directly to users and groups. Further it introduces the concept of group hierarchy

Table 2. Formal OT-RBAC Model Definitions

Basic Sets and Functions
- U, G, R, S (finite set of users, groups, roles and subjects respectively)
- HS, OP_{HS} (finite set of Hadoop services and operations respectively)
- ES, OB (finite set of ecosystem services and objects respectively)
- OP, Tag (finite set of object operations and object tags respectively)
- directUG : $U \rightarrow 2^G$, mapping each user to a set of groups, equivalently UGA $\subseteq$ U × G
- [**]directUR : $U \rightarrow 2^R$, mapping each user to a set of roles, equivalently URA $\subseteq$ U × R
- directGR : $G \rightarrow 2^R$, mapping each group to a set of roles, equivalently GRA $\subseteq$ G × R
- [**]GH $\subseteq$ G×G, a partial order relation $\succeq_g$ on G
- object-tag $\subseteq$ OB×Tag, relation between object and object tags
- OBJECT-PRMS = $2^{ES \times (OB \cup Tag) \times OP}$, set of data and service object permissions
- HS-PRMS = $2^{HS \times OP_{HS}}$, set of Hadoop services permissions

[††]**Role Permission Assignments**
- $PA_{HS} \subseteq$ R×HS-PRMS, mapping roles to Hadoop service permissions. Alternatively,
  hs_{prms} : (r:R) $\rightarrow 2^{HS-PRMS}$, defined as $hs_{prms}(r) = \{p \mid (r,p) \in PA_{HS}\}$
- $PA_{ES} \subseteq$ R×OBJECT-PRMS, mapping roles to object permissions. Alternatively,
  es_{prms} : (r:R) $\rightarrow 2^{OBJECT-PRMS}$, defined as $es_{prms}(r) = \{p \mid (r,p) \in PA_{ES}\}$

Hadoop Cross Services Access
- $PA_{HS-HS} \subseteq$ HS × HS-PRMS, mapping Hadoop service to Hadoop service access.
  Alternatively, $hs\text{-}hs_{prms}$: (hs:HS) $\rightarrow 2^{HS-PRMS}$, defined as
  $hs\text{-}hs_{prms}(hs) = \{p \mid (hs,p) \in PA_{HS-HS}\}$

[††]**Effective Roles of Users (Derived Functions)**
- effectiveGR : $G \rightarrow 2^R$, defined as
  $$effectiveGR(g_i) = directGR(g_i) \cup \left(\bigcup_{\forall g \in \{g_j \mid g_i \succeq_g g_j\}} effectiveGR(g) \right)$$
- effectiveR : $U \rightarrow 2^R$, defined as
  $$effectiveR(u) = directUR(u) \cup \left(\bigcup_{\forall g \in directUG(u)} effectiveGR(g) \right)$$

[††]**Effective Roles and Permissions of Subjects**
- userSub : $S \rightarrow U$, mapping each subject to its creator user
- effectiveR : $S \rightarrow 2^R$, mapping of subject s to a set of roles. It is required that :
  effectiveR(s) $\subseteq$ effectiveR(userSub(s))
- $effectiveHS_{prms}$: $S \rightarrow 2^{HS-PRMS}$, defined as $effectiveHS_{prms}(s) = \bigcup_{\forall r \in effectiveR(s)} hs_{prms}(r)$
- $effectiveES_{prms}$: $S \rightarrow 2^{OBJECT-PRMS}$, defined as $effectiveES_{prms}(s) = \bigcup_{\forall r \in effectiveR(s)} es_{prms}(r)$

Ecosystem Service Object Operation Decision
A subject s $\in$ S is allowed to perform an operation op $\in$ OP on an object ob $\in$ OB in
ecosystem service es $\in$ ES if the effective object permissions of subject s include permissions
to object ob or to tag t $\in$ Tag associated with object ob. Formally,
(es,ob,op) $\in$ $effectiveES_{prms}$ (s) $\vee$
($\exists$ t) [(ob,t) $\in$ object-tag $\wedge$ (es,t,op) $\in$ $effectiveES_{prms}$ (s)]

[**] and [††] highlight new and modified components respectively with respect to HeAC

which results in roles inheritance and eases administrative responsibilities of the
security administrator. Also including group hierarchy makes OT-RBAC model
easier to fit into attributes based models where role is one of the other attributes.
In such case group hierarchy can be very useful in attributes inheritance offering

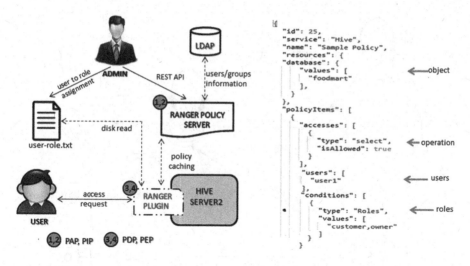

Fig. 4. Proposed Implementation in Apache Ranger and Sample JSON Policy

convenient administration by assigning or removing multiple attributes to users with single administrative operation [35].

The proposed OT-RBAC model presents a novel approach for adding attributes to RBAC (besides NIST strategies [27]), by introducing object tags. The model represents object permissions (OBJECT-PRMS) as union of permissions on attribute values (reflected as tags) associated with objects and regular permissions as discussed in RBAC [33]. In the following section, we propose an implementation approach for OT-RBAC using open-source Apache Ranger.

5   Proposed Implementation

One approach to implement OT-RBAC model is by extending open-source Apache Ranger which provides centralized security administration to multiple Hadoop ecosystem services. It offers REST API to create security policies which are enforced using plugins appended to each secured service. These plugins intercept a user access request, and check against policies cached sporadically from policy server to make access decisions. Apache Ranger 0.5 and above provide extensible framework to add new authorization functionalities by offering context enricher and condition evaluator hooks. Context enricher is a Java class which appends user access request with additional information used for policy evaluation. Condition evaluator enables a security architect to add custom conditions to policies. These hooks can be used to extend plugins to enforce OT-RBAC.

Proposed Apache Ranger architecture for Hive service authorization is shown in Fig. 4. Users and groups are stored in LDAP, which are synced to Ranger policy manager to create policies. A text file is added which stores current users to roles assignment. This file is used by context enricher implemented, to add roles

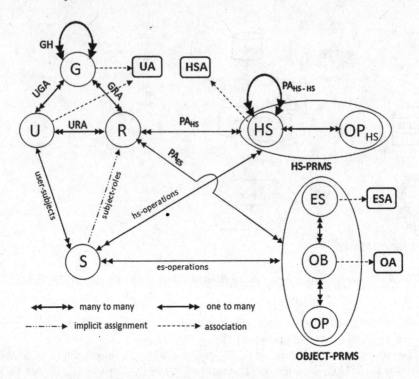

Fig. 5. Adding Attributes to OT-RBAC model

of user to access request along with objects and actions. A condition evaluator should also be implemented to include roles in policy used for evaluation. A sample policy in JSON format is shown in Fig. 4. This policy includes roles in condition which specifies the roles allowed to perform select operation on table foodmart. Hive service definition should be updated with new context and condition hooks information using REST API. Access decision and enforcement is done in Ranger plugin embedded with Hive service whereas policy administration and information is through central policy server as shown in Fig. 4. Similar implementation approach can be adopted in other ecosystem services also. This proposed implementation requires roles addition at two places, one in text file and other in policy conditions which requires extra effort by administrator.

6 Attributes Based Extensions to OT-RBAC

We outline some approaches for adding attributes in OT-RBAC model to achieve finer-grained access control. OT-RBAC model incorporates tags for objects, which is further generalized by introducing set of object attributes along with attributes for other entities. As shown in Fig. 5, UA is a set of attributes for users and groups, and OA is a set of attributes for data and service objects. HSA and ESA are set of attributes for HS and ES. An attribute is a function which takes

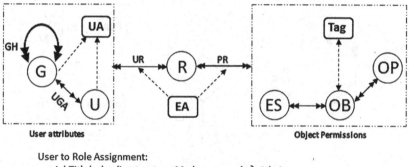

Fig. 6. Dynamic Roles and Object Permissions in OT-RBAC

as input an entity and returns values from a specified range [24]. Attribute-based authorization policies are used to determine access permissions of users on services and objects. With group hierarchy, senior groups inherit attributes from junior groups [20], and a user assigned to senior groups gets all attributes of group besides its direct attributes. A set of environment attributes is also added to incorporate contextual information (like access time, threat level) in policies.

We outline how an attribute enhanced OT-RBAC model, along the lines of Fig. 5, can incorporate NIST proposed strategies [27] for adding attributes in RBAC, i.e., Dynamic Roles, Attribute Centric and Role Centric. We discuss these in context of objects permissions assignment. These approaches can be similarly applied to Hadoop services permissions assignment also.

6.1 Dynamic Roles

Dynamic Roles approach considers user and environment attributes to determine roles of a user. This automated approach require rules defined using a policy language [8] composed of attributes and resulting roles. The roles of the user will change based on the user's current attributes as well as current environment attributes. As shown in Fig. 6, OT-RBAC model can be configured to achieve dynamic roles assignment to users based on the direct or inherited attributes through group memberships [20]. We can further extend the use of attributes for dynamic permissions assignment to roles based on object tags, environment attribute values and operations.

As in Fig. 6, user u_1 with attribute jobTitle value director and environment attribute optMode value normal can be assigned Admin role, which can change to role Faculty when attribute optMode changes to emergency. Similarly, permission

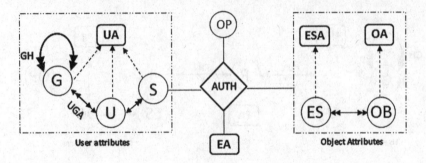

Authorization$_{write}$ (s:S, es:ES, ob:OB) :: effective$_{jobTitle}$(s) = director ∧ access(s,es) = True
∧ name(es) = hdfs ∧ tag(ob)= PII ∧ optMode = normal

Authorization$_{read}$ (s:S, es:ES, ob:OB) :: effective$_{jobTitle}$(s) = professor ∧ access(s,es) = True
∧ name(es) = hdfs ∧ tag(ob)= PCI ∧ optMode = emergency.

Fig. 7. Attribute Centric Approach in OT-RBAC

containing operation write on object ob with tag value PII can be assigned to role Admin which can change to role Faculty when tag changes to PCI.

6.2 Attribute Centric

In this approach, access decision is based on attributes of entities (role is also an attribute) where authorization policies comprise attributes of subjects, objects or environment [23,24,42]. To configure OT-RBAC with attribute centric strategy, boolean authorization functions are defined using propositional logic formula for each operation in OP which specify policy if subject s can perform operation op on object ob in ecosystem service es under some environment attributes.

As shown in Fig. 7, authorization policy is defined stating that subject s with effective attribute jobTitle value director is allowed to perform write on object ob with attribute tag value PII in ecosystem service es with name hdfs when environment attribute optMode is normal. It should be noted that object ob must belong to ecosystem service es and subject must be allowed to access es (expressed by access(s,es)) before performing any operation on object in es. Similar authorization policy for read operation can be defined by administrators.

6.3 Role Centric

In this approach the maximum permissions (avail_prms) are assigned to user through roles assignment (similar to RBAC [33]) but the final set of permissions (final_prms) is dependent on the attributes of entities. Permission Filtering boolean functions are defined based on the attributes, which are checked for each permission in avail_prms set available to users via roles, to determine the final_prms set assigned to the users as discussed in [25].

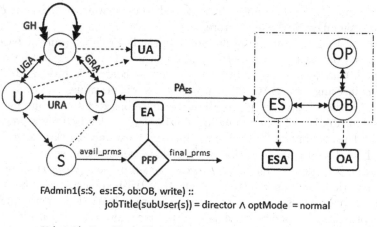

FAdmin1(s:S, es:ES, ob:OB, write) ::
    jobTitle(subUser(s)) = director ∧ optMode = normal

FAdmin2(s:S, es:ES, ob:OB, read) ::
    jobTitle(subUser(s)) = faculty ∧ tag(ob) = PCI

Fig. 8. Role Centric Approach in OT-RBAC

Assume user u_1 assigned to role Admin then u_1 gets permissions (avail_prms) of writing to hdfs service file customer and reading a file having PII tag. These permissions are checked against filter functions selected using target functions discussed in [25]. As shown in Fig. 8, filter function FAdmin1 is invoked to check if first permission is in final_prms set. The function checks if creator user of s has jobTitle attribute with value director and optMode is normal to avail this permission. If it returns true, the permission will be included in final set (final_prms). Similar filter function can be called for other permissions also.

7   Related Work

Several papers [6,7,14,19,32,36,43] discuss security threats and solutions in Hadoop ecosystem. Recently, Gupta et al. [18] presented a multi-layer authorization framework for Hadoop ecosystem, which covers several access control enforcement points and demonstrates their application using Apache Ranger. Access control using cryptography based on proxy re-encryption [31] provides approach for delegated access to Hadoop cluster. A security model for G-Hadoop framework using public key and SSL is presented in [46]. Security and privacy concerns of MapReduce applications are discussed in [15]. Ulusoy et al. [39,40] proposed approaches for fine grained access control for MapReduce systems. Privacy issues in Big Data are addressed in [13,29,37,38].

Risk aware information disclosure in [9] can be used for Hadoop Data lake. Secure information access model via data services [11] can be applied for Hadoop data services. HDFS can use data access protection using data distribution and swapping in [16]. Vimercati et al. [41] discuss confidentiality of outsourced data.

Colombo et al. [12] also proposed fine-grained context-aware access control features for MongoDB NoSQL datastore.

Risk based access using classification [10] studies role assignment based on risk factors. Contextual attributes in location aware ABAC in [21] can be applied in Hadoop. Classification of data object based on content is presented in [44]. Policy engineering for ABAC [26] can be used to define values based on risk or context. Another promising approach in attribute based data sharing has been presented in [45]. Use of role mining in [28] can be extended to determine roles of users based on attributes. A research roadmap on trust and Big Data is presented in [34]. Trust based Data ingestion or processing can use models in [30].

Hu et al. [22] presented a general access control model for Big Data processing frameworks. The paper introduces chain of trust among several entities to authorize access request. The work provides a preliminary document which can be conceptualized to specific systems like Hadoop. However, the authors do not address details particular to the Hadoop ecosystem.

8   Conclusion and Future Work

In this paper we present first formalized access control model called HeAC for Hadoop ecosystem. Besides the regular permissions including objects and operations, this model also includes object attribute values (represented as tags) in object permissions. We further extended HeAC model to propose Object-Tagged RBAC model (OT-RBAC) which preserves role based permission assignment and presents a novel approach for adding object attributes to RBAC. We proposed an implementation approach for introducing roles in open-source Apache Ranger using context enricher and condition evaluators. We additionally draft some extensions to OT-RBAC by adding attributes to provide fine grained access policies. We outline OT-RBAC model to support NIST strategies for including attributes using Dynamic Roles, Attribute Centric and Role Centric.

For future work, we plan to develop pure attribute based access control models for fine grained access to Hadoop ecosystem resources. Also, since the Hadoop data lake is used by multiple tenants it would be interesting to introduce data ingestion security into the system to secure data at HDFS data nodes level.

Acknowledgement. Sincere gratitude is extended to James Benson, Technology Research Analyst at Institute for Cyber Security, UTSA, for his useful comments. This research is partially supported by NSF Grants CNS-1111925, CNS-1423481, CNS-1538418, DoD ARL Grant W911NF-15-1-0518 and by The Texas Sustainable Energy Research Institute at University of Texas at San Antonio.

References

1. Apache Hadoop. http://hadoop.apache.org/
2. Apache Knox. https://knox.apache.org/

3. Apache Ranger. http://ranger.apache.org/
4. Apache Sentry. https://sentry.apache.org/
5. Data Age 2025: The Evolution of Data to Life-Critical. https://www.idc.com/
6. Big Data: Securing Intel IT's Apache Hadoop Platform (2016). http://www.intel.com/content/dam/www/public/us/en/documents/white-papers/big-data-securing-intel-it-apache-hadoop-platform-paper.pdf
7. Securing Hadoop: Security Recommendations for Hadoop Environments (2016). https://securosis.com/assets/library/reports/Securing_Hadoop_Final_V2.pdf
8. Al-Kahtani, M.A., Sandhu, R.: A model for attribute-based user-role assignment. In: Proceedings of IEEE ACSAC, pp. 353–362 (2002)
9. Armando, A., Bezzi, M., Metoui, N., Sabetta, A.: Risk-based privacy-aware information disclosure. IJSSE 6(2), 70–89 (2015)
10. Badar, N., Vaidya, J., Atluri, V., Shafiq, B.: Risk based access control using classification. In: Al-Shaer, E., Ou, X., Xie, G. (eds.) Automated Security Management, pp. 79–95. Springer, Cham (2013)
11. Barhamgi, M., Benslimane, D., Oulmakhzoune, S., Cuppens-Boulahia, N., Cuppens, F., Mrissa, M., Taktak, H.: Secure and privacy-preserving execution model for data services. In: Salinesi, C., Norrie, M.C., Pastor, Ó. (eds.) CAiSE 2013. LNCS, vol. 7908, pp. 35–50. Springer, Heidelberg (2013). doi:10.1007/978-3-642-38709-8_3
12. Colombo, P., Ferrari, E.: Complementing MongoDB with advanced access control features: concepts and research challenges. In: Proceedings of SEBD 2015 (2015)
13. Colombo, P., Ferrari, E.: Privacy aware access control for Big Data: a research roadmap. Big Data Res. 2(4), 145–154 (2015)
14. Das, D., O'Malley, O., Radia, S., Zhang, K.: Adding security to Apache Hadoop. Hortonworks, IBM (2011)
15. Derbeko, P., Dolev, S., Gudes, E., Sharma, S.: Security and privacy aspects in mapreduce on clouds: a survey. Comput. Sci. Rev. 20, 1–28 (2016)
16. Di Vimercati, S.D.C., Foresti, S., Paraboschi, S., Pelosi, G., Samarati, P.: Protecting access confidentiality with data distribution and swapping. In: Proceedings of IEEE BdCloud, pp. 167–174 (2014)
17. Ferraiolo, D.F., Sandhu, R., Gavrila, S., Kuhn, D.R., Chandramouli, R.: Proposed NIST standard for role-based access control. ACM TISSEC 4(3), 224–274 (2001)
18. Gupta, M., Patwa, F., Benson, J., Sandhu, R.: Multi-layer authorization framework for a representative Hadoop ecosystem deployment. In: Proceedings of ACM SACMAT (2017, to appear). 8 pages
19. Gupta, M., Patwa, F., Sandhu, R.: POSTER: access control model for the Hadoop ecosystem. In: Proceedings of ACM SACMAT (2017, to appear). 3 pages
20. Gupta, M., Sandhu, R.: The GURA$_G$ administrative model for user and group attribute assignment. In: Chen, J., Piuri, V., Su, C., Yung, M. (eds.) NSS 2016. LNCS, vol. 9955, pp. 318–332. Springer, Cham (2016). doi:10.1007/978-3-319-46298-1_21
21. Hsu, A.C., Ray, I.: Specification and enforcement of location-aware attribute-based access control for online social networks. In: Proceedings of ACM ABAC 2016, pp. 25–34 (2016)
22. Hu, V.C., Grance, T., Ferraiolo, D.F., Kuhn, D.R.: An access control scheme for Big Data processing. In: Proceedings of IEEE CollaborateCom, pp. 1–7 (2014)
23. Hu, V.C., Kuhn, D.R., Ferraiolo, D.F.: Attribute-based access control. IEEE Comput. 48(2), 85–88 (2015)

24. Jin, X., Krishnan, R., Sandhu, R.: A unified attribute-based access control model covering DAC, MAC and RBAC. In: Cuppens-Boulahia, N., Cuppens, F., Garcia-Alfaro, J. (eds.) DBSec 2012. LNCS, vol. 7371, pp. 41–55. Springer, Heidelberg (2012). doi:10.1007/978-3-642-31540-4_4

25. Jin, X., Sandhu, R., Krishnan, R.: RABAC: role-centric attribute-based access control. In: Kotenko, I., Skormin, V. (eds.) MMM-ACNS 2012. LNCS, vol. 7531, pp. 84–96. Springer, Heidelberg (2012). doi:10.1007/978-3-642-33704-8_8

26. Krautsevich, L., Lazouski, A., Martinelli, F., Yautsiukhin, A.: Towards attribute-based access control policy engineering using risk. In: Bauer, T., Großmann, J., Seehusen, F., Stølen, K., Wendland, M.-F. (eds.) RISK 2013. LNCS, vol. 8418, pp. 80–90. Springer, Cham (2014). doi:10.1007/978-3-319-07076-6_6

27. Kuhn, D.R., Coyne, E.J., Weil, T.R.: Adding attributes to role-based access control. IEEE Comput. 43(6), 79–81 (2010)

28. Lu, H., Hong, Y., Yang, Y., Duan, L., Badar, N.: Towards user-oriented RBAC model. J. Comput. Secur. 23(1), 107–129 (2015)

29. Lu, R., Zhu, H., Liu, X., Liu, J.K., Shao, J.: Toward efficient and privacy-preserving computing in Big Data era. IEEE Netw. 28(4), 46–50 (2014)

30. Moyano, F., Fernandez-Gago, C., Lopez, J.: A conceptual framework for trust models. In: Fischer-Hübner, S., Katsikas, S., Quirchmayr, G. (eds.) TrustBus 2012. LNCS, vol. 7449, pp. 93–104. Springer, Heidelberg (2012). doi:10.1007/978-3-642-32287-7_8

31. Nunez, D., Agudo, I., Lopez, J.: Delegated access for Hadoop clusters in the cloud. In: Proceedings of IEEE CloudCom, pp. 374–379 (2014)

32. OMalley, O., Zhang, K., Radia, S., Marti, R., Harrell, C.: Hadoop security design. Technical report, Yahoo Inc. (2009)

33. Sandhu, R.S., Coyne, E.J., Feinstein, H.L., Youman, C.E.: Role-based access control models. IEEE Comput. 29(2), 38–47 (1996)

34. Sänger, J., Richthammer, C., Hassan, S., Pernul, G.: Trust and Big Data: a roadmap for research. In: Proceedings of IEEE DEXA, pp. 278–282. IEEE (2014)

35. Servos, D., Osborn, S.L.: HGABAC: towards a formal model of hierarchical attribute-based access control. In: Cuppens, F., Garcia-Alfaro, J., Zincir Heywood, N., Fong, P.W.L. (eds.) FPS 2014. LNCS, vol. 8930, pp. 187–204. Springer, Cham (2015). doi:10.1007/978-3-319-17040-4_12

36. Sharma, P.P., Navdeti, C.P.: Securing big data Hadoop: a review of security issues, threats and solution. IJCSIT 5, 2126–2131 (2014)

37. Soria-Comas, J., Domingo-Ferrer, J.: Big Data privacy: challenges to privacy principles and models. Data Sci. Eng. 1(1), 21–28 (2016)

38. Tene, O., Polonetsky, J.: Big Data for all: privacy and user control in the age of analytics. Nw. J. Tech. Intell. Prop. 11, xxvii (2012)

39. Ulusoy, H., Colombo, P., Ferrari, E., Kantarcioglu, M., Pattuk, E.: GuardMR: fine-grained security policy enforcement for MapReduce systems. In: Proceedings of ACM ASIACCS, pp. 285–296 (2015)

40. Ulusoy, H., Kantarcioglu, M., Pattuk, E., Hamlen, K.: Vigiles: fine-grained access control for MapReduce systems. In: Proceedings of IEEE Big Data Congress, pp. 40–47 (2014)

41. Vimercati, S.D.C.D., Foresti, S., Paraboschi, S., Pelosi, G., Samarati, P.: Shuffle index: efficient and private access to outsourced data. ACM TOS 11(4), 19 (2015)

42. Wang, L., Wijesekera, D., Jajodia, S.: A logic-based framework for attribute based access control. In: Proceedings of ACM FMSE, pp. 45–55 (2004)

43. White, T.: Hadoop: The Definitive Guide. O'Reilly Media, Inc., Sebastopol (2012)

44. Wrona, K., Oudkerk, S., Armando, A., Ranise, S., Traverso, R., Ferrari, L., McEvoy, R.: Assisted content-based labelling and classification of documents. In: Proceedings of IEEE ICMCIS, pp. 1–7 (2016)
45. Yu, S., Wang, C., Ren, K., Lou, W.: Attribute based data sharing with attribute revocation. In: Proceedings of ACM ASIACCS, pp. 261–270 (2010)
46. Zhao, J., Wang, L., Tao, J., Chen, J., Sun, W., Ranjan, R., Kołodziej, J., Streit, A., Georgakopoulos, D.: A security framework in G-Hadoop for Big Data computing across distributed cloud data centres. JCSS **80**(5), 994–1007 (2014)

Identification of Access Control Policy Sentences from Natural Language Policy Documents

Masoud Narouei[✉], Hamed Khanpour, and Hassan Takabi

Department of Computer Science and Engineering,
University of North Texas, Denton, TX, USA
{Masoudnarouei,Hamedkhanpour}@my.unt.edu, Hassan.Takabi@unt.edu

Abstract. Access control mechanisms are a necessary and crucial design element to any application's security. There are a plethora of accepted access control models in the information security realm. However, attribute-based access control (ABAC) has been proposed as a general model that could overcome the limitations of the dominant access control models (i.e., role-based access control) while unifying their advantages. One issue with migrating to an ABAC model is the information that needs to be encoded in the model is typically buried within existing natural language artifacts, hence difficult to interpret. This requires processing natural language documents and extracting policies from those documents. Software requirements and policy documents are the main sources of declaring organizational policies, but they are often huge and consist of a lot of general descriptive sentences that lack any access control content. Manually processing these documents to extract policies and then using them to build a model is a laborious and expensive process. This paper is the first step towards a new policy engineering approach for ABAC by processing policy documents and identifying access control contents. We take advantage of multiple natural language processing techniques including pointwise mutual information to identify access control policy sentences within natural language documents. We evaluate our approach on documents from different domains including conference management, education, and healthcare. Our methodology effectively identifies policy sentences with an average recall and precision of 90% on all datasets, which bested the state-of-the-art by 5%.

Keywords: Access control policy · Attribute-based access control · Policy engineering · Natural language processing

1 Introduction

A fundamental management responsibility is securing information and information systems. Almost all of the applications that deal with safety, privacy, defense and even finance include some form of access control, which regulates who or what can view or use resources in a computing environment. Access control policies (ACP) are those rules that a corporation exerts in order to control access to

© IFIP International Federation for Information Processing 2017
Published by Springer International Publishing AG 2017. All Rights Reserved
G. Livraga and S. Zhu (Eds.): DBSec 2017, LNCS 10359, pp. 82–100, 2017.
DOI: 10.1007/978-3-319-61176-1_5

its information assets. They determine which activities are allowed for authorized users, controlling every attempt by a user for accessing system resources. They also reveal a lot of information about the internal processes within an organization, which sometimes mirror the structure of the organization. The risks of not using adequate access control policies range from inconvenience to critical loss or corruption of data. A big challenge for security administrators while implementing an access control model is to properly define ACPs, especially in large corporations. Most of these corporations have high-level requirement specification documents that specify how information access is manipulated and who, under which circumstances, can access what asset. All US federal agencies are required to provide information security by the "Federal Information Security Act of 2002" [2], and policy documentation is part of that requirement. Although it is not necessary for private industry to prepare such documentation, the remarkable costs affiliated with recent cyber-attacks have forced many corporations to record their security policies. In addition, recording security policies aid corporations in transitioning from access control lists into a more robust access control model such as Attribute-based Access Control (ABAC) with more ease. In this paper, we refer to these documents as natural language access control policies (NLACPs), which are described as "statements governing management and access of enterprise objects. They are human expressions that can be translated to machine-enforceable access control policies" [10].

An issue with NLACPs is they are usually declared in human understandable terms, are unstructured, and also may be ambiguous, hence it is difficult to encode them directly in a machine-enforceable form. In our previous work [19], we addressed this issue by proposing semantic role labeling technique, where we were able to process sentences and extract necessary elements such as subject, object and action to create machine-enforceable ACPs. However, prior to identifying ACP elements, one has to first process the unstructured NLACP documents in order to identify those sentences that express policies. NLACPs are often huge and consist of many sentences. Several of these sentences are general descriptive sentences that lack any access control policy content (non-ACP sentences). The process of manually sifting through NLACPs to extract the buried ACPs is very laborious and error-prone.

This paper is the first step towards our ultimate goal of building an ABAC system from existing information. Current ABAC solutions try to convert already existing policies in the form of ACL [32], RBAC, etc. to an equivalent ABAC model. However, the previous work ignores an important source of information, NLACP documents, in the process of building an ABAC model. In this paper, we aim to propose a new technique to solve this issue in order to make it easier for corporations to extract ACPs from NLACPs. We will limit our discussion to identifying ACP sentences and separating them from other non-informative, irrelevant sentences. In the future, we aim to use the extracted ACP sentences to build a new ABAC model. Our proposed technique takes advantage of four different types of features in order to come up with a final discriminating feature set. These features include pointwise mutual information (PMI) features, security features, syntactic complexity features and dependency features. To the best of our knowledge, this is the first report that elaborates on the effectiveness of using four different feature vectors to extract policies.

The contributions of this paper are as follow:

- We take the first step towards proposing a new policy engineering approach for ABAC by processing NLACP documents and extracting policy contents.
- We introduce a new technique to effectively distinguish ACP sentences from non-ACP sentences.
- We introduce four different types of features in order to come up with a final discriminating feature set.
- We introduce the first publicly available policy dataset to encourage more research on this area.

The rest of this paper is organized as follows: We begin with a discussion of related work in Sect. 2. Section 3 will discuss some background concepts and then Sect. 4 will present our methodology. The experiments are presented in Sect. 5, and finally, conclusion and future works wraps up the paper.

2   Related Work

Previous work in the literature took advantage of predefined patterns and combining machine learning approaches to find access control policy sentences. Xiao *et al.* proposed Text2Policy, which employs shallow parsing techniques with finite state transducers to match a sentence into one of four access control patterns [30]. One example of such a pattern is *Modal Verb in Main Verb Group*, which helps recognize that the main verb contains a modal verb, which leads to identifying the sentence as an ACP sentence. If the matching is successful, it uses the annotated portions of the sentence to extract the subject, action, and object from the sentence. Text2Policy does not need a labeled data set. However, it misses ACPs that do not follow one of its four semantic patterns. It is reported that only 34.4% of the identified ACP sentences followed one of Text2Policy's patterns [25].

Slankas *et al.* proposed access control rule extraction (ACRE), a supervised learning approach that uses an ensemble classifier to determine whether a sentence expresses an ACR or not [25]. Their ensemble classifier is composed of a k-nearest neighbors (k-NN) classifier, a naïve bayes classifier, and a support vector machine classifier. To determine how to use these classifiers, they calculated a threshold value for the ratio of the computed distance to the neighbors compared to the number of words in the sentence. If the k-NN classifier's ratio for a sentence is below a threshold value of 0.6, they return the k-NN classifier's predication. Otherwise, they output a majority vote of the three classifiers. Even though this approach performs very well, the k-NN classifier suffers from a slow processing time since it has to compare each sentence to all other sentences to make a decision.

In our previous work, we proposed semantic role labeling (SRL) to automatically extract ACP elements from unrestricted natural language documents, define roles, and build an RBAC model [20]. We did not attempt to identify ACP sentences, but instead used the already extracted sentences by [25] and left implementing the ACP sentence identification step for future work. In this work, we propose our methodology for ACP sentence identification.

The problem of mining ABAC policies from natural language documents has not been studied in the literature. However, there are few works for deriving ABAC policies from request logs. This problem was first investigated in [31]. The authors present an algorithm for mining ABAC policies from logs and attribute data. The algorithm iterates over tuples in the user-permission relation extracted from the log, uses selected tuples as seeds for constructing candidate rules, and attempts to generalize each candidate rule to cover additional tuples in the user-permission relation. Finally, it selects the highest quality candidate rules for inclusion in the generated policy. In [16], the authors proposed a multi-objective evolutionary approach for learning ABAC policies from sets of authorized and denied access requests. They used the same ABAC language and the same case studies of the [31]. They presented a strategy for learning policies by learning single rules, each one focused on a subset of requests, an initialization of the population; a scheme for diversity promotion and for early termination. Most recently, we introduced a policy engineering framework for ABAC where we were able to identify access control policy sentences using deep learning techniques and convert some ACP sentences to ABAC policies using semantic role labeling technique [18]. There are also some other work for inferring RBAC policies from logs for less expressive access control models (e.g., RBAC [7]). Usage of evolutionary techniques for inferring RBAC rules explaining the observed actions in environments with tree-structured role hierarchies was also proposed in [9].

3   Background

This section provides background with regards to PMI, syntactic complexity, ML, and ABAC with the motivation behind using these techniques.

3.1   Pointwise Mutual Information

Pointwise mutual information (PMI) has been extensively applied in information retrieval (IR) and text based applications. PMI was introduced by Turney [27] as an unsupervised learning algorithm for finding synonyms based on statistical information. It gives an estimation of semantic similarity between two words based on word co-occurrence[1] on a very large corpus, e.g., Web, Wikipedia. Given two words t1 and t2, their PMI score is defined as Eq. 1:

$$PMI(t_1, t_2) = \log_2 \frac{P(t_1, t_2)}{P(t_1) * P(t_2)} \tag{1}$$

---

[1] Frequent occurrence of two words from a text corpus alongside each other in a certain order.

where $P(t_1, t_2)$ is the probability of $P(t_1)$ and $P(t_2)$ under the joint distribution and $P(t_1)$, $P(t_2)$ is the probability of each word independently. Thus, if $P(t_1)$ and $P(t_2)$ are independent, $P(t_1, t_2) = P(t_1)P(t_2)$, and PMI is $log(1) = 0$, meaning that $P(t_1)$ and $P(t_2)$ share no information.

We take advantage of PMI scores in order to extract informative features that are relevant to the current task. This will also narrow down the number of features significantly as will be described in Sect. 4.2.

3.2 Measures of Syntactic Complexity

ACP sentences usually contain more complex structures such as clauses than non-ACP sentences (e.g., *"If by chance a student is employed at a particular clinic or health care institution for any reason, that student may not be placed at that clinic or institution for any of their clinical practical."* compared to non-ACP content such as *"All Health providers and staff."* Hence, we evaluate syntactic complexity of written English texts to better discriminate ACP sentences from non-ACP contents. Although many different measures have been suggested for characterizing syntactic complexity, previous literature only considered a small set since not many computational tools are available and manual analysis is laborious. Recently Lu [13] proposed a computational system that uses 14 different measures to analyze syntactic complexity in second language writing. These 14 syntactic complexity measures are chosen from a large number of measures that were discussed in [22,29].

Table 1. The 14 syntactic complexity measures used in previous research [13]

Mean length of clause	MLC	# of words / # of clauses
Mean length of sentence	MLS	# of words / # of sentences
Mean length of T-unit	MLT	# of words / # of T-units
Sentence complexity ratio	C/S	# of clauses / # of sentences
T-unit complexity ratio	C/T	# of clauses / # of T-units
Complex T-unit ratio	CT/T	# of complex T-units / # of T-units
Dependent clause ratio	DC/C	# of dependent clauses / # of clauses
Dependent clauses per T-unit	DC/T	# of dependent clauses / # of T-units
Coordinate phrases per clause	CP/C	# of coordinate phrases / # of clauses
Coordinate phrases per T-unit	CP/T	# of coordinate phrases / # of T-units
Sentence coordination ratio	T/S	# of T-units / # of sentences
Complex nominals per clause	CN/C	# of complex nominals / # of clauses
Complex nominals per T-unit	CN/T	# of complex nominals / # of T-units
Verb phrases per T-unit	VP/T	# of verb phrases / # of T-units

This system takes input as a plain text file and then counts the frequency of the following nine structures:

- Words (W)
- Sentences (S)
- Verb phrases (VP)
- Clauses (C)
- T-units (T)
- Dependent clauses (DC)
- Complex T-units (CT)
- Coordinate phrases (CP)
- Complex nominals (CN)

Using these calculated frequencies, the system produces 14 indices of syntactic complexity, presented in Table 1. We use these 14 features to measure syntactic complexity of sentences, which will help classification.

3.3 Machine Learning

Machine learning (ML) is a method of data analysis that automates building an analytical model. Using algorithms that iteratively learn from data, ML allows computers to find hidden insights without being explicitly programmed. ML algorithms are often categorized as being supervised or unsupervised. Supervised algorithms can apply what has been learned in the past to new data. Unsupervised algorithms, however, draw inferences from new data. In this work, we have a labeled dataset and hence we use supervised ML algorithms. We will take advantage of two ML algorithms, naïve bayes and support vector machines. The naïve bayes algorithm is based on conditional probabilities. It uses Bayes' Theorem, to find the probability of an event occurring given the probability of another event that has already occurred. Using these probabilities, the algorithm calculates the probability of any sentence being either ACP or non-ACP and then makes the final decision based on the higher probability. We also employ support vector machine, which is a supervised ML algorithm that constructs a hyperplane or set of hyperplanes in a high-dimensional space that is used for classification. The algorithm plots all sentences in a n-dimensional space (where n is number of features) with the value of each feature being the value of a particular coordinate. Then, the algorithm finds the hyper-plane(s) that best separates ACP and non-ACP sentences.

3.4 Attribute-Based Access Control

Attribute-based access control (ABAC) is an access control model wherein the access control decisions are made based on a set of attributes, associated with the requester, the environment, and/or the resource itself. An attribute is a property expressed as a name:value pair that can capture identities and access control lists (DAC), security labels, clearances and classifications (MAC), and roles (RBAC). ABAC was proposed as a general model that could overcome the limitations of the dominant access control models (i.e., discretionary-DAC, mandatory-MAC, and role-based-RBAC) while unifying their advantages. Our

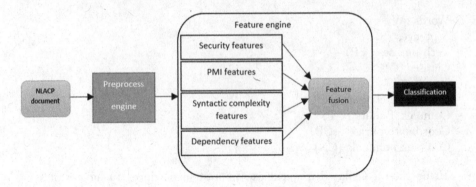

Fig. 1. Overview of the proposed system

central contribution is to take the first step towards proposing a new policy engineering approach for ABAC by processing policy documents and extracting those sentences that exhibit ACP content. We then use these ACP sentences to extract ABAC policies.

4   The Proposed Methodology

In order to effectively distinguish ACP sentences from non-ACP sentences, we take advantage of different NLP and ML techniques. Our proposed system is composed of a set of components that extract various types of features. The overview of the system is presented in Fig. 1. In the next sections, we will describe each of these components in detail.

4.1   Preprocess Engine

Figure 2 presents part of a large NLACP document[2]. It is obvious that there are many non-relevant contents such as titles, tables, etc. that need to be removed. As these formal NLACP documents are usually expressed in PDF format, the first step is to read each PDF document. For this purpose we used Apache PDFBox [1] toolkit, which extracts texts and ignore the other contents such as tables. In order to parse the extracted text, we fed it into Stanford Corenlp toolkit [15]. The tool split the text by sentence boundaries where each sentence was on a separate line and ended by a period. As many of the extracted sentences are not statements (e.g., titles), we introduce the following equation to filter out everything other than sentences.

$$Ratio(sent) = \frac{Capitals(sent)}{Tokens(sent)}$$

---

[2] http://policy.unt.edu/sites/default/files/07.022_AdministrativeEntrySearches UniversityResidenceHalls_2012.pdf.

Where *Capitals* stand for the number of capital letters in the sentence (*sent*) and *Tokens* counts for the number of tokens in each sentences. If this ratio is less than 0.6, we consider the sentence for further processing. We used different ratios but 0.6 gave us the most accurate results. We also limited ourselves to sentences with more than 15 characters, which helped us remove more incomplete sentences. After this step, the following four sentences are extracted:

- *The University respects its resident students' reasonable expectation of privacy in their rooms and makes every effort to ensure privacy in university residences.*
- *However, in order to protect and maintain the property of the university and the health and safety of the university's students, the university reserves the right to enter and/or search student residence hall rooms in the interest of preserving a safe and orderly living and learning environment.*
- *Designated university officials are authorized to enter a residence hall room unaccompanied by a resident student to conduct room inspections under the following conditions.*
- *To perform reasonable custodial, maintenance, and repair services.*

Next, the extracted sentences were fed to feature engine in order to extract discriminating features.

Policies of the University of North Texas	
	Chapter 7
07.022 Administrative Entry and Searches of University Residence Halls	Student Affairs

Policy Statement. UNT respects its resident students' reasonable expectation of privacy in their rooms and makes every effort to ensure privacy in university residences. However, in order to protect and maintain the property of the university and the health and safety of the university's students, the university reserves the right to enter and/or search student residence hall rooms in the interest of preserving a safe and orderly living and learning environment.

Application of Policy. Resident students.

Procedures and Responsibilities.

    1. Administrative room inspections

        a. Designated university officials are authorized to enter a residence hall room unaccompanied by a resident student to conduct room inspections under the following conditions:

            i. To perform reasonable custodial, maintenance, and repair services.

Fig. 2. Part of a NLACP document

4.2 Feature Engine

Feature engine is responsible for extracting various types of features and creating the final feature vector. The following sections will describe each sub-component in detail.

Security Features. Kong *et al.* proposed a system called AUTOREB that was able to infer the mobile app security behaviors using apps' reviews from different users with an average accuracy of 94% [12]. As their problem is quite similar to ours in terms of classifying security sentences, we decided to take advantage of their proposed methodology to extract more discriminating features. Our adapted methodology is described as follows:

First, we ranked all words in ACP sentences based on their frequencies of occurrence and chose the top 15 most frequent keywords. These 15 keywords were considered as the initial set of our security features. Then, the new keywords were chosen from those that had a high co-occurrence with current keywords in our feature set. If this co-occurrence exceeded a threshold, we added the new keyword into the feature set. To avoid repetitive calculations, at each round we only considered those keywords that were added in the previous round. This process was repeated until no more new keywords were added. In the case of applying the same methodology on any other dataset, the same process can be applied to identify the keywords relevant to that dataset.

PMI Features. N-grams have been extensively used for text classification as a good syntactic feature. However, they often result in large and sparse feature vectors. By taking advantage of PMI scores, we can limit ourselves to only those words that are informative and correlated to the current task. This will improve the performance of classification algorithm and speed it up considerably. To have an accurate calculation of PMI scores, a large corpus is required. In the original arrangement of PMI by Turney [27], it is required to consult the Web for counting words. However, there are some disadvantages with using the Web. The Web is always changing and the search scheme of commercial search engines is always changing too, which makes it hard to maintain the functionality of the system. For computing semantic relations, it has been shown that Wikipedia is more reliable and effective than a search engine and covers more concepts than WordNet [23]. In order to extract PMI scores, we used two different setups. For the first one, we gathered pre-calculated PMI data through Semilar project [24]. This data was calculated using whole Wikipedia text (as of Jan 2013). For the second setup, we calculated PMI scores from access control contexts as mentioned briefly in Sect. 3.1 so that it will be more adaptable to our approach. We gathered over 10 MB of text consisting all of our datasets and then calculated PMI scores based on the formula described in Sect. 3.1. To calculate relevant keywords, we adopt the same process that was described in previous section. Starting with the same initial set of 15 keywords, we calculated the PMI score of all words in our dataset. For each setup, we used the corresponding set of PMI scores. If this PMI score was more than a threshold, we added the other word to the feature set too. After the first round, we came up with a new set of features in addition to the initial 15 keywords. Then in the second iteration, we repeated the process using newly found keywords in the previous step. We followed this process until no more features were added to our feature set.

Syntactic Complexity Features. This component computes complexity measures based on the explanations in Sect. 3.2. For each sentence a feature vector of 23 values were created.

Dependency Relations. Marneffe *et al.* described a methodology for extracting typed dependency parses of English sentences [4]. A dependency parse demonstrates dependencies among single words. It also labels dependencies with grammatical relations, such as subject or indirect object. Consider the sentence: "Bills on ports and immigration were submitted by Senator Brownback, Republican of Kansas." The corresponding typed dependency tree is drawn in Fig. 3.

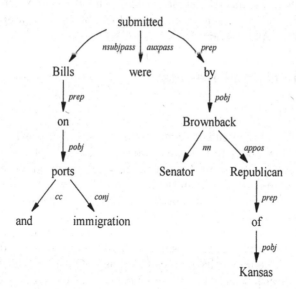

Fig. 3. Typed dependency parse for the sentence "Bills on ports and immigration were submitted by Senator Brownback, Republican of Kansas."

Typed dependency parses of ACP sentences capture inherent relations occurring in ACP sentences that can be critical in discriminating them from non-ACP contents. Using this structure, we extracted the ratio of occurrences of each of the following attributes to the length of sentence and used them as features:

– Subject
– object
– auxiliary verb
– verb

Feature Fusion. This component stacks all the four feature types in a single, long feature vector that were used for classification. For PMI and security features, there are some redundancies in extracted keywords that were removed too.

4.3   Classification

After building the final feature vector for each sentence and creating the final dataset consisting of all vectors, we train a predictive model using ML algorithms. Since our training set has fairly little data, we need to choose a classifier with high bias according to machine learning theory [14]. For this work, we chose naïve bayes as there are many theoretical and empirical results that naïve bayes does well in such circumstances (e.g. [6,21]). We also employed support vector machines classifier as it has consistently achieved good performance on text categorization tasks (see [11]). The created model will be used to categorize unseen sentences to either ACP or non-ACP.

5   Experiments and Results

5.1   Dataset(s)

The access policy domain suffers from a scarcity of publicly available data for researchers. To help alleviate this issue, we created a dataset to serve the dual purpose of (1) making our evaluation of the proposed method more robust, and (2) providing the research community with more data, which will in turn allow other researchers to evaluate their work both more comprehensively and in direct comparison to others.

We constructed our dataset from real-world policy documents from the authors' home institution. To do this, we gathered over 430 policy documents in PDF format from the university policy office[3], as well as policy documents from the university's health science center[4]. The documents described security access authorizations for a wide variety of departments, including humanresources, information technology, risk management services, faculty affairs, administration, intellectual property, technology transfer, and equity development, among others. Altogether, these documents were comprised of more than 21,000 sentences. Since manually labeling the sentences is a labor-intensive process, in this work we limited our data to a randomly selected subset of 1,000 sentences from the full dataset.

---

[3] https://policy.unt.edu/policy-alphabetical/a.
[4] https://app.unthsc.edu/policies/Home/ByChapter.

The sentences were annotated for the presence of ACP content by two Ph.D. students studying cybersecurity, who are familiar with access control policies and the contexts in which they occur. They were also provided a coding guide that told them what should be considered as an access control policy sentence and what should not. The first author of this paper adjudicated any discrepancies in the annotations after discussing them with both annotators. We computed kappa on the annotations, finding $\kappa = 0.75$ between the two annotators. The final annotated dataset is comprised of 455 ACP sentences and 545 non-ACP sentences. This dataset is available upon request.

5.2 Evaluation Criteria

In order to evaluate results, we use recall, precision, and the F_1 measure. Precision is the fraction of ACP sentences that are relevant, while recall is the fraction of ACP sentences that are retrieved. To compute these values, the classifier's predictions are categorized into four categories. True positives (TP) are correct predictions. True negatives (TN) are sentences that we correctly predicted as not an ACP sentence. False positives (FP) are sentences that were mistakenly identified as an ACP sentence. Finally, false negatives (FN) are ACP sentences that we failed to correctly predict as an ACP sentence. Using these values, precision is calculated using $P = \frac{TP}{TP+FP}$ and recall using $R = \frac{TP}{TP+FN}$. To have an effective model, a high value for both precision and recall is required. Lower recall means the approach could more likely miss ACP sentences while a lower precision implies that the approach could more likely identify non-ACP sentences as ACP sentences. We define F_1 as the harmonic mean of precision and recall, giving an equal weight to both elements. F_1 measure is calculated using the $F_1 = 2 \times \frac{P \times R}{P+R}$ respectively.

5.3 Experimental Results

After performing the preprocessing, the dataset is divided into 70% training (700 sentences) and 30% testing sets (300 sentences). In order for both datasets to be independent and identically distributed, stratification was performed to make sure the distribution of both classes (ACP and non-ACP) are the same in both

Table 2. Obtained results using two classifiers (naïve bayes and support vector machines (SVM)) alongside comparison with baseline and the implemented ensemble classifier

Methodology	Precision (%)	Recall (%)	F_1 (%)
ZeroR	30	54	39
Ensemble	74	70	72
SVM	76	76	76
Naïve Bayes	82	80	80

train and test sets. Then, the training set is fed to feature engine for feature extraction. The initial set of seeds for all datasets are presented in Table 4. All the experiments were performed using weka toolkit [8], which is a collection of ML algorithms for data mining tasks. We used SMO (support vector machine's implementation in weka) as well as naïve bayes classifier for classification task. We trained both classifiers using the training set. The performance of the system on the testing dataset is reported in Table 2. The second row shows the baseline results generated using weka's ZeroR classifier. The ZeroR algorithm selects the majority class in the dataset and uses it to make all predictions.

Table 3. Study document set statistics

Dataset	Domain	# of sentences	# of ACP sentences
iTrust for ACRE	Healthcare	1160	550
iTrust for Text2policy	Healthcare	471	418
IBM Course Management	Education	401	169
CyberChair	Conference Mgmt	303	139
Collected ACP Documents	Multiple	142	114

The third row shows the comparison with the ACRE system proposed by Slankas et al. [25]. In their ensemble classifier, an object is assigned to the class most common among its k nearest neighbors. The classifier used a modified version of Levenshtein distance as distance metric. The details of this classifier was presented in Sect. 2. We were not able to receive the source codes for their ensemble classifier and hence we implemented it based on the description provided in their paper as well as the main author's thesis [26]. We did our best to make sure that our implemented ensemble classifier is similar to their methodology. Finally, the fourth and fifth rows present results obtained using our proposed method. As Table 2 shows, our methodology was able to outperform the majority baseline and previous work by a huge margin.

In order to have an exact comparison between our proposed methodology and the ACRE system, we performed experiments on the same datasets that they used in their paper. These datasets were manually labeled by Slankas et al. [25] and consisted of five sections, described as follows:

- **iTrust for ACRE.** iTrust [17] is an open source healthcare application that consists of 40 use cases plus additional non-functional requirements. *iTrust for ACRE* was extracted directly from the project's wiki.
- **iTrust for Text2policy.** The second version of iTrust that was taken from the documentations used by Xiao et al. [30].
- **IBM Course Management.** Eight use cases from the IBM Course Registration System [3].
- **CyberChair.** CyberChair documents [28], which has been used by over 475 different conferences and workshops.

– **Collected ACP Documents.** A combined document of 142 sentences that were collected by Xiao *et al.* [30].

More information about these dataset(s) can be found in Table 3. The initial seeds for each dataset is presented in Table 4. Using these seeds, each dataset was independently fed to the feature engine and experiments were performed. The obtained results are presented in Table 5. As the table presents, applying our implemented system on all five dataset(s) achieved a macro average of 90% F_1 while the ACRE system achieved 85%.

Table 4. The top 15 frequent keywords for each dataset

UNT Policies	will, must, may, health, student, shall, information, employee, should, science, review, center, policy, university, staff
Collected	can, access, read, if, customer, subject, assign, information, paper, resident, use, patient, medical, may, reps
CyberChair	reviewer, paper, submit, author, assign, number, must, expertise, should, indicate, process, base, overview, topic, abstract
IBM	system, student, course, professor, schedule, registrar, offering, case, use, semester, information, offering, delete, will, select
iTrustforACRE	patient, office, hcp, can, name, date, message, list, appointment, information, user, lab, representative, view, office
iTrustfortext2policy	patient, view, can, message, representative, name, list, system, hcp, date, appointment, data, user, present, administrator

In order to extract effective PMI scores, we used both pre-calculated PMI values using Wikipedia and also our self-calculated PMI values based on security context. In order to come up with the best threshold, we used ten-fold cross validation on the train set to evaluate the performance using features generated by each threshold value. Figure 4(a) presents the performance using pre-calculated PMI values from Wikipedia. The best set of features were extracted using a threshold value of 0.4. Figure 4(b) shows the performance using self-calculated PMI values. The best performance (83%) was obtained using a threshold value of 0.8. As these figures indicate, our self-generated PMI scores yield a higher performance, which seems reasonable considering the fact that they were generated from the security context while pre-calculated PMI values were generated using public Wikipedia article, which are general to any method but not specific to our domain. Hence, we used our self-calculated PMI scores with the threshold of 0.8 as the output of PMI component. For security features, the best results were obtained using a threshold value of 15 co-occurrences in the whole context, as Fig. 5 shows, hence we used this threshold. It is obvious that the performance decreases considerably after a threshold of 20 co-occurrences since there are not many words that occur together more than 20 times in the dataset.

Table 5. Comparison with ACRE system

Dataset	ACRE			Proposed system		
	Precision (%)	Recall (%)	F_1 (%)	Precision (%)	Recall (%)	F_1 (%)
iTrust for ACRE	90	86	88	89	90	90
iTrust for Text2policy	96	99	98	97	99	98
IBM Course Management	83	92	87	94	93	93
CyberChair	63	64	64	79	74	77
Collected ACP Documents	83	96	89	91	93	92
Average	83	87	85	90	90	90

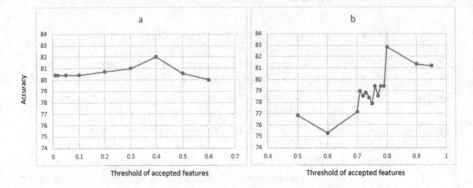

Fig. 4. Threshold analysis using PMI values from Wikipedia (a) and self-calculated PMI values (b)

Table 6 shows the final number of features generated by each component while performing the first experiment (policy documents). Overall, 750 features were extracted after combining all of the features together and removing duplicates. These features were used to build a final feature vector for each sentence. For all features except complexity and dependency features that had their own calculated values, the presence or absence of the features in the corresponding sentence was considered.

Our dataset consists of ACP and non-ACP sentences both expressed in security context. Even distinguishing an ACP sentence from a non-ACP sentence was sometimes difficult as expressed by our labelers. Our proposed system was able to distinguish between these sentences by 80% F_1 while the state-of-the-art

Table 6. The number of different feature values

Feature type	# of features
Security features	73
Policy PMI	721
Syntactic complexity	23
Dependency features	4

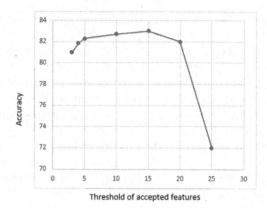

Fig. 5. Threshold analysis of co-occurrences of security features

gains around 72% F_1. The strength of our methodology comes from incorporating both lexical and semantic features. As all sentences are expressed in ACP
context, using just lexical approach is not sufficient. However, using syntactic
complexity features improves the results as ACP sentences usually consist of
clauses and more complex structures compared to non-ACP sentences. These
structures convey lots of information but are usually ignored using only lexical
approaches. The ensemble classifier that was used in the previous work considered the syntax of sentence (using a modified version of Levenshtein distance
for comparing words) to identify ACP sentences. This approach performed very
well on their reported experiments, but while analyzing policy documents, both
syntactic and semantic features proved more fruitful.

As we mentioned in the previous section, we were unable to get their codes,
and hence we implemented their methodology in Java language using Stanford
Corenlp package. On average, our implementation reported 84% average precision on all five datasets (iTrust for ACRE, iTrust for Text2policy, IBM Course
Management, CyberChair, Collected ACP Documents) while their reported average precision was 81%. The F_1, however, was lower as we got around 81% while
they reported 84%. The main problem with ensemble classifier was the speed
since for each instance, the k-NN classifier needs to compute the distance to all
other sentences. This resulted in an average execution time of about 5–6 h on
the five datasets while our proposed methodology runs for less than a minute
given the features, or about 20 min for extracting features depending on the size
of dataset.

6 Discussion

There are several threats to validity of experiments including lack of represen-
tativeness of datasets, identifying a threshold for features and human factors
for determining correct ACP sentences from NLACPs. The five datasets used in

previous literature covered mostly limited grammars and many of their policies were of similar structure and form, not representing the diversity of policies in real-world. To reduce the threat, we evaluated our approach on policy documents from authors' home institution. These documents covered a large variety of policies ranging from human resources, information technology, risk management services, faculty affairs, administration, intellectual property, technology transfer, and equity development, among others. To further reduce this threat, additional evaluation needs to be done to choose a more representative sample of dataset, instead of choosing sentences randomly. Choosing the proper threshold is another issue as it determines the quality of extracted keywords. To reduce this threat, we examined different threshold values from different ranges. A final threat include human factors for determining correct ACP sentences from NLACPs. To reduce the human factor threats, the sentences were annotated for the presence of ACP content by two Ph.D. students studying cybersecurity, who are familiar with ACPs and the contexts in which they occur. The co-author of this paper adjudicated any discrepancies in the annotations after discussing them with both annotators.

To further evaluate our method's performance on ACP sentences that were previously used in the literature, we performed another experiment. We randomly sampled 250 ACP sentences from the four dataset(s) that were previously described (iTrust, IBM course registration system, cyberchair and collected documents). We also gathered 250 sentences that carry no ACP information from Microsoft research paraphrase corpus [5]. This dataset contains 5,800 pairs of sentences that were extracted from news sources on the web, alongside human annotations indicating whether each pair capture a semantic equivalence relationship. Only one sentence has been extracted from any given news article. Using 10-fold cross validation, our proposed system was able to correctly identify ACP sentences with an accuracy of 94%, which shows applicability of this method in the wild.

7   Conclusion and Future Work

ABAC is a promising alternative to traditional models of access control (i.e., DAC, MAC and RBAC) that is drawing attention in both recent academic literature and industry. However, the cost of developing ABAC policies can be a significant obstacle for organizations to migrate from traditional access control models to ABAC. In this paper, we took the first step towards a new policy engineering approach for ABAC by processing policy documents and extracting access control contents. We took advantage of multiple natural language processing techniques including pointwise mutual information to identify access control policy sentences. Experimental results yielded an average 90% F_1, which bested the state-of-the-art by 5%. In future, we plan to extend our work to a comprehensive policy engineering framework that includes extracting ABAC policies from ACP sentences.

References

1. Apache pdfbox. https://pdfbox.apache.org/index.html
2. Federal information security management act of 2002. Title III of the E-Government Act of 2002 (2002)
3. Ibm course registration requirements (2004)
4. De Marneffe, M.C., MacCartney, B., Manning, C.D., et al.: Generating typed dependency parses from phrase structure parses. In: Proceedings of LREC, Genoa, vol. 6, pp. 449–454 (2006)
5. Dolan, B., Brockett, C., Quirk, C.: Microsoft research paraphrase corpus (2005). Accessed 29 Mar 2008
6. Forman, G., Cohen, I.: Learning from little: comparison of classifiers given little training. In: Boulicaut, J.-F., Esposito, F., Giannotti, F., Pedreschi, D. (eds.) PKDD 2004. LNCS, vol. 3202, pp. 161–172. Springer, Heidelberg (2004). doi:10.1007/978-3-540-30116-5_17
7. Gal-Oz, N., Gonen, Y., Yahalom, R., Gudes, E., Rozenberg, B., Shmueli, E.: Mining roles from web application usage patterns. In: Furnell, S., Lambrinoudakis, C., Pernul, G. (eds.) TrustBus 2011. LNCS, vol. 6863, pp. 125–137. Springer, Heidelberg (2011). doi:10.1007/978-3-642-22890-2_11
8. Hall, M., Frank, E., Holmes, G., Pfahringer, B., Reutemann, P., Witten, I.H.: The weka data mining software: an update. ACM SIGKDD Explor. Newsl. 11(1), 10–18 (2009)
9. Hu, N., Bradford, P.G., Liu, J.: Applying role based access control and genetic algorithms to insider threat detection. In: Proceedings of the 44th Annual Southeast Regional Conference, pp. 790–791. ACM (2006)
10. Hu, V.C., Ferraiolo, D., Kuhn, R., Friedman, A.R., Lang, A.J., Cogdell, M.M., Schnitzer, A., Sandlin, K., Miller, R., Scarfone, K., et al.: Guide to attribute based access control (abac) definition and considerations (draft). NIST special publication 800(162) (2013)
11. Joachims, T.: Text categorization with Support Vector Machines: learning with many relevant features. In: Nédellec, C., Rouveirol, C. (eds.) ECML 1998. LNCS, vol. 1398, pp. 137–142. Springer, Heidelberg (1998). doi:10.1007/BFb0026683
12. Kong, D., Cen, L., Jin, H.: Autoreb: automatically understanding the review-to-behavior fidelity in android applications. In: Proceedings of the 22nd ACM SIGSAC Conference on Computer and Communications Security, pp. 530–541. ACM (2015)
13. Lu, X.: Automatic analysis of syntactic complexity in second language writing. Int. J. Corpus Linguist. 15(4), 474–496 (2010)
14. Manning, C.D., Raghavan, P., Schütze, H.: Probabilistic information retrieval. In: Introduction to Information Retrieval, pp. 220–235 (2009)
15. Manning, C.D., Surdeanu, M., Bauer, J., Finkel, J.R., Bethard, S., McClosky, D.: The stanford corenlp natural language processing toolkit. In: ACL (System Demonstrations), pp. 55–60 (2014)
16. Medvet, E., Bartoli, A., Carminati, B., Ferrari, E.: Evolutionary inference of attribute-based access control policies. In: Gaspar-Cunha, A., Henggeler Antunes, C., Coello, C.C. (eds.) EMO 2015. LNCS, vol. 9018, pp. 351–365. Springer, Cham (2015). doi:10.1007/978-3-319-15934-8_24
17. Meneely, A., Smith, B., Williams, L.: itrust electronic health care system: a case study (2011)
18. Narouei, M., Khanpour, H., Takabi, H., Parde, N., Nielsen, R.: Towards a top-down policy engineering framework for attribute-based access control. In: Proceedings of

the 22nd ACM Symposium on Access Control Models and Technologies. ACM (2017)

19. Narouei, M., Takabi, H.: Automatic top-down role engineering framework using natural language processing techniques. In: Akram, R.N., Jajodia, S. (eds.) WISTP 2015. LNCS, vol. 9311, pp. 137–152. Springer, Cham (2015). doi:10.1007/978-3-319-24018-3_9

20. Narouei, M., Takabi, H.: Towards an automatic top-down role engineering approach using natural language processing techniques. In: Proceedings of the 20th ACM Symposium on Access Control Models and Technologies, pp. 157–160. ACM (2015)

21. Ng, A.Y., Jordan, M.I.: On discriminative vs. generative classifiers: a comparison of logistic regression and naive bayes. Adv. Neural Inf. Process. Syst. **2**, 841–848 (2002)

22. Ortega, L.: Syntactic complexity measures and their relationship to l2 proficiency: a research synthesis of college-level l2 writing. Appl. Linguist. **24**(4), 492–518 (2003)

23. Ponzetto, S.P., Strube, M.: Knowledge derived from wikipedia for computing semantic relatedness. J. Artif. Intell. Res. (JAIR) **30**, 181–212 (2007)

24. Rus, V., Lintean, M.C., Banjade, R., Niraula, N.B., Stefanescu, D.: Semilar: The semantic similarity toolkit. In: ACL (Conference System Demonstrations), pp. 163–168. Citeseer (2013)

25. Slankas, J., Xiao, X., Williams, L., Xie, T.: Relation extraction for inferring access control rules from natural language artifacts. In: Proceedings of the 30th Annual Computer Security Applications Conference, pp. 366–375. ACM (2014)

26. Slankas, J.B.: Implementing database access control policy from unconstrained natural language text (2015)

27. Turney, P.D.: Mining the web for synonyms: PMI-IR versus LSA on TOEFL. In: Raedt, L., Flach, P. (eds.) ECML 2001. LNCS, vol. 2167, pp. 491–502. Springer, Heidelberg (2001). doi:10.1007/3-540-44795-4_42

28. Van De Stadt, R.: Cyberchair: A web-based groupware application to facilitate the paper reviewing process. arXiv preprint arXiv:1206.1833 (2012)

29. Wolfe-Quintero, K., Inagaki, S., Kim, H.Y.: Second Language Development in Writing: Measures of Fluency, Accuracy, & Complexity. No. 17, University of Hawaii Press, Honolulu (1998)

30. Xiao, X., Paradkar, A., Thummalapenta, S., Xie, T.: Automated extraction of security policies from natural-language software documents. In: Proceedings of the ACM SIGSOFT 20th International Symposium on the Foundations of Software Engineering, p. 12. ACM (2012)

31. Xu, Z., Stoller, S.D.: Mining attribute-based access control policies from logs. In: Atluri, V., Pernul, G. (eds.) DBSec 2014. LNCS, vol. 8566, pp. 276–291. Springer, Heidelberg (2014). doi:10.1007/978-3-662-43936-4_18

32. Xu, Z., Stoller, S.D.: Mining attribute-based access control policies. IEEE Trans. Dependable Secure Comput. **12**(5), 533–545 (2015)

Fast Distributed Evaluation of Stateful Attribute-Based Access Control Policies

Thang Bui, Scott D. Stoller[(✉)], and Shikhar Sharma

Department of Computer Science, Stony Brook University, Stony Brook, NY, USA
stoller@cs.stonybrook.edu

Abstract. Separation of access control logic from other components of applications facilitates uniform enforcement of policies across applications in enterprise systems. This approach is popular in attribute-based access control (ABAC) systems and is embodied in the XACML standard. For this approach to be practical in an enterprise system, the access control decision engine must be scalable, able to quickly respond to access control requests from many concurrently running applications. This is especially challenging for stateful (also called history-based) access control policies, in which access control requests may trigger state updates. This paper presents an policy evaluation algorithm for stateful ABAC policies that achieves high throughput by distributed processing, using a specialized multi-version concurrency control scheme to deal with possibly conflicting concurrent updates. The algorithm is especially designed to achieve low latency, by minimizing the number of messages on the critical path of each access control decision.

1 Introduction

Separation of access control logic from other components of applications facilitates uniform enforcement of policies across applications in enterprise systems. This approach is adopted in the ISO standard for access control in open systems [13] and the XACML standard[1]. Servers that run the access control policy evaluation algorithm and provide access control decisions to applications are called *policy decision points* (PDPs) in XACML terminology. In this paper, we refer to them simply as *servers*, since we do not discuss other kinds of server.

For this approach to be practical in an enterprise system, the policy evaluation algorithm must be scalable, able to quickly respond to access control requests from many concurrently running applications. To scale beyond the capacity of a single server, distributed policy evaluation algorithms are needed,

---

This material is based on work supported in part by NSF Grants CNS-1421893, and CCF-1414078, ONR Grant N00014-15-1-2208, AFOSR Grant FA9550-14-1-0261, and DARPA Contract FA8650-15-C-7561. Any opinions, findings, and conclusions or recommendations expressed in this material are those of the authors and do not necessarily reflect the views of these agencies.

[1] http://www.oasis-open.org/committees/xacml/.

© IFIP International Federation for Information Processing 2017
Published by Springer International Publishing AG 2017. All Rights Reserved
G. Livraga and S. Zhu (Eds.): DBSec 2017, LNCS 10359, pp. 101–119, 2017.
DOI: 10.1007/978-3-319-61176-1_6

to coordinate concurrent processing of requests on multiple servers. This is relatively straightforward if the policy and the information it references are static. However, this is challenging for *stateful* (also called *state-modifying*, *dynamic*, or *history-based*) access control policies, in which access control requests may trigger state updates, i.e., updates to information referenced by the policy. The classical examples of stateful access control policies are dynamic separation-of-duty (DSOD) policies, such as the Chinese wall policy [5] and DSOD in role-based access control (RBAC) [2]. Another classic category of stateful access control policies are usage control policies [20], such as policies that limit the number of times a user can view a video or the number of videos that a user with a particular type of subscription can view each month. The research literature contains numerous additional examples of stateful access control policies, policy models, and policy evaluation algorithms [3,4,6,8,10–12,14,17–19,22]. In the context of Attribute-Based Access Control (ABAC), the updated state is typically attribute data.

The main challenge in distributed policy evaluation algorithms for stateful policies is ensuring serializability, as in concurrent transaction processing in databases [21]. Processing of each access control request, including its reads of attribute data and its updates to attribute data, should be serializable with respect to processing of other requests. Since concurrent requests may read or write the same attribute data, a concurrency control mechanism is needed to ensure this.

To illustrate the importance of serializability in this context, consider a typical Chinese wall policy in which companies A and B are in the same conflict of interest (COI) class, so user who has accessed documents of one them cannot access documents of the other. When a server allows a request for access to documents of either company, it updates a user attribute to reflect this. Suppose a devious user concurrently submits an access request for a document of company A to one server, and an access request for a document of company B to another server. In a non-serializable execution in which both requests are evaluated in the initial state (where the user has not accessed any documents), both requests could be permitted, violating the intended policy. In a serializable execution, the result must be equivalent to a serial execution, where one of the requests sees the effect of the update performed by the other request, causing the second request to be denied, as it should be.

A straightforward approach to this problem is to use a distributed replicated database that supports serializability for multi-row transactions, and to evaluate each request in a transaction. However, this requirement eliminates well-known scalable NoSQL databases, such as Bigtable [7], Cassandra[2], and MongoDB[3], which achieve scalability in part by supporting only single-row transactions. Master-slave replication in SQL databases, such as MySQL[4], allows multi-row transactions, but has limited scalability, because all read-write transactions must

---

[2] http://cassandra.apache.org/.

[3] https://www.mongodb.com/.

[4] https://dev.mysql.com/.

be submitted to a single master server, and provides inadequate consistency guarantees, because slaves can return slightly out-of-date data. Multi-phase commit protocols, such as in Oracle, IBM DB2, and Microsoft SQL Sever, allow multi-row transactions and ensure serializability, but are less scalable.

Decat *et al.* present a distributed policy evaluation algorithm for stateful ABAC policies [8] that is more scalable than multi-phase commit protocols, by exploiting the fact that evaluation of an ABAC request involves at most two objects (i.e., two rows), typically called the *subject* and the *resource*. Their algorithm uses a specialized scheme for optimistic concurrency control [21, Sect. 15.5]. Their experimental results demonstrate that their algorithm scales well in terms of throughput. However, their algorithm incurs a significant increase in latency, since processing of each request involves a chain of 6 messages (including the messages to and from the client).

This paper presents a new distributed policy evaluation algorithm for stateful ABAC policies. The algorithm is called FACADE (Fast Access Control Algorithm with Distributed Evaluation). It uses a specialized scheme for multiversion timestamp ordering concurrency control [21, Sect. 15.6] that simultaneously achieves low latency by minimizing the length of the message chain on the critical path (i.e., the message chain ending with the result sent to the client). Low latency is of obvious importance for interactive applications: developers struggle to keep the latency of the application's core functionality within limits acceptable to users, especially for multi-tier enterprise applications, where many requests involve processing by multiple servers (web servers, application servers, database servers, etc.), and latency contributions from non-core functionality such as access control are acceptable only if they are low. Low latency is also important for batch applications. These applications often process large amounts of data, hence requiring many access control checks. If the latency of these checks is not kept very low, the repeated delays in the core application processing will cause poor system utilization. Reducing the number of messages per request has the additional benefit of reducing the required network capacity.

FACADE processes read-only requests differently than read-write requests, in contrast to Decat *et al.*'s algorithm, which processes all requests the same way. This, together with use of multiversion timestamp ordering concurrency control, enables FACADE to use especially short message chains for read-only requests. Multiversion timestamp ordering concurrency control has the desirable property that *read-only requests never abort*. This helps FACADE use a shorter critical path than Decat *et al.*'s algorithm for read-only requests.

FACADE also uses shorter message chains than Decat *et al.*'s algorithm for read-write requests. This is achieved partly by use of multiversion concurrency control and partly by specialization to requests that update at most one object. This specialization is motivated by the observation that in every stateful policy given as an example in every paper cited above, each request updates the state of at most one object. FACADE can be extended to handle requests that update two objects, but that extension is not described in this paper.

FACADE is more flexible than Decat *et al.*'s algorithm, in that FACADE allows an object to be a subject and a resource, while Decat *et al.*'s algorithm requires the sets of subjects and resources to be disjoint [8, Sect. 3.4]

We ran experiments, described in Sect. 3, to compare the performance of FACADE and Decat *et al.*'s algorithm. Our experiments show that FACADE has significantly lower average latency, uses significantly fewer network messages per request, and has slightly higher throughput than Decat *et al.*'s algorithm in many cases.

2   Algorithm

System Architecture. We adopt the system architecture in [8]. There are two types of hosts: *clients* and *servers*. Each client runs applications and a small client-side stub that interacts with the access control servers. Each server runs three kinds of processes: a *coordinator*, which receives requests from clients and is responsible for concurrency control, a *database*, which stores a copy of the attribute data used by the policy, and one or more *workers*, which evaluate requests based on the access control policy and send the result to the coordinator and/or client.

Each worker reads attribute data from the co-located replica of the database. Workers never update the database. The set of objects is partitioned across the set of coordinators. Thus, for each object x, there is a unique coordinator, denoted coord(x), responsible for x; we also say that coord(x) *manages* x. Only coord(x) submits updates of x to the master database (this is done using a standard database connector, such as ODBC or JDBC, regardless of whether the master database is on the same server or a different server). Coordinators never read the database.

Multiversion Timestamp Ordering Concurrency Control. Before presenting our algorithm, we briefly review multiversion timestamp ordering concurrency control [21, Sect. 15.6], with a change in terminology: we refer to "requests" in place of "transactions". A sequence of *versions* is associated with each data item. In FACADE, we treat each attribute of each object as a data item. Each version v has a value v.value, a write timestamp v.wts (the timestamp of the request that created v), and a read timestamp v.rts (the largest timestamp of any request that successfully read v). Each request req is assigned a timestamp req.ts. Let v denote the most recent version of x.attr whose timestamp is at most req.ts. A read of x.attr by req returns v.value. A write by req requires a conflict check: if req.ts < v.rts, then req aborts and restarts; if req.ts == v.wts, then the value of v is overwritten; otherwise (if req.ts > v.rts), a new version of x.attr is created. Note that reads never cause aborts, and read-only transactions always commit.

To support conflict checking, each coordinator maintains a data structure containing the read timestamp and write timestamp of every version of an attribute created during the coordinator's current session (i.e., since the coordinator process started running). This data structure does not store the value of each version, since it is not needed for conflict checking. Entries for old versions can be garbage-collected; details are straightforward and omitted. Although this data structure has some information overlap with cachedUpdates, we keep the two data structures separate for clarity, because they serve different purposes.

This data structure is accessed using two functions. getVersion(x,attr,ts) returns the most recent version of x.attr written at or before ts; if no such version exists, it returns a special version v with v.wts = 0 and v.rts = 0, representing the last version written in the previous session (any timestamp guaranteed to precede all timestamps generated in the current session is safe; 0 is a convenient choice). addVersion(x,attr,ts) creates and stores a version of x.attr with write timestamp and read timestamp equal to ts.

Database. To avoid use of a heavyweight multi-phase commit protocol in the database, we assume a database that supports master-slave (also called primary-secondary) replication, in which updates are committed at one replica, called the *master* or *primary*, and the updates are visible at the other replicas, called *slaves* or *secondaries*, within a known time limit, called the *database latency*. This assumption is satisfied by the replication schemes in popular databases, such as primary-secondary replication in MongoDB and master-slave replication in MySQL. Loose bounds on the database latency are sufficient: the size of the database latency has little effect on FACADE's performance, mainly affecting how long updates are cached by coordinators. Since distributed concurrency control is provided by the coordinators, it does not matter what, if any, centralized concurrency control scheme is used by the master replica of the database.

FACADE masks the database latency in the same way as Decat *et al.*'s algorithm. Each coordinator maintains a LRU cache of recent committed updates to objects it manages, and it piggybacks on each request (when forwarding the request to a coordinator or worker) the cached updates for objects it manages that are involved in the request. Each cached update specifies a write timestamp as well as an attribute and its new value. A cached update is never evicted before the current time exceeds the update's write timestamp plus the database latency. The cache is accessed using the function cachedUpdates(x), which returns the set of cached updates to x.

FACADE needs to store multiple versions of objects in the database. This can easily be done in any database, by including a "version" column in the database schema. Our implementation using MySQL works this way.

Request Objects. We model requests as objects with fields subject, resource, ts (timestamp), cachedUpdates[i] (i = 1 and i = 2 for piggybacked cached updates to subject and resource, respectively), worker (worker selected to evaluate this request), and evalResult (result of evaluating the request, described below).

Policy Language. FACADE is independent of the details of the policy language. Any ABAC policy language can be used, provided it can express updates. For example, XACML can be used, with updates expressed as obligations, as in

[8,20]. Details of the policy language are abstracted behind an interface containing a single function evaluateRequest(policy,request) that returns an EvalResult object with these fields: decision (permit or deny), readAttr[i] (i = 1 and i = 2 for the set of attributes of the subject and resource, respectively, read during evaluation of the request), updatedObj (the index of the updated object, if any, otherwise −1), rdonlyObj (if updatedObj > 0, this is the index of the other object, otherwise −1), and updates (set of attribute-value pairs, specifying updates to updatedObj). The index values are interpreted as: 1 = subject, 2 = resource. evaluateRequest evaluates the request using attribute values current as of req.ts, reading values from req.cachedUpdates when they exist, otherwise reading values from the database using queries with timestamp req.ts.

Bounds on Attribute Accesses. Our algorithm can exploit bounds on attribute read and written by requests, when available, to improve performance. In particular, for a request r, for each object x that might be accessed by r (namely, the subject and resource), the client stub provides (1) a lower bound on the set of attributes of x that will definitely be read by r, (2) an upper bound on the set of attributes of x that might be read by r, and (3) an upper bound on the set of attributes of x that might be updated by the request. It is always safe to use the trivial bounds, i.e., the empty set for (1) and the set of all attributes for (2) and (3). When tighter bounds are available for (1), the algorithm can sometimes use them to conclude that a request definitely conflicts with an in-progress request r, without waiting to learn the exact set of attributes read by r. When tighter bounds are available for (2) and (3), the algorithm can sometimes use them to conclude that two requests involving the same object access disjoint sets of attributes and hence cannot conflict, without waiting to learn the exact sets of attributes they accessed. Note that these situations arise only in the typically small fraction of cases that two concurrent requests access the same object, and at least one of the requests is not known to be read-only.

Tighter bounds can often be obtained from basic knowledge about the request and the policy. The code or rules defining these bounds could be written manually for small systems or generated by a straightforward static analysis of the access control policy, based on the types of object and type of action in each rule and the names of the attributes read and written by each rule. For example, consider an access control system for an online video service, in which requests to play a video are subject to usage control to limit the number of views, and all other requests (browsing the video catalog, paying for a video, account maintenance, etc.) are not. In this system, a client can identify a request as read-only if the resource type is not "video" or the action is not "play".

These bounds are provided by defining (possibly using trivial bounds) the following policy-specific functions, where x is req.subject or req.resource.

– defReadAttr(x, req) is a set of attributes of x definitely read by req.
– mightReadAttr(x, req) is an upper bound on the set of attributes of x that might be read by req (including definitely read attributes).
– mightWriteAttr(x, req) is an upper bound on the set of attributes of x that might be updated by req.

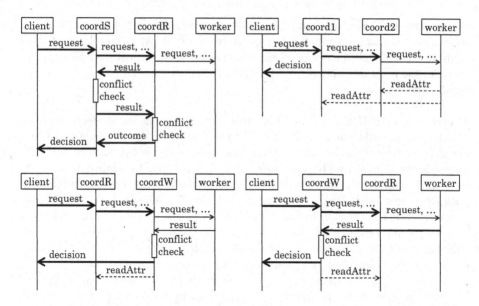

Fig. 1. Sequence diagrams. Top left: Decat *et al.*'s algorithm. Top right: FACADE for read-only request. Thick and thin solid lines are non-local and local messages, respectively, on the critical path. Dashed lines are messages not on the critical path. Bottom left: FACADE for read-write request, when client correctly predicts a read-only object. Bottom right: FACADE for read-write request, when client incorrectly predicts a read-only object.

Sequence Diagrams. We give brief overviews of Decat *et al.*'s algorithm and our algorithm, focusing on the message patterns shown in the sequence diagrams in Fig. 1. The sequence diagrams show the common case in which the request does not restart due to a conflict and the two objects accessed by the request are managed by coordinators on different servers. Accesses to the database are not shown; they are the same for Decat *et al.*'s algorithm and FACADE.

Overview of Decat et al.'s Algorithm. In Decat *et al.*'s algorithm, the client sends the request to coordS, the coordinator for the subject of the request. coordS updates data structures used for conflict detection and then forwards the request (with piggybacked cached committed updates) to coordR, the coordinator for the resource of the request, which does the same and then forwards to the request to a worker on the same server. The worker evaluates the request and then sends the result to coordS. coordS checks for conflicts involving the subject; specifically, it checks whether any attribute of the subject read by the request was updated after it forwarded the request to coordR (any such update was not piggybacked on the request and hence might not have been used in its evaluation). If there is no conflict, it forwards the result to coordR, which performs a similar conflict check and, if there is no conflict, commits the updates (if any) to the resource, and then sends the outcome of the conflict check to coordS. coordS commits the updates to the subject and then sends the decision to the client. If either coordinator detects a conflict, the request is restarted. After coordS sends the result to coordR and before it receives the outcome of coordR's conflict check, it treats the request's updates to the subject specially, as *tentative updates*; for details, see [8].

Overview of FACADE for Read-Only Requests. The client sends the request to coord1, the coordinator for one of the objects accessed by the request (either one is fine). coord1 updates data structures used for conflict detection and then forwards the request (with piggybacked cached committed updates) to coord2, the coordinator for the other object accessed by the request. coord2 updates its data structures and forwards the request to the worker. The worker evaluates the request, sends the decision to the client, and sends the sets of read attributes of the subject and resource to their respective coordinators, which update the read timestamps of the read versions. It is safe for the worker to send the decision directly to the client, because read-only requests never abort in FACADE.

Note that this message pattern is used for any request that turns out to be read-only, regardless of whether this is known in advance, i.e., regardless of whether mightWriteAttr is empty for either object involved in the request.

Overview of FACADE for Read-Write Requests. When the client sends the request to the coordinator for an object not updated by the request, we say that the client *correctly predicts a read-only object* for the request. This is guaranteed if mightWriteAttr returns an empty set for at least one object involved in the request, and has 50% probability otherwise. It is preferable for the client to send the request to such a coordinator, denoted coordR, because the worker sends the evaluation result to the coordinator for the updated object, denoted coordW, and that result message is local if the worker is co-located with coordW, which happens if coordR receives the request from the client and forwards it to coordW. If mightWriteAttr returns a non-empty set for both objects, then the client arbitrarily selects a coordinator to which to send the request. If that turns out to be coordW, we say that the client *incorrectly predicts a read-only object* for the request. The only consequence is that the worker's result message is a network message instead of a local message.

When the client correctly predicts a read-only object for the request, the client sends the request to the coordinator for that object, denoted coordR. coordR updates data structures used for conflict detection and then forwards the request (with piggybacked cached committed updates) to coordW. The worker evaluates the request and sends the result, including the decision and the sets of read and written attributes of the subject and resource, to coordW. coordW checks for conflicts; specifically, it checks whether any attribute updated by this request was read by a request with a later timestamp. Even if there is no conflict yet, a conflict could arise later, involving a request with a later timestamp that has already been forwarded and might read the attribute. A set of such requests, called "pending might read requests", is associated with each version of an attribute. The worker waits until there are no such pending might read requests and then checks for conflicts again. If there is no conflict, it commits the updates, sends the decision to the client, and sends the set of read attributes of the other object to the other coordinator.

When the client incorrectly predicts a read-only object for the request, the message pattern is the same, except that coordW receives the request first and then forwards it to coordR, and the evaluation result message from the worker to coordW is a network message instead of a local message.

Handling of Requests Known to be Read-Only. A request req is *known to be read-only* iff mightWriteAttr(req.subject, req) and mightWriteAttr(req.resource, req) are empty. Handling of requests known to be read-only is described separately from handling of other requests, for ease of understanding, although the two are similar in places, and the code for them is integrated in our implementation. Handling of requests known to be read-only follows the pseudocode in Fig. 2. The pseudocode syntax is generally Python-like, except we denote tuples using angle brackets instead of parentheses. Implicitly, coarse-grain locking is used to ensure that coordinators process each incoming message atomically, i.e., without interruption by processing of other messages (as an optimization, finer-grained locking could be used).

Handling of Read-Write Requests. Handling of other requests follows the pseudocode in Figs. 3 and 4.

Liveness. The algorithm presented in the pseudocode is deadlock-free: the inequality on timestamps in the **await** statement in Fig. 4 ensures that two requests cannot be stuck waiting for each other. However, it can starve some read-write requests. For example, a long stream of reads to an attribute x.attr can cause the condition in the **await** statement in Fig. 4 to remain true for a long time, causing a pending update to x.attr to starve. The underlying reason is that FACADE gives precedence to reads over writes, in the sense that reads never abort, and writes can be aborted due to conflicting reads.

To counter-balance this, and thereby help prevent starvation of writes, we modify the algorithm to delay reads in two cases (these modifications are not reflected in the pseudocode). (1) After a coordinator c receives $\langle$"request", req, 1$\rangle$ from a client, if req might update req.obj[1], c delays processing of incoming requests that potentially conflict with req (temporarily storing them in a queue) until c determines the outcome (commit or restart) of the current execution of req, at which time c processes the delayed requests normally. An incoming request req2 potentially conflicts with req if req2 might read an attribute that req might update. (2) After a coordinator c receives an evaluation result message $\langle$"result", req$\rangle$ that includes updates to an object x managed by c, while c is waiting for the **await** condition to become true, c delays processing of incoming requests that potentially conflict with those updates until c determines the outcome (commit or restart) for req, at which time c processes the delayed requests normally An incoming request req2 potentially conflicts with the updates if req2 might read one of the updated attributes. Note that these two kinds of delays

1. client:
for read-only requests, the coordinator order (subject first or resource first) does
not affect correctness or performance. arbitrarily do subject first.
req.obj[1], req.obj[2] = req.subject. req.resource
send ⟨"request", req, 1⟩ **to** coord(req.obj[1])

2. coordinator: on receiving ⟨"request", req, 1⟩:
x = req.obj[1]
req.ts = now()   # now() returns the current date-time.
for attr **in** defReadAttr(x, req):
  getVersion(x, attr, req.ts).rts = req.ts
for attr **in** mightReadAttr(x, req) - defReadAttr(x, req):
  getVersion(x, attr, req.ts).pendingMightReads.add(⟨req.id, req.ts⟩)
req.cachedUpdates[1] = cachedUpdates(x)
send ⟨"request", req, 2⟩ **to** coord(req.obj[2])

3. coordinator: on receiving ⟨"request", req, 2⟩:
x = req.obj[2]
for attr in defReadAttr(x, req):
  getVersion(x, attr, req.ts).rts = req.ts
for attr **in** mightReadAttr(x, req) - defReadAttr(x, req):
  getVersion(x, attr, req.ts).pendingMightReads.add(⟨req.id, req.ts⟩)
select worker w to evaluate this request
req.worker = w
req.cachedUpdates[2] = cachedUpdates(x)
send req **to** w

4. worker: on receiving req:
req.evalResult = evaluateRequest(policy, req)
send ⟨"decision", req.id, evalResult.decision⟩ **to** req.client
send ⟨"readAttr", req, 1⟩ **to** coord(req.subject)
send ⟨"readAttr", req, 2⟩ **to** coord(req.resource)

5. coordinator: on receiving ⟨"readAttr", req, i⟩:
x = req.subject if i==1 else req.resource
for attr **in** mightReadAttr(x, req) - defReadAttr(x, req):
  v = getVersion(x, attr, req.ts)
  v.pendingMightReads.remove(⟨req.id, req.ts⟩)
  if attr in req.evalResult.readAttr[i]:
    v.rts = req.ts

Fig. 2. Handling of requests known to be read-only.

cannot lead to deadlock (i.e., to circular wait), because the delayed requests are younger than req.

Decat *et al.*'s algorithm can also starve requests. It gives precedence to writes over reads, in the sense that writes never abort, and reads can be aborted because of conflicting writes. Consequently, long streams of writes can starve read-only

1. client:
if either object is known to be read-only, send req to its coordinator first.
if isEmpty(mightWriteAttr(req.obj[2], req)):
  req.obj[1], req.obj[2] = 2, 1
else:
  req.obj[1], req.obj[2] = 1, 2
send ⟨"request", req, 1⟩ **to** coord(req.obj[1])

2. coordinator: on receiving ⟨"request", req, 1⟩:
x = req.obj[1]
req.ts = now()   # now() returns the current date-time.
for attr **in** mightReadAttr(x, req)
  v = getVersion(x,attr,req.ts)
  v.pendingMightReads.add(⟨req.id, req.ts⟩)
req.cachedUpdates[1] = cachedUpdates(x)
send ⟨"request", req, 2⟩ **to** coord(req.obj[2])

3. coordinator: on receiving ⟨"request", req, 2⟩:
x = req.obj[2]
for attr **in** mightReadAttr(x,req)
  v = getVersion(x,attr,req.ts)
  v.pendingMightReads.add(⟨req.id, req.ts⟩)
select worker w to evaluate this request
req.worker = w
req.cachedUpdates[2] = cachedUpdates(x)
send req **to** w

4. worker: on receiving req:
req.evalResult = evaluateRequest(policy, req)
if req.updatedObj == -1:
  # req is read-only.
  send ⟨"decision", req.id, req.evalResult.decision⟩ to req.client
  send ⟨"readAttr", req, 1⟩ **to** coord(req.subject)
  send ⟨"readAttr", req, 2⟩ **to** coord(req.resource)
else:
  # req updated an object.
  send ⟨"result", req⟩ **to** coord(req.obj[req.updatedObj])

Fig. 3. Handling of requests not known to be read-only, part 1.

and read-write requests. Their algorithm does not incorporate any mechanism to compensate for this. This is probably acceptable for workloads in which writes are infrequent relative to reads.

Optimizations. Our implementation incorporates the following optimizations that are not reflected in the pseudocode. (1) If the same coordinator is responsible for both objects involved in a request, then the coordinator performs the processing for both objects together, without sending itself a message in between.

5. coordinator: on receiving ⟨"result", req⟩:
# req updates an object that this coordinator is responsible for.
x = req.obj[req.updatedObj]
conflict = checkForConflicts()
if not conflict:
  # wait for relevant pending reads to complete. await(*expr*) blocks until *expr* is true.
  await (∀⟨attr,val⟩∈ req.updates. ∀⟨id,ts⟩∈ getVersion(x, attr, req.ts).pendingMightReads.
          id == req.id or ts < req.ts)
  conflict = checkForConflicts()
  if not conflict:
    commit req.evalResult.updates to the database with write timestamp req.ts
    # cache the updates, and store the new versions for conflict checking
    for (attr, val) **in** req.evalResult.updates:
      cachedUpdates(x).add(⟨attr, val, req.ts⟩)
      addVersion(x,attr,req.ts)
    # update read timestamps
    for attr **in** mightReadAttr(x,req)
      v = getVersion(x,attr,req.ts)
      v.pendingMightReads.remove(⟨req.id, req.ts⟩)
      if attr in req.readAttr[req.updatedObj]:
        v.rts = req.ts
    # send decision to client
    send ⟨"decision", req.id, req.decision⟩ **to** req.client
    # send read attributes to coordinator for read-only object
    roCoord = coord(req.subject) if req.evalResult.rdonlyObj==1 else coord(req.resource)
    send ⟨"readAttr", req, req.evalResult.rdonlyObj⟩ **to** roCoord
  else:
    restart(req)
else:
  restart(req)

coordinator: on receiving ⟨"restart", req⟩:
remove entries for req from all pendingMightReads sets
restart processing of req, as if it were newly received from client

def checkForConflicts():
  for (attr, val) **in** req.updates:
    # note: if x.attr has not been read or written in this session, then
    # v is the special version with v.rts=0 and v.wts=0.
    v = getVersion(x, attr, req.ts)
    if v.rts > req.ts:
      return true
  return false

def restart(req):
  remove entries for req from all pendingMightReads sets
  # tell the other coordinator to restart processing of this request
  roCoord = coord(req.subject) if req.evalResult.rdonlyObj==1 else coord(req.resource)
  send ⟨"restart", req⟩ **to** roCoord

Fig. 4. Handling of requests not known to be read-only, part 2.

(2) The **await** statement in Fig. 4 waits for all relevant pending reads to complete before checking whether any of them conflict with the pending update. As an optimization, when each relevant pending read completes, the coordinator immediately checks whether it conflicts with the pending update, and if so, immediately restarts the request performing the update. (3) To reduce the number of database queries, workers piggyback data read from the database on messages sent to coordinators, and coordinators add it to the data structure that caches recent committed updates. Note that caching of attribute data is done only by coordinators, not workers, because a coordinator performs all updates to objects it manages and hence knows when cached data is stale (relative to a specified request timestamp).

Fault-Tolerance. Like Decat *et al.* in [8], we focus in this paper on scalability and leave detailed consideration of fault-tolerance for future work. We briefly sketch how to extend our algorithm to tolerate crash failures. A fault-monitoring service is needed to detect crashes and restart crashed processes. Requests that were in-progress at the time of a crash might be dropped. If a client does not receive a decision for a request in a reasonable amount of time, the client can re-submit the request with the same identifier. If the request is read-only, the worker simply re-evaluates it in the current state. If the request performs updates, the worker checks whether the request already committed, and if so, re-sends the original decision, otherwise re-evaluates the request in the current state. To support this, when a coordinator commits the attribute updates for a request, it also inserts a record containing the request id and decision in a request log table. The worker looks up the request id in this table before evaluating a request.

3   Evaluation

Implementation. We implemented FACADE in DistAlgo [15, 16], an extension of Python with high-level communication and synchronization constructs. The DistAlgo compiler[5] translates DistAlgo into Python for execution. We also implemented Decat *et al.*'s algorithm in DistAlgo, to allow a performance comparison of the algorithms, not influenced by the performance of different programming language implementations (Decat *et al.*'s implementation is in Scala). Our implementations of both algorithms are publicly available[6]. The experimental platform consists of three desktop PCs with Intel Core 2 Quad processors (two at 2.83 GHz, one at 2.66 GHz), with Gigabit Ethernet NICs connected to a Gigabit Ethernet switch, and running Windows 10 64-bit, Python 3.6, DistAlgo 1.0.9, and MySQL 5.7.17.

Workload. The workload consists of pseudorandom sequences of requests. The same seeds, hence the same workload, are used for corresponding experiments with the two algorithms. Configuration parameters for each experiment include:

---

[5] http://sourceforge.net/projects/distalgo/files/.

[6] http://www.cs.stonybrook.edu/~stoller/software/.

- nClient: number of clients. This is also the maximum number of concurrent requests, since each client sends a request and waits for the response before sending the next request.
- nWorker: number of workers per coordinator.
- nObj: number of objects in database. We use objects with 10 attributes, two of which are mutable (i.e., might be updated by access control policy rules).
- nRequest: total number of requests (split evenly among the clients).
- pWrite: probability that a request is read-write; other requests are read-only.
- pSameCoord: probability that the two objects involved in a request have the same coordinator. As discussed below, we emulate experiments with nCoord coordinators using our platform with 2 coordinators by setting pSameCoord = 1/nCoord.
- wrongWrite: flag controlling accuracy of client's prediction of written objects. wrongWrite = 0 means completely accurate. wrongWrite = 1 means the prediction always includes an object not written by the request.
- wrongAttr: flag controlling accuracy of client's prediction of accessed attributes. wrongAttr = 0 means completely accurate. wrongAttr = 1 means the predictions of read and written attributes contain all attributes and all mutable attributes, respectively.

Latency. To evaluate how the performance, primarily latency, of FACADE would depend on the number of coordinators in a system, we ran experiments analogous to the latency experiments in [8, Sect. 3.4, Fig. 9]. We use nClient = 1, like they do, to measure the intrinsic latency of the algorithm, in the absence of contention. In their experiment, latency is measured instead as a function of the actual number of coordinators. However, the number of coordinators affects the latency only indirectly, by affecting the probability that the same coordinator is responsible for the two objects involved in the request. For clarity, we measure the latency directly as a function of this probability, by making pSameCoord a workload parameter, as described above. This also allows us to use a smaller platform for the experiments. Values of the other fixed workload parameters in these experiments are nWorker = 1, nObj = 1000, nRequest = 5000 and pWrite = 0.1. For FACADE, we repeat the experiments for each of the four possible combinations of values of wrongWrite and wrongAttr. Figure 5 shows average latency per request and average number of network messages sent per request for FACADE and Decat *et al.*'s algorithm. When pSameCoord is 0.5 or less, corresponding to deployments with 2 or more coordinators, FACADE has lower latency and sends fewer network messages than Decat *et al.*'s algorithm. FACADE's lower latency stems from using fewer network messages and fewer database queries (due to optimization (3)). Deployments in large systems would probably use around 10 coordinators, as in Decat *et al.*'s experiments. This corresponds to pSameCoord = 0.1, for which the average latency of FACADE is *less than half* the average latency of Decat *et al.*'s algorithm (37.7 ms compared to 91.1 ms), and the average network messages per request is 3.8 for FACADE vs. 5.6 for Decat *et al.*'s algorithm. This is true regardless of whether accurate

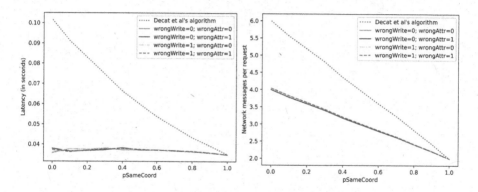

Fig. 5. Average latency per request (left) and average number of network messages per request (right) as a function of pSameCoord.

prediction of accessed attributes and written objects is possible. More generally, we see that incorrect prediction of accessed attributes and written objects have negligible effect on these results. We also see that the average latency of FACADE is almost independent of pSameCoord; this is because local processing time accounts for much of the latency, and the average number of network messages per request changes less for FACADE than Decat et al.'s algorithm.

Throughput. To evaluate throughput, we ran experiments analogous to the performance experiments in [8, Sect. 4.4, Fig. 13]. To determine the maximum throughput of each algorithm, we ran experiments with increasing numbers of clients, until the throughput plateaus. For each value of nClient, we ran experiments with increasing numbers of workers, until throughput plateaus. We then used the value of nWorkers determined for the largest value of nClient in experiments with all smaller values of nClient, since we wanted only one workload parameter to vary in the final results. For FACADE with wrongWrite = 0 and wrongAttr = 0, we found nClient = 23 and nWorker = 4 provided the maximum throughput of 344 requests/second, with mean latency of 65.5 ms. For Decat et al.'s algorithm, we found nClient = 19 and nWorker = 14 provided the maximum throughput of 318 requests/second, with mean latency of 79.5 ms. Values of the other fixed workload parameters are nObj = 1000, nRequest = 5000, pWrite = 0.1 and pSameCoord = 0.1. Figure 6 shows average throughput as a function of nClient for Decat et al.'s algorithm and FACADE. For FACADE, average throughput is shown for each of the four possible combinations of values of wrongWrite and wrongAttr. We see that FACADE achieves higher maximum throughput than Decat et al.'s algorithm in most cases in these experiments. We also see that FACADE's throughput is more sensitive than its latency to the accuracy of predictions of accessed attributes and written objects.

Local Processing Time. The CPU time per request for coordinators is similar for FACADE and Decat et al.'s algorithm. The CPU time per request for workers is roughly double for FACADE compared to Decat et al.'s algorithm, due

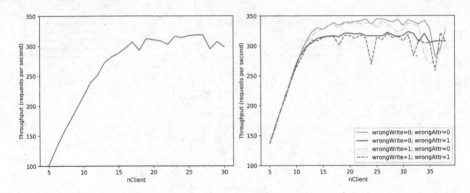

Fig. 6. Average throughput as a function of nClient for Decat *et al.*'s algorithm (left) and FACADE (right).

to versioning and piggybacking data read from the database on messages to coordinators (i.e., optimization (3)). Local processing is a significant fraction of the overall latency (and throughput is relatively low in absolute terms), because Python is relatively slow. If both algorithms were implemented in a faster language such as C++, local processing would be a smaller part of the overall latency, and the ratio of average latency for FACADE to average latency for Decat *et al.*'s algorithm would be even smaller than in our experiments.

Performance with Different Write Probabilities. To evaluate the effect of pWrite on performance, we also ran the latency experiments and throughput experiments (described in the *Latency* and *Throughput* paragraphs above, respectively) with pWrite = 0.0 (i.e., all requests are read-only) and pWrite = 0.2. We consider pWrite = 0.1 to be a realistic value and pWrite = 0.2 to be on the high side of the realistic range. pWrite = 0.0 is a natural boundary value to consider; it is also the best case for both algorithm's performance. For the latency experiments, the results with pWrite = 0.0 and pWrite = 0.2 are almost the same as those described above for pWrite = 0.1, because writes have little effect on performance when there are no conflicts, and there are no conflicts in experiments with only one client. For the throughput experiment with pWrite = 0.0, for FACADE, we found nClient = 24 and nWorker = 8 provided the maximum throughput of 412 requests/second, with mean latency of 56.6 ms; for Decat *et al.*'s algorithm, we found nClient = 24 and nWorker = 9 provided the maximum throughput of 373 requests/second, with mean latency of 62.0 ms. For throughput experiment with pWrite = 0.2, for FACADE, we found nClient = 24 and nWorker = 2 provided the maximum throughput of 303 requests/second, with mean latency of 77.4 ms; for Decat *et al.*'s algorithm, we found nClient = 25 and nWorker = 4 provided the maximum throughput of 283 requests/second, with mean latency of 86.8 ms. Thus, FACADE's maximum throughput is 11%, 8%, and 7% higher than Decat *et al.*'s algorithm's maximum throughput when pWrite = 0.0, 0.1, and 0.2, respectively, and FACADE has lower latency in all three experiments.

Performance with More Conflicts. To evaluate the effect of a higher conflict rate on performance, we also ran the throughput experiments with an unrealistically small number of objects; decreasing nObj is the simplest way to increase the conflict rate. Specifically, we reduced nObj from 1000 (a more realistic value) to 200 (an unrealistically small value) for these experiments. Other workload parameters, including nClient and nWorker, are the same as described above for the throughput experiments. For FACADE with wrongWrite = 0 and wrongAttr = 0, the number of restarts due to conflicts increased from 1 to 16, throughput decreased from 344 to 295 requests/second, and average latency increased from 65.5 to 75.9 ms. For Decat *et al.*'s algorithm, the number of restarts due to conflicts increased from 1 to 6, throughput decreased from 318 to 305 requests/second, and average latency increased from 79.5 to 81.7 ms. Although FACADE is more sensitive than Decat *et al.*'s algorithm to this change, FACADE's performance is still competitive, with 3% lower throughput and 7% lower latency than Decat *et al.*'s algorithm.

4   Related Work

Decat *et al.*'s work in [8] is the most closely related and is discussed in previous sections.

Chadwick describes a distributed architecture for a XACML-based stateful policy framework, consisting of multiple policy decision points (PDPs) interacting with a centralized database containing the mutable state [6]. Each PDP locks all relevant rows in the database before evaluating a request. The design has limited scalability, due to the centralized database and locking.

Alzahrani *et al.* describe a similar distributed architecture [1], without committing to a specific approach to storage of the state. They briefly mention a few alternatives, e.g., in a centralized database, or replicated at or partitioned among the PDPs, but do not discuss any of them in detail.

Dhankhar *et al.* consider evaluation of stateful distributed XACML policies. Different PDPs have different policies, and the policies can refer to each other [9]. Concurrency control is provided by a centralized lock manager. Each PDP locks all relevant attributes before evaluating a request. The centralized lock manager limits scalability of their design.

Kelbert and Pretschner describe a fault-tolerant decentralized infrastructure for enforcement of usage control policies [14]. They rely on the database, Cassandra (see Foonote 2), for concurrency control. As mentioned in Sect. 1, Cassandra provides serializability only for single-row transactions, so their system does not support serializable evaluation of requests involving attributes of two objects.

Weber *et al.* present a framework for stateful access control policies in distributed systems based on weakly consistent replication of the state, as provided by eventually consistent data stores [22]. In contrast, our design is based on the traditional notion of strong consistency. When weak consistency is acceptable, it potentially allows more fault-tolerance and scalability. They do not present a completed implementation or any performance results.

Acknowledgments. We thank M. Decat for explaining the details of [8].

References

1. Alzahrani, A., Janicke, H., Abubaker, S.: Decentralized XACML overlay network. In: Proceedings of the 10th IEEE International Conference on Computer and Information Technology (CIT 2010), pp. 1032–1037. IEEE Computer Society (2010)
2. American National Standards Institute (ANSI), International Committee for Information Technology Standards (INCITS): Role-based access control. ANSI INCITS Standard, pp. 359–2004, February 2004
3. Becker, M.Y.: Specification and analysis of dynamic authorisation policies. In: Proceedings 22nd IEEE Computer Security Foundations Symposium (CSF), pp. 203–217. IEEE Computer Society (2009)
4. Becker, M.Y., Nanz, S.: A logic for state-modifying authorization policies. ACM Trans. Inf. Syst. Secur. **13**(3), 20:1–20:28 (2010)
5. Brewer, D.F.C., Nash, M.J.: The Chinese wall security policy. In: Proceedings of the 1989 IEEE Symposium on Security and Privacy, pp. 206–214. IEEE Computer Society (1989)
6. Chadwick, D.: Coordinated decision making in distributed applications. Inf. Secur. Tech. Rep. **12**, 147–154 (2007)
7. Chang, F., Dean, J., Ghemawat, S., Hsieh, W.C., Wallach, D.A., Burrows, M., Chandra, T., Fikes, A., Gruber, R.E.: Bigtable: a distributed storage system for structured data. ACM Trans. Comput. Syst. **26**(2), 4:1–4:26 (2008)
8. Decat, M., Lagaisse, B., Joosen, W.: Scalable and secure concurrent evaluation of history-based access control policies. In: Proceedings of the 31st Annual Computer Security Applications Conference (ACSAC 2015), pp. 281–290. ACM (2015)
9. Dhankhar, V., Kaushik, S., Wijesekera, D., Nerode, A.: Evaluating distributed XACML policies. In: Proceedings of the 4th ACM Workshop On Secure Web Services (SWS 2007), pp. 99–110. ACM (2007)
10. Edjlali, G., Acharya, A., Chaudhary, V.: History-based access control for mobile code. In: Proceedings of the 5th ACM Conference on Computer and Communications Security (CCS 1998), pp. 38–48. ACM (1998)
11. Gama, P., Ribeiro, C., Ferreira, P.: A scalable history-based policy engine. In: Proceedings of the 7th IEEE International Workshop on Policies for Distributed Systems and Networks (POLICY 2006), pp. 100–112. IEEE Computer Society (2006)
12. Gay, R., Mantel, H., Sprick, B.: Service automata. In: Barthe, G., Datta, A., Etalle, S. (eds.) FAST 2011. LNCS, vol. 7140, pp. 148–163. Springer, Heidelberg (2012). doi:10.1007/978-3-642-29420-4_10
13. ISO/IEC: Information technology – open systems interconnection – security frameworks for open systems: access control framework. ISO/IEC Standard 10181–3:1996, International Organization for Standardization (2006)
14. Kelbert, F., Pretschner, A.: A fully decentralized data usage control enforcement infrastructure. In: Malkin, T., Kolesnikov, V., Lewko, A.B., Polychronakis, M. (eds.) ACNS 2015. LNCS, vol. 9092, pp. 409–430. Springer, Cham (2015). doi:10.1007/978-3-319-28166-7_20
15. Liu, Y.A., Stoller, S.D., Lin, B.: From clarity to efficiency for distributed algorithms. ACM Trans. Program. Lang. Syst. **39**(3), 395–410 (2017)

16. Liu, Y.A., Stoller, S.D., Lin, B., Gorbovitski, M.: From clarity to efficiency for distributed algorithms. In: Proceedings of the 2012 ACM International Conference on Object Oriented Programming Systems Languages and Applications (OOPSLA), pp. 395–410. ACM Press, October 2012

17. Lobo, J., Ma, J., Russo, A., Lupu, E., Calo, S., Sloman, M.: Refinement of history-based policies. In: Balduccini, M., Son, T.C. (eds.) Logic Programming, Knowledge Representation, and Nonmonotonic Reasoning. LNCS (LNAI), vol. 6565, pp. 280–299. Springer, Heidelberg (2011). doi:10.1007/978-3-642-20832-4_18

18. Martinelli, F., Matteucci, I., Mori, P., Saracino, A.: Enforcement of U-XACML history-based usage control policy. In: Barthe, G., Markatos, E., Samarati, P. (eds.) STM 2016. LNCS, vol. 9871, pp. 64–81. Springer, Cham (2016). doi:10.1007/978-3-319-46598-2_5

19. Nguyen, D., Park, J., Sandhu, R.S.: A provenance-based access control model for dynamic separation of duties. In: Eleventh Annual International Conference on Privacy, Security and Trust (PST 2013), pp. 247–256. IEEE Computer Society (2013)

20. Park, J., Sandhu, R.: The $ucon_{abc}$ usage control model. ACM Trans. Inf. Syst. Secur. **7**(1), 128–174 (2004)

21. Silberschatz, A., Korth, H.F., Sudarshan, S.: Database System Concepts, 6th edn. McGraw-Hill, New York (2011)

22. Weber, M., Bieniusa, A., Poetzsch-Heffter, A.: Access control for weakly consistent replicated information systems. In: Barthe, G., Markatos, E., Samarati, P. (eds.) STM 2016. LNCS, vol. 9871, pp. 82–97. Springer, Cham (2016). doi:10.1007/978-3-319-46598-2_6

Privacy

Gaussian Mixture Models for Classification and Hypothesis Tests Under Differential Privacy

Xiaosu Tong[1], Bowei Xi[2(✉)], Murat Kantarcioglu[3], and Ali Inan[4]

[1] Amazon, Seattle, WA, USA
xiaosutong@gmail.com
[2] Department of Statistics, Purdue University, West Lafayette, IN, USA
xbw@purdue.edu
[3] Department of Computer Science, University of Texas at Dallas, Dallas, TX, USA
muratk@utdallas.edu
[4] Department of Computer Engineering,
Adana Science and Technology University, Adana, Turkey
ainan@adanabtu.edu.tr

Abstract. Many statistical models are constructed using very basic statistics: mean vectors, variances, and covariances. Gaussian mixture models are such models. When a data set contains sensitive information and cannot be directly released to users, such models can be easily constructed based on noise added query responses. The models nonetheless provide preliminary results to users. Although the queried basic statistics meet the differential privacy guarantee, the complex models constructed using these statistics may not meet the differential privacy guarantee. However it is up to the users to decide how to query a database and how to further utilize the queried results. In this article, our goal is to understand the impact of differential privacy mechanism on Gaussian mixture models. Our approach involves querying basic statistics from a database under differential privacy protection, and using the noise added responses to build classifier and perform hypothesis tests. We discover that adding Laplace noises may have a non-negligible effect on model outputs. For example variance-covariance matrix after noise addition is no longer positive definite. We propose a heuristic algorithm to repair the noise added variance-covariance matrix. We then examine the classification error using the noise added responses, through experiments with both simulated data and real life data, and demonstrate under which conditions the impact of the added noises can be reduced. We compute the exact type I and type II errors under differential privacy for one sample z test, one sample t test, and two sample t test with equal variances. We then show under which condition a hypothesis test returns reliable result given differentially private means, variances and covariances.

Keywords: Differential privacy · Statistical database · Mixture model · Classification · Hypothesis test

© IFIP International Federation for Information Processing 2017
Published by Springer International Publishing AG 2017. All Rights Reserved
G. Livraga and S. Zhu (Eds.): DBSec 2017, LNCS 10359, pp. 123–141, 2017.
DOI: 10.1007/978-3-319-61176-1_7

1   Introduction

Building a model over a data set is often a straightforward task. However, when the data set contains sensitive information, special care has to be taken. Instead of having direct access to data, the users are provided with a sanitized view of the database containing private information, either through perturbed individual records or perturbed query responses.

From users' perspective, knowing the responses to their queries are perturbed, users may not want to directly query the output of a complex model. Many statistical models are constructed using very basic statistics. Knowing the values of means, variances and covariances, or equivalently the sums, the sums of squares and the sums of cross products, users can build least square regression models, conduct principal component analysis, construct hypothesis tests, and construct Bayesian classifiers under Gaussian mixture models, etc. Although the basic statistics (e.g., means, variances and covariances) satisfy differential privacy guarantee, the complex models constructed using these basic statistics may no longer meet the differential privacy guarantee.

We notice it is up to the users to decide how to query a database and how to further utilize the queried results. Building statistical models using the perturbed basic statistics provides quick initial estimates. If the results based on the perturbed query responses are promising, users can then proceed to improve the accuracy of the results.

In this article, our goal is to understand the impact of differential privacy mechanism for the mixture of Gaussian models. Gaussian mixture models refer to the case where each model follows multivariate Gaussian distribution. Hence users only need to obtain the mean vector and the variance-covariance matrix for each class. Out of all the statistical techniques that can be applied to Gaussian mixture models without further querying the database, we focus on building a classifier or performing a hypothesis test with the noisy responses. Through extensive experiments and theoretical discussions, we show when the classifiers and tests work reliably under privacy protection mechanism, in particular, differential privacy.

k-anonymity [17,18,20] and differential privacy [5] are two major privacy preserving models. Under k-anonymity model the perturbed individual records are released to the users, while under differential privacy model the perturbed query responses are released to the users. Recent work pointed out the two privacy preserving models are complimentary [3]. Main contributions of this article could be summarized as follows:

1. We provide theoretical results on the type I and type II errors under differential privacy for several hypothesis tests. We also show when a hypothesis test returns reliable result under differential privacy mechanism.
2. We propose a heuristic algorithm to repair the noise added variance-covariance matrix, which is no longer positive definite and cannot be directly used in building a Bayesian classifier.

3. We examine the classification error for the multivariate Gaussian case through experiments. The experiments demonstrate when the impact of the added noise can be reduced.

The rest of the paper is organized as follows. Section 1.1 provides a brief overview of differential privacy mechanism. Related work is discussed in Sect. 2. Section 3 provides theoretical results for hypothesis tests under differential privacy. In Sect. 4 we provide an algorithm to repair the noise added variance-covariance matrix, and study the classification error through extensive experiments. Section 5 concludes our discussion.

1.1 Differential Privacy

Let $D = \{X_1, \cdots, X_d\}$ be a d-dimensional database with n observations, where the domain of each attribute X_i is continuous and bounded. We are interested in building Gaussian mixture models over database D. One only needs to compute the expected values of each attribute X_i and the variance-covariance matrix, $\Sigma_{ij} = cov(X_i, X_j) = E[(X_i - \mu_i)(X_j - \mu_j)]$, where $\mu_i = E(X_i)$. More details follow in Sect. 4. Users obtain the values of μ_is and Σ_{ij}s by querying the database D. The query results are perturbed according to differential privacy mechanism. Next we briefly review differential privacy.

Given a set of queries $Q = \{Q_1, ..., Q_q\}$, Laplace mechanism for differential privacy adds Laplace noise with parameter λ to the actual value. λ is determined by privacy parameter ε and sensitivity $S(Q)$. Here, ε is a pre-determined parameter, selected by the database curator, while sensitivity $S(Q)$ is a function of the query Q. Hence differential privacy mechanism minimizes the risk of identifying individual records from a database containing sensitive information.

Sensitivity is defined over sibling databases, which differ in only one observation.

$$S(Q) = \max_{\forall \text{ sibling databases } D_1, D_2} \sum_{i=1}^{q} |Q_i^{D_1} - Q_i^{D_2}| \tag{1}$$

That is, sensitivity of Q is the maximum difference in the L_1 norm of the query response caused by a single record update. Sensitivities for standard queries, such as sum, mean, variance-covariance are well established [6].

Once ε and $S(Q)$ are known, λ is set such that $\lambda \geq S(Q)/\varepsilon$. Then for each query Q, the database first computes the actual value Q^D in D, then adds Laplace noise to obtain the noisy response R^D, and return R^D to users: $R^D = Q^D + r$, where $r \sim \text{Laplace}(\lambda)$. There have been many work on sensitivity analysis. For querying mean, variance and covariance, we use the sensitivity results as in [22] in this article. Later in the experimental studies, the Laplace noises are added according to the results in [22]. Although there are other techniques to satisfy differential privacy (e.g., exponential mechanism [16]), for the three basic queries needed to build Gaussian mixture models, we leverage the Laplace mechanism discussed above.

2   Related Work

Gaussian mixture models are widely used in practice [4,9]. Differential privacy mechanism [5] models the database as a statistical database. Random noises are added to the responses to user queries. The magnitude of random noise is proportional to the privacy parameter ε and the sensitivity of the query set. Different formulations of differential privacy have been proposed. One definition of sensitivity consider sibling data sets that have the same size but differ in one record [5,7]. Other studies have sibling data sets through insertion of a new record sets [6]. We follow the formulation in [5] in this article.

Classification under differential privacy has received some attention. In [8], Friedman et al. built a decision tree, a method of ID3 classification, through recursive queries retrieving the information gain. Jagannathan et al. [10] built multiple random decision trees using sum queries. [1] proposed perturbing the objective function before optimization for empirical risk minimization. The lower bounds of the noisy versions of convex optimization algorithms were studied. Privacy preserving optimization is an important component in some classifiers, such as regularized logistic regression and support vector machine (SVM). [12] extended the results in [1], and also proposed differentially private optimization algorithms for convex empirical risk minimization. [19] proposed a privacy preserving mechanism for SVM.

In [14] every component in a mixture population follows a Gaussian mixture distribution. A perturbation matrix was generated based on a gamma distribution. Gamma perturbations were included in the objective function as multipliers, and a classifier was learned through maximizing the perturbed objective function. On the other hand, we consider classifiers that can be constructed using very basic statistics, i.e., means, variances and covariances, and show how their performance is affected by the added noises. In this article, we present Bayes classifiers based on Gaussian mixture models by querying the mean vector and the variance-covariance matrix for each class.

[2] proposed an algorithm using a Markov Chain Monte Carlo procedure to produce principal components that satisfy differential privacy. It is a differentially private lower rank approximation to a semi-positive definite matrix. Typically the rank k is much smaller than the dimension d. [11] also proposed an algorithm to produce differentially private low rank approximation to a positive definite matrix. [15] focused on producing recommendations under differential privacy. In [15], the true ratings were perturbed. A variance-covariance matrix was computed using the perturbed ratings; noises were added to the resulting matrix; then a low rank approximation to the noise added matrix was computed. Compared to the existing work, we focus on the scenario where all the variables are used to learn a variance-covariance matrix and the subsequent classifier and dimension reduction is not needed.

[13] proposed the differentially private M-estimators, such as sample quantiles, maximum likelihood estimator, based on the perturbed histogram. Our work has a different focus. We examine the classifiers and hypothesis tests constructed using the differentially private sample means, variances and covariances.

[21] derived rules for how to adjust sample sizes to achieve a pre-specified power for Pearson's χ^2 test of independence and the test of sample proportion. For the second test, when sample size is reasonably large, the sample proportion is approximately normally distributed. [21] developed sample size adjustment results based on the approximate normal distribution. Our work provides theoretical results to compute the exact type I and type II errors for one sample z test, one sample t test, and two sample t test. Both type I and type II errors are functions of ε and n. Hence with a known ε value users can obtain a minimum sample size required to achieve a pre-specified power while the exact type I error is controlled by a certain upper bound.

3   Hypothesis Tests Under Differential Privacy

Differential privacy mechanism has a big impact on hypothesis tests because the test statistic is now created using the noise added query results, and hypothesis tests often apply to data with smaller sample size than classification. Next we provide the distributions for the noise added test statistic under the null value and an alternative value.

Only when we know the true λs for the Laplace noises, we can numerically compute the exact p-value given a noise added test statistic. The true λs are unknown to the users querying a database. Hence in this section we examine a more realistic scenario: A rejection region is constructed using the critical values from a Gaussian distribution or a t distribution as usual, users compute a test statistic using the noise added mean and variance, and make a decision. The exact type I and type II errors can be computed numerically for likely ε values, which provide a reliability check of the test for users. Here we show for what sample size the exact type I and type II errors are close to those without the added noises. We consider the most commonly used hypothesis tests: the one sample z test, the one sample t test, the two sample t test with equal variance.

For the two sample t test with unequal variances, the degrees of freedom for the standard test is also affected by the added Laplace noises. To construct a rejection region and compute the exact type I and type II errors merits more effort in this case. It is part of our future work.

3.1   One Sample z Test

Assume n samples $Y_1, Y_2, ..., Y_n$ i.i.d $\sim N(\mu, \sigma^2)$, where σ^2 is known. The null hypothesis is $H_0 : \mu = \mu_0$. We consider the common two-sided alternative hypothesis $H_a : \mu \neq \mu_0$ or the one-sided $H_a : \mu > \mu_0$ and $H_a : \mu < \mu_0$.

The test statistic is based on the noise added sample mean. $\bar{Y}^a = \bar{Y} + r$, where $r \sim \text{Laplace}(\lambda)$. The test statistic under differential privacy is

$$Z = \frac{\bar{Y}^a - \mu_0}{\sigma/\sqrt{n}}.$$

$\bar{Y}^a$ follows a Gaussian-Laplace mixture distribution, $\mathrm{GL}(\mu, \sigma^2, n, \lambda)$. It has the cumulative distribution function (CDF) $F_a(y|\mu)$ as follows.

$$
\begin{aligned}
F_a(y|\mu) = {} & \Phi\left(\frac{y-\mu}{\sigma/\sqrt{n}}\right) + \frac{1}{2}\exp\{\frac{y-\mu}{\lambda} + \frac{\sigma^2}{2n\lambda^2}\}\Phi\left(-\frac{y-\mu}{\sigma/\sqrt{n}} - \frac{\sigma}{\lambda\sqrt{n}}\right) \\
& - \frac{1}{2}\exp\{\frac{-y+\mu}{\lambda} + \frac{\sigma^2}{2n\lambda^2}\}\Phi\left(\frac{y-\mu}{\sigma/\sqrt{n}} - \frac{\sigma}{\lambda\sqrt{n}}\right),
\end{aligned}
\tag{2}
$$

where $\Phi(\cdot)$ is the CDF of the unit Gaussian distribution.

We can easily derive the distribution of the test statistic under the null value and an alternative value by re-scaling $\bar{Y}^a$. However for the one sample z test the computation of the exact type I and type II errors can be done in a simpler fashion. Here and for the rest of this section we show the exact type I and type II errors for the two-sided alternative $H_a : \mu \neq \mu_0$. The results for the one-sided alternatives can be derived similarly.

Let α be the significance level of the test. Let $z_{\frac{\alpha}{2}}$ be the $(1 - \frac{\alpha}{2})$ quantile of the unit Gaussian distribution (i.e., the upper quantile). α and β are the type I and type II errors for the standard test, without the added Laplace noise. For the test under differential privacy, we have the exact type I error, α^a, and type II error, β^a, as follows.

$$
\begin{aligned}
\alpha^a & = P\left(\left|\frac{\bar{Y}^a - \mu_0}{\sigma/\sqrt{n}}\right| > z_{\frac{\alpha}{2}}|H_0\right) = 1 - F_a\left(\mu_0 + z_{\frac{\alpha}{2}}\frac{\sigma}{\sqrt{n}} \mid \mu_0\right) + F_a\left(\mu_0 - z_{\frac{\alpha}{2}}\frac{\sigma}{\sqrt{n}} \mid \mu_0\right) \\
& = \alpha + e^{\frac{-z_{\frac{\alpha}{2}}\sigma}{\lambda\sqrt{n}} + \frac{\sigma^2}{2n\lambda^2}}\Phi\left(z_{\frac{\alpha}{2}} - \frac{\sigma}{\lambda\sqrt{n}}\right) - e^{\frac{z_{\frac{\alpha}{2}}\sigma}{\lambda\sqrt{n}} + \frac{\sigma^2}{2n\lambda^2}}\Phi\left(-z_{\frac{\alpha}{2}} - \frac{\sigma}{\lambda\sqrt{n}}\right),
\end{aligned}
$$

$$
\begin{aligned}
\beta^a = {} & P\left(\left|\frac{\bar{Y}^a - \mu_0}{\sigma/\sqrt{n}}\right| < z_{\frac{\alpha}{2}}|H_a\right) = F_a\left(\mu_0 + z_{\frac{\alpha}{2}}\frac{\sigma}{\sqrt{n}} \mid \mu_a\right) - F_a\left(\mu_0 - z_{\frac{\alpha}{2}}\frac{\sigma}{\sqrt{n}} \mid \mu_a\right) \\
= {} & \beta + \frac{1}{2}\exp\{\frac{-z_{\frac{\alpha}{2}}\sigma}{\lambda\sqrt{n}} + \frac{\mu_0 - \mu_a}{\lambda} + \frac{\sigma^2}{2n\lambda^2}\} \times \Phi\left(-z_{\frac{\alpha}{2}} + \frac{\mu_0 - \mu_a}{\sigma/\sqrt{n}} + \frac{\sigma}{\lambda\sqrt{n}}\right) \\
& + \frac{1}{2}\exp\{\frac{z_{\frac{\alpha}{2}}\sigma}{\lambda\sqrt{n}} - \frac{\mu_0 - \mu_a}{\lambda} + \frac{\sigma^2}{2n\lambda^2}\} \times \Phi\left(-z_{\frac{\alpha}{2}} + \frac{\mu_0 - \mu_a}{\sigma/\sqrt{n}} - \frac{\sigma}{\lambda\sqrt{n}}\right) \\
& - \frac{1}{2}\exp\{\frac{\sigma^2}{2n\lambda^2} + \frac{\mu_0 - \mu_a}{\lambda} - \frac{z_{\frac{\alpha}{2}}\sigma}{\lambda\sqrt{n}}\} + \frac{1}{2}\exp\{\frac{\sigma^2}{2n\lambda^2} + \frac{\mu_0 - \mu_a}{\lambda} + \frac{z_{\frac{\alpha}{2}}\sigma}{\lambda\sqrt{n}}\} \\
& - \frac{1}{2}\exp\{\frac{z_{\frac{\alpha}{2}}\sigma}{\lambda\sqrt{n}} + \frac{\mu_0 - \mu_a}{\lambda} + \frac{\sigma^2}{2n\lambda^2}\} \times \Phi\left(z_{\frac{\alpha}{2}} + \frac{\mu_0 - \mu_a}{\sigma/\sqrt{n}} + \frac{\sigma}{\lambda\sqrt{n}}\right) \\
& - \frac{1}{2}\exp\{\frac{-z_{\frac{\alpha}{2}}\sigma}{\lambda\sqrt{n}} - \frac{\mu_0 - \mu_a}{\lambda} + \frac{\sigma^2}{2n\lambda^2}\} \times \Phi\left(z_{\frac{\alpha}{2}} + \frac{\mu_0 - \mu_a}{\sigma/\sqrt{n}} - \frac{\sigma}{\lambda\sqrt{n}}\right).
\end{aligned}
$$

3.2   One Sample t Test

Assume n samples $Y_1, Y_2, ..., Y_n$ i.i.d $\sim N(\mu, \sigma^2)$, where σ^2 is unknown. The null hypothesis is $H_0 : \mu = \mu_0$. The common alternative hypotheses are $H_a : \mu \neq \mu_0$, $H_a : \mu > \mu_0$, or $H_a : \mu < \mu_0$. Suppose users query the sample mean and the sample variance. Then the test statistic involves two noise added sample statistics,

$$T^a = \frac{\bar{Y}^a - \mu_0}{S^a/\sqrt{n}},$$

where $Y^a = \bar{Y} + r_1$ with $r_1 \sim$ Laplace(λ_1), and $S^a = \sqrt{S^2 + r_2}$ with $r_2 \sim$ Laplace(λ_2).

To obtain the distribution of the test statistic under either the null value or an alternative value, we re-write the test statistic as

$$T^a = \frac{Z^a}{X^a}, \quad \text{where } Z^a = \frac{\bar{Y}^a - \mu}{\sigma/\sqrt{n}} + \frac{\mu - \mu_0}{\sigma/\sqrt{n}} \text{ and } X^a = \sqrt{(S^a)^2/\sigma^2}.$$

We obtain the distribution of Z^a by rescaling a Gaussian-Laplace mixture distribution. Similarly we obtain the distribution of X^a based on a Chi-Square-Laplace mixture distribution. Let $F_Z(z)$ be the CDF of Z^a and $f_X(x)$ be the PDF of X^a.

$$F_Z(z|\mu) = \Phi(z - \delta) + \frac{1}{2}\exp\{\frac{\sigma(z-\delta)}{\lambda_1\sqrt{n}} + \frac{\sigma^2}{2n\lambda_1^2}\} \times \Phi\left(-(z-\delta) - \frac{\sigma}{\lambda_1\sqrt{n}}\right)$$
$$- \frac{1}{2}\exp\{-\frac{\sigma(z-\delta)}{\lambda_1\sqrt{n}} + \frac{\sigma^2}{2n\lambda_1^2}\} \times \Phi\left((z-\delta) - \frac{\sigma}{\lambda_1\sqrt{n}}\right), \qquad (3)$$

where $\delta = \frac{\mu_0 - \mu}{\sigma/\sqrt{n}}$. δ equals to 0 under the null and does not equal to 0 under the alternative. The distribution of X^a does not depend on the mean.

$$f_X(x) = [\ 2xf_g(x^2|\frac{n-1}{2}, \theta_0) + \frac{\sigma^2 x}{\lambda_2}e^{-\frac{\sigma^2 x^2}{\lambda_2}}(\frac{\theta_1}{\theta_0})^{\frac{n-1}{2}}F_g(x^2|\frac{n-1}{2}, \theta_1)$$
$$- xe^{-\frac{\sigma^2 x^2}{\lambda_2}}(\frac{\theta_1}{\theta_0})^{\frac{n-1}{2}}f_g(x^2|\frac{n-1}{2}, \theta_1) + \frac{\sigma^2 x}{\lambda_2}e^{\frac{\sigma^2 x^2}{\lambda_2}}(\frac{\theta_2}{\theta_0})^{\frac{n-1}{2}}(1 - F_g(x^2|\frac{n-1}{2}, \theta_2))$$
$$- xe^{\frac{\sigma^2 x^2}{\lambda_2}}(\frac{\theta_2}{\theta_0})^{\frac{n-1}{2}}f_g(x^2|\frac{n-1}{2}, \theta_2)\ ]\ /\ [1 - \frac{1}{2}(\frac{\theta_2}{\theta_0})^{\frac{n-1}{2}}] \qquad (4)$$

where $\theta_0 = \frac{2}{n-1}$, $\theta_1 = \frac{2}{n-1-2\sigma^2/\lambda_2}$, $\theta_2 = \frac{2}{n-1+2\sigma^2/\lambda_2}$, and F_g and f_g are the CDF and PDF of a gamma distribution respectively.

The distribution of the test statistic T^a given mean μ is

$$F_T(t|\mu) = \begin{cases} \int_0^\infty F_Z(tx|\mu)f_X(x)\,dx & t \geq 0 \\ \int_0^\infty (1 - F_Z(tx|\mu))\,f_X(x)\,dx & t < 0 \end{cases} \qquad (5)$$

Let $t_{\frac{\alpha}{2}, n-1}$ be the $(1 - \frac{\alpha}{2})$ quantile of a t distribution with $n-1$ degrees of freedom. The exact type I and type II errors can be computed numerically. Again we just show α^a and β^a under the two sided alternative. Similarly we can obtain the revised errors for the one sided alternatives.

$$\alpha^a = P\left(|T^a| > t_{\frac{\alpha}{2}, n-1}\big|\mu = \mu_0\right) = 1 - F_T(t_{\frac{\alpha}{2}, n-1}|\mu_0) + F_T(-t_{\frac{\alpha}{2}, n-1}|\mu_0),$$

$$\beta^a = P\left(|T^a| < t_{\frac{\alpha}{2}, n-1}\big|\mu = \mu_a\right) = F_T\left(t_{\frac{\alpha}{2}, n-1}|\mu_a\right) - F_T\left(-t_{\frac{\alpha}{2}, n-1}|\mu_a\right).$$

3.3   Two Sample t Test with Equal Variance

Assume n_1 samples $Y_1^1, Y_2^1, ..., Y_{n_1}^1$ i.i.d $\sim N(\mu_1, \sigma^2)$, n_2 samples $Y_1^2, Y_2^2, ..., Y_{n_2}^2$ i.i.d $\sim N(\mu_2, \sigma^2)$, where σ^2 is unknown. The null hypothesis is $H_0 : \mu_1 - \mu_2 = 0$. The common alternative hypotheses are $H_a : \mu_1 - \mu_2 \neq 0$, $H_a : \mu_1 - \mu_2 > 0$, or $H_a : \mu_1 - \mu_2 < 0$.

Suppose users query the sample means and the sample variances. Then the test statistic involves multiple noise added sample statistics.

$$T^a = \frac{\bar{Y}_1^a - \bar{Y}_2^a}{S^a \sqrt{\frac{1}{n_1} + \frac{1}{n_2}}},$$

where $\bar{Y}_1^a = \bar{Y}_1 + r_1$, $\bar{Y}_2^a = \bar{Y}_2 + r_2$, and $S^a = \sqrt{\frac{(n_1-1)(S_1^2+r_3)+(n_2-1)(S_2^2+r_4)}{n_1+n_2-2}}$, with $r_i \sim \text{Laplace}(\lambda_i)$, $i = 1 \sim 4$. We re-write the test statistic as

$$T^a = \frac{Z^a}{X^a}, \text{ where } Z^a = \frac{\bar{Y}_1^a - \bar{Y}_2^a - (\mu_1 - \mu_2)}{\sigma \sqrt{\frac{1}{n_1} + \frac{1}{n_2}}} + \frac{(\mu_1 - \mu_2)}{\sigma \sqrt{\frac{1}{n_1} + \frac{1}{n_2}}} \text{ and } X^a = \frac{S^a}{\sigma}.$$

Since the Laplace noises are added independently, we can then obtain the distribution of the numerator by convoluting Gaussian and Laplace distributions. The distribution of X^a is based on convolution of chi-square and Laplace distributions. The distributions of Z^a and X^a depend on the Laplace noise parameters λ_i, $i = 1 \sim 4$. We obtain their distributions under two separate cases. Let $v = n_1 + n_2 - 2$. Let $\delta = \frac{\mu_1 - \mu_2}{\sigma \sqrt{\frac{1}{n_1} + \frac{1}{n_2}}}$. δ equals to 0 under H_0 and is non-zero under H_a.

Distribution of Z^a, $\lambda_1 \neq \lambda_2$: We have the CDF

$$F_Z(z|\mu_1 - \mu_2) = \Phi(z - \delta) - \frac{\lambda_2^2}{2(\lambda_1^2 - \lambda_2^2)} e^{\tau_2(z-\delta) + \frac{\tau_2^2}{2}} (1 - \Phi(z - \delta + \tau_2))$$

$$+ \frac{\lambda_2^2}{2(\lambda_1^2 - \lambda_2^2)} e^{\frac{\tau_2^2}{2} - \tau_2(z-\delta)} \Phi(z - \delta - \tau_2) + \frac{\lambda_1^2}{2(\lambda_1^2 - \lambda_2^2)} e^{\frac{\tau_1^2}{2} + \tau_1(z-\delta)} (1 - \Phi(z - \delta + \tau_1))$$

$$- \frac{\lambda_1^2}{2(\lambda_1^2 - \lambda_2^2)} e^{\frac{\tau_1^2}{2} - \tau_1(z-\delta)} \Phi(z - \delta - \tau_1) \tag{6}$$

where $\tau_1 = \sigma \sqrt{\frac{1}{n_1} + \frac{1}{n_2}}/\lambda_1$, and $\tau_2 = \sigma \sqrt{\frac{1}{n_1} + \frac{1}{n_2}}/\lambda_2$.

Distribution of Z^a, $\lambda_1 = \lambda_2$: We have the CDF

$$F_Z(z|\mu_1 - \mu_2) = \Phi(z - \delta) - \left(\frac{1}{2} + \frac{\tau(z - \delta)}{4} - \frac{\tau^2}{4}\right) e^{\frac{\tau^2}{2} - \tau(z-\delta)} \Phi(z - \delta - \tau)$$

$$- \frac{\tau}{4\sqrt{2\pi}} e^{\frac{\tau^2}{2} - \tau(z-\delta) - \frac{(z-\delta-\tau)^2}{2}} + \frac{\tau}{4\sqrt{2\pi}} e^{\frac{\tau^2}{2} + \tau(z-\delta) - \frac{(z-\delta+\tau)^2}{2}}$$

$$+ \left(\frac{1}{2} - \frac{\tau(z - \delta)}{4} - \frac{\tau^2}{4}\right) e^{\frac{\tau^2}{2} + \tau(z-\delta)} (1 - \Phi(z - \delta + \tau)) \tag{7}$$

where $\tau = \sigma\sqrt{\frac{1}{n_1} + \frac{1}{n_2}}/\lambda_1$.

Distribution of X^a, $\lambda_3 \neq \lambda_4$: It does not depend on $\mu_1 - \mu_2$. Note $v = n_1 + n_2 - 2$. We have the PDF

$$
\begin{aligned}
f_X(x) = &[2x f_G(x^2; \tfrac{v}{2}, \theta_0) + \frac{b_2^2}{b_2^2 - b_1^2} e^{-b_1 x^2} (b_1 x)(\frac{\theta_1}{\theta_0})^{\frac{v}{2}} F_G(x^2; \tfrac{v}{2}, \theta_1) \\
&- \frac{b_2^2 x}{b_2^2 - b_1^2} e^{-b_1 x^2} (\frac{\theta_1}{\theta_0})^{\frac{v}{2}} f_G(x^2; \tfrac{v}{2}, \theta_1) - \frac{b_1^2}{b_2^2 - b_1^2} e^{-b_2 x^2} (b_2 x)(\frac{\theta_2}{\theta_0})^{\frac{v}{2}} F_G(x^2; \tfrac{v}{2}, \theta_2) \\
&+ \frac{b_1^2 x}{b_2^2 - b_1^2} e^{-b_2 x^2} (\frac{\theta_2}{\theta_0})^{\frac{v}{2}} f_G(x^2; \tfrac{v}{2}, \theta_2) + \frac{b_2^2}{b_2^2 - b_1^2} e^{b_1 x^2} (b_1 x)(\frac{\theta_3}{\theta_0})^{\frac{v}{2}} (1 - F_G(x^2; \tfrac{v}{2}, \theta_3)) \\
&- \frac{b_2^2 x}{b_2^2 - b_1^2} e^{b_1 x^2} (\frac{\theta_3}{\theta_0})^{\frac{v}{2}} f_G(x^2; \tfrac{v}{2}, \theta_3) - \frac{b_1^2}{b_2^2 - b_1^2} e^{b_2 x^2} (b_2 x)(\frac{\theta_4}{\theta_0})^{\frac{v}{2}} (1 - F_G(x^2; \tfrac{v}{2}, \theta_4)) \\
&+ \frac{b_1^2 x}{b_2^2 - b_1^2} e^{b_2 x^2} (\frac{\theta_4}{\theta_0})^{\frac{v}{2}} f_G(x^2; \tfrac{v}{2}, \theta_4)]/[1 - \frac{b_2^2}{2(b_2^2 - b_1^2)} (\frac{\theta_3}{\theta_0})^{\frac{v}{2}} + \frac{b_1^2}{2(b_2^2 - b_1^2)} (\frac{\theta_4}{\theta_0})^{\frac{v}{2}}]
\end{aligned}
$$

where $\tau_1 = \sigma\sqrt{\frac{1}{n_1} + \frac{1}{n_2}}/\lambda_1$, $\tau_2 = \sigma\sqrt{\frac{1}{n_1} + \frac{1}{n_2}}/\lambda_2$, $b_1 = \frac{(n_1+n_2-2)\sigma^2}{(n_1-1)\lambda_3}$, $b_2 = \frac{(n_1+n_2-2)\sigma^2}{(n_2-1)\lambda_4}$, $\theta_0 = \frac{2}{n_1+n_2-2}$, $\theta_1 = \frac{2}{n_1+n_2-2-2b_1}$, $\theta_2 = \frac{2}{n_1+n_2-2-2b_2}$, $\theta_3 = \frac{2}{n_1+n_2-2+2b_1}$, $\theta_4 = \frac{2}{n_1+n_2-2+2b_2}$.

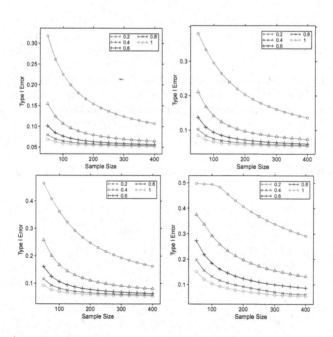

Fig. 1. Exact type I errors for increasing sample size n and five εs: 0.2, 0.4, 0.6, 0.8, and 1. Top left is one sample z test; top right is one sample t test; bottom left is two sample t test with equal sample size and equal variance; bottom right is two sample t test with unequal sample sizes and equal variance.

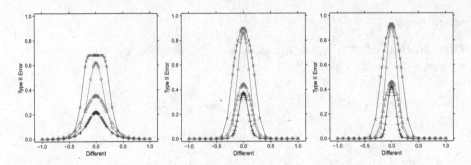

Fig. 2. Red line X is for one sample z test; blue line o is for one sample t test; pink line triangle is for two sample t test with equal sample size and equal variance; green line + is for two sample t test with unequal sample size and equal variance. $n = 50$. Left: $\varepsilon = 0.2$; Middle: $\varepsilon = 0.6$; Right: $\varepsilon = 1$. (Color figure online)

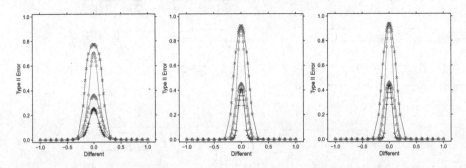

Fig. 3. Red line X is for one sample z test; blue line o is for one sample t test; pink line triangle is for two sample t test with equal sample size and equal variance; green line + is for two sample t test with unequal sample size and equal variance. $n = 100$. Left: $\varepsilon = 0.2$; Middle: $\varepsilon = 0.6$; Right: $\varepsilon = 1$. (Color figure online)

Distribution of X^a, $\lambda_3 = \lambda_4$: Again, it does not depend on $\mu_1 - \mu_2$. We have the PDF

$$
\begin{aligned}
f_X(x) = [\ & 2xf_G(x^2; \frac{v}{2}, \theta_0) + (\frac{b^2x^3 + bx}{2})e^{-bx^2}(\frac{\theta_1}{\theta_0})^{\frac{v}{2}}F_G(x^2; \frac{v}{2}, \theta_1) \\
& - (\frac{2x + bx^3}{2})e^{-bx^2}(\frac{\theta_1}{\theta_0})^{\frac{v}{2}}f_G(x^2; \frac{v}{2}, \theta_1) - (\frac{b^2x}{2})e^{-bx^2}(\frac{\theta_1}{\theta_0})^{\frac{v+2}{2}}F_G(x^2; \frac{v+2}{2}, \theta_1) \\
& + (\frac{bx}{2})e^{-bx^2}(\frac{\theta_1}{\theta_0})^{\frac{v+2}{2}}f_G(x^2; \frac{v+2}{2}, \theta_1) + (\frac{bx - b^2x^3}{2})e^{bx^2}(\frac{\theta_2}{\theta_0})^{\frac{v}{2}}(1 - F_G(x^2; \frac{v}{2}, \theta_2)) \\
& - (\frac{2x - bx^3}{2})e^{bx^2}(\frac{\theta_2}{\theta_0})^{\frac{v}{2}}f_G(x^2; \frac{v}{2}, \theta_2) + (\frac{b^2x}{2})e^{bx^2}(\frac{\theta_2}{\theta_0})^{\frac{v+2}{2}}(1 - F_G(x^2; \frac{v+2}{2}, \theta_2)) \\
& - (\frac{bx}{2})e^{bx^2}(\frac{\theta_2}{\theta_0})^{\frac{v+2}{2}}f_G(x^2; \frac{v+2}{2}, \theta_2)\ ]/[\ 1 - \frac{1}{2}(\frac{\theta_2}{\theta_0})^{\frac{v}{2}} - \frac{b}{4}(\frac{\theta_2}{\theta_0})^{\frac{v+2}{2}}\ ]
\end{aligned}
$$

where $b = 2\sigma^2/\lambda_3$, $\theta_0 = \frac{2}{n_1 + n_2 - 2}$, $\theta_1 = \frac{2}{n_1 + n_2 - 2 - b}$, and $\theta_2 = \frac{2}{n_1 + n_2 - 2 + b}$.

Given the Laplace noise parameters λ_i, we select the CDF and PDF of Z^a and X^a respectively. The distribution of the test statistic T^a given the value of $\mu_1 - \mu_2$ follows Eq. 5. Let $t_{\frac{\alpha}{2},v}$ be the $(1 - \frac{\alpha}{2})$ quantile of a t distribution with v degrees of freedom. The exact type I and type II errors again can be computed numerically. We show α^a and β^a under the two sided alternative. Similarly we can obtain the revised errors for the one sided alternatives. Let $\delta = \mu_1 - \mu_2$.

$$\alpha^a = P\left(|T^a| > t_{\frac{\alpha}{2},v} \mid \delta = 0\right) = 1 - F_T(t_{\frac{\alpha}{2},v} \mid \delta = 0) + F_T(-t_{\frac{\alpha}{2},v} \mid \delta = 0),$$

$$\beta^a = P\left(|T^a| < t_{\frac{\alpha}{2},v} \mid \delta \neq 0\right) = F_T(t_{\frac{\alpha}{2},v} \mid \delta \neq 0) - F_T(-t_{\frac{\alpha}{2},v} \mid \delta \neq 0).$$

3.4 Experimental Evaluation

To examine when the exact type I and II errors are less reliable, we run a set of experiments and provide the results in the following tables and figures. For all the experiments we set $\alpha = 0.05$, increase sample size n from 50 to 400 by steps of 25, and examine five ε values, 0.2, 0.4, 0.6, 0.8 and 1. $\lambda = 1/(n_i \varepsilon)$.

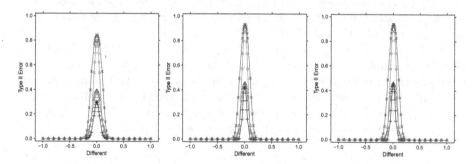

Fig. 4. Red line X is for one sample z test; blue line o is for one sample t test; pink line triangle is for two sample t test with equal sample size and equal variance; green line + is for two sample t test with unequal sample size and equal variance. $n = 200$. Left: $\varepsilon = 0.2$; Middle: $\varepsilon = 0.6$; Right: $\varepsilon = 1$. (Color figure online)

In Table 1, we show the exact type I errors for selected sample size n: 50, 100, 200, 300, and 400. Figure 1 shows the exact type I errors for the tests with increasing sample size n. As sample size increases and ε becomes larger, the exact type I errors is approaching $\alpha = 0.05$. Considering the exact type I error only, when users construct a test statistic with noise added mean and variance, the sample size needs to 100 or larger to provide a reliable result for moderate to small noise. For large noise, i.e. $\varepsilon \leq 0.2$, the sample size needs to be 400 or larger for a reliable test.

Figures 2, 3, 4 and 5 show the type II errors with noise added query results for selected n: 50, 100, 200, 400 and ε: 0.2, 0.6, 1. Hypothesis tests often operate with far less samples than classification, since the test is always significant for

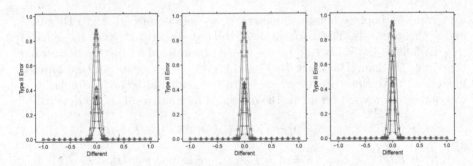

Fig. 5. Red line X is for one sample z test; blue line o is for one sample t test; pink line triangle is for two sample t test with equal sample size and equal variance; green line + is for two sample t test with unequal sample size and equal variance. $n = 400$. Left: $\varepsilon = 0.2$; Middle: $\varepsilon = 0.6$; Right: $\varepsilon = 1$. (Color figure online)

large dataset. For the tests considered in this article, the type I errors based on noise added query results decrease sharply as sample size increases. Type II error depends on the difference between the true value and the hypothesized value. The type II error under differential privacy also improves significantly as sample size increases and ε becomes larger.

Notice users cannot know how much noises are added to the query results. Small noises can cause major distortion to the test results. We must apply differential privacy query results with caution in hypothesis tests. Often users have only a handful or a few dozen samples in a test, the direct noise addition makes the test result unreliable. With very small datasets, users need the clean query results or direct access to the raw data for a reliable output.

4   Differentially Private Bayesian Classifier for Gaussian Mixture Models

Let database $D = \{X_1, \ldots, X_d, W\}$, where W is a binary class label, $Dom(W) = \{w_1, w_2\}$, and each X_i, $1 \leq i \leq d$ is a continuous attribute. A Bayesian classifier has the following decision rule:

Assign a record $\mathbf{x}$ to w_1 if $P(w_1|\mathbf{x}) > P(w_2|\mathbf{x})$; otherwise assign it to w_2.

The probabilities $P(w_i|\mathbf{x})$ can be calculated as: $P(w_i|\mathbf{x}) = p(\mathbf{x}|w_i)P(w_i)/p(\mathbf{x})$. If $p(\mathbf{x}|w_i)$ follows multivariate Gaussian distribution, it is known as the Gaussian mixture model [4]. For each class w_i, its mean μ_i and the variance-covariance matrix Σ_i of $p(\mathbf{x}|w_i) \sim N(\mu_i, \Sigma_i)$ are estimated from the data set D. For binary case, the Bayes error (i.e., the classification error) is calculated as [4]:

$$\text{Bayes Error} = \int_{\mathscr{R}_1} p(\mathbf{x}|w_2)P(w_2)d\mathbf{x} + \int_{\mathscr{R}_2} p(\mathbf{x}|w_1)P(w_1)d\mathbf{x}.$$

Table 1. (a) Z test type I error with added noises. $\sigma = 0.5$. (b) One sample t test type I error with added noises. $\sigma = 0.4$. (c) Two sample t test with equal sample size type I error with added noises. $\sigma_1 = \sigma_2 = 0.35$. $n_1 = n_2 = n$. (d) Two sample t test with unequal sample size type I error with added noises. $\sigma_1 = \sigma_2 = 0.2$. $n_1 = n$ and $n_2 = 1.1n$.

n	$\varepsilon = 0.2$	$\varepsilon = 0.4$	$\varepsilon = 0.6$	$\varepsilon = 0.8$	$\varepsilon = 1$	n	$\varepsilon = 0.2$	$\varepsilon = 0.4$	$\varepsilon = 0.6$	$\varepsilon = 0.8$	$\varepsilon = 1$
a						b					
50	0.3177	0.1542	0.1011	0.0794	0.0689	50	0.3805	0.2109	0.1373	0.1023	0.0841
100	0.2251	0.1070	0.0762	0.0647	0.0594	100	0.2953	0.1415	0.0935	0.0746	0.0657
200	0.1542	0.0794	0.0631	0.0573	0.0546	200	0.2035	0.0968	0.0711	0.0618	0.0575
300	0.1239	0.0697	0.0587	0.0548	0.0531	300	0.1606	0.0813	0.0639	0.0578	0.0549
400	0.1070	0.0647	0.0565	0.0536	0.0523	400	0.1363	0.0734	0.0603	0.0559	0.0537
c						d					
50	0.4645	0.2576	0.1612	0.1157	0.0924	50	0.4977	0.3748	0.2726	0.1975	0.1518
100	0.3609	0.1673	0.1057	0.0815	0.0701	100	0.4944	0.2920	0.1844	0.1319	0.1039
200	0.2472	0.1108	0.0774	0.0653	0.0597	200	0.4066	0.1976	0.1236	0.0922	0.0744
300	0.1936	0.0907	0.0681	0.0601	0.0564	300	0.3376	0.1557	0.1001	0.0714	0.0602
400	0.1627	0.0805	0.0634	0.0575	0.0542	400	0.2917	0.1320	0.0871	0.0613	0.0549

$\mathscr{R}_1$ is the region where records are labeled as class 1, and $\mathscr{R}_2$ is the region where records are labeled as class 2.

In this article we examine Bayes error for Gaussian mixture models under differential privacy protection. The database D only needs to return the following for users to build a Bayesian classifier:

- The sample size in D, which has sensitivity 0,
- The proportions of the two classes, i.e., $P(w_1)$ and $P(w_2)$,
- For each category, mean μ_i and variance-covariance Σ_i of the multivariate Gaussian distribution for $p(\mathbf{x}|w_i)$.

Bounded variables fit well into differential privacy mechanism. With unbounded variables one very large or small record can cause a significant increase the sensitivity. Notice Gaussian distribution is unbounded. Hence we work with truncated Gaussian distribution over interval $[\mu - 6\sigma, \mu + 6\sigma]$, a probability range of 0.999999998. Truncated Gaussian has density $I_{\{\mu - 6\sigma \leq x \leq \mu + 6\sigma\}}(x) \frac{f(x)}{\Phi(6) - \Phi(-6)}$.

4.1   Repair Noise Added Variance-Covariance Matrix

Let $\hat{\Sigma} = (\hat{\sigma}_{ij})_{d \times d}$ be the sample variance-covariance matrix. When users query variances and covariances separately, independent Laplace noises are added to every element of $\hat{\Sigma}$. Let $A = (r_{ij})_{d \times d}$ be the matrix of independent Laplace noises, where $r_{ij} = r_{ji}$. The returned query result is $\Sigma_Q = \hat{\Sigma} + A$.

Σ_Q is the noise added variance-covariance matrix, which is the results that users can easily obtain to test their model. Σ_Q is still symmetric but seize to be positive definite. In order to have a valid variance-covariance matrix, we repair the noise added variance-covariance matrix, and obtain a positive definite matrix Σ_+ close to Σ_Q, since $\hat{\Sigma}$ is not disclosed to users under differential privacy.

Let (l_j, e_j), $j = 1, ..., d$ be the eigenvalue and eigenvector pairs of Σ_Q, where the eigenvalues follow the decreasing order, $l_1 \geq l_2 \geq ... \geq l_d$. The last several eigenvalues of Σ_Q are negative. Let $l_k, ..., l_d$ be the negative eigenvalues. The positive definite matrix Σ_+ has eigenvalue and eigenvector pairs as the following: $(l_1, e_1), ..., (l_{k-1}, e_{k-1}), (l_k^+, e_k), ..., (l_d^+, e_d)$. We keep the eigenvectors, and use an optimization algorithm to search over positive eigenvalues to find a Σ_+ that minimizes the determinant of $\Sigma_+ - \Sigma_Q$.

$$(l_k^+, ..., l_d^+) = \mathrm{argmin}\, |\Sigma_+ - \Sigma_Q|.$$

Let $E_j = e_j e_j'$, $j = 1, ..., d$. We have

$$\Sigma_+ - \Sigma_Q = \sum_{j=k}^{d} (l_j^+ - l_j) E_j.$$

Therefore we perform a fine grid search over wide intervals to obtain positive eigenvalues that

$$(l_k^+, ..., l_d^+) = \mathrm{argmin}_{\{w_k > 0, ..., w_d > 0\}} \Big| \sum_{j=k}^{d} (w_j - l_j) E_j \Big|.$$

4.2   Experimental Evaluation

We have conducted extensive experiments in this section. We consider binary classification scenario. To understand how differential privacy affects the Bayes error, we do not want to introduce any other errors. Note Gaussian distribution may not represent the underlying data accurately. To avoid additional errors due to modeling real data distribution inaccurately, we generate data sets from

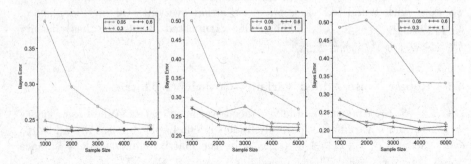

Fig. 6. Small training sample LDA Bayes error. Left: 2 dimension; Middle: 5 dimension; Right: 10 dimension.

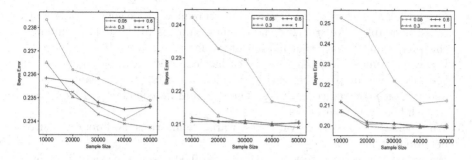

Fig. 7. Large training sample LDA Bayes error. Left: 2 dimension; Middle: 5 dimension; Right: 10 dimension.

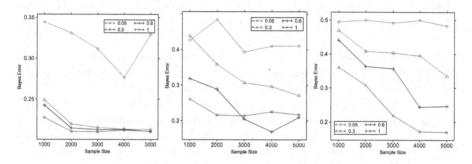

Fig. 8. Small training sample QDA Bayes error. Left: 2 dimension; Middle: 5 dimension; Right: 10 dimension.

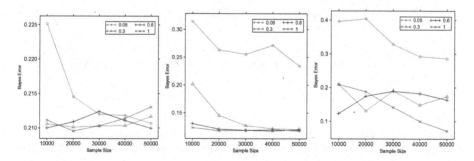

Fig. 9. Large training sample QDA Bayes error. Left: 2 dimension; Middle: 5 dimension; Right: 10 dimension.

known Gaussian mixture parameters. The parameters are estimated from real life data in two experiments, and synthetic in the rest.

In Eq. 4, if the two Gaussian distributions have the same variance-covariance matrix, we perform a linear discriminant analysis (LDA). If the two Gaussian distributions have different variance-covariance matrices, we perform a quadratic discriminant analysis (QDA). Every experimental run has the following steps.

1. Given the parameters of the Gaussian mixture models, we generate a training set of n samples. We truncate the training samples to the $\mu \pm 6\sigma$ interval, throwing away samples that fall out of the interval.
2. Using the truncated training set which has less than n samples, given a pre-specified ε, we compute the sensitivity values according to [22], sample means and variance-covariance matrices. Then we add independent Laplace noises to each Gaussian component.
3. We repair the noise added variance-covariance matrices, and obtain positive definite matrices.
4. We generate a separate test data set of size 50,000 using the original parameters without the noises, and report the effectiveness of the Gaussian mixture models using the noise added sample means and the positive definite matrices from the previous step. Test data set of size 50,000 is chosen to make sure that the estimated Bayes errors are accurate.

Experiment 1. We set $\mu_1 = 0.75 \times 1_d$ and $\mu_2 = 0.25 \times 1_d$, where 1_d is a d-dimensional vector with elements all equal to 1. The two d-dimensional Gaussian distributions have the same variance-covariance matrix Σ, where $\sigma_{ii} = 0.8^2$ and $\sigma_{ij} = 0.5 \times 0.8^2$. The prior is $p_1 = p_2 = 0.5$. We pool the two classes to estimate the sample variance-covariance matrix. We compute the sensitivity for variances and covariances adjusted to the range of the pooled data. The sample means and the sensitivity values for sample means are computed. We run the experiments in 2-dimension, 5-dimension, and 10-dimension, $d = 2, 5, 10$. We have four ε values, $\varepsilon = 0.05, 0.3, 0.6, 1$. Meanwhile we gradually increase the training set size.

Table 2. True LDA and QDA Bayes errors

	2-D	5-D	10-D		2-D	5-D	10-D
LDA Bayes error	0.2351	0.2100	0.1996	QDA Bayes error	0.2105	0.1170	0.0589

Using the prespecified parameter values, we have the true LDA classification rule, following Eq. 4. We generate 5 million samples using the prespecified parameter values without truncation, using the true LDA classification rules to estimate Bayes error. We take the average Bayes error of four such runs as the actual LDA Bayes error, shown in Table 2.

Figures 6 and 7 show the Bayes error under differential privacy for LDA experiment in increasing dimensions. For each combination (ε, n, d), we perform five runs. The average Bayes error of five runs is shown on the figures.

When two classes have the same variance-covariance matrix, the LDA Bayes error in general is not significantly affected by the noise added query results used in the classifier. For ε from 0.3 to 1, several thousand training samples are sufficient to return a preliminary Bayes error estimate which is very close to the actual LDA Bayes error. For this special case, we can obtain a fairly accurate idea about how well the LDA classifier performs using the noise added query results.

Experiment 2. We set $\mu_1 = 0.75 \times 1_d$ and $\mu_2 = 0.25 \times 1_d$. We set $\Sigma_1 = I_d$, where I_d is a d-dimensional identity matrix, and set Σ_2 as the one used in Experiment 1. We set the prior as $p_1 = p_2 = 0.5$. The sample means, variances, covariances, and the sensitivity values are computed. Again, we run the experiments in 2-dimension, 5-dimension, and 10-dimension, $d = 2, 5, 10$. We have four ε values, $\varepsilon = 0.05, 0.3, 0.6, 1$. Meanwhile we gradually increase the training set size.

Using the prespecified parameter values, we have the true QDA classification rule, following Eq. 4. We generate 5 million samples using the prespecified parameter values without truncation, using the true QDA classification rules to estimate Bayes error. We take the average Bayes error of four such runs as the actual QDA Bayes error, shown in Table 1.

Figures 8 and 9 show the Bayes error rate for QDA experiment in increasing dimensions. For each combination (ε, n, d), we perform five runs. The average Bayes error of five runs is shown on the figures.

When two classes have different variance-covariance matrices, dimensionality has a large impact on the Bayes error estimates obtained under differential privacy. For ε from 0.3 to 1, 2 dimensional experiment shows that three thousand training samples is sufficient to return a reasonable estimate of the actual Bayes error. 5 dimensional experiment needs 40,000 training samples to eliminate the impact of the added noises. 10 dimensional experiment needs even more training samples to return a reasonable estimate of the Bayes error under differential privacy.

Experiment 3. We used the Parkinson data set from the UCI Machine learning repository (https://archive.ics.uci.edu/ml/datasets/Parkinsons). We computed the mean and variance-covariance matrix of each class in the Parkinson data and used these parameters in our Gaussian mixture models. In all of the experiments, we set $\varepsilon = 0.6$. For the Parkinson data, the majority class equals to 75.38% of the total. There are 197 observations and 21 numerical variables besides the class label. Without differential privacy mechanism, directly using the sample estimates, the Bayes error is less than 0.01. On the other hand, the Gaussian mixture models with increasing sample sizes under differential privacy have Bayes error decreasing from 0.246 to 0.198. The Bayes error 0.198 is obtained from 50,000 training samples. The above results confirm that direct noise addition to Gaussian mixture parameters could cause significant distortion in higher dimensional space when two classes have different variance-covariance matrices. As dimensionality increases, we need a very large number of training samples to reduce the impact of the added noises.

Experiment 4. We also used the Adult data set from the UCI Machine learning repository (https://archive.ics.uci.edu/ml/datasets/Adult). The Adult data is much larger than the Parkinson data, with 32,561 observations. We used all the numerical variables in this experiment, i.e., 6 variables. We computed the mean and variance-covariance matrix of each class in the Adult data and used these parameters in our Gaussian mixture models. Again we set $\varepsilon = 0.6$. For the Adult data, the majority class equals to 75.92% of the total, similar to the Parkinson data. Without differential privacy mechanism, directly using the sample estimates, the Bayes error is 0.0309. With 50,000 training samples, the Gaussian mixture model under differential privacy has the Bayes error equal to 0.0747. The impact of the added noises is less severe for this lower dimensional data. Training sample size around 50,000 provides a reasonable result.

5   Summary

In this article we examine the performance of Bayesian classifier using the noise added mean and variance-covariance matrices. We also study the exact type I and type II errors under differential privacy for various hypothesis tests. In the process we identify an interesting issue associated with random noise addition: The variance-covariance matrix without the added noise is positive definite. However simply adding noise can only return a symmetric matrix, which is no longer positive definite. Consequently the query result cannot be used to construct a classifier. We implement a heuristic algorithm to repair the noise added matrix.

This is a general issue for random noise addition. Users may simply assemble basic query results without directly querying a complex statistic. Then adding noises causes the assembled result to no longer satisfy certain constraints. The query results need to be further modified in order to be used in subsequent studies.

References

1. Chaudhuri, K., Monteleoni, C., Sarwate, A.D.: Differentially private empirical risk minimization. J. Mach. Learn. Res. **12**, 1069–1109 (2011)
2. Chaudhuri, K., Sarwate, A.D., Sinha, K.: A near-optimal algorithm for differentially-private principal components. J. Mach. Learn. Res. **14**, 2905–2943 (2013)
3. Clifton, C., Tassa, T.: On syntactic anonymity and differential privacy. Trans. Data Priv. **6**(2), 161–183 (2013)
4. Duda, R.O., Hart, P.E., Stork, D.G.: Pattern Classification, 2nd edn. Wiley, New York (2001)
5. Dwork, C.: Differential privacy. In: Bugliesi, M., Preneel, B., Sassone, V., Wegener, I. (eds.) ICALP 2006. LNCS, vol. 4052, pp. 1–12. Springer, Heidelberg (2006). doi:10.1007/11787006_1
6. Dwork, C.: Differential Privacy: A Survey of Results. In: Agrawal, M., Du, D., Duan, Z., Li, A. (eds.) TAMC 2008. LNCS, vol. 4978, pp. 1–19. Springer, Heidelberg (2008). doi:10.1007/978-3-540-79228-4_1
7. Dwork, C., McSherry, F., Nissim, K., Smith, A.: Calibrating noise to sensitivity in private data analysis. In: Halevi, S., Rabin, T. (eds.) TCC 2006. LNCS, vol. 3876, pp. 265–284. Springer, Heidelberg (2006). doi:10.1007/11681878_14

8. Friedman, A., Schuster, A.: Data mining with differential privacy. In: KDD 2010: Proceedings of the 16th ACM SIGKDD International Conference on Knowledge Discovery and Data Mining, New York, NY, USA, pp. 493–502. ACM (2010)

9. Fukunaga, K.: Introduction to Statistical Pattern Recognition, 2nd edn. Academic Press Professional Inc., San Diego (1990)

10. Jagannathan, G., Pillaipakkamnatt, K., Wright, R.N.: A practical differentially private random decision tree classifier. In: ICDM Workshops, pp. 114–121 (2009)

11. Kapralov, M., Talwar, K.: On differentially private low rank approximation. In: SODA 2013: Proceedings of the Twenty-Fourth Annual ACM-SIAM Symposium on Discrete Algorithms, New Orleans, Louisiana, pp. 1395–1414. SIAM (2013)

12. Kifer, D., Smith, A., Thakurta, A.: Private convex empirical risk minimization and high-dimensional regression. J. Mach. Learn. Res. **23**, 1–41 (2012)

13. J. Lei. Differentially private M-estimators. In: Advances in Neural Information Processing Systems, pp. 361–369 (2011)

14. Pathak, M.A., Raj, B.: Large margin Gaussian mixture models with differential privacy. IEEE Trans. Dependable Secure Comput. **9**(4), 463–469 (2012)

15. McSherry, F., Mironov, I.: Differentially private recommender systems: building privacy into the Netflix prize contenders. In: KDD 2009: Proceedings of the 15th ACM SIGKDD International Conference on Knowledge Discovery and Data Mining, Paris, France, pp. 627–636. ACM (2009)

16. McSherry, F., Talwar, K.: Mechanism design via differential privacy. In: FOCS 2007: 48th Annual IEEE Symposium on Foundations of Computer Science, Providence, Rhode Island, pp. 94–103. IEEE (2007)

17. Samarati, P., Sweeney, L.: Generalizing data to provide anonymity when disclosing information. In: Proceedings of the 17th ACM SIGACT-SIGMOD-SIGART Symposium on Principles of Database Systems (PODS), Seattle, WA, USA, 1–3 June 1998

18. Samarati, P.: Protecting respondents identities in microdata release. IEEE Trans. Knowl. Data Eng. **13**(6), 1010–1027 (2001)

19. Rubinstein, B., Bartlett, P.L., Huang, L., Taft, N.: Learning in a large function space: privacy-preserving mechanisms for SVM learning. J. Priv. Confidentiality **4**(1), 65–100 (2012)

20. Sweeney, L.: k-anonymity: a model for protecting privacy. Int. J. Uncertainty Fuzziness Knowl. Based Syst. **10**(5), 557–570 (2002)

21. Vu, D., Slavkovic, A.: Differential privacy for clinical trial data: preliminary evaluations. In: IEEE 13th International Conference on Data Mining Workshops, Los Alamitos, CA, USA, pp. 138–143. IEEE (2009)

22. Xi, B., Kantarcioglu, M., Inan, A.: Mixture of Gaussian models and Bayes error under differential privacy. In: Proceedings of the First ACM Conference on Data and Application Security and Privacy, pp. 179–190. ACM (2011)

Differentially Private K-Skyband Query Answering Through Adaptive Spatial Decomposition

Ling Chen[1]([⊠]), Ting Yu[2], and Rada Chirkova[1]

[1] Department of Computer Science, North Carolina State University, Raleigh, USA
{lchen10,rychirko}@ncsu.edu
[2] Qatar Computing Research Institute, Hamad Bin Khalifa University, Doha, Qatar
tyu@qf.org.qa

Abstract. Given a set of multi-dimensional points, a k-skyband query retrieves those points dominated by no more than k other points. k-skyband queries are an important type of multi-criteria analysis with diverse applications in practice. In this paper, we investigate techniques to answer k-skyband queries with differential privacy. We first propose a general technique BBS-Priv, which accepts any differentially private spatial decomposition tree as input and leverages data synthesis to answer k-skyband queries privately. We then show that, though quite a few private spatial decomposition trees are proposed in the literature, they are mainly designed to answer spatial range queries. Directly integrating them with BBS-Priv would introduce too much noise to generate useful k-skyband results. To address this problem, we propose a novel spatial decomposition technique k-skyband tree specially optimized for k-skyband queries, which partitions data adaptively based on the parameter k. We further propose techniques to generate a k-skyband tree over spatial data that satisfies differential privacy, and combine BBS-Priv with the private k-skyband tree to answer k-skyband queries. We conduct extensive experiments based on two real-world datasets and three synthetic datasets that are commonly used for evaluating k-skyband queries. The results show that the proposed scheme significantly outperforms existing differentially private spatial decomposition schemes and achieves high utility when privacy budgets are properly allocated.

Keywords: k-skyband query · Differential privacy · Adaptive spatial decomposition

1 Introduction

Given a set of multi-dimensional points, a k-skyband query [30] identifies the set of points that are *dominated* by at most k other points. A point p dominates another point q if p is at least as good as q on all dimensions and strictly better

G. Livraga and S. Zhu (Eds.): DBSec 2017, LNCS 10359, pp. 142–163, 2017.
DOI: 10.1007/978-3-319-61176-1_8

than q in at least one dimension. A k-skyband query is a generalization of a skyline query [5,7,25]: when k is 0, a k-skyband query is just a skyline query. As an important type of preference queries, skyband queries and their variants [15,30] have wide applications in practice, e.g., location-based services [22] and service recommendations [23].

Similar to other data analysis tasks, directly releasing the results of k-skyband queries over sensitive data of individuals could result in privacy breach. For example, the presence (absence) of one point may cause a large set of points to be removed from (included in) the k-skyband results. Thus, by analyzing the output of k-skyband queries, an adversary may infer the presence or absence of an individual in the dataset, which could be very sensitive. Due to such potential privacy risks, a data owner may be reluctant to share k-skyband query results with collaborators or the public, even if such sharing could bring significant benefits.

In this paper, we develop techniques to answer k-skyband queries with differential privacy [10,11]. Unlike syntactic approaches such as k-anonymity [16,31], differential privacy provides a provable strong privacy guarantee that the output of a computation is insensitive to any particular individual. That is, an adversary has limited ability to make inference about whether an individual is present or absent in the dataset.

We first propose a general technique BBS-Priv, which accepts any differentially private spatial decomposition tree as input and leverages data synthesis to generate private k-skyband results. Specifically, in a spatial decomposition tree, an internal node contains the coordinates of a region, the number of data points within the region (referred as point count), and pointers to its child nodes (i.e., subregions) at the lower level. Given a spatial decomposition tree, such as private quad-tree or kd-tree [9], BBS-Priv adopts the branch-and-bound paradigm to progressively traverse nodes for dominance checking, and prunes internal nodes that do not contain k-skyband points, i.e., there is no need to access all the partitions. When reaching a node that could not be further pruned, BBS-Priv generate approximate k-skyband results using synthesized points based on the node's point count.

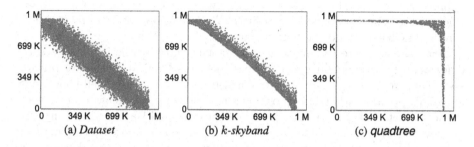

Fig. 1. Comparison of true and private k-skyband results with anti-correlated synthetic dataset when $k = 200$.

As several techniques have been proposed in the literature to generate private spatial decomposition trees [9], it seems that we can directly combine them with BBS-Priv to answer k-skyband queries privately. Unfortunately, such a straight-forward approach would significantly distort k-skyband query results. Figure 1a shows an example synthetic dataset following an anti-correlated distribution, which is commonly used in skyline query evaluation. Figure 1b shows the true k-skyband results, and Fig. 1c shows the private k-skyband results when combining BBS-Priv with a differentially private quadtree [9]. We see that private quadtree fails to sufficiently capture the properties of the dataset that are important to k-skyband queries, producing k-skyband results significantly different from the true results. There are two major reasons for such a poor performance. First, in k-skyband queries, the regions close to the upper-right corner are much more important than those lower-left regions, since these regions contain points with preferred values in all dimensions. Thus, it is much more desirable to accurately capture data distributions in upper-right regions than the lower-left ones. Existing spatial decomposition schemes are designed for spatial range queries and thus all regions are treated equally, and thus are not suitable to answer k-skyband queries. Second, existing schemes achieve differential privacy by perturbing the point count in each region. After such perturbation, some empty regions' point counts may become positive. If these regions are close to the upper-right corner, these noisy points would distort the k-skyband results significantly.

Based on the above observations, we develop k-skyband tree, a novel spatial decomposition technique, which partitions data space adaptively based on the parameter k. The insight of k-skyband tree is that not all the regions contribute equally to the k-skyband results (e.g., points in dominance regions do not contribute to the k-skyband results at all), and thus finer and more accurate decompositions should be performed on the regions that are likely to contain k-skyband results. Built on this insight, when choosing a splitting point to partition a region, k-skyband tree finds an appropriate upper-right region ne that contains more than k points, which guarantees that the points in the lower-left region sw can be safely pruned. The upper-right region ne gets finer decompositions in subsequent splittings. We further present a suite of techniques to publish a k-skyband tree privacy, and propose a post-processing technique to improve its accuracy by suppressing data synthesis in those empty partitions whose noisy point counts become positive. We can then feed BBS-Priv with the private k-skyband tree to compute private k-skyband results.

For evaluation, we conduct experiments over three synthetic datasets with different distributions and two real-world datasets [1,2], and compare private k-skyband trees with private quad-tree and private kd-tree, two well-known differentially private spatial decomposition schemes. Our results show that for synthetic datasets k-skyband tree outperforms quadtree and kd-tree when sufficient privacy budgets are allocated ($\epsilon > 0.5$). Further, our proposed technique significantly outperforms private quadtree and kd-tree in the two real datasets for all studied privacy budgets (ϵ ranging from 0.1 to 2.0). One key observation from our experiments is that, though the three synthetic datasets are commonly used in skyline query evaluation, they unfortunately do not capture the actual data distributions in real applications where k-skyband queries matter most. We observe that real datasets tend to have fan-shaped data distributions, where dominant points spread sparsely in dominance regions (e.g., the ne region in a 2-D space), while other inferior points more densely spread over other regions. Our scheme is adaptive enough to capture such distributions while existing spatial decomposition schemes fail to do so.

2 Preliminaries

2.1 Differential Privacy

Differential privacy [10] is a formal privacy model that guarantees the output of a query function to be insensitive to any particular record in the data set.

Definition 1 *(ε-differential privacy): Given any pair of neighboring databases D and D' that differ in at most one individual record, a randomized algorithm A is ϵ-differentially private iff for any $S \subseteq Range(A)$:*

$$Pr[A(D) \in S] \le Pr[A(D') \in S] * e^\epsilon$$

The parameter ϵ is often referred as the privacy budget in differential privacy, as it directly affects the level of privacy protection. Obviously, the smaller ϵ, the harder to distinguish between D and D' from the output of A, and thus the stronger the privacy protection.

The most common strategy to achieve ϵ-differential privacy is to add noise to the output of a function. The magnitude of the noise is calibrated by the privacy budget ϵ and the sensitivity of the query function f, which is defined as the maximum difference between the outputs of the query function f on any pair of neighboring databases:

$$\Delta f = \max_{D,D'} \parallel f(D) - f(D') \parallel_1$$

There are two common approaches for achieving ϵ-differential privacy: Laplace mechanism [12] and Exponential mechanism [26].

Laplace Mechanism: The output of a query function f is perturbed by adding noise from the Laplace distribution with probability density function $pdf(x|b) = \frac{1}{2b} \exp(-\frac{|x|}{b})$, $b = \frac{\Delta f}{\epsilon}$. The following randomized mechanism A_l satisfies ϵ-differential privacy:

$$A_l(D) = f(D) + Lap(\frac{\Delta f}{\epsilon})$$

Exponential Mechanism: This mechanism returns an output that is close to the optimum, with respect to a quality function. A quality function $q(D, r)$ assigns a score to all possible outputs $r \in R = range(f)$, and outputs closer to the true output receive higher scores. A randomized mechanism A_e that outputs $r \in R$ with probability

$$Pr[A_e(D) = r] \propto exp(\frac{\epsilon q(D, r)}{2 \Delta q})$$

satisfies ϵ-differential privacy, where Δq is the sensitivity of the quality function.

Differential privacy has two properties: *sequential composition* and *parallel composition*. Sequential composition is that given n independent randomized mechanisms $A_1, A_2, \ldots, A_n$ where A_i ($1 \le i \le n$) satisfies ϵ_i-differential privacy, a sequence of A_i over the dataset D satisfies ϵ-differential privacy, where $\epsilon = \sum_1^n \epsilon_i$. Parallel composition is that given n independent randomized mechanisms $A_1, A_2, \ldots, A_n$ where A_i ($1 \le i \le n$) satisfies ϵ-differential privacy, a sequence of A_i over a set of *disjoint data sets* D_i satisfies ϵ-differential privacy.

2.2   K-Skyband Queries

Given a d-dimensional data set D, a k-skyband query returns all the points in D that are dominated by at most k other points in D. The dominance relationship in k-skyband queries is defined as follows:

Definition 2 (Dominance). *Given two d-dimensional points $p = (u_1, \ldots, u_d)$ and $q = (v_1, \ldots, v_d)$, if for all $i = 1, \ldots, d$, $u_i \succeq v_i$ and $\exists j, u_j \succ v_j$, we say that p dominates q ($p \succ q$), where $\succ$ denotes better than and $\succeq$ denotes better than or equal to.*

In k-skyband queries, k represents the thickness of the skyband results, and a skyline query [5,30] is a special case of k-skyband queries when $k = 0$. k-skyband queries can be answered by extending algorithms for skyline queries, such as Branch-and-Bound Skyline (BBS) [30]. BBS is an efficient algorithm built on top of any spatial decomposition. A spatial decomposition is a hierarchical (tree) decomposition of a geometric space into smaller regions. In a spatial decomposition tree, an internal node stores the coordinates of a region, the number of points in that region (referred to as point count) and pointers to its child nodes, while the leaf nodes are individual data points.

Given a spatial decomposition tree, BBS traverses the nodes for dominance checking, and prunes internal nodes that are determined to contain no skyline points, i.e., not all the points will be accessed. It can be easily adapted to answer k-skyband queries by excluding a region if it is already dominated by k other points. In this paper, we focus on how to adapt BBS to answer k-skyband queries with differential privacy.

3   Approaches

3.1   BBS-Priv

BBS-Priv is inspired by the BBS algorithm [30]. BBS maintains a set S to keep track of all the k-skyband points discovered so far during the algorithm, and accesses the nodes in a spatial decomposition tree starting from the root node that covers the whole region. When a node n is accessed, if BBS finds more than k points in S that dominate n, n is pruned. Otherwise, BBS (1) inserts n into S if n is a leaf node that represents a point, or (2) expands n by accessing n's child nodes if n is an internal node. For n's child nodes, distances are computed according to L_1 norm, i.e., the *maxdist* of a point is the sum of its coordinates and the *maxdist* of an internal node is the *maxdist* of its upper-right corner point. These child nodes are inserted into a max heap that sorts nodes based on their *maxdist*, so that nodes with higher *maxdist* are accessed earlier. The intuition is that nodes with larger *maxdist* cannot be dominated by those with smaller *maxdist*, and thus BBS only needs to check the dominance relationship between the current accessed node and each point in S. The access order based on the max heap guarantees that points inserted into S are k-skyband points. BBS continues to access nodes from the max heap until the heap is empty, and returns points in S as the k-skyband results.

In BBS-Priv, the input is a differentially private spatial decomposition tree, whose leaf nodes are not individual points but a region. We can simply adapt BBS such that when reaching a leaf node e, if e is not pruned by points in S, we uniformly generate points in the region according to e's point count, treat each of them as a child of e, and continue. Due to space limit, we omit the detailed pseudocode of BBS-Priv.

Privacy Analysis. It is easy to see that BBS-Priv only conducts post processing of a spatial decomposition T. As long as T is constructed with differential privacy, BBS-Priv would provide the same privacy guarantee.

3.2 Differentially Private K-Skyband Tree

Although BBS-Priv can be combined with any existing differentially private spatial decomposition trees, as we will show in Sect. 4 later, the resulting k-skyband results are often highly distorted compared with the true results. The reason is that existing differentially private spatial decomposition schemes aim to capture the distribution of all the data. For example, in existing schemes, it is common that a dense region (with high point count) will be further partitioned. However, for k-skyband queries, we may not need to do so if that area is already dominated by more than k points. Similarly, for a sparse region, if it is close to the upper-right corner, we may still need to continue the partition as it is likely to contain k-skyband results and we need to better capture their distributions. Based on this observation, we propose a novel spatial decomposition algorithm k-skyband tree specifically tailored to answer k-skyband queries, and will show further how to build k-skyband tree with differential privacy. For simplicity and easy explanation, we present our scheme for handling spatial data (2-D data). It can be easily extended to handle multi-dimensional data, which we omit due to space limit.

K-Skyband Tree. The insight of k-skyband tree is to perform finer and more accurate decompositions on the regions that are more important for k-skyband results, i.e., the regions close to the upper-right corner. Note that when we say "upper-right corner", it is relative to the input dataset instead of to the whole space. Therefore, the partition of space must be dependent on the dataset. Specifically, k-skyband tree chooses a splitting point s such that there are more than k data points in the the upper-right region ne and all the data points falling into the dominance region sw can be excluded from the computation of k-skyband results. Details of k-skyband tree is presented at Appendix A. The main point of k-skyband tree is not to make query answering more efficient. Instead, it would guide fine-grained decomposition towards those regions that are likely to contain k-skyband results, so that when we add noise later to satisfy differential privacy, the distribution of points in those regions is preserved better, reducing the distortion of the true results.

Differentially Private K-Skyband Tree. To make the process of building k-skyband tree satisfy differential privacy, we need to revise the algorithm in the following major steps. First, as the tree will reveal the split points for each region (i.e., the coordinates of internal nodes), we need to make the splitting process differentially private. Specifically, we leverage the Exponential Mechanism to choose private values for the splitting point. For s_x (s_y), we divide the possible output range, $[x_{min}, x_{max}]([y_{min}, y_{max}])$, into intervals based on the ranks of the true data points, and assign higher probability to the intervals closer to s_x (s_y). Once an interval is chosen based on the Exponential Mechanism, a value is uniformly sampled from the interval to be the private value of the splitting point.

Second, the point count of each node in the tree will reveal the number of points in each region. To achieve differential privacy, we adopt the laplace mechanism to add noise. Specifically, we add Laplacian noise to the point count of each node, protecting the true count of the points falling into the split regions.

The pseudo code of building differentially private k-skyband tree is shown in Algorithm 1 and the function *splitByK* is shown in Algorithm 3 (in Appendix A). Besides the region r, k in k-skyband queries, and the max height h, the algorithm accepts as input the privacy budgets $\epsilon_0, \ldots, \epsilon_h$ for each level of the spatial decomposition tree and the split rate α used for computing the budget of choosing splitting points. In Algorithm 3, we apply the Exponential Mechanism to obtain noisy values for s_x and s_y (Lines 14–15), and use the obtained random values to form the noisy splitting point. In Algorithm 1, Line 1 and Lines 17–18 correspond to adding Laplacian noise to point counts.

The algorithm starts by adding Laplacian noise to the point count of the input region r, and adds the root node into the queue Q for splitting (Lines 1–2). The algorithm splits the node recursively until there are no nodes left in Q. For each node n to be split (Line 4), the algorithm first computes the budget ϵ_c for obtaining noisy count and k' based on k and ϵ_c (Line 8). The reason why we use $k' = k + 1 + \frac{\sqrt{2}}{\epsilon_c}$ instead of $k + 1$ is because (1) by adding Laplacian noise

---

Algorithm 1. Differentially Private k-skyband tree

Input: A region r, k in k-skyband queries, max height of the tree h, privacy budgets
for each level of the tree $\epsilon_0, \ldots, \epsilon_h$, split rate α
Output: A differentially private spatial decomposition tree T
1: $r.ncount = r.count + \text{Lap}(\frac{1}{\epsilon_0 * (1-\alpha)})$
2: $Q.\text{enqueue}(r)$
3: **while** Q is not empty **do**
4:     $n = Q.\text{dequeue}()$
5:     **if** isLeaf(n, h) **then**
6:         updateNoisyCount(n)
7:         **continue** // back to Line 3
8:     $l = n.level$, $\epsilon_c = \epsilon_{l+1} * (1 - \alpha)$, $k' = k + 1 + \frac{\sqrt{2}}{\epsilon_c}$
9:     **if** $n.ncount > k'$ and $n.parent.midPointSplit$ is $false$ **then**
10:         $N = \text{splitByK}(n, k', \epsilon_l * \alpha)$
11:         **if** $N.ne = n$ **then**
12:             $N = \text{splitByMidPoint}(n)$
13:             $n.midPointSplit = true$, $\epsilon_c = \epsilon_{l+1}$
14:     **else**
15:         $N = \text{splitByMidPoint}(n)$
16:         $n.midPointSplit = true$, $\epsilon_c = \epsilon_{l+1}$
17:     **for** $c \in N$ **do**
18:         $c.ncount = c.count + \text{Lap}(\frac{1}{\epsilon_c})$
19:     $Q.\text{enqueue}(N.ne, N.nw, N.se)$
20:     **if** $N.ne.ncount \leq k$ **then**
21:         $Q.\text{enqueue}(N.sw)$
22: **return** N

---

with mean 0 to k, there is 50 % of the chance that we would obtain a noisy value smaller than $k + 1$; (2) if the noisy count of ne is smaller than $k + 1$, then sw cannot be pruned and we need to further split sw; (3) the standard deviation of Laplacian noise based on ϵ_c is $\frac{\sqrt{2}}{\epsilon_c}$ and adding this standard deviation to k to obtain k' (i.e., making the count of ne to be slightly larger than k) can make the noisy count of ne more likely larger than k; Based on k', the algorithm splits the node n using the corresponding budget in the level of the node n (Line 10). After splitting, for each split node, the algorithm computes the noisy count based on the count budget (Lines 17–18). If the node is split using midpoint as quadtree, the splitting budget is saved and the count budget is updated (Lines 11–13 and 15–16), and all the children nodes are set to use midpoint split (Lines 13 and 16). If the noisy count of ne is less than or equal to k, the dominance region sw needs to be further split (Lines 20–21).

When a node's level reaches the maximum height of the tree (h), the node is considered as a leaf node and no further split is applied (Line 5). Also, if the noisy count of the node is too small (e.g., less than 8), further splitting the node will caused the counts of child nodes to be distorted significantly by the Laplacian noise, and thus the algorithm considers the node as a leaf node.

In this case, if the leaf node is not at the maximum level, the remaining budgets allocated for the rest of the levels are used to recompute the noisy count [9] (Line 6). The algorithm terminates when there are no nodes left in Q.

Obtaining Splitting Point with Differential Privacy. To protect the values of the splitting point, the algorithm leverages the Exponential Mechanism (EM) [26] to sample the noisy values.

Let $L = \{x_1, \ldots, x_m\}$ be a set of m values in ascending order in some domain range $[lo, hi]$, and let x_s be the desired value of the splitting point. Let $rank(x)$ denote the rank of x in L, representing the number of items in L that are smaller than x. The quality function fed into the EM [9] is:

$$q(L, x) = -|rank(x) - rank(x_s)|,$$

The EM returns x with $Pr[EM(L) = x] \propto e^{-\frac{\epsilon}{2}|rank(x) - rank(x_s)|}$ based on this quality function. Since all values x between two consecutive values in L have the same rank, they are equally likely to be chosen, which can be implemented using uniformly random sampling between the two consecutive values in L. Accordingly, EM can be implemented by choosing an output from the interval $I_k = [x_k, x_{k+1})$ with probability proportional to $|I_k|e^{-\frac{\epsilon}{2}|k - rank(x_s)|}$. Once an interval I_k is chosen, EM then returns a uniformly random value in I_k.

Privacy Analysis. Due to space limit, we provide next only a sketch of the proof of the privacy guarantee of private k-skyband tree. Note that, the construction process of k-skyband tree tree falls into the category of hybrid spatial decomposition as defined in [9], where in the first few levels it uses data-dependent split (i.e., SplitByK (Algorithm 3)) and in the remaining levels it uses data-independent split (i.e., through midpoints). Therefore the analysis of k-skyband tree's privacy guarantee is very similar to that in [9].

To construct a k-skyband tree with differential privacy, we need to combine the privacy guarantees of both tree structures and point counts. Note that adding or deleting a single data point changes the counts of all the nodes on the path from the root to the leaf containing that data point, and it could also affect the node splitting in the levels where data-dependent split is used. For a k-skyband tree with max height h, our algorithm assigns the privacy budget $\epsilon_0, \ldots, \epsilon_h$ for the nodes at level $0, \ldots, h$. To protect both node splitting results and point counts, for nodes at the level i $(0 \le i \le h)$, our algorithms uses the Exponential Mechanism with budget $\alpha\epsilon_i$ to obtain noisy splitting points and adds Laplacian noise with budget $(1 - \alpha)\epsilon_i$ to obtain noisy point counts. For the levels where data-independent split is used, there is no need to protect node splitting, and the budget for splitting is also used for computing noisy counts. If nodes n_1 and n_2 are not on the same root-to-leaf path, their point counts are independent of each other, and knowing the noisy counts of n_1 does not affect the privacy guarantees of n_2. Thus, based on the parallel composition property (Sect. 2.1), k-skyband tree at the level i satisfies ϵ_i-differential privacy. Further, based on the sequence composition property of differential privacy in Sect. 2.1, for $\epsilon = \epsilon_0 + \ldots + \epsilon_h$, the whole k-skyband tree satisfies ϵ-differential privacy.

Budget Allocation Strategy. For budget allocation, we follow the geometric scheme proposed in [9]. Let ϵ_i denote the budget for level i of the tree, and the budgets for each level is computed using $\epsilon_{i+1} = 2^{\frac{1}{3}}\epsilon_i$. The intuition is that for nodes at the levels closer to root, the point counts are larger and more resistant to noise, and more budgets should be allocated for levels close to leaves.

For a level i where data-dependent split is used, we allocate $\alpha\epsilon_i$ to compute the noisy splitting point and $(1-\alpha)\epsilon_i$ to obtain noisy point counts. Previous study [9] shows that a small portion of budget is enough for splitting, and we set $\alpha = 10\%$ based on our empirical evaluations. Such allocation is also consistent with the study on private kd-tree [9], which shows that finding a splitting point using Exponential Mechanism requires less budget than adding Laplacian noise to point counts.

Post Processing. Post processing is commonly used in differential privacy to improve utility [10,13]. For example, [9,19] leverages post-processing to improve count query accuracy. However, the utility of k-skyband queries do not solely depend on the accuracy of point counts, and the existing post-processing techniques optimized for count queries do not work properly for k-skyband queries. The reason is that k-skyband tree uses data-dependent split at the first few levels, and certain dominance regions of the noisy splitting point (i.e., sw regions) are excluded in the computation of k-skyband results. If we adjust the noisy counts of the nodes using existing post-processing techniques, the noisy counts of some ne regions may become less than k, making the corresponding sw regions not qualified for pruning. Due to such inconsistency between the region splitting and noisy counts, the properties of the dataset that are important to k-skyband queries cannot be sufficiently captured, further distorting the k-skyband results significantly.

Empirically, we observed that the major factor affecting the utility of private k-skyband results is to synthesize data points for the regions whose noisy counts are positive but their true counts are zero. If the regions of these nodes are close to the upper-right point (x_{max}, y_{max}), then the private k-skyband results are significantly distorted since the data points in these regions could dominate the data points in any other region.

To smooth such errors caused by data synthesis with Laplacian noise, we propose a novel post-processing technique. The insight is to not synthesize points for half of the empty leaf nodes, as Laplacian noise has a 50% of probability to be positive (or negative) and could turn half of the empty leaf nodes to have positive noisy counts. However, the number of empty leaf nodes partially reveals the data distribution of the data set. Therefore, instead of directly using the number of empty leaf nodes, we use the number of leaf nodes whose noisy count are negative to approximate the number of empty leaf nodes whose noisy count are positive.

Let C be the set of nodes whose true point counts are 0. Denote C_p as the subset of nodes in C whose noisy point counts are positive, and denote C_n as all the leaf nodes (not only those in C) whose noisy point counts are negative.

As Laplacian noise has a 50% probability to be positive, the expected size of C_p is $E[|C_p|] = \frac{|C|}{2}$. C_n includes two types of nodes: (1) the empty leaf nodes whose noisy counts are negative ($C_{n,1}$) and (2) the leaf nodes whose point counts are positive but noisy counts become negative ($C_{n,2}$). Ideally, we should use $|C_{n,1}|$ to approximate $|C_p|$. But $|C_{n,2}|$ is usually a small number since we do not split nodes whose noisy counts are too small, we can ignore $|C_{n,2}|$ and directly use $|C_n|$ to approximate $|C_p|$.

Based on the analysis, with the private k-skyband tree, we compute the number of leaf nodes whose noisy counts are negative ($|C_n|$), sort the leaf nodes with noisy positive counts in L_C in ascending order, and set the noisy count of the first $|C_n|$ nodes in L_C to zero. In this way, we can reduce the error of data synthesis in the nodes in C_p.

Another alternative approach is to directly compute $|C|$, and set the noisy count of the first $\frac{|C|}{2}$ nodes in L_C to zero. However, such approach relies on the true count of leaf nodes, which requires further privacy protection that consumes another portion of the privacy budget. Thus, we choose the approach based on $|C_n|$.

4   Evaluations

In our evaluations, we compare the performances of three techniques: k-skyband tree, kd-tree, and quadtree. For k-skyband tree, we limit the max tree height to be 7, and set the noisy count threshold to 8 to stop splitting. In this way, the max height allows k-skyband tree to get fine enough decompositions and the noisy count threshold prevents the noisy count of some nodes becoming too small and too sensitive to noise. For quadtree, we limit the max tree height to be 7, same as k-skyband tree. For kd-tree, we use the hybrid tree with the default parameters from the existing work [9], which was shown to perform the best for answering count queries. We conduct experiments with privacy budgets ranging from 0.1 to 1.0 and k ranging from 0 to 200; for each budget and each k, we apply the techniques on each dataset 10 times and report their average.

Datasets. The evaluations are carried out over three synthetic datasets that are commonly used for evaluating many interesting variations of skyline queries. They follow independent, correlated and anti-correlated distributions respectively [5]. Each of these datasets contains 10,000 points. For these synthetic datasets, we normalize the values to be in the range $[0, 1000000] \times [0, 1000000]$, and assume the larger values to be preferred in each dimension. Besides synthetic datasets, we also conduct experiments over two real-world datasets: NBA [2] and forest cover type [1]. The NBA dataset includes the statistics of all NBA players from 1997 to 2016, and there are 8645 points in total. In the NBA dataset, we want to find out NBA players who can score high points (in the range $[0, 3156]$) and get many rebounds (in the range $[0, 1449]$). The second dataset is uniformly sampled from the forest cover type dataset, which provides basic information for forested lands in the United States. It contains about 50,000 records and has

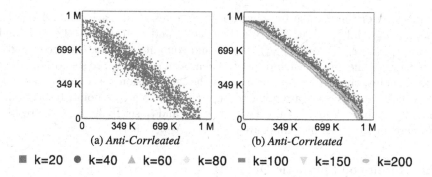

Fig. 2. Illustration of the synthetic dataset following anti-correlated distributions and their k-skyband results when k ranges from 20 to 200. (Color figure online)

been used to evaluate skyline query answering schemes [35]. From this dataset, we want to find out those forests located in uninhabited areas, i.e. those with high elevations (in the range [1859, 3858]) and long distance to roadways (in the range [0, 7117]), since those areas might exist some rare or endangered animal species for research.

The distributions of the synthetic datasets and the real datasets as well as their true skyband query results are shown in Figs. 2 and 5. When k increases, the k-skyband expands toward the area containing less preferred points. Figures 2b and 5c–d show different k-skybands on different datasets when k increases from 0 to 200. If $k_1 > k_2$, k_1-skyband results are the super set of k_2-skyband results, and visually a new stripe in a different color is added when k increases from k_2 to k_1.

Utility Metric. We use $F1$-measure to examine the similarity between the true k-skyband results S_t and the private k-skyband results S_p. To compute $F1$-measure, we first define false positives and false negatives based on the distance among points in S_p and S_t. Intuitively, with differential privacy, we could not guarantee that any true k-skyband results are returned. Instead, if a private skyband point is close to a real skyband point, then we say it is a hit (a true positive). Otherwise, it is a false positive. Similarly, if a true skyband point is not hit by any private skyband point, then it is counted as a false negative. More formally, given t_x and t_y, a point q in S_p is a *true positive (TP)* if there exists a point p in S_t such that $|p.x - q.x| \leq t_x$ and $|p.y - q.y| \leq t_y$; otherwise, we say q is a *false positive (FP)*. Similarly, a point p in S_t is a *false negative (FN)* if there exists no point q in S_p such that $|p.x - q.x| \leq t_x$ and $|q.y - p.y| \leq t_y$. Here for simplicity we use t_x and t_y instead of a radius to quantify the threshold of distance between a skyband point and its true positives. We refer to t_x (t_y) as the error tolerance threshold in x (y) dimension. Based on the counts of TP, FP and FN, we can compute the precision and recall and further derive $F1$-measure.

$$Precision = \frac{TP}{TP + FP} \qquad Recall = \frac{TP}{TP + FN} \qquad F1 = \frac{2 \times Precision \times Recall}{Precision + Recall}$$

For each of the five datasets, we obtain the ranges for each dimension (i.e., $[x_{min}, x_{max}]$ and $[y_{min}, y_{max}]$). We then set t_x and t_y to be 1%, 3%, 5%, and 7% of $(x_{max} - x_{min})$ and $(y_{max} - y_{min})$, and compute $F1$-measure accordingly. Note that for the 2-dimension dataset, when both t_x and t_y are set to 1% of the data domains, the error tolerance rate in the 2-dimension space becomes 1% × 1%, which requires a private point to be very close to a true point. In Sects. 4.1 and 4.2, we show the $F1$-measure with both t_x and t_y set to 3%, and the results of different error tolerance rates are shown in Appendix B.

4.1   Results on Synthetic Datasets

Figure 3 shows the $F1$-measure results for the anti-correlated synthetic datasets. Due to space limit, we omit the results for the normal and correlated distributions. In the figure, the x-axis shows different values for k and the y-axis shows the values of $F1$-measure.

Impacts of Datasets. For the anti-correlated dataset, quadtree has very poor $F1$-measure scores (about 0.1). Unlike the independent and correlated datasets, the true k-skyband points are concentrated in the center areas instead of the areas close to the upper-right point, as shown in Fig. 2b. Thus, when quadtree fails to capture the properties of those regions that contribute most to k-skyband results, the resulting private k-skyband points are still in the regions close to the optimal point (as shown in Fig. 4a), causing high false positives and high false negatives, leading to $F1$-measure scores. kd-tree performs better than k-skyband tree when ϵ is less than or equal to 0.5. For example, when ϵ is 0.1, k-skyband tree uses very little budget for splitting the region and cannot capture the data distribution precisely. The resulting k-skyband is similar to the one shown in Fig. 4a. However, when ϵ becomes larger, which allows enough budget for splitting the region, k-skyband tree performs much better than the other two trees (shown in Figs. 4a and b), and produces the private k-skyband (shown in Fig. 4c) that is very similar to the true k-skyband. Though both k-skyband tree and kd-tree are data dependent and will adaptively conduct finer decompositions in dense regions, k-skyband tree focuses the decompositions on the regions that are most important to k-skyband queries (i.e., the upper-right regions), while kd-tree treats each region equally and does not provide fine enough decompositions in the upper-right regions, which explains the accuracy gap between kd-tree and k-skyband tree, especially when ϵ becomes bigger. For the independent and correlated datasets, the results are similar.

Impacts of k. Generally, $F1$-measure improves with the increase of k, and k-skyband tree is better than the other two approaches when budgets are >0.5. The reason is that when k is small (<20), k-skyband query results intend to contain only a few data points. For privacy protection, a relatively large amount of noise is needed to hide the exact locations of these data points, causing high distortion of actual query results. When k increases, more and more data points are contained in k-skyband query results. k-skyband tree would more accurately

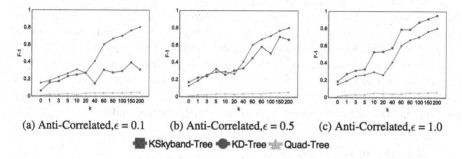

(a) Anti-Correlated,$\epsilon = 0.1$ (b) Anti-Correlated,$\epsilon = 0.5$ (c) Anti-Correlated,$\epsilon = 1.0$

KSkyband-Tree   KD-Tree   Quad-Tree

Fig. 3. Comparing $F1$-measure of k-skyband tree, kd-tree and quadtree on the anti-correlated dataset when k ranges from 0 to 200 and ϵ ranges from 0.1 to 1.0.

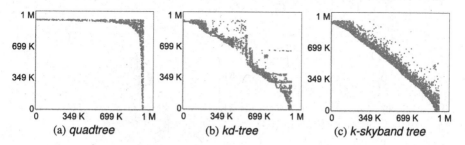

(a) *quadtree*   (b) *kd-tree*   (c) *k-skyband tree*

Fig. 4. Private k-skyband results of the anti-correlated dataset when $\epsilon = 2$ and $k = 200$.

capture the distribution of skyband points, even if their exact locations are protected, which explains the increased utility. In this case, producing an accurate k-skyband is very difficult because the corresponding nodes with small number of points are sensitive to the added noise; when k becomes larger, the number of points in k-skyband becomes larger, and the corresponding nodes with larger number of points are more resistant to the added noise.

Summary. From the synthetic datasets, it does not seem k-skyband tree offers significant advantages over kd-tree or quadtree (depending on which datasets we look at). However, we note that these three datasets are widely used in past work to evaluate the *efficiency* and *scalability* of algorithms to compute exact skyline/k-skyband points. In this work however we focus on the accuracy of differentially privacy algorithms. We argue that none of the distributions in the three synthetic datasets are representative of practical datasets where k-skyband queries are meaningful and useful. For example, in the independent dataset, all the points evenly spread in the whole region, which means k-skyband points are as common as any other non-skyband points. Similarly, for the correlated dataset, it implies that if a point is superior in one dimension, it also tends to be so in the other. In that case, there would be no need to have k-skyband queries over multiple dimensions, as we only need to query points superior in one dimension and their superiority in the other dimension is implicitly ensured.

The anti-correlated dataset goes to another extreme: if a point is superior at one dimension, it must be poor at the other, which also renders k-skyband queries over multi-dimensions unnecessary. Essentially k-skyband queries are to find *unusual* points who are good at both dimensions. Unusual points mean they cannot be as common as other points (inferior at both dimension) as in the independent and correlated datasets, and, on the other hand, they do exist (i.e., superior in both dimensions), not as in the anti-correlated dataset.

4.2   Results on Real Datasets

Figures 6a–f show the $F1$-measure results for the two real datasets when varying k under different privacy budget ϵ.

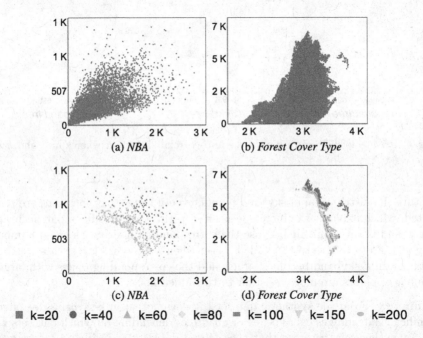

Fig. 5. Illustration of NBA and Forest Cover Type datasets and their k-skyband results when k ranges from 20 to 200. (Color figure online)

The first thing we notice is that the distributions of the two real datasets (shown in Fig. 5) do not resemble any of the three synthetic datasets (shown in Fig. 2). Most of the points are "ordinary". They are not good at either dimension (i.e., they are largely concentrated around regions that are inferior in both dimensions, and k-skyband points instead spread out sparsely: we do have points that are superior in one dimension or in both dimensions, but they are not as concentrated as those "ordinary" points.

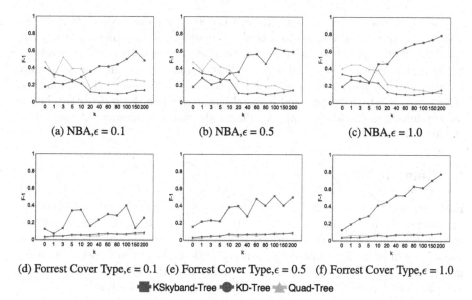

(a) NBA,$\epsilon = 0.1$ (b) NBA,$\epsilon = 0.5$ (c) NBA,$\epsilon = 1.0$

(d) Forrest Cover Type,$\epsilon = 0.1$ (e) Forrest Cover Type,$\epsilon = 0.5$ (f) Forrest Cover Type,$\epsilon = 1.0$

KSkyband-Tree   KD-Tree   Quad-Tree

Fig. 6. Comparing F1-measure of 3 techniques based on k-skyband tree, kd-tree and quadtree on the two real datasets of NBA player stats and Forest Cover Type when k ranges from 0 to 200 and ϵ ranges from 0.1 to 1.0.

From the F1-measure results, for all ϵ values, we can see that k-skyband tree clearly outperforms the other two approaches when $k > 20$ in the NBA dataset; in the forest cover type dataset, k-skyband tree achieves the best performance for all k values. The major reason is that both kd-tree and quadtree focus on splitting more in dense areas, which unfortunately in the real datasets correspond to regions that are not likely to contain k-skyband results. On the other hand, for the regions containing k-skyband points, since they are sparse, kd-tree and quadtree can only generate very course-grained partitions close to the optimal point or along each dimension. The consequence is that during data synthesizing phase, many points will be generated quite near the optimal point or the x and y axes. These synthesized points will be very likely to be included in the private k-skyband query results, which are far different from the real results. As a contrast, k-skyband tree will quickly prune out those dense but not interesting regions (from k-skyband queries' point of view) and split more in regions that likely contain k-skyband points even if these regions are not dense.

Summary. In general, k-skyband tree outperforms the other two trees for both real datasets for all ϵ values when k is reasonably large ($k > 20$). In the forest cover type dataset, k-skyband tree are better than the other two trees for all k values. Such results show that k-skyband tree achieves high utility not only in synthetic datasets used for evaluating various skyline computations, but also in real-world datasets used in multi-criteria decision making.

For these real-world datasets, we can observe that the desired data points usually are in the sparse areas that contain small number of points, and most of the data points are condensed in the areas that represent less desirable values. For example, in the NBA dataset, the best players spread in the areas that represent high scoring or high rebounding, and only a few are in the region that represents both high scoring and high rebounding; while the rest of the players are condensed in the areas along the diagonals from the upper-right corners to the lower-left corners.

5   Related Work

Early works to ensure privacy of released data were based on syntactic approaches such as k-anonymity [16,31] and ℓ-diversity [24]. However, these approaches only satisfy syntactic privacy notions, and cannot provide formal guarantees of privacy as differential privacy. Differential privacy ensures that no matter what knowledge or power an adversary has, the adversary cannot infer an individual's presence in a dataset from the randomized output.

In this work, our goal is to perform k-skyband queries under differential privacy. Initial efforts on differential privacy [10–12,14,21] focused on the theoretical proof of its feasibility on various data analysis tasks, e.g., histogram [3,19,36]. More recent work has focused on practical applications of differential privacy for privacy-preserving data publishing, such as data publishing based on private spatial decompositions. Inan et al. [20] proposed a differentially private technique to build data-partitioning index structures in the context of private record matching, which uses an approximate mean as a surrogate for median (on numerical data) to build kd-tree. Recent works [9,37] also proposed several private spatial decompositions, such as quadtree, kd-tree and PrivTree, building the noisy trees with effective budget allocation strategies. These differentially private data publishing techniques are specifically crafted for answering range count queries. However, synthesizing the dataset based on the spatial decompositions and applying BNL to compute k-skyband results cannot capture the accurate results. Data synthesis on the partitions whose true counts are zero but becomes positive after adding noise would introduce too much unnecessary noise for k-skyband results. Unlike these approaches generating a tree for answering k-skyband queries with different k values, our technique generates a private tree for each k value (i.e., choosing the ne regions based on k). Our technique optimizes the data decomposition for k-skyband queries and suppresses data synthesis on partitions whose noisy counts become positive from zero with post-processing techniques. Evaluation results demonstrate the superiority of our space decomposition optimized based on k over the general space decompositions proposed by the existing works.

Differentially private cluster analysis has also been studied in prior work. Zhang et al. [38] proposed differentially private model fitting based on genetic algorithms and McSherry [27] introduced the PINQ framework, both of which have been applied to achieve differentially privacy for k-means clustering.

Nissim et al. [29] proposed the sample-aggregate framework that calibrates the noise magnitude according to the smooth sensitivity of a function. Their framework can be applied to k-means clustering under the assumption that the dataset is well-separated. Chen et al. [8] proposed several techniques that achieve differential privacy for WaveCluster [32,33], which can capture spatial information to detect clusters with complex shapes, e.g. concave shapes. Leveraging private clustering analysis for computing k-skyband results would easily miss some partitions that contain a small number of k-skyband points, since these partitions are too sparse to form clusters.

Another important line of prior work focuses on privacy-preserving database queries over sensitive data distributed among multiple parties. Recently, the advances in the theory of secure multiparty computations [6,17,18] proved that comparison, addition, and multiplication (XOR and AND) can be computed securely with reasonable computation cost. Based on these primitive protocols, a line of research has focused on developing efficient secure multi-party communication protocols for various database queries, such as set operations [4,28], top-k queries [34]. These protocols focus on protecting the privacy of the data among multiple parties and only the final query results are released to the public, while our work presents an approach to release the query results without compromising individual privacy of the individuals. What's more, there are no existing secure protocols for k-skyband queries, and building such protocols is not trivial.

6   Conclusion

In this paper we have addressed the problem of k-skyband queries with differential privacy. We propose a general technique BBS-Priv that accepts any space decomposition tree with differential privacy as input and selective performs data synthesis for interested tree nodes to compute private k-skyband results. To improve the query accuracy, we further devise a new space decomposition tree k-skyband tree specifically optimized for k-skyband queries, which partitions the space based on k other than median value or midpoint of each dimension. We further present a suite of techniques to publish a k-skyband tree satisfying differential privacy, and propose a post-processing technique to improve accuracy by suppressing data synthesis on those partitions whose noisy counts become positive from zero. In the future, we will investigate under differential privacy other categories of multi-criteria decision making queries, such as top-k dominating queries.

A   Algorithm of k-Skyband Tree

Algorithm 2 shows the detailed steps of generating k-skyband tree. Given a region $\langle (x_{min}, x_{max}), (y_{min}, y_{max}) \rangle$, k-skyband tree first inserts the input region node r into a queue Q (Line 1), and then removes a node n from Q for splitting (Line 3). If the height of n reaches the maximum height h, n is considered as a leaf node (Line 4) and k-skyband tree continues to process a new node from Q (back to

Line 2). If the point count of n is larger than $k+1$, k-skyband tree uses a function $SplitByK$ (Algorithm 3) to choose a splitting point $s = (s_x, s_y)$ based on k (Line 7), such that the upper-right region $ne = \langle (s_x, s_y), (x_{max}, y_{max}) \rangle$ contains more than k data points. When $SplitByK$ cannot find such a ne (i.e., ne is the same as n at Line 8), k-skyband tree uses the midpoint of each dimension as the splitting point (same as quadtree) (Line 9). If the point count of n is smaller than $k+1$ (not possible to find ne whose point count is larger than $k+1$), k-skyband tree also uses the midpoint of each dimension as the splitting point (same as quadtree) (Line 11). After splitting, the dominance region of s, $sw = \langle (x_{min}, s_x), (y_{min}, s_y) \rangle$, is considered as a leaf node and no further split is required. The splitting terminates when there are no more nodes in Q to be split (Line 2).

Algorithm 2. k-skyband tree

Input: A region r, k in k-skyband queries, the max height of the tree h
Output: A spatial decomposition tree T
 1: Q.enqueue(r)
 2: **while** Q is not empty **do**
 3:     $n = Q$.dequeue()
 4:     **if** isLeaf(n, h) **then**
 5:         **continue** // back to Line 2
 6:     **if** $n.count > k + 1$ **then**
 7:         $N =$ splitByK($n, k, -1$) // -1 means no noise added
 8:         **if** $N.ne = n$ **then**
 9:             $N =$ splitByMidPoint(n)
10:     **else**
11:         $N =$ splitByMidPoint(n)
12:     Q.enqueue($N.ne, N.nw, N.se$)
13:     **if** $N.ne.count \leq k$ **then**
14:         Q.enqueue($N.sw$)
15: **return** N

To obtain an upper-right region with more than k points, we propose an efficient algorithm shown in Algorithm 3. Given a region n, the algorithm uses two max heaps (H_x and H_y) to sort the data points within n (Line 2), where H_x (H_y) sorts the points based on their x (y) coordinates. The algorithm accesses a point n_x from H_x and a point n_y from H_y, and uses the x coordinate of n_x and the y coordinate of n_y as a new splitting point $s = \langle s_x, s_y \rangle$ (Line 8). Then n_x and n_y are put into a set P (Line 9), which are later checked to see whether they fall into the upper-right region ne split based on s. The reason is that the y (x) coordinate of n_x (n_y) may be smaller than s_y (s_x), and thus n_x (n_y) may not fall into ne. To compute the point count of ne, we only need to check the points in P: each $c \in P$ is checked to see whether it falls into ne (Lines 10–12). If so, then c is moved from P to S, which holds all the points that dominate the current split point (Line 12); otherwise, c remains in P and will be checked again when a new splitting point is formed in the next iteration. Also, due to the way split points are generated, if a point dominates an early split point, it will also dominate later ones. That is why we can safely put that point into S. The algorithm terminates when S contains more than k points (Line 4), i.e., splitting by s guarantees that the upper-right region contains more than k points. The non-private k-skyband tree directly returns the split regions (Lines 16–17) and skip the steps for obtaining the private values for s (Lines 14–15).

B   Results with Different Error Tolerance Rates

A real skyband point is considered to be hit by a private skyline point if the private skyband point is close enough to the real skyband point. Error tolerance rates quantitatively define how close they should be in order to be considered as a hit. Larger error tolerance rates mean more loose constraints on the distance between the private and the real skyband points, and thus make $F1$-measure become better. In other words, when the error tolerance rates become larger, the privacy technique provides better guarantee in utility. We compute results of $F1$-measure by varying error tolerance rates to observe the impacts of error tolerance rates on the performance.

Algorithm 3. SplitByK

Input: A region n, k in k-skyband queries, ϵ_s privacy budgets for splitting
Output: A set N that contains four children nodes sw,se,nw,ne
1: $S = \emptyset$, $P = \emptyset$
2: H_x.insert($n.data$), H_y.insert($n.data$)
3: $s_x = n.x_{max}$, $s_y = n.y_{max}$
4: **while** $S.size < k + 1$ **do**
5:     **if** H_x is empty *or* H_y is empty **then**
6:         **break**
7:     $n_x = H_x$.remove(), $n_y = H_y$.remove()
8:     $s_x = n_x.x$, $s_y = n_y.y$
9:     P.add(n_x), P.add(n_y)
10:    **for** $c \in P$ **do**
11:        **if** $c.x \geq s_x$ and $c.y \geq s_y$ **then**
12:            S.add(c), P.remove(c)
13: **if** $\epsilon_s > 0$ **then**
14:    $s_x = EM(s_x, \epsilon_s)$
15:    $s_y = EM(s_y, \epsilon_s)$
16: $N = \{sw, se, nw, ne\} = n$.split($s_x, s_y$)
17: **return** N

Figure 7 shows the $F1$-measure results for the two real-world datasets when the error tolerance rate t_x and t_y ranges from 1% to 7%. In each figure, the x-axis shows different values for the error tolerance rates and the y-axis shows the values of $F1$-measure. Due to space limit, we choose to show the results with k set to 40 and ϵ set to 1.0 as the representative results. Clearly, k-skyband tree performs much better than the other two trees in all the error tolerance rates, and its $F1$-measure improves significantly when t_x and t_y reach 7%.

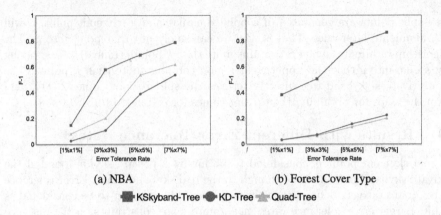

(a) NBA   (b) Forest Cover Type

KSkyband-Tree   KD-Tree   Quad-Tree

Fig. 7. F1-measure of 3 techniques based on k-skyband tree, kd-tree and quadtree on the real datasets for $k = 40$, $\epsilon = 1.0$, and error tolerance rate ranges from 1% to 7% on each dimension.

References

1. http://kdd.ics.uci.edu/databases/covertype/covertype.html
2. Nba players statistics. http://www.hoopsstats.com/basketball/fantasy/nba/playerstats
3. Barak, B., Chaudhuri, K., Dwork, C., Kale, S., McSherry, F., Talwar, K.: Privacy, accuracy, and consistency too: a holistic solution to contingency table release (2007)
4. Blanton, M., Aguiar, E.: Private and oblivious set and multiset operations. In: ASIACCS (2012)
5. Borzsony, S., Kossmann, D., Stocker, K.: The skyline operator. In: ICDE (2001)
6. Cachin, C.: Efficient private bidding and auctions with an oblivious third party. In: CCS (1999)
7. Chen, L., Gao, S., Anyanwu, K.: Efficiently evaluating skyline queries on RDF databases. In: ESWC (2011)
8. Chen, L., Yu, T., Chirkova, R.: Wavecluster with differential privacy. In: CIKM (2015)
9. Cormode, G., Procopiuc, C., Srivastava, D., Shen, E., Yu, T.: Differentially private spatial decompositions. In: ICDE (2012)
10. Dwork, C.: Differential privacy: a survey of results. In: TAMC (2008)
11. Dwork, C., Lei, J.: Differential privacy and robust statistics. In: STOC (2009)
12. Dwork, C., McSherry, F., Nissim, K., Smith, A.: Calibrating noise to sensitivity in private data analysis. In: TCC (2006)
13. Dwork, C., Roth, A.: The algorithmic foundations of differential privacy. Found. Trends Theor. Comput. Sci. **9**, 211–407 (2014)
14. Feldman, D., Fiat, A., Kaplan, H., Nissim, K.: Private coresets. In: STOC (2009)
15. Feng, X., Gao, Y., Jiang, T., Chen, L., Miao, X., Liu, Q.: Parallel k-skyband computation on multicore architecture. In: APWeb (2013)
16. Ghinita, G., Zhao, K., Papadias, D., Kalnis, P.: A reciprocal framework for spatial k-anonymity. Inf. Syst. **35**(3), 299–314 (2010)
17. Gordon, D.S., Carmit, H., Katz, J., Lindell, Y.: Complete fairness in secure two-party computation. In: STOC (2008)

18. Harnik, D., Naor, M., Reingold, O., Rosen, A.: Completeness in two-party secure computation: a computational view. In: STOC (2004)
19. Hay, M., Rastogi, V., Miklau, G., Suciu, D.: Boosting the accuracy of differentially private histograms through consistency. PVLDB **3**, 1021–1032 (2010)
20. Inan, A., Kantarcioglu, M., Ghinita, G., Bertino, E.: Private record matching using differential privacy. In: EDBT (2010)
21. Kasiviswanathan, S.P., Lee, H.K., Nissim, K., Raskhodnikova, S., Smith, A.: What can we learn privately? In: FOCS (2008)
22. Kodama, K., Iijima, Y., Guo, X., Ishikawa, Y.: Skyline queries based on user locations and preferences for making location-based recommendations. In: International Workshop on Location Based Social Networks (2009)
23. Levandoski, J.J., Mokbel, M.F., Khalefa, M.E.: Preference query evaluation over expensive attributes. In: CIKM (2010)
24. Machanavajjhala, A., Kifer, D., Gehrke, J., Venkitasubramaniam, M.: L-diversity: privacy beyond k-anonymity. ACM Trans. Knowl. Discov. Data **1**(1) (2007). http://doi.acm.org/10.1145/1217299.1217302
25. Magnani, M., Assent, I., Mortensen, M.L.: Taking the big picture: representative skylines based on significance and diversity. VLDB J. **23**(5), 795–815 (2014)
26. McSherry, F., Talwar, K.: Mechanism design via differential privacy. In: FOCS (2007)
27. McSherry, F.: Privacy integrated queries: an extensible platform for privacy-preserving data analysis. Commun. ACM **53**(9), 19–30 (2010)
28. Freedman, M.J., Nissim, K., Pinkas, B.: Efficient private matching and set intersection. In: Cachin, C., Camenisch, J.L. (eds.) EUROCRYPT 2004. LNCS, vol. 3027, pp. 1–19. Springer, Heidelberg (2004). doi:10.1007/978-3-540-24676-3_1
29. Nissim, K., Raskhodnikova, S., Smith, A.: Smooth sensitivity and sampling in private data analysis. In: STOC (2007)
30. Papadias, D., Tao, Y., Fu, G., Seeger, B.: Progressive skyline computation in database systems. ACM Trans. Database Syst. **30**(1), 41–82 (2005)
31. Samarati, P., Sweeney, L.: Protecting privacy when disclosing information: k-anonymity and its enforcement through generalization and suppression. In: Proceedings IEEE Security & Privacy (1998)
32. Sheikholeslami, G., Chatterjee, S., Zhang, A.: Wavecluster: a multi-resolution clustering approach for very large spatial databases. In: VLDB (1998)
33. Sheikholeslami, G., Chatterjee, S., Zhang, A.: Wavecluster: a wavelet-based clustering approach for spatial data in very large databases. VLDB J. **8**(3–4), 289–304 (2000)
34. Vaidya, J., Clifton, C.: Privacy-preserving top-k queries. In: ICDE (2005)
35. Valkanas, G., Papadopoulos, A.N., Gunopulos, D.: Skydiver: a framework for skyline diversification. In: EDBT (2013)
36. Xu, J., Zhang, Z., Xiao, X., Yang, Y., Yu, G., Winslett, M.: Differentially private histogram publication. VLDB J. **22**(6), 797–822 (2013)
37. Zhang, J., Xiao, X., Xie, X.: Privtree: a differentially private algorithm for hierarchical decompositions. In: SIGMOD (2016)
38. Zhang, J., Xiao, X., Yang, Y., Zhang, Z., Winslett, M.: Privgene: differentially private model fitting using genetic algorithms. In: SIGMOD (2013)

Mutually Private Location Proximity Detection with Access Control

Michael G. Solomon[1($\boxtimes$)], Vaidy Sunderam[1], Li Xiong[1], and Ming Li[2]

[1] Department of Mathematics and Computer Science,
Emory University, Atlanta, USA
{msolo01,lxiong,vss}@emory.edu, michael@solomonconsulting.com
[2] Department of ECE, University of Arizona, Tucson, USA
ming.li@arizona.edu

Abstract. Mobile application users want to consume location-based services without disclosing their locations and data owners (DO) want to provide different levels of service based on consumer classifications, sometimes without disclosing areas of interest (AOI) locations to all users. Both actors want to leverage location-based services utility without sacrificing privacy. We propose a protocol that supports queries from different classifications of users, such as subscribers/non-subscribers, or internal/external personnel, and imposes embedded fine-grained access control without disclosing user or DO location information. We use Ciphertext Policy Attribute-Based Encryption (CP-ABE) and Hidden Vector Encryption (HVE) to provide flexible access control and mutually private proximity detection (MPPD). Our protocol minimizes expensive cryptographic operations through the use of location mapping with compressed Gray codes, each representing multiple locations. Our protocol encrypts AOI locations using HVE, and then encrypts AOI information using CP-ABE with an expressive access policy. Our protocol's use of these two encryption methods allows DOs to define a single set of AOIs that can be accessed by sets of users, each with potentially different access permissions. A separate service provider (SP) processes queries without divulging location information of the user or any DO provided AOI.

1 Introduction

Mobile applications increasingly incorporate location awareness into services they provide. Many such services rely on location to provide context-sensitive results, such as proximity alerts when a consumer approaches some defined area of interest. Today's smart phones, tablets, and other mobile devices make accurate location-sensing a commonly used feature. However, consumers of such features are becoming increasingly concerned over the loss of privacy resulting

---

Research supported by AFOSR DDDAS grant FA9550-17-1-0006 and NSF TWC grant CNS-1618932.

G. Livraga and S. Zhu (Eds.): DBSec 2017, LNCS 10359, pp. 164–184, 2017.
DOI: 10.1007/978-3-319-61176-1_9

from ongoing location disclosure. Consumers want to use location-based services without sacrificing their privacy. Likewise, Data providers or data owners (DO) of areas of interest (AOI) want to provide differing levels of service based on consumer classifications, without disclosing the locations of all AOIs to all consumers. For example, DOs may provide services of greater value to paid subscribers than to users who consume services for free. Alternatively, DOs may keep specific AOI locations private from some users in the interest public safety or to avoid public panic, as in the case of areas of active criminal activity or ongoing health hazards. DOs may want to restrict access to, and even awareness of, AOIs to consumers based on access permissions and location.

We propose a protocol that allows DOs to define a single set of AOIs with consumer access rules that limits AOI proximity alerts to specific groups of consumers. The locations of defined AOIs are kept private and not disclosed to consumers until they are in proximity to individual AOIs. The protocol fundamentally provides proximity detection based on a consumer's current location, without disclosing any consumer's location. Our protocol supports mutual privacy (privacy for the data provider and the consumer), along with embedded access control that allows DOs to control proximity alerts by consumer type.

1.1 Motivation

Consider Mary, a guest at the "Fun Times" amusement park. Mary wants to make the most of her day and desires to minimize time spent waiting in line for attractions or shows. Mary has subscribed to the "Fun Times InTheKnow app" premium service that sends information to her smart phone about nearby attractions and shows with short wait times. Bob is also in the "Fun Times" park and has the "InTheKnow" app, but Bob did not subscribe to the premium service. Bob only receives general information about attraction wait times for attractions that are somewhat close to his current location. Both Mary and Bob want to receive helpful information without disclosing their locations to "Fun Times." On the other hand, "Fun Times" wants to provide location-sensitive services to Mary and Bob without publishing all of the "Fun Times" AOIs. If "Fun Times" published all of their AOIs, no subscribers would need to pay for the subscription to the premium service. "Fun Times" protects their revenue stream by keeping their AOIs private. "Fun Times" can limit the information they provide to subscribers, and even define different subscriber levels based on subscription fees paid. "Fun Times" can also use this service to direct their employees to areas of the park that need cleaning, servicing, or even crowd management. Other use cases for such a location privacy preserving framework could include providing epidemiological related alerts or criminal activity investigation proximity with precision granularity based on clearance and/or "need to know".

1.2   Existing and Potential Solutions

A naive approach to location privacy could be to simply stop the sensing or release of physical location. However, this results in the loss of utility (e.g. navigation, lost device tracking, etc.) Alternatively, applications could introduce novel ways of exchanging and processing location information that protects the location owner's privacy without requiring user action. Application layer location privacy efforts generally focus on one of four main areas, each with its own drawbacks: location perturbation (limited accuracy), access control (limited privacy/utility trade-off), private information retrieval (PIR) (lack of DO privacy), and encryption (performance cost). Each type of approach attempts to provide the ability for users to consume location based services without divulging their exact locations to any untrusted service provider (SP) or any DO, and in some cases, allow the DO to keep its AOI location data private as well. Existing approaches assume equal access to proximity detection for all AOIs. Traditional access controls require a trusted authority to make decisions at run time.

Existing MPPD solutions largely focus on location as the only criteria for determining proximity to a defined area. In some cases, additional attributes should be considered, such as security level or subscriber status, before providing proximity responses. A useful MPPD protocol should allow a single set of AOIs to service a large group of diverse users. Such a protocol would only provide proximity alerts for users that are both near a defined AOI and meet requirements set by the data owner. This is normally accomplished by separating the MPPD functionality from the authorization phase. This requires a third party to examine each user and AOI attributes to determine if a proximity alert is allowed, based on data owner policy. In this scenario, the third party possesses substantial information about users and AOIs.

1.3   Our Contributions

We propose a novel method of providing Mutually Private Proximity Detection (MPPD) with embedded fine-grained access control. Our protocol provides fine-grained access control that is embedded in the encryption technique. That is, decryption is only successful when attempted with an authorized user's key. Our method does not disclose location data to any user, DO, or SP. Our protocol introduces a novel use of two existing techniques, Ciphertext Policy Attribute-Based Encryption (CP-ABE) to provide fine-grained access control based on descriptive consumer attributes, and Hidden Vector Encryption (HVE) to efficiently determine user proximity to AOIs. Neither CP-ABE, nor HVE, alone solve the problem of MPPD, but when both are integrated into a new protocol, work together to efficiently provide MPPD. To the best of our knowledge, there is only one other proposed protocol [24] that controls access to AOIs based on access rules the data owner supplies, along with MPPD. However, this protocol requires distance calculations for determining proximity to each AOI, which can be costly. Our protocol substantially reduces computational overhead and increases scalability by using HVE with compressed AOI location tokens to determine if a user's location overlaps one or more AOIs, along with CP-ABE to provide flexible fine-grained access control.

Our proposed protocol is novel in that it provides fine-grained flexible consumer access control, minimizes computational load on devices with limited processing power (eg. mobile devices), and also provides a high level of privacy guarantees for both users and DOs. The protocol supports these services and reduces the DO administrative workload by incorporating two state of the art partial solutions, drawing on the existing strengths of each approach (CP-ABE and HVE). Using this new protocol, DOs can create a single set of AOIs, along with access policies for each AOI, that restricts user consumption of proximity alerts based on DO defined descriptive attributes. Users (consumers) must possess the attributes necessary to satisfy a DO defined access policy for each AOI to receive proximity alerts for that AOI. All of this functionality is provided without requiring any third party to access unencrypted location information to make access authorization decisions. Using CP-ABE, DOs can define the granularity of user proximity alert consumption based on its own defined access policies. The use of HVE provides the actual proximity determination of a user location to an AOI without disclosing either location to any party. The combination of CP-ABE and HVE provides a flexible and scalable approach to implementing MPPD with controlled access based on user classifications.

2   Related Work

Existing private proximity detection solutions generally fall into the following four areas, each with its own drawbacks: location perturbation (limited accuracy), access control (limited privacy/utility trade-off), private information retrieval (PIR) (lack of DO privacy), and encryption (performance cost). Current private proximity detection solutions, with the exception of [24], provide results based on location alone, and do not consider other attributes. Results filtering is left to a trusted third party that can examine attributes and learn user locations.

2.1   Location Perturbation and Transformation

Location perturbation schemes use obfuscated or perturbed user location data. K-anonymity is the most common technique to limit the ability to determine any user location. Kim [21,23] proposed two different approaches to cloaking locations. One weakness of such schemes is that attackers can use the same cloaking techniques as authorized users to generate and analyze shared cloaked locations. Another weakness of these schemes is that the Location Based Server (LBS) only responds with answers of the same, or lower, precision of the obfuscated user location. Yiu [30] proposed a collection of schemes to strengthen the shared knowledge weakness by partitioning data using different criteria, but these techniques are still vulnerable to attacks based on a priori knowledge and brute-force attacks. Hossain [18] proposed shear-based spatial perturbation schemes.

Recent work in cloaking [3,29] extends differential privacy [11] and provides greater semantic privacy protection. However, all perturbation schemes reduce location data precision.

2.2 Access Control

Another approach is via structured access control techniques. Bugiel [7] proposed FlaskDroid, a fine-grained Mandatory Access Control (MAC) approach for Android devices. Li [24] proposed Privacy-preserving Location Query Protocol (PLQP). PLQP allows queries to access location information while still upholding user privacy and is efficient enough to operate on mobile devices. Lu [25] proposed Secure and Privacy-preserving Opportunistic Computing (SPOC) framework for health care data, which requires close proximity of a patient and medical personnel to grant PHI access. Fawaz [13] proposed more fine-grained access control for sharing location data with third-party apps. Access control methods only provide binary access control with limited granularity for privacy protection, i.e. users can either grant or block access, and these approaches require a trusted third party to access locations to make access decisions.

2.3 Private Information Retrieval

PIR [10] allows users to issue DO queries without the DO learning the query's content. These techniques build private spatial indexes that utilize PIR operations, which can provide efficient spatial query processing while the underlying PIR scheme provides privacy. Hengartner [17] uses PIR to hide location information from an untrusted server and uses trusted computing to guarantee that the PIR algorithms are only performing the intended operations. Khoshgozaran [20] proposed two location privacy approaches that eliminate the need for an LBS anonymizer. Ghinita [14] uses PIR techniques to support Nearest-Neighbor (NN) queries without requiring a trusted third party. This approach uses cryptography to protect user locations and increases performance with data mining techniques to eliminate redundant calculations. PIR techniques can be efficient and provide a high level of privacy guarantees, but they assume that the AOI data is public, and provide no privacy for DOs.

2.4 Encryption

Other approaches use encryption for location data to provide privacy, requiring varying degrees of communication and computation in the encrypted space to determine proximity. Encryption techniques can be viewed as symmetric PIR, that is, PIR plus the restriction of AOI privacy, which is the main focus of our research. Although encryption introduces overhead, techniques that use it can be more resistant to attackers, even those with prior knowledge. Proposed encryption techniques can loosely be grouped into three general categories: spatial data encryption, novel encoding/notation, and homomorphic encryption schemes.

Spatial data encryption approaches include Khoshgozaran [19], who proposed a grid-based scheme that uses group shared symmetric keys, allowing users to query nearby cell contents and then locally decrypt location details for items encrypted with their group shared key. Wang's [28] approach performs a geometric range search on encrypted spatial data.

Other approaches depend on novel location encoding or notation. Ghinita [15] proposed a method based on HVE to protect user locations. Calderoni [8, 26] proposed a new compact data structure, Spatial Bloom Filter (SBF), to privately store AOI locations, and the Paillier cryptosystem to protect AOIs from disclosure while still allowing comparison with encoded (but unencrypted) user locations. Kim [22] proposed Hilbert Curve Transformation (HCT), which encodes locations using a Hilbert Curve and then encrypts the result using AES.

And finally, other approaches depend on homomorphic encryption to provide location protection. Elmehdwi [12] proposed Secure k-Nearest Neighbors (SkNN), which uses the Paillier cryptosystem and protocols to support private k-NN queries. Choi [9] also uses Paillier cryptosystem, along with Order Preserving Encryption (OPE) [1] to detect proximity between proximity zones. Sedenka [27] uses the Paillier cryptosystem as well, but incorporates garbled circuits to define three protocols to privately calculate distances between any two points using different coordinate systems on a spherical surface.

None of these encryption based techniques provides access control They only address private proximity detection. To the best of our knowledge, only one other proposed technique addresses MPPD and access control. Li [24] proposes a MPPD solution that does include fine-grained access control by using CP-ABE [4], along with Paillier cryptosystem. Li's use of Pailler is similar to the secure distance calculations of Elmehdwi's approach, and is the most similar to our approach. We compare our framework to Li's in the experiments section.

Encryption techniques are promising and can offer good privacy guarantees for both users and DOs, but at a cost. Encryption requires computation overhead which can be a drawback for devices with limited processing power.

3   Problem Setting and Preliminaries

In this section, we define our problem setting and privacy model, then introduce the two cryptographic primitives that we use in our protocol.

3.1   Framework Model

Our location-based proximity detection model is based on users traveling through real space represented by a two-dimensional grid. At any one point in time, a user occupies exactly 1 grid cell. On demand, users can query an encrypted database of AOIs, using their own encrypted locations, to determine if their current location overlaps any AOI. The DO defines multiple AOIs, encrypts them, and supplies them to the SP. The SP determines access authorization using encrypted data, and then carries out the calculations on the encrypted data if access is authorized to determine if the user's current location is in proximity to one or more AOIs. The SP responds to the user with a list of AOIs that overlap the cell that encloses the user's location. Cells can represent any physical size. The only restriction is that the DO and all users must use the same cell granularity when converting AOI or user locations into grid coordinates.

Our protocol uses an architecture of four distinct entities. They are:

- Data Owner/Provider (DO) - Defines/encrypts AOI locations/access policies
- Key Generator (KG) - Generates CP-ABE and HVE keys and tokens.
- Service provider (SP) - Provides computation services for users to determine user proximity to any AOI
- User - user with ability to determine current location (GPS or other means)

The function of each step of the protocol is explained in a later section.

3.2   Privacy Model

Our protocol provides the following privacy guarantees:

- DOs learn nothing
- Users learn nothing except when in proximity to an AOI
- SPs learn nothing

All DOs learn nothing about user locations and users learn nothing about AOI locations (defined by a DO), except in the case when their current location overlaps one or more AOIs. Even when a user learns that he or she overlaps an AOI, the complete dimensions of the AOI are undisclosed. A user could travel around the grid in a structured attempt to learn AOI locations by remembering cells with AOI overlaps. Our protocol does not protect eventual AOI location disclosure from such an attack. SPs only learn that a user's location overlaps one or more AOIs, but do not know any actual user or AOI location, and thus cannot infer any physical location information. SPs can learn that two or more users are in proximity to one another, when those users overlap the same AOIs, but again without any AOI or user location information, cannot determine any physical location information.

Protocol extensions to protect AOI locations from deliberate user brute-force attacks are left for future work, but could be implemented at the KG by having the KG detect patterns of such attacks and responding to the user accordingly. The KG does know the location of a user when that user requests an HVE key, but the KG is trusted, and thus does not share user location with any other entity and does not have access to any AOI information.

The KG does not return the generated HVE encrypted user location to the user. Instead, the KG sends the encrypted user location to the SP. The SP only has access to encrypted AOI and user locations. It can only learn that a user's location overlaps one or more AOIs, but does not know what location the overlap represents in the grid. The SP could infer user to user proximity when multiple users are in proximity to the same AOI within a short period of time. But the

SP still would not know the location of any user or AOI. Although HVE is a public key cryptography scheme and would provide the possibility for a malicious service provider to attempt to build a fake encrypted AOI list, our framework limits the distribution of the HVE public keys. This practice limits the ability to encrypt AOI locations to only trusted entities, thereby removing the ability to use dictionary type attacks to generate imposter AOI lists.

3.3   Ciphertext Policy Attribute Based Encryption

A CP-ABE scheme provides fine-grained access control over data [4]. CP-ABE associates a user with a descriptive attribute set to generate a secret key, SK. Only users whose attributes match the encryption access policy can decrypt the data. CP-ABE provides the ability for a set of users with identical attributes to share data (i.e. able to decrypt the same ciphertext) without having to share keys or have any awareness of the existence of other users with the same attributes. Further, any user's attributes may satisfy the requirements of many different ciphertext access policies. CP-ABE expressive access policies free ciphertext from being bound to a static set of attributes. Users that possess different attributes may be able to decrypt a ciphertext, as long as each user's attributes satisfies the DO supplied access policy. To encrypt a message M using CP-ABE, the encryptor provides an access policy, expressed as a boolean expression containing selected attributes and values for M. Figure 1 shows an access policy in a tree structure. M is then encrypted based on the access structure, T. Decryptors generate SK based on their attributes. A decryptor is only able to decrypt ciphertext, CT, when her SK satisfies the access policy used to encrypt M. Unauthorized users cannot decrypt CT even if they collude and combine disjoint attributes.

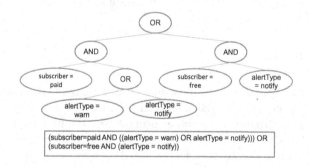

(subscriber=paid AND ((alertType = warn) OR alertType = notify))) OR (subscriber=free AND (alertType = notify))

Fig. 1. CP-ABE access policy tree

CP-ABE defines the following four essential functions:

1. Setup(): Input security parameter, output public key (PK), for encryption, and master key (MK), to generate user secret keys.

2. Encrypt: Input message M, access structure T, public key PK, output cipher-text CT.
3. KeyGen: Input set of user's attributes SX and MK, output secret key SK for SX.
4. Decrypt: Input CT, SK. If SK satisfies access structure in CT, return M, else return NULL.

3.4 Hidden Vector Encryption

Hidden Vector Encryption (HVE) [6] is an extension of an anonymous identity based encryption (IBE) [5] scheme. With IBE, the keys used for encryption and decryption are based on identities and attributes. HVE allows an attribute string that is associated with the ciphertext or the user secret key to contain wildcards. Thus, HVE provides a searchable encryption scheme that supports conjunctive equality, range and subset queries. An HVE attribute is represented as a vector of elements with a value of 0, 1, or a wildcard (represented as "*" and often referred to as a "don't care" value). The wildcard in an HVE attribute matches the values 0 or 1 in comparison operations. An HVE comparison of a search predicate S and a ciphertext C evaluates as True if the attribute vector I used to encrypt C contains the same values as S for all positions that are not "*".

HVE defines the following four essential functions:

1. Setup(): Input security parameter, output public key (PK), for encryption, and master key (MK).
2. KeyGen: Input MK and a string y in 0, 1, *, output secret key SK for y.
3. Encrypt: Input public key PK, message M, attribute string x in 0, 1, output ciphertext CT.
4. Query: Input CT, SK. If SK satisfies attribute string x in CT, return M, else return NULL.

4 Protocol Description

In this section, we present our proposed protocol for MPPD with access control by combining HVE and CP-ABE. We call the protocol PrivProxABE, which stands for Private Proximity detection using ABE. We first define AOI types and user attributes which will be used to enforce access control and then present the details for each of the steps of the protocol.

4.1 AOI and User Attributes

Suppose "Fun Times" defines four types of AOIs, each with a different color designator. Table 1 lists the AOI types and what each one represents.

These AOI types are simply examples of what our protocol can represent. AOI types can be of any type and any number. The AOI type definition is left up to the specific implementation definition, as long as each AOI is uniquely

Table 1. AOI Types

Color	Who can access	Alert type
Red	Paid subscribers	Warn
Blue	Paid subscribers	Notify
Green	Paid subscribers	Approach
Yellow	Free users	Notify

identified with a character label. AOIs can overlap by sharing one or more cells. In such cases, users would receive a response to a proximity query indicating that their current location places them within more than one AOI. Using our example, "Fun Times" wants to provide some value to users who use their mobile app, but reserve the more detailed information for premium subscribers to their service. Users of the "InTheKnow" app that have paid for the premium service can receive "warn", "notify", and "approach" alerts. The first two alert types could be used to provide guidance for areas to avoid, while the third alert type could inform users of areas that would be beneficial to visit. Users that have not paid for the premium subscription will only receive "notify" alerts. Notice that there are two different colors for the "notify" alert. Premium subscribers will receive more specific information about "notify" alerts, while free users will only receive general messages. This approach allows "Fun Times" to define different classifications of users, each of which receives different proximity alerts.

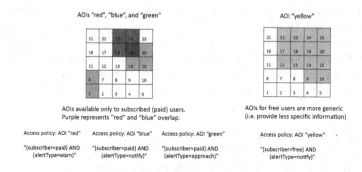

Fig. 2. Paid subscriber and free user AOIs (Color figure online)

Figure 2 shows how each AOI accessible by both paid and free subscribers is defined on a grid and the access policy associated with each AOI. Our example includes four users to demonstrate the protocol's flexibility. Table 2 shows each user and their associated descriptive attributes used to generate each user's secret key. A user's attributes must satisfy (match) an AOI's access policy to decrypt and access the AOI.

To determine proximity to an AOI, a user sends the current encrypted location to a service provider, which then determines if that user is within any cell

Table 2. Users and Descriptive Attributes

User	Subscriber	AOI Types	Can decrypt AOIs
Alice	free	warn, notify	yellow
Bart	free	approach	none
Mary	paid	approach, notify	blue, green
Daniel	paid	warn, notify	red, blue

defined as an AOI, all without ever learning any location information from the user or data provider. Users request a secret key from a key generator based on descriptive attributes. The attribute-based key provides the ability for the service provider to assess accessibility for each AOI. To continue our example, four users request proximity alerts for the amusement park defined AOIs.

We call the technique PrivProxABE (MPPD using ABE). The protocol is made up of four basic phases:

I Setup - DO Initializes protocol state
II InitAOIs - DO Encrypts AOIs with access policy
III InitUserLoc - User encrypts current user location
IV Query - User initiates location proximity query

4.2 Setup

Algorithm 1 shows the steps in the "Setup" phase and refers to Fig. 3. In the "Setup" phase, the DO calls the ABEsetup() method on the KG to generate the ABE public key, PK_{ABE}. The KG sends PK_{ABE} to the DO and the SP. The DO also calls the HVEsetup() function on the KG to generate the HVE public key, PK_{HVE}, and returns PK_{HVE} to the DO. The DO keeps PK_{ABE} and PK_{HVE} secret, (i.e. the DO does not share either key with any other entity.)

Algorithm 1. PrivProxABE Protocol - Setup

---

1. DO calls ABEsetup() on KG. KG sends PK_{ABE} to DO and SP.
2. DO calls HVEsetup() on KG. KG sends PK_{HVE} to DO.

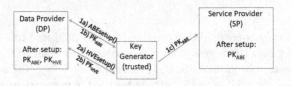

Fig. 3. PrivProxABE Setup phase

4.3   Encrypting AOIs with Access Policy

Algorithm 2 shows the steps in the "InitAOIs" phase and refers to Fig. 4. Our evaluation assumes a static set of AOIs. If AOIs change frequently, this phase would be revisited to initialize any new or changes AOIs. Initializing individual AOIs has a small impact on performance, and would only impact overall performance in relation to the number of frequently update AOIs.

Algorithm 2. PrivProxABE Protocol - InitAOIs

---

3. DO encrypts encoded AOI cells (encoded using Gray encoding) and keeps CT_{HVE}.

4a. DO encrypts AOI label and other descriptive AOI info using access policy to get CT_{ABE}.

4b. DO sends CT_{ABE} and CT_{HVE} to SP.

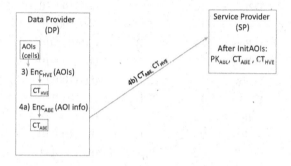

Fig. 4. PrivProxABE InitAOIs phase

Encode AOI Locations. In the "InitAOIs" phase, the DO maps cell IDs to coordinates, (x,y), and generates Gray codes for the x and y values. The complete Gray code for a cell ID is the concatenation of the ordered pair's $GrayCode(x)$ and $GrayCode(y)$. Table 3 shows the Gray codes for the "red", "blue", "green", and "yellow" AOIs presented in Fig. 2.

We use Gray encoding to reduce the number of stored values to represent AOIs. In many cases, using Gray codes dramatically reduces the number of distinct values necessary to represent groups of cells that comprise AOIs. To reduce the number of required comparisons to determine AOI proximity in user queries, we compress the Gray codes into search tokens that contain wildcards, as proposed in [15]. If two Gray code values differ by only a single digit, we replace the digit with a wildcard, "*", allowing the new single search token to represent two individual Gray codes, or Cell IDs. The iterative compression process continues until no two remaining search tokens differ by a single digit, potentially resulting in a small number of search tokens for each AOI.

Table 3. AOI location encoding - step 1

AOI color	Cell ID	Cell coord	Gray code
Red	18	(3,3)	0001000110
Red	19	(4,3)	0011000110
Red	23	(3,4)	0001000111
Red	24	(4,4)	0011000111
Blue	14	(4,3)	0011000010
Blue	15	(5,3)	0011100010
Blue	19	(4,4)	0011000110
Blue	20	(5,4)	0011100110
Green	1	(1,1)	0000100001
Green	6	(1,2)	0000100011
Yellow	9	(4,2)	0011000011
Yellow	10	(5,2)	0011100011
Yellow	12	(2,3)	0001100010
Yellow	13	(3,3)	0001000010
Yellow	14	(4,3)	0011000010
Yellow	15	(5,3)	0011100010
Yellow	17	(2,4)	0001100110
Yellow	18	(3,4)	0001000110
Yellow	19	(4,4)	0011000110
Yellow	20	(5,4)	0011100110
Yellow	22	(2,5)	0001100111
Yellow	23	(3,5)	0001000111
Yellow	24	(4,5)	0011000111
Yellow	25	(5,5)	0011100111

For example, the Gray codes for cell 18 (0001000110) and cell 19 (0011000110) only differ by the value in position 2. Therefore, we can combine the two Gray codes into a single token, replacing the value in position 2 with a wildcard "*", (00*1000110). The wildcard represents a "don't care" value, since the token matches any value in position 2. The Gray codes for cell 23 and 24 also differ by only a single digit and can be represented with the token (00*1000111). Notice that the two resulting tokens differ by only a single digit and can also be combined into a single token (00*100011*). In this way, a single token can represent four cells. Table 4 shows the result of the compression process for the"red", "blue", "green", and "yellow" AOIs presented in Fig. 2.

Table 4. AOI location encoding - step 2

AOI color	Compressed Gray code token(s)
red	00*100011*
blue	0011*00*10
green	00001000*1
yellow	0011*00*1*, 0001*00*10, 0001*00111

Encrypt Encoded AOIs. After compressing each of the Gray encoded AOI locations, each AOI is represented by one or more compressed tokens. The DO encrypts each compressed token with HVE using PK_{HVE}, returning CT_{HVE}. The DO then encrypts the AOI label and any additional AOI information for a single AOI with CP-ABE using PK_{ABE}, returning CT_{ABE}. The DO then sends CT_{ABE} and CT_{HVE} for all AOIs to the SP.

4.4 Encrypting User Location

Algorithm 3 shows the steps in the "InitUserLoc" phase and refers to Fig. 5. If the user is new (eg. not a user known to the KG), the KG authenticates the user and records the authorized user attributes in the KG user table. The KG uses the user attributes to generate a secret key (SK_{ABE}), and returns SK_{ABE} to the user. The user then calls HVEencrypt(currentLocation) on the KG to encrypt the user's current location using HVE, generating Loc_{HVE}. The KG returns Loc_{HVE} to the user. The user then generates a randomIdentifier for use in the query phase. The randomIdentifier detaches queries from user identities.

Algorithm 3. PrivProxABE Protocol - InitUserLoc

5. User calls ABEkeyGen(userID) on KG. The KG authenticates new users and generates a CP-ABE secret key (SK_{ABE}) based on the user's attributes, and returns SK_{ABE} to the user.

6. User calls HVEencrypt(currentLocation) on KG. KG sends the encrypted location (Loc_{HVE}) to the user. The user generates a randomIdentifier.

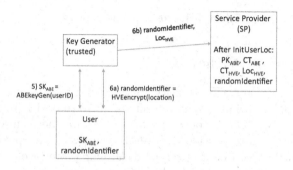

Fig. 5. PrivProxABE InitUserLoc phase

4.5   Querying Proximity to AOIs

Algorithm 4 shows the steps in the "Query" phase and refers to Fig. 6. The HVEdecrypt(Loc$_{HVE}$) function, run on the SP, iterates through each AOI and attempts to decrypt CT$_{HVE}$. If HVE decryption is successful, that means that the user's location shares 1 cell defined by the AOI. The SP returns a list of all successfully decrypted HVE ciphertexts to the user. The user runs ABEdecrypt(SK$_{ABE}$, CT$_{ABE}$) to attempt to decrypt each AOI returned from the SP. Successful ABE decryption means the user's attributes satisfy the AOI's access policy for an AOI that overlaps the current user's location. The service provider returns a list of encrypted AOIs that overlap the user's location. The user then attempts to decrypt each AOI using CP-ABE. Any successfully decrypted AOI means that the user's location overlaps the AOI and the user is authorized to consume the AOI's information.

Algorithm 4. PrivProxABE Protocol - Query

---

7. User calls HVEdecrypt() on SP. SP attempts to decrypt all AOIs using user's Loc$_{HVE}$.

8. SP returns a list of all successfully decrypted (HVE) AOIs.

9. User calls ABEdecrypt(SK$_{ABE}$,CT$_{ABE}$) for each AOI returned from the SP to determine if the user is authorized to consume proximity alerts for each AOI.

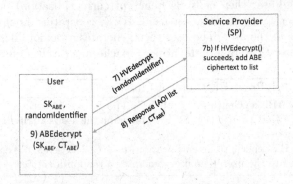

Fig. 6. PrivProxABE Query phase

5   Security and Privacy

In this section, we analyze the security and privacy guarantees of our proposed protocol. One of the primary requirements of this protocol is to maintain the privacy of all actors. The proposed protocol relies on the security guarantees of CP-ABE and HVE to provide the overall security of our approach. In each scheme, the SK is necessary to decrypt data. Our protocol ensures that only trusted entities, users and the KG, have access to any SK. The following list explains the privacy guarantees with respect to each architectural entity.

- Data Owner/Provider (DO) - The DO is the only entity with access to the unencrypted AOI data. It does not share unencrypted AOI locations with any other entity to protect AOI locations. The DO does not have access to any other information generated or stored by the KG or any user, and thus cannot learn any information about user locations.
- Key Generator (KG) - The KG (trusted entity) generates SKs, but never has access to AOIs and cannot decrypt any AOI data without colluding with another entity. The KG does learn user locations as it generates Loc_{HVE} for a user's current location, but never shares this data with any entity other than the user and does not have access to any information, such as AOI locations, with which to correlate user locations. The KG could remember user locations to build user trajectories, but our protocol defines the KG as a trusted entity and we expect that the KG will not exceed its authority.
- Service provider (SP) - The SP never has access to unencrypted AOI or user locations, and cannot access the decryption keys to decrypt any ciphertext. The SP never learns any information about AOI or user locations. The SP does learn that a user is in proximity to one or more AOIs when the locations overlap, but does not have any information to imply actual user/AOI locations. The SP can determine when users are close to one another when they are in proximity to the same AOI(s). The service provider cannot attempt a dictionary type attack by building a fake AOI list since it lacks PK_{HVE} which is necessary to encrypt AOI locations.
- User - The user only divulges actual location to the KG. Since the SP carries out all calculations on encrypted data, the SP never learns the user's location, and as stated previously, the DO never has access to any user location information. The user never has access to any AOI location and only learns general AOI information when the SP discloses that the user is in proximity to one or more AOIs. The user learns that the current location overlaps the reported AOIs, but does not immediately learn the dimensions of any AOI. Through a process of strategically visiting multiple cells, a user could build a map of AOI locations based on proximity alerts. However, a user could only construct meaningful map entries from AOI data the user is authorized to decrypt. One method to restrict malicious users from building even a partial AOI map in a short period of time could be to set a threshold of maximum speed at which a user could realistically travel from cell to cell. If a user attempts to exceed this threshold, he would be deemed malicious. Hallgren et al. [16] propose a protocol, MaxPace, based on the concept of using travel speed thresholds to detect malicious users. Implementing protection from this type of attack is left for future work.

Our protocol protects both AOI and user locations from disclosure, and only allows users to eventually build an AOI map after determined actions resulting in recording history of cells visited and proximity alerts.

6 Experiments

In this section, we evaluate our protocol in terms of computational performance. We do not evaluate communication overhead in this analysis, as it is expected that communication overhead is similar between the approaches we used in out evaluation. In future work we will expand the evaluation to measure communication overhead. We implemented a prototype of our framework and Li's [24] similar framework and ran multiple tests to evaluate performance as input data size varied. We ran each test by first defining access policies, half of which are satisfied by test user attributes. We then define a set of AOIs, each with one of the test access policies. The test user then issues location proximity queries, each with a random user location. The times reported in the results represent the average time to resolve a single location proximity query. All tests were run on a single computer with 16GB memory and an Intel Core i7 2.4GHz CPU running Ubuntu Linux 16.04. We chose to run all tests on a single computer to focus on computation load. Tests were run for varying numbers of AOIs, ranging from 10 to 1500, using grid sizes from 100×100, up to 1500×1500, and various AOI shapes. Figure 7 shows the 11 different AOI shapes used in the performance evaluation. We defined 9 basic shapes, 1 irregular shape, and a standard symmetric AOI represented as a square when using grid cells and as a circle for circle-based AOIs. For Li's Paillier based implementation, we constructed each AOI using multiple circles to approximate the shapes built from grid cells. The only AOI type that can be generally approximated with a single circle is the square AOI. All other AOI shape definitions require multiple circles, which results in more computation.

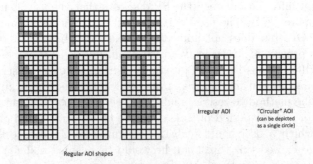

Irregular AOI

"Circular" AOI
(can be depicted
as a single circle)

Regular AOI shapes

Fig. 7. AOI shapes used in tests

Both frameworks were implemented in Python 2.7, using the Charm [2] framework for rapid cryptosystem prototyping. The primary purpose of our prototype implementations is to demonstrate the viability and advantages of our proposed protocol and to compare its relative performance to similar proposed protocol that is less scalable, as opposed to providing a deployment-ready solution. Our experiments showed that varying the grid sizes only had a minor affect

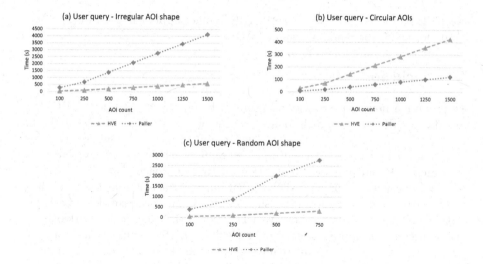

Fig. 8. Experimental results

on performance - much less than varying the number of AOIs. For this reason, we chose to present results of tests all run using a grid size of 250×250. The unit size can represent any physical distance. In this way, the association of a physical measurement with the unit size defines the precision of the approach. That is, a smaller unit size results in higher precision.

Figure 8 shows the performance time of our scheme compared against Paillier-based protocol [24] with varying number of AOIs of different shapes. Our tests show that the choice of HVE for AOI encoding and encryption provides superior scalability over Paillier cryptosystem distance calculations. Li's CP-ABE/Paillier approach works efficiently for circular AOIs, but lacks scalability for more complex AOI shapes. In our evaluations, we represented Li's circular AOIs as square regions. Although squares only approximate circular regions, they provide the most compact structure when using Gray code compression. Figure 8(a) shows that CP-ABE/Paillier performs better than our CP-ABE/HVE framework when all AOIs are represented as circles only. This is because our HVE approach requires a collection of cells to define each AOI. Although using Gray codes with token compression to combine multiple tokens, the Paillier distance calculation for a single circle is still more efficient than an HVE token match.

However, when AOI shapes are not circular, as is desirable to represent distinct regions in the real world of various shapes, the CP-ABE/Paillier approach which requires multiple circles to represent a single AOI exhibits slower performance due to the additional overhead. Our approach can efficiently represent irregular AOI shapes and still provide good performance. Figure 8(b) shows the performance when using AOIs with irregular shapes. The last set of test we ran randomly selected AOI shapes from 10 pre-defined shapes, including circular AOIs. Even when selecting some regular shapes, our approach using CP-ABE and HVE performs better than the P-ABE/Paillier approach. Figure 8(c) shows the performance for randomly shaped AOIs.

7   Conclusion

We propose a novel framework and protocol based on CP-ABE and HVE that provides MPPD with embedded access control for different classifications of users. With our framework, DOs can define and maintain a single set of AOIs and grant access to AOI information based on user attributes. We implemented our framework and protocol, along with another approach that uses CP-ABE and Paillier cryptosystem, and showed that our approach is more scalable when using AOIs defined as non-circular shapes. Our framework provides a basis for implementers to develop scalable MPPD services that minimizes workload on DOs or users. This approach can provide flexible MPPD services to meet a wide variety of client needs, without requiring a trusted third party to examine user locations to determine proximity.

Future work toward refining our MPPD with access control solution include hardening the protocol against attack, optimizing the HVE token compression algorithm, and reducing the communication overhead between framework components. Each of these enhancements will make our protocol more secure and scalable, and easier to deploy to mobile devices with limited resources.

References

1. Agrawal, R., Kiernan, J., Srikant, R., Xu, Y.: Order preserving encryption for numeric data. In: Proceedings of the 2004 ACM SIGMOD International Conference on Management of Data, pp. 563–574. ACM (2004)
2. Akinyele, J.A., Garman, C., Miers, I., Pagano, M.W., Rushanan, M., Green, M., Rubin, A.D.: Charm: a framework for rapidly prototyping cryptosystems. J. Cryptographic Eng. **3**(2), 111–128 (2013)
3. Andrés, M.E., Bordenabe, N.E., Chatzikokolakis, K., Palamidessi, C.: Geo-indistinguishability: differential privacy for location-based systems. In: Proceedings of the 2013 ACM SIGSAC Conference on Computer & Communications Security, pp. 901–914. ACM (2013)
4. Bethencourt, J., Sahai, A., Waters, B.: Ciphertext-policy attribute-based encryption. In: 2007 IEEE symposium on security and privacy (SP 2007), pp. 321–334. IEEE (2007)
5. Boneh, D., Crescenzo, G., Ostrovsky, R., Persiano, G.: Public key encryption with keyword search. In: Cachin, C., Camenisch, J.L. (eds.) EUROCRYPT 2004. LNCS, vol. 3027, pp. 506–522. Springer, Heidelberg (2004). doi:10.1007/978-3-540-24676-3_30
6. Boneh, D., Waters, B.: Conjunctive, subset, and range queries on encrypted data. In: Vadhan, S.P. (ed.) TCC 2007. LNCS, vol. 4392, pp. 535–554. Springer, Heidelberg (2007). doi:10.1007/978-3-540-70936-7_29
7. Bugiel, S., Heuser, S., Sadeghi, A.R.: Flexible and fine-grained mandatory access control on android for diverse security and privacy policies. In: Usenix Security, pp. 131–146 (2013)

8. Calderoni, L., Palmieri, P., Maio, D.: Location privacy without mutual trust: the spatial bloom filter. Comput. Commun. **68**, 4–16 (2015)
9. Choi, S., Ghinita, G., Bertino, E.: Secure mutual proximity zone enclosure evaluation. In: Proceedings of the 22nd ACM SIGSPATIAL International Conference on Advances in Geographic Information Systems, pp. 133–142 (2014)
10. Chor, B., Kushilevitz, E., Goldreich, O., Sudan, M.: Private information retrieval. J. ACM (JACM) **45**(6), 965–981 (1998)
11. Dwork, C.: Differential privacy: a survey of results. In: Agrawal, M., Du, D., Duan, Z., Li, A. (eds.) TAMC 2008. LNCS, vol. 4978, pp. 1–19. Springer, Heidelberg (2008). doi:10.1007/978-3-540-79228-4_1
12. Elmehdwi, Y., Samanthula, B.K., Jiang, W.: Secure k-nearest neighbor query over encrypted data in outsourced environments. In: 2014 IEEE 30th International Conference on Data Engineering, pp. 664–675. IEEE (2014)
13. Fawaz, K., Shin, K.G.: Location privacy protection for smartphone users. In: Proceedings of the 2014 ACM SIGSAC Conference on Computer and Communications Security, pp. 239–250. ACM (2014)
14. Ghinita, G., Kalnis, P., Khoshgozaran, A., Shahabi, C., Tan, K.L.: Private queries in location based services: anonymizers are not necessary. In: Proceedings of the 2008 ACM SIGMOD International Conference on Management of Data, pp. 121–132. ACM (2008)
15. Ghinita, G., Rughinis, R.: A privacy-preserving location-based alert system. In: Proceedings of the 21st ACM SIGSPATIAL International Conference on Advances in Geographic Information Systems, pp. 432–435. ACM (2013)
16. Hallgren, P., Ochoa, M., Sabelfeld, A.: Maxpace: Speed-constrained location queries. In: 2016 IEEE Conference on Communications and Network Security (CNS), pp. 136–144. IEEE (2016)
17. Hengartner, U.: Hiding location information from location-based services. In: 2007 International Conference on Mobile Data Management, pp. 268–272. IEEE (2007)
18. Hossain, A.A., Lee, S.J., Huh, E.N.: Shear-based spatial transformation to protect proximity attack in outsourced database. In: 2013 12th IEEE International Conference on Trust, Security and Privacy in Computing and Communications (TrustCom), pp. 1633–1638. IEEE (2013)
19. Khoshgozaran, A., Shahabi, C.: Private buddy search: enabling private spatial queries in social networks. In: International Conference on Computational Science and Engineering, CSE 2009, vol. 4, pp. 166–173. IEEE (2009)
20. Khoshgozaran, A., Shahabi, C.: Private information retrieval techniques for enabling location privacy in location-based services. In: Bettini, C., Jajodia, S., Samarati, P., Wang, X.S. (eds.) Privacy in Location-Based Applications. LNCS, vol. 5599, pp. 59–83. Springer, Heidelberg (2009). doi:10.1007/978-3-642-03511-1_3
21. Kim, H.I., Chang, J.W.: k-nearest neighbor query processing algorithms for a query region in road networks. J. Comput. Sci. Technol. **28**(4), 585–596 (2013)
22. Kim, H.I., Hong, S., Chang, J.W.: Hilbert curve-based cryptographic transformation scheme for spatial query processing on outsourced private data. Data Knowl. Eng. **104**, 32–44 (2015)
23. Kim, H.I., Kim, Y.K., Chang, J.W.: A grid-based cloaking area creation scheme for continuous lbs queries in distributed systems. J. Convergence **4**(1), 23–30 (2013)
24. Li, X.Y., Jung, T.: Search me if you can: privacy-preserving location query service. In: 2013 Proceedings of the IEEE INFOCOM, pp. 2760–2768 (2013)
25. Lu, R., Lin, X., Shen, X.: Spoc: a secure and privacy-preserving opportunistic computing framework for mobile-healthcare emergency. IEEE Trans. Parallel Distrib. Syst. **24**(3), 614–624 (2013)

26. Palmieri, P., Calderoni, L., Maio, D.: Spatial bloom filters: enabling privacy in location-aware applications. In: Lin, D., Yung, M., Zhou, J. (eds.) Inscrypt 2014. LNCS, vol. 8957, pp. 16–36. Springer, Cham (2015). doi:10.1007/978-3-319-16745-9_2
27. Šeděnka, J., Gasti, P.: Privacy-preserving distance computation and proximity testing on earth, done right. In: Proceedings of the 9th ACM Symposium on Info, Computer and Comm Security, ASIA CCS 2014, pp. 99–110 (2014)
28. Wang, B., Li, M., Wang, H.: Geometric range search on encrypted spatial data. IEEE Trans. Info Forensics Secur. **11**(4), 704–719 (2016)
29. Xiao, Y., Xiong, L.: Protecting locations with differential privacy under temporal correlations. In: Proceedings of the 22nd ACM SIGSAC Conference on Computer and Communications Security, pp. 1298–1309. ACM (2015)
30. Yiu, M.L., Ghinita, G., Jensen, C.S., Kalnis, P.: Enabling search services on outsourced private spatial data. VLDB J. **19**(3), 363–384 (2010)

Privacy-Preserving Elastic Net for Data Encrypted by Different Keys - With an Application on Biomarker Discovery

Jun Zhang, Meiqi He, and Siu-Ming Yiu[✉]

Department of Computer Science, The University of Hong Kong,
Pokfulam Road, Pok Fu Lam, Hong Kong
{jzhang3,mqhe,smyiu}@cs.hku.hk

Abstract. Elastic net is a popular linear regression tool and has many important applications, in particular, finding genomic biomarkers for cancers from gene expression profiles for personalized medicine (elastic net is currently the most accurate prediction method for this problem). There is an increasing trend for organizations to store their data (e.g. gene expression profiles) in an untrusted third-party cloud system in order to leverage both its storage capacity and computational power. Due to the privacy concern, data must be stored in its encrypted form. While there are quite a number of privacy-preserving data mining protocols on encrypted data, there does not exist one for elastic net. In this paper, we propose the first privacy-preserving elastic net protocol using two non-colluding servers. Our protocol is able to handle expression profiles encrypted from multiple medical units using different encryption keys. Thus, collaboration between multiple medical units are made possible without jeopardizing the privacy of data records. We formally prove that our protocol is secure and implemented the protocol. The experimental results show that our protocol runs reasonably fast, thus can be applied in practice.

Keywords: Privacy-preserving elastic net · Multiple encryption keys · Encrypted gene expression profiles · Biomarker discovery

1 Introduction

We motivate our study based on the following biomarker discovery application. The current cancer treatment based on doctors' empirical knowledge can be described as "one-size-fits-all" - almost all the patients diagnosed with the same cancer will receive similar treatment. Under this situation, some patients are likely to be under-treated while others be over-treated. Even worse, not all patients will benefit from the treatment, a proportion of them may suffer from severe side effects. By contrast, personalized medicine aims at treating patients differently with different drugs at the right dose [1]. To achieve personalized treatment for cancer, we need *biomarkers* (i.e. a set of genes) to predict

© IFIP International Federation for Information Processing 2017
Published by Springer International Publishing AG 2017. All Rights Reserved
G. Livraga and S. Zhu (Eds.): DBSec 2017, LNCS 10359, pp. 185–204, 2017.
DOI: 10.1007/978-3-319-61176-1_10

a patient's response to anticancer drugs (e.g. sensitivity and resistance). With the advert of bioinformatics technology, we are able to make use of data mining and statistical methods to discover biomarkers from genomic data. Cancer Genome Project (CGP) [2] and Cancer Cell Line Encyclopedia (CCLE) [3] are examples showing the analysis results for discovering biomarkers using genomic features derived from human tumor samples against drug responses. A typical input for genomic features is a *gene expression profile* which is a vector recording the degrees of activation of different genes. The number of genes can be up to tens of thousands. A patient's response is usually measured by GI50 value (log of drug concentration for 50% growth inhibition) [4]. Given n gene expression profiles (e.g. from n patients) of which the dimension is m ($n \ll m$), the task is to perform regression analysis between gene expression profiles and GI50 values. Elastic net regression was found to be the most accurate predictor [5] among existing approaches.

Why encrypted by different keys? Due to the huge volume of medical records and DNA information, there is an increasing trend for medical units to make use of a third-party cloud system to store the records as well as to leverage its massive computational power to analyze the data. It is well recognized that genomic information such as DNA is particularly sensitive and must be well protected [6]. The privacy of gene expression has been overlooked until Schadt et al. pointed out that gene contents can be inferred based on expression profiles alone [7]. Even worse, some expression data is strongly correlated to important personal indexes such as body mass index and insulin levels. It is likely that an entire profile can be derived and linked to a specific individual. Therefore, gene expression profiles stored in cloud should also be encrypted. As medical units need to *retrieve* expression profiles when implementing personalized treatment, profiles from different medical units would be encrypted using *different keys* to avoid leaking the details of the records to other medical units.

It is important for different medical units to combine their datasets in order to increase the size of n (the number of patients) for accurate predication. Collaborative data mining on encrypted data is a promising direction for medical units to "share" data for more accurate prediction without jeopardizing the privacy of the data. The problem to be tackled in this paper is to design a privacy-preserving elastic net protocol to predict biomarkers based on gene expression profiles encrypted by different keys and GI50 values. Our goal is to ensure that the cloud learns nothing about the patients' expression profiles beyond what is revealed by the final result of elastic net regression.

Difficulties of the problem: There is no existing work for elastic net or lasso (another popular linear regression model) [8] while most of the work was designed for ordinary least square (OLS) and ridge regression. The difficulty lies on the fact that unlike solving OLS and ridge regression, the state-of-the-art solution (e.g. glmnet [9]) for elastic net is based on an iterative algorithm, which requires information of one training sample in each iteration. It is not clear how to perform these iterations if all data records are encrypted. Other existing solutions (e.g. Least Angle Regression (LARS) [10], computing the Euclidean projection onto a closed convex set [11] and using proximal stochastic dual coordinate descent [12]) suffer from a similar problem.

Ideas of our proposed solution: Instead of using the iterative algorithms to solve the elastic net problem directly, it has been proved that elastic net regression can be reduced to support vector machine (SVM) [13]. An identical solution as glmnet [9] up to a tolerance level can be obtained with a solver for SVM. Our main idea is to transform the encrypted training dataset of elastic net to that of SVM, based on which we compute the *gram matrix*[1]. Then the gram matrix will be used as input to a modern SVM solver. Once obtaining the solution to SVM, we reconstruct the solution to elastic net. We make sure that the cloud server cannot recover patients' expression profiles based on the gram matrix, for which we provide a security proof in this paper.

Roughly speaking, there are two ways to achieve privacy preserving SVM. One is perturbation based approach. Data sent to the cloud is perturbed by a random transformation [14], which considers only one user (i.e. medical unit). The other is cryptography based approach, such as secret sharing [15], Oblivious Transfer (OT) [15,16] and Fully Homomorphic Encryption (FHE) [15,17]. The cryptography based approach provides a higher level of privacy compared to perturbation based approach, but incurs higher computation/communication overhead. Most of the previous work focused on distributed databases [15,16,18–20], while we consider a centralized *outsourced encrypted database under multiple keys*. Liu et al. proposed a secure protocol based on FHE for outsourced encrypted SVM [17], but it requires the users to be online during the whole process. It has been proved that completely non-interactive multiple party computation cannot be achieved in the single server setting when user-server collusion might exist [21]. Thus, we need at least *two non-colluding servers* [22] if we want to keep the medical units offline. This two non-colluding servers model makes sense in the practical community (e.g. [22,23]). For example, we set up two cloud servers which belong to Amazon Web Services (AWS) cloud service and Google Cloud Platform (GCP) respectively. Considering the consequences of legal action for breach of contract and bad reputation, it is reasonable to assume that they will not collude. According to [24], each user can secret-share its data among the two non-colluding servers. Then the two servers compute on the shares of the input interactively and send the shares of the result to the users to reconstruct the final output. Although the secret sharing based approach is better in terms of computation cost, it incurs higher communication cost [25] and cannot deal with data encrypted under multiple keys. Moreover, oblivious transfer focuses on the single key setting, which is not suitable for the case of *multiple keys*. Consequently, we focus on homomorphic encryption based approach in this paper. There indeed exists a multikey FHE primitive that allows computation on data encrypted under multiple keys [26]. However, its efficiency is still far from practice and it requires interactions among all the medical units during the decryption phase. Peter et al. came up with a scheme that transforms the ciphertexts under different keys into those under the same

---

[1] A matrix that contains dot product of any two training samples.

key [27], incurring a huge amount of interactions between the servers. To reduce communication overhead, proxy re-encryption [30] can be utilized to transform ciphertexts [28, 29]. However, the amount of interactions is still heavy. Because they used partially homomorphic encryption - if the underlying cryptosystem is additively homomorphic, they need joint work between the two servers to compute multiplication and vice versa. To further reduce the *communication overhead*, we utilize a framework to enable additively homomorphic encryption to support one multiplication [31]. We choose the BCP Cryptosystem [32] as the underlying additively homomorphic encryption and modify it to support multikey additive homomorphism. In this way, we successfully remove the need to transform the ciphertexts to those under the same key, while it is a must in [27–29]. To remove the constraint that medical units need to be online during decryption phase, we divide a medical unit's secret key s into two shares s_1 and s_2, and distribute them to the servers. Final decryption is obtained after two rounds of partial decryption.

To summarize, our contributions include the following:

(1) We construct a homomorphic cryptosystem that supports one multiply operation under single key and multiple add operations under both single key and different keys. Compared with the BCP cryptosystem, our scheme only doubles the encryption time. With 1024-bit security parameter, an add operation takes less than 1 ms while a multiply operation takes about 16 ms. The size of ciphertext increases linearly from 6138 B to 26 KB with the number of involved users increasing from 2 to 100. Overall speaking, the proposed scheme is practical.

(2) We propose the first privacy preserving protocol to solve elastic net on gene expression profiles encrypted by different encryption keys for cancer biomarker discovery, which encourages cooperation between medical units. Through reduction from elastic net to SVM, we demonstrate how to train SVM securely based on the gram matrix. The solution to elastic net is reconstructed based on the solution to SVM. Moreover, our solution can allow users (medical units) to stay offline except for the initialization phase.

(3) We evaluate our scheme on a real database[2] for drug sensitivity in cancer cell lines [33]. Moreover, our scheme can also be used to solve lasso, based on a similar reduction from lasso to SVM [34].

2   Model Description

In this paper, we propose a collaborative model for privacy preserving biomarker discovery for anticancer drugs using encrypted expression profiles extracted from the tumor samples of patients. As shown in Fig. 1, the involved parties are patients, medical units, certified institution and the cloud.

(1) Patients (P). Cancer patients go to the medical units to receive personalized treatment. We list six patients here labeled as $\{P_1, P_2 \cdots, P_6\}$.

---

[2] http://www.cancerrxgene.org, accessed on 10 Aug 2016.

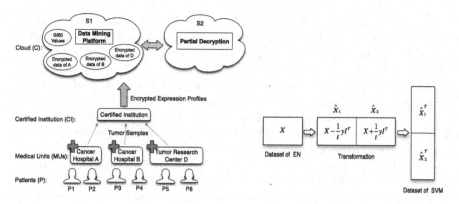

Fig. 1. System model for genomic bio-marker discovery through collaborative data mining.

Fig. 2. Dataset transformation from elastic net to SVM

(2) Medical Units (MUs). There are different MUs (e.g. cancer hospitals, tumor research centers) in our model. Each MU is able to extract tumor samples from the patients, observe the effect of 72 h of anticancer drug treatment on them, and upload the GI50 values to the cloud. On the other hand, MU sends the tumor samples to the certified institution.

(3) Certified Institution (CI). CI is responsible to perform gene expression profiling. CI encrypts the gene expression profiles from different MUs with different encryption keys, and sends the encrypted profiles to the cloud. Only the MU that holds the correct private key can decrypt the encrypted profiles.

(4) Cloud (C). It consists of two non-colluding servers S_1 and S_2, which is responsible for storage and massive computation.

Threat model: CI is a trusted party. S_1 and S_2 are both honest-but-curious, and they are non-colluding. There might exist collusion between a MU and S_1. However, none of the medical units will collude with S_2. We consider two types of potential attacks: (i) attacker at one MU tries to know the expression profiles of other MUs (ii) attacker at S_1 or S_2 in the cloud aims at recovering gene expression profiles through observing the input, intermediate or final results.

3 Preliminaries

3.1 Elastic Net Regression

Let the input dataset be $\{(x_i, y_i)\}_{i=1}^{n}$, where each $x_i \in R^m$ is a column vector representing a gene expression profile, and $y_i \in R$ is the GI50 value.[3] Let $X \in R^{n \times m}$ be a matrix containing all gene expression profiles (the transposed

---

[3] GI50 denotes the log of the drug concentration for 50% growth inhibition.

i-th row of X is x_i) and the column vector $y \in R^n$ (i-th element of y is y_i) be the responses, the goal of linear regression analysis is to find a column vector $\beta \in R^m$ such that y_i can be approximated by $\tilde{y}_i = \beta^T x_i$. The ordinary least squares (OLS) regression works by minimizing the residual sum of squares $\min_\beta ||X\beta - y||_2^2$. There are some situations where OLS is not a good solution, for example, when m is large or the columns of X are highly correlated. One way to handle this problem is to introduce a penalization term. Ridge regression uses l_2-norm penalization ($||\beta||_2^2$), while lasso regression uses l_1-norm penalization ($|\beta|_1$). Ridge regression cannot produce a sparse model. By contrast, owing to the nature of l_1 penalty, lasso is able to generate a sparse model. Nevertheless, lasso has some limitations - it selects at most n variables in the $n \ll m$ case, picks out only one variable from a group of correlated variables not caring which one is selected (the robustness issue: we want to identify all related variables). In our application, since $n \ll m$ and genes may be highly correlated, lasso regression is not the ideal method in this situation and elastic net penalty ($\lambda_1|\beta|_1 + \lambda_2||\beta||_2^2$) is introduced, which is a convex combination of the lasso and ridge penalty [10]. It performs well under the situation of $n \ll m$ and correlated variables. The elastic net regression can be represented as follows.

$$\min_{\beta \in R^m} ||X\beta - y||_2^2 + \lambda||\beta||_2^2 \qquad such\ that\ |\beta|_1 \leq t \qquad (1)$$

where $\lambda > 0$ is the l_2-regularization constant and $t > 0$ is the l_1-norm budget.

3.2   Support Vector Machine with Squared Hinge Loss

Given that we have a dataset $\{(x_i, y_i)\}_{i=1}^n$ where $x_i \in R^m$ and $y_i \in \{+1, -1\}$, we aim at finding a separating plane $w^T x + b = 0$ ($w \in R^m$) to classify the training samples into two classes. There exists many eligible separating planes. For sake of robustness, support vector machine maximizes the margin ($\frac{1}{||w||}$) between two classes, which is equivalent to minimize $||w||^2$. However, sometimes the training dataset is linearly inseparable. One solution is to allow SVM to make mistakes on some samples. We use the squared hinge loss $max(0, 1 - y_i(w^T x_i + b))^2$ to measure the error of sample x_i, which need to be minimized. Therefore, the linear SVM with squared hinge loss can be represented as follows.

$$\min_w \frac{1}{2}w^T w + C\sum_{i=1}^n max(0, 1 - y_i(w^T x_i + b))^2 \qquad (2)$$

where C is the penalty parameter of the error term. The above is the primal form of SVM, which is often solved in its dual form:

$$\min_{\alpha_i \geq 0} \quad f(\alpha) = \alpha^T Q\alpha + \frac{1}{2C}\sum_{i=1}^n \alpha_i^2 - 2\sum_{i=1}^n \alpha_i \qquad (3)$$

where $\alpha \in R^n$ and each α_i is the coefficient for x_i. Q is a $n \times n$ matrix with $Q_{ij} = y_i y_j x_i^T x_j$. Gram matrix K is defined as $K = x_i^T x_j$. Once we get α by solving (3), we can further compute $w = \sum_{i=1}^n \alpha_i x_i y_i$.

3.3 Reduction from Elastic Net to SVM

Zhou et al. demonstrated that elastic net regression can be reduced to SVM [13]. They do not include any bias item b (they assume that the separating hyperplane passes through the origin). After a series of transformations, (1) and (3) can be changed to (4) and (5) respectively.[4] We do not provide the transformation steps (Please refer to [13] for details).

$$\min_{\hat{\beta}_i > 0} ||\hat{Z}\hat{\beta}||_2^2 + \lambda \sum_{i=1}^{2m} \hat{\beta}_i^2 \qquad \sum_{i=1}^{2m} \hat{\beta}_i = 1 \qquad (4)$$

where $\hat{Z} = [\hat{X}_1, -\hat{X}_2] \in R^{n \times 2m}$, $\hat{X}_1 = X - \frac{1}{t}yI^T$ and $\hat{X}_2 = X + \frac{1}{t}yI^T$ ($I \in R^m$ is an identity vector).

$$\min_{\alpha_i > 0} ||Z(\frac{\alpha}{|\alpha^*|_1})||_2^2 + \lambda \sum_{i=1}^{2m} (\frac{\alpha_i}{|\alpha^*|_1})^2 \qquad \sum_{i=1}^{2m} \frac{\alpha_i}{|\alpha^*|_1} = 1 \qquad (5)$$

where $Z = y_i x_i$, $C = \frac{1}{2\lambda}$ and α^* is the optimal solution. Comparing (4) and (5), we notice that they have similar form except for two differences. The first one is that the class labels in elastic net are real valued but binary in SVM. As shown in Fig. 2, to transform the training dataset X of elastic net to that of SVM, we compute $\hat{X}_1$ as subtracting each column of X by $\frac{1}{t}y$ and calculate $\hat{X}_2$ as adding each column of X by $\frac{1}{t}y$, then concatenate $\hat{X}_1$ and $\hat{X}_2$ together and transpose it. The first m training samples of SVM are of class $+1$, and the remaining are of class -1. The second difference is that they have different scale. The optimal solution $\hat{\beta}^*$ can be represented by the optimal solution α^* as $\hat{\beta}^* = \frac{\alpha^*}{|\alpha^*|_1}$. Finally, the optimal solution β to elastic net (see (1)) can be recovered from $\hat{\beta}$ according to $\beta = t \times (\hat{\beta}_{1\cdots m} - \hat{\beta}_{m+1\cdots 2m})$, where t is the l_1-norm budget and $\hat{\beta}_{i\cdots j}$ denotes a vector consisting of elements of $\hat{\beta}$ from index i to j.

4   Our Scheme

Fully homomorphic encryption can be used to compute arbitrary polynomial functions over encrypted data. However, the high computational complexity and communication cost preclude its use in practice. If only focusing on those operations of interest to the target application, more practical homomorphic encryption schemes are possible. For example, Zhou and Wornell [35] proposed an integer vector encryption scheme which supports addition, linear transformation and weighted inner product on ciphertexts. Nevertheless, reduction from elastic net to SVM leads to changes of the training dataset. To be specific, one gene expression profile of a patient across all genes (i.e. one row) is a training sample of elastic net. But one training sample of SVM see Fig. 2 can be considered as gene expression values of a particular gene among all the patients (i.e. one column).

---

[4] Here we use $\hat{\beta} \in R^{2m}$ to differentiate from $\beta \in R^m$, β can be derived from $\hat{\beta}$.

Therefore, if we encrypt gene expression profiles using the integer vector encryption, there is no way to construct ciphertexts for the training dataset of SVM. As a result, we restrict our attention to cryptosystems encrypting one element of the profile at a time instead of encrypting the whole profile. Recall that we can use gram matrix as input to train SVM (see Sect. 3.2) and the basic operation to compute gram matrix is the dot product of two samples, it requires a ciphertext to support one multiply operation and multiple add operations. Indeed, the BGN cryptosystem [36] can compute one multiplication on ciphertexts using the bilinear maps. However, it does not support multikey homomorphism. In our setting of collaborative data mining in the cloud, the training dataset of elastic net is horizontally partitioned (different units holding different records with the same set of attributes) while the training dataset of SVM is vertically partitioned (records are partitioned into different parts with different attributes after the transformation). In order to train SVM on encrypted training dataset, we thus need a cryptosystem that *supports one multiply operation under single key and multiple add operations under both single key and different keys*. In this paper, we try to let medical units stay offline except for the initialization phase. Specifically, we use secret sharing to authorize one server (i.e. S_1) to decrypt the encrypted gram matrix without knowing the secret key of any medical unit.

4.1 Building Blocks

Framework to Enable One Multiplication on Cihphertexts. Catalano and Fiore [31] showed a framework to enable existing additively homomorphic encryption schemes (i.e. Paillier, ElGamal) to compute multiplication on ciphertexts. We use $E()$ to denote the underlying additively homomorphic encryption. The idea is to transform a ciphertext $E(x_{ij})$ into "multiplication friendly". To be specific, we use $\mathcal{E}(x_{ij}) = (x_{ij} - b_{ij}, E(b_{ij}))$ (where b_{ij} is a random number) to represent the "multiplication friendly" ciphertext. Given two "multiplication friendly" ciphertexts $\mathcal{E}(x_{11}) = (x_{11} - b_{11}, E(b_{11}))$ and $\mathcal{E}(x_{21}) = (x_{21} - b_{21}, E(b_{21}))$, we compute multiplication as $\mathcal{E}(x_{11}x_{21}) = (\alpha_1, \beta_1, \beta_2)$.

$$\alpha_1 = E[(x_{11} - b_{11})(x_{21} - b_{21})]E(b_{11})^{x_{21}-b_{21}}E(b_{21})^{x_{11}-b_{11}}$$
$$= E(x_{11}x_{21} - b_{11}b_{21}) \tag{6}$$
$$\beta_1 = E(b_{11}) \qquad \beta_2 = E(b_{21}) \tag{7}$$

To decrypt $\mathcal{E}(x_{11}x_{21})$, we will add $b_{11}b_{21}$ to the decryption of α where b_{11}, b_{21} is retrieved from β_1, β_2. The addition of two ciphertexts after multiplication works by adding the α components and concatenating the β components. Therefore, the β component will grow linearly with additions after performing a multiplication. To remove this constraint, two non-colluding servers are used to store $\mathcal{E} = (x_{ij} - b_{ij}, E(b_{ij}))$ and b_{ij} respectively. In this way, S_1 can throw away the β component after performing a multiplication, because S_2 will operate on the b_{ij}'s in plaintext. Therefore, the ciphertext contains only the α component after performing a multiplication. This framework has a nice property that it *inherits the multikey homomorphism* of the underlying additively homomorphic encryption.

Given two safe primes p and q, we compute $N = pq$, $g = -a^{2N}$ ($a \in \mathbb{Z}_{N^2}^*$), the secret key $s \in [1, \frac{N^2}{2}]$, the public key $(N, g, h = g^s)$.

Encryption: To encrypt plaintext $m \in \mathbb{Z}_N$, we select a random $r \in [1, \frac{N}{4}]$ and generate the ciphertext $E(m) = (C_m^{(1)}, C_m^{(2)})$ as below:

$$C_m^{(1)} = g^r \bmod N^2 \quad and \quad C_m^{(2)} = h^r(1 + mN) \bmod N^2 \tag{8}$$

Decryption:

$$t = \frac{C_m^{(2)}}{(C_m^{(1)})^s} \qquad m = \frac{t - 1 \bmod N^2}{N} \tag{9}$$

Proxy Re-encryption: If the secret key s is divided into two shares s_1, s_2 such that $s = s_1 + s_2$, then we can use s_1 to partially decrypt $E(m)$ to $E(m)' = (C_m^{(1)'}, C_m^{(2)'})$, which can be considered as a ciphertext under key s_2.

$$C_m^{(1)'} = C_m^{(1)} \qquad C_m^{(2)'} = \frac{C_m^{(2)}}{C_m^{(1)^{s_1}}} \tag{10}$$

Single Key Homomorphism: Supposed that we have two plaintexts m_1, m_2 and their ciphertexts $E(m_1) = (C_{m_1}^{(1)}, C_{m_1}^{(2)})$ and $E(m_2) = (C_{m_2}^{(1)}, C_{m_2}^{(2)})$ under the same key s. The ciphertext $E(m_1 + m_2)$ can be computed as $E(m_1 + m_2) = (C_{m_1}^{(1)} C_{m_2}^{(1)}, C_{m_1}^{(2)} C_{m_2}^{(2)})$.

Fig. 3. The BCP cryptosystem

Multikey Homomorphism of the BCP Cryptosystem. The BCP cryptosystem (also known as Modified Paillier Cryptosystem) is an additively homomorphic encryption under single key [32]. We briefly review the BCP cryptosystem in Fig. 3 and discuss how to modify it to support multikey homomorphism at the expense of expanding the ciphertext size. Supposed that $E(m_a) = (C_{m_a}^{(1)}, C_{m_a}^{(2)})$ is under key s_a, $E(m_b) = (C_{m_b}^{(1)}, C_{m_b}^{(2)})$ is under key s_b, then $E^{(ab)}(m_a + m_b)$ where $E^{(ab)}$ denotes a ciphertext related to key s_a and s_b can be computed as

$$E^{(ab)}(m_a + m_b) = (C_{m_a}^{(1)}, C_{m_b}^{(1)}, C_{m_a}^{(2)} C_{m_b}^{(2)}) \tag{11}$$

The ciphertext size *only depends on the number of involved medical units (i.e. keys)*. There are two MUs with key s_a and s_b respectively in this example, the addition of their ciphertexts is a 3-tuple. If n MUs cooperate together, the addition of their ciphertexts should be a $(n+1)$-tuple. To decrypt $E^{(ab)}(m_a + m_b)$, the secret key s_a and s_b are required.

$$t = \frac{C_{m_a}^{(2)} C_{m_b}^{(2)}}{(C_{m_a}^{(1)})^{s_a} (C_{m_b}^{(1)})^{s_b}} \qquad m_a + m_b = \frac{t - 1 \bmod N^2}{N} \tag{12}$$

Incorporating the above modified the BCP cryptosystem to the framework that enables additively homomorphic encryption to support one multiplication, we obtain our final encryption scheme $\mathcal{E}_{BCP}$.[5]

---

[5] $\mathcal{E}$ denotes the framwork and BCP denotes the underlying cryptosystem.

Gram Matrix Computation. Gram matrix K is defined as $K_{ij} = \langle x_i, x_j \rangle = x_i^T x_j$ where x_i and x_j are any two training samples (see Sect. 3.2). Recall that the original training dataset X of elastic net regression is transformed to the training dataset $\hat{X}$ of SVM during the reduction process (see Sect. 3.3), we use $\hat{X} = \{\hat{x}_i\}_{i=1}^{2m}$ to denote the transformed dataset. After dataset transformation, the horizontally partitioned dataset of the elastic net is converted to vertically partitioned dataset of SVM. The gram matrix $K(\hat{X})$ of $\hat{X}$ is computed as follows.

$$K(\hat{X}) = \begin{bmatrix} \langle \hat{x}_1, \hat{x}_1 \rangle & \langle \hat{x}_1, \hat{x}_2 \rangle & \cdots & \langle \hat{x}_1, \hat{x}_{2m} \rangle \\ \langle \hat{x}_2, \hat{x}_1 \rangle & \langle \hat{x}_2, \hat{x}_2 \rangle & \cdots & \langle \hat{x}_2, \hat{x}_{2m} \rangle \\ \vdots & \vdots & \ddots & \vdots \\ \langle \hat{x}_{2m}, \hat{x}_1 \rangle & \langle \hat{x}_{2m}, \hat{x}_2 \rangle & \cdots & \langle \hat{x}_{2m}, \hat{x}_{2m} \rangle \end{bmatrix} \tag{13}$$

For ease of description, we firstly consider the case of two medical units denoted as MU_A and MU_B. Assume that the cloud store n gene expression profiles, among which n_A records are from MU_A and n_B records are from MU_B. Then in the transformed dataset of SVM, for each training sample, the first n_A elements are encrypted under key s_A, the remaining n_B elements are encrypted under key s_B. Assume that we have two training samples $\hat{x}_1$ and $\hat{x}_2$ of SVM, their dot product $\langle \hat{x}_1, \hat{x}_2 \rangle$ can be computed as follows and their ciphertexts are denoted as $\mathcal{E}_{BCP}(\hat{x}_1) = (\hat{x}_1 - b_1, E(b_1))$, $\mathcal{E}_{BCP}(\hat{x}_2) = (\hat{x}_2 - b_2, E(b_2))$.

$$\langle \hat{x}_1, \hat{x}_2 \rangle = \sum_{i=1}^{n_A} \hat{x}_{1i}\hat{x}_{2i} + \sum_{i=n_A+1}^{n} \hat{x}_{1i}\hat{x}_{2i} \tag{14}$$

Supposed that the ciphertext of $\sum_{i=1}^{n_A} \hat{x}_{1i}\hat{x}_{2i}$ and $\sum_{i=n_A+1}^{n} \hat{x}_{1i}\hat{x}_{2i}$ are α_A and α_B respectively,[6] then S_1 will compute α_A, α_B as follows. The computation of α_A or α_B only requires single key homomorphism.

$$\alpha_A = E(\sum_{i=1}^{n_A} \hat{x}_{1i}\hat{x}_{2i} - b_{1i}b_{2i}) = (C_A^{(1)}, C_A^{(2)}) \tag{15}$$

$$\alpha_B = E(\sum_{i=n_A+1}^{n_A+n_B} \hat{x}_{1i}\hat{x}_{2i} - b_{1i}b_{2i}) = (C_B^{(1)}, C_B^{(2)}) \tag{16}$$

As α_A and α_B are encrypted under different keys, adding them together requires multikey homomorphism.

$$E(\langle \hat{x}_1, \hat{x}_2 \rangle - \langle \hat{b}_1, \hat{b}_2 \rangle) = \alpha_A + \alpha_B = (C_A^{(1)}, C_B^{(1)}, C_A^{(2)} C_B^{(2)}) \tag{17}$$

Keep Medical Units Offline. We leverage $\mathcal{E}_{BCP}$'s proxy re-encryption property, which inherits from the underlying the BCP cryptosystem (see (10)). To keep MU_A and MU_B offline, we split the secret key s of each involved medical

---

[6] S_1 will abandon the β_1 and β_2 component after a multiplication.

unit into two shares. Specifically, we have $s_A = s_{A_1} + s_{A_2}$ and $s_B = s_{B_1} + s_{B_2}$. S_1 holds s_{A_1}, s_{B_1} and S_2 holds s_{A_2}, s_{B_2}. To compute $\langle \hat{x}_1, \hat{x}_2 \rangle$, S_1 will firstly decrypt (17) partially.

$$C_A^{(1)'} = C_A^{(1)} \qquad C_B^{(1)'} = C_B^{(1)} \tag{18}$$

$$C_A^{(2)'} C_B^{(2)'} = \frac{C_A^{(2)} C_B^{(2)}}{(C_A^{(1)})^{s_{A_1}} (C_B^{(1)})^{s_{B_1}}} \tag{19}$$

Then S_1 will send $C_A^{(1)}$ and $C_B^{(1)}$ to S_2. S_2 will compute and return $(C_A^{(1)})^{s_{A_2}}$, $(C_B^{(1)})^{s_{B_2}}$, $\langle \hat{b}_1, \hat{b}_2 \rangle$ to S_1 afterwards. Finally, S_1 is able to decrypt $E(\langle \hat{x}_1, \hat{x}_2 \rangle - \langle \hat{b}_1, \hat{b}_2 \rangle)$ completely and get $\langle \hat{x}_1, \hat{x}_2 \rangle$ in plaintext.

$$C_A^{(2)''} C_B^{(2)''} = \frac{C_A^{(2)'} C_B^{(2)'}}{(C_A^{(1)'})^{s_{A_2}} (C_B^{(1)'})^{s_{B_2}}} \tag{20}$$

$$\langle \hat{x}_1, \hat{x}_2 \rangle = \frac{(C_A^{(2)''} C_B^{(2)''} - 1) \bmod n^2}{n} + \langle \hat{b}_1, \hat{b}_2 \rangle \tag{21}$$

The above shows how to compute $\langle \hat{x}_1, \hat{x}_2 \rangle$ based on two ciphertexts $\mathcal{E}_{BCP}(\hat{x}_1)$ and $\mathcal{E}_{BCP}(\hat{x}_2)$. Similarly, we can compute each element of the gram matrix $K_{ij} = \langle \hat{x}_i, \hat{x}_j \rangle = \hat{x}_i^T \hat{x}_j$ based on the ciphertexts $\mathcal{E}_{BCP}(\hat{x}_i)$ and $\mathcal{E}_{BCP}(\hat{x}_j)$. Observing that the gram matrix in (13) is symmetric, we can only compute the upper triangular half of it. In the end, S_1 gets the gram matrix K in plaintext. If there are more than two medical units, we can easily extend (14), (15) and (17) to handle the case of multiple medical units. The size of ciphertext of $\langle \hat{x}_1, \hat{x}_2 \rangle$ increases linearly with the number of involved medical units. Likewise, the communication overhead also increases linearly during the decryption phase.

4.2 Our Construction

Given the encrypted gene expression profiles $\mathcal{E}_{BCP}(X)$ derived from multiple medical units, the cloud runs privacy preserving elastic net on it to discover biomarkers to predict a patient's response to anticancer drugs. As it is not clear how to design a privacy preserving protocol based on iterative algorithms to solve elastic net. We resort to reduction to shift our attention from elastic net to SVM. In Algorithm 1, we firstly demonstrate how to transform the encrypted dataset of elastic net to that of SVM (see Sect. 3.3). It is easy to perform such transformation on the dataset in plaintext. However, once it is encrypted, we need to rely on the homomorphic properties of our cryptosystem to finish the transformation. Next, we compute the encrypted gram matrix $\mathcal{E}_{BCP}(K)$ of the transformed training dataset (see Sect. 4.1). Gram matrix plays a role as intermediate dataset based on which SVM model can be generated correctly without breaching the privacy of patients' gene expression profiles. In order to keep medical units offline, we authorize S_1 to decrypt $\mathcal{E}_{BCP}(K)$. Based on K, we train SVM and obtain the solution α. Finally, we use α to reconstruct β, which is the solution to elastic net.

Algorithm 1. Protocol for Privacy-preserving Elastic Net.

Input: Encrypted dataset $\mathcal{E}_{BCP}(X) = \{\mathcal{E}_{BCP}(x_i)\}_{i=1}^n$, where $x_i \in Z^m$ and response vector in plaintext $y \in R^n$; l_1-norm budget t and l_2-regularization parameter λ.
Output: Solution β.

1. **Dataset Transformation:** Compute $\mathcal{E}_{BCP}(\hat{X}_1^T)$ and $\mathcal{E}_{BCP}(\hat{X}_2^T)$, where $\hat{X}_1 = X - \frac{1}{t}yI^T$ and $\hat{X}_2 = X + \frac{1}{t}yI^T$. Given t and y, we might need to scale $\mathcal{E}_{BCP}(X)$ to make sure operations are run on integer domain. The encrypted training dataset of SVM is $\mathcal{E}_{BCP}(\hat{X}) = [\mathcal{E}_{BCP}(\hat{X}_1), \mathcal{E}_{BCP}(\hat{X}_2)]^T$ and the class labels $\hat{y} \in Z^{2m}$ where $\hat{y}_i = +1$ if $i \in [1, m]$ and $y_i = -1$ if $i \in [m + 1, 2m]$.
2. **Gram Matrix Computation:** Compute first the encrypted gram matrix $\mathcal{E}(K)$ of SVM based on $\mathcal{E}_{BCP}(\hat{X})$, then authorize S_1 to decrypt $\mathcal{E}(K)$ and get the gram matrix K in clear.
3. **Train SVM:** Solve the dual optimization problem of SVM (see (3)) based on gram matrix K and $C = \frac{1}{2\lambda}$ to get SVM's solution α (Please refer to the Appendix if readers are interested in how to train SVM).
4. **Reconstruct elastic net's solution** β based on α.

Model Assessment. In Algorithm 1, there are two parameters: l_1-norm constraint t and l_2-regularization parameter λ. It is not known beforehand which t and λ are best for the elastic net. For different combinations of (t, λ), the predictive power of the derived solution varies. We do "grid search" on t and λ using k-fold cross validation [37] to assess the goodness-of-fit of our model under different parameters. The grid-search is straightforward. We specify the range of t and λ respectively. Then we try various pairs of (t, λ). As it might be time-consuming to do a complete grid-search, we recommend using a coarse grid first. Once a "better" region is identified, we will conduct a finer grid search on that region. We divide our training dataset $\mathcal{E}_{BCP}(X)$ into k subsets satisfying $\mathcal{E}_{BCP}(X) = \mathcal{E}_{BCP}(X_1) \cup \cdots \cup \mathcal{E}_{BCP}(X_k)$, $\mathcal{E}_{BCP}(X_i) \cap \mathcal{E}_{BCP}(X_j) = \emptyset$ $(i \neq j)$. We use $\mathcal{E}_{BCP}(X_i)$ where $i \in [1, k]$ as the validation set and the remaining $k - 1$ subsets as the training dataset each time. In order to measure the performance of regression, we choose Rooted Mean Squared Error (RMSE). An RMSE value closer to 0 indicates the regression model is more useful for prediction. In the setting of k-fold cross validation, we need to compute the average of k RMSE values. Supposed that there are d samples in the validation set, the predicted GI50 value of gene expression profile x_i is $\tilde{y}_i$ and the true value is y_i, then RMSE is computed as

$$RMSE = \sqrt{(\frac{1}{d}\sum_{i=1}^{d}(\tilde{y}_i - y_i)^2)} \tag{22}$$

Recall that each gene expression profile is encrypted, we can compute the ciphertext of the predicted GI50 value $\tilde{y}_i$ as $\mathcal{E}_{BCP}(\tilde{y}_i) = (\beta^T(x_i - b_i), \beta^T E(b_i))$ where β is the solution to elastic net. To get $\tilde{y}_i$ in plaintext, we make the two non-colluding servers work together. S_1 reveals β to S_2. S_2 return $\beta^T b_i$ to S_1. Then S_1 computes $\tilde{y}_i = \beta^T(x_i - b_i) + \beta^T b_i = \beta^T x_i$. For each (t, λ) pair, S_1 computes RMSE with each predicted value $\tilde{y}_i$. Finally, S_1 will pick the optimal (t, λ) which achieves the smallest RMSE and get the optimal solution β^*.

5   Security Analysis

We consider the honest-but-curious model, meaning that all the medical units, S_1 or S_2 will follow our protocol but try to gather information about the inputs of MUs. There might exist collusion between a medical unit and S_1. We analyze the security of our model with the Real and Ideal paradigm and Composition Theorem [38]. The main idea is to use a simulator in the ideal world to simulate the view of a semi-honest adversary in the real world. If the view in the real world is computationally indistinguishable from the view in the ideal world, then the protocol is believed to be secure. According to the Composition Theorem, the entire scheme is secure if each step is proved to be secure. Due to page limit, the Proof of Theorem 1 is given in the Appendix.

Theorem 1. *In Algorithm 1, it is computationally infeasible for S_1 to distinguish the gene expression profiles encrypted under multiple keys as long as $\mathcal{E}_{BCP}$ is semantically secure and the two servers are non-colluding.*

Theorem 2. *No encryption scheme is secure against known-sample attack if dot products are revealed.*

Proof: We define known-sample attack as an attacker obtaining the plaintexts of a set of records of the encrypted database but not knowing the correspondence between the plaintexts and the encrypted records. According to [39], no encryption scheme is secure against known-sample attack if distance information is revealed. As distance computation can be decomposed into dot products, revealing dot products equals to revealing distance. Given n encrypted samples whose dimension is m, if an attacker knows the plaintexts of m linearly independent samples, the attacker can obtain the plaintext of any encrypted samples even without the decryption key. The idea is to construct m linear equations, whose unique solution corresponds to the desired sample.

Fortunately, in the following theorem, we show that it is impossible for the attackers to make use of Theorem 2 to launch the attack.

Theorem 3. *S_1 cannot reconstruct gene expression profile of a patient with gram matrix K known, considering the impossibility that an attacker collects enough samples of SVM to launch the attack mentioned in Theorem 2.*

Proof: According to Sect. 3.3, one gene expression profile of a patient across all genes (i.e. one row) is a training sample of elastic net. But the training sample of SVM can be considered as gene expression values of a particular gene among all the patients (i.e. one column). If S_1 colludes with MU_1, it only brings minor advantage that some of the elements from MU_1 of a training sample are revealed. Unless the attacker cracks our cryptosystem and obtains all the private keys of the involved medical units, he cannot set up linear equations to launch known-sample attack.

6   Experimental Evaluation

The configuration of our PC is Windows 7 Enterprise 64-bit Operating System with Intel(R) Core(TM) i5 CPU (4 cores), 3.4 GHz and 16 GB memory. We use a public database for drug sensitivity in cancer cell lines [33]. To provide platform independence, we use Java to implement our scheme together with open-source IDE. We use BigInteger class to process big numbers which offers all basic operations we need. We utilize SecureRandom class to produce a cryptographically strong random number. As for the generation of safe prime numbers, we use the probablePrime method provided by BigInteger class. The probability that a BigInteger returned by this method is composite does not exceed 2^{-100}. The performance of our scheme depends heavily on the size of modulus N, and the number of additions and multiplications performed. During the initialization phase, public and private key pair are generated. The runtime of generating a key pair varies with the bit length of N, as it depends a lot on the random number generator. A typical value for N is 1024 and it takes about 2 s in average to generate one key pair. We firstly compare the encryption time of two training samples when using the BCP Cryptosystem and our proposed cryptosystem $\mathcal{E}_{BCP}$. We vary the dimension of each sample from 1000 to 10000. The bit length of modulus N is set to 1024 and 1536 respectively (following the same setting as in [27]). As shown in Fig. 4, the encryption time scales linearly as the dimension increases. We use $E(x)$ and $(x - b, E(b))$ where b is a random number to denote the ciphertext of x under BCP and $\mathcal{E}_{BCP}$ cryptosystem respectively. Leveraging the framework proposed in [31], the encryption time of $\mathcal{E}_{BCP}$ doubles that of the BCP Cryptosystem. The additional encryption time is caused by generating the random number b. Moreover, we measure the time to compute dot product of two encrypted training samples. We focus on vertically partitioned dataset of SVM. To facilitate understanding, we encrypt the first half (belonging to Alice) of a sample using secret key s_A and the second half (belonging to Bob) using secret key s_B. We show the runtime of dot product computation on the ciphertexts in Fig. 5. Similarly, time to calculate dot product increases linearly with the dimension of samples. For vectors of dimension m, one dot product operation includes m multiplications and $m - 1$ additions, among which one addition is multikey homomorphic. It takes only 1 ms to run a multikey homomorphic addtition. For operations under single key, addition is much faster than multiplication. With a 1024-bit modulus, the runtime of additions is less than 1 s. The runtime of multiplications varies from 16 s to 185 s with the dimension of a sample increasing from 1000 to 10000. Therefore, multiplications are the bottleneck of dot product

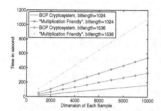 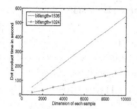 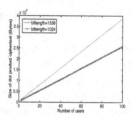

Fig. 4. Encryption time   **Fig. 5.** Dot product time   **Fig. 6.** Ciphertext size

computation. To decrypt one encryted dot product, it takes 285 ms and 572 ms with and without secret sharing separately. Recall that the multikey homomorphism property is achieved at the expense of expanding ciphertext size, we also measure the effect of the number of involved users on the increase of ciphertext size. As shown in Fig. 6, the ciphertext size increases linearly from 6138 B to 26 KB when the number of involved users increasing from 2 to 100.

The public database for drug sensitivity in this paper consists of 1002 cancer cell lines, 265 anticancer drugs. For each drug, GI50 values of around 300 to 1000 cell lines are available. As for gene expression profiling, it contains the RMA-normalized expression values of 17737 genes of 1018 cell lines. We preprocess them in MATLAB, keeping those cell lines that belong to the intersection of gene expression profiles and GI50 values. For example, considering drug PD-0325901, we get 843 expression profiles and GI50 values. As our cryptosystem only supports operations on the integer domain, we need to preprocess the database. To be specific, we first select a system parameter p to represent the number of bits for the fractional part of expression values. We next multiply each expression value by 10^p to get its integer value. Then, we need to divide each element of the gram matrix by 10^{2p} to remove the influence of scaling up. After running our privacy-preserving elastic net, we successfully pick out 165 genomic biomarkers.

Comparison with existing schemes: We focus on the homomorphic encryption based schemes [17,27–29], of which the setting is outsourced encrypted database under multiple keys. According to the experimental evaluation of [31], using their proposed framework to enable one multiplication on additively homomorphic ciphertexts outperforms the BGV homomorphic encryption [40] (in terms of ciphertext size, time of encryption/decryption/homomorphic operations). As shown in the experiments above, modification to the BCP Cryptosystem for multikey homomorphism only doubles the encryption time. Therefore, addition and multiplication can be run more efficiently in our scheme compared to [17], which deals with only two users (i.e. keys). Besides, they require the users to be online while we keep the users offline in this paper. Under two non-colluding servers model, the schemes in [27–29] can be used to compute addition/multiplication. The main drawback of their schemes is that they have to transform the ciphertexts under multiple keys to those under the same key, which is a heavy workload for the cloud server. Moreover, computing multiplication incurs interactions between the two servers. By contrast, our cryptosystem enables calculating multiplication without interactions.

7   Discussion and Conclusions

In practical scenarios, a gene expression profile typically has dimension of order 10^4. We can only collect hundreds of patients' profiles of different cancers. If we store gram matrix K in the memory, it is of order 10^8 in our case, which requires a lot of memory. Keerthi et al. [41] proposed to restrict the support vectors to some subset of *basis vectors* $\mathcal{J} \subset \{1, \cdots, n\}$ in order to reduce the memory requirement. This method requires $O(|\mathcal{J}|n)$ space where $|\mathcal{J}| \ll n$. However, the derived α using this method can be different from the one we get using the entire gram matrix K. It is a trade-off between accuracy and efficiency. For genomic biomarker discovery, it is obviously more important to pick out accurate biomarkers. Therefore, it makes sense to maintain the gram matrix in memory. Furthermore, each element of the gram matrix can be calculated independently. To accelerate the computation of gram matrix, we can utilize existing parallel computing frameworks.

To conclude, in this paper, by assuming the existence of two non-colluding servers, we proposed a privacy preserving collaborative model to conduct elastic net regression through reduction to SVM on encrypted gene expression profiles and GI50 values of anticancer drugs. To compute the gram matrix on ciphertexts, we successfully construct a cryptosystem that supports one multiply operation under single key and multiple add operations under both single key and different keys. Besides, we use secret sharing to allow one of the cloud server to get the gram matrix. Our scheme keeps the medical units offline and is proved to be secure in the semi-honest model or even if a medical unit colludes with one cloud server. The experimental results highlight the practicability of our scheme. The proposed protocol could also be applied to other applications that use elastic net or lasso for linear regression. Our future work is to extend our scheme to malicious adversaries (either S_1 or S_2 is malicious). One promising direction is to use commitment scheme [42] and zero-knowledge protocols.

Acknowledgements. This work is supported in part by RGC CRF (Project No. CityU C1008-16), Hong Kong and National High Technology Research and Development Program of China (No. 2015AA016008).

Appendix

Train SVM: We do not include any bias item in this paper, according to Sect. 3.3. Therefore, efficient dual coordinate descent method [43] can be used, of which the main idea is to optimize one variable once at a time, reducing memory requirements. However, coordinate descent methods are inherently sequential and hard to parallelize. To utilize the parallel properties of SVM, we also seek to find an solution which can be parallelized [44]. As shown in [45], optimizing on either the primal problem or the dual problem is in fact equivalent. Linear SVM can be considered as non-linear SVM with a linear kernel k and an associated

Reproducing Hilbert Space $\mathcal{H}$.

$$\min_{f\in\mathcal{H}} \frac{1}{2}\|f\|_{\mathcal{H}}^2 + C\sum_{i=1}^{n} max(0, 1 - y_i f(x_i))^2 \tag{23}$$

According to the *representer theorem*, the *minimizer* of (23) can be represented by $f^*(x) = \sum_{i=1}^{n} \theta_i k(x_i, x)$. Note that these coefficients θ_i are different from the Lagrange multipliers α_i in standard SVM literature. The relationship between θ_i and α_i is $\theta_i = y_i\alpha_i$. (23) can be rewritten as

$$\min_{\theta} \frac{1}{2}\theta^T K\theta + C\sum_{i=1}^{n} max(0, 1 - y_i\theta^T K_i)^2 \tag{24}$$

where K is also the gram matrix with $K_{ij} = x_i^T x_j$ and K_i is the i_{th} row of K. The Newton optimization algorithm of the above problem can be expressed as dense linear algebra operations. When combined with highly optimized libraries such as Intel's MKL for multicores, Jacket, and CuBLAS for GPUs, we can largely speedup the training of SVM.

Proof of Theorem 1. We discuss the security of each step in Algorithm 1.

Step 1 Dataset Transformation: Given $\mathcal{E}_{BCP}(X)$, it requires homomorphic addition to compute $\mathcal{E}_{BCP}(\hat{X}_1)$ and $\mathcal{E}_{BCP}(\hat{X}_2)$. Therefore, we need to prove the security of addition over ciphertexts against a semi-honest adversary $\mathcal{A}_{S_1}^{\text{SH}}$ in the real world. We set up a simulator $\mathcal{F}^{\text{SH}}$ in the ideal world to simulate the view of $\mathcal{A}_{S_1}^{\text{SH}}$. Considering one operation $\mathcal{E}_{BCP}(X_{ij} + \frac{1}{t}y)$, The view of $\mathcal{A}_{S_1}^{\text{SH}}$ in this step includes input $\{\mathcal{E}_{BCP}(X_{ij}), \mathcal{E}_{BCP}(\frac{1}{t}y)\}$ and output $\mathcal{E}_{BCP}(X_{ij} + \frac{1}{t}y)$. Without loss of generality, we assume that simulator $\mathcal{F}^{\text{SH}}$ computes $\mathcal{E}_{BCP}(m_1)$ and $\mathcal{E}_{BCP}(m_2)$ where $m_1 = 1$ and $m_2 = 2$. Then the simulator computes $\mathcal{E}_{BCP}(m_1 + m_2)$ and returns $\{\mathcal{E}_{BCP}(m_1), \mathcal{E}_{BCP}(m_2), \mathcal{E}_{BCP}(m_1 + m_2)\}$ to $\mathcal{A}_{S_1}^{\text{SH}}$. Since the view of $\mathcal{A}_{S_1}^{\text{SH}}$ are ciphertexts generated under $\mathcal{E}_{BCP}$ cryptosystem and $\mathcal{A}_{S_1}^{\text{SH}}$ has no knowledge of the private key. If $\mathcal{A}_{S_1}^{\text{SH}}$ could distinguish the real world from the ideal world, then it indicates $\mathcal{A}_{S_1}^{\text{SH}}$ is able to distinguish ciphertexts generated by $\mathcal{E}_{BCP}$, which contradicts to the assumption that $\mathcal{E}_{BCP}$ is semantically secure. Therefore, $\mathcal{A}_{S_1}^{\text{SH}}$ is computationally infeasible to distinguish the real world from the ideal world.

For the case where a medical unit (denoted as MU_1) colludes with S_1, we use $\mathcal{A}_{(S_1, MU_1)}^{\text{SH}}$ to denote the corresponding adversary. $\mathcal{A}_{(S_1, MU_1)}^{\text{SH}}$ cannot learn anything beyond gene expression values of MU_1.

Step 2 Gram Matrix Computation: Recall that the basic operation of gram matrix computation is dot product of two training samples $\hat{x}_1$ and $\hat{x}_2$. The security of addition has been proved in step 1. As for the security of multiplication, computing α component is implemented over ciphertexts on S_1 (see (6)). Similar to the proof above, we can prove the security of multiplication with the real and ideal paradigm. Moreover, the multikey homomorphic addition of the BCP

Cryptosystem is based on ciphertexts (see (11)). As long as the BCP Cryptosystem is semantically secure, the multikey homomorphic addition is secure. As for decrypting the encrypted dot product, S_1 interacts with S_2. S_1 sends C_{A_1} and C_{B_1} to S_2 (see (18)). S_2 returns $(C_{A_1})^{s_{A_2}}$, $(C_{B_1})^{s_{B_2}}$ to S_1. Based on the hardness of computing discrete logarithm, it is infeasible for S_1 to deduce s_{A_2} or s_{B_2}. Therefore, S_1 cannot recover s_A or s_B.

For the case of collusion between S_1 and MU_1, $\{\hat{x}_{1i}\}_{i=1}^{n_a}$ and $\{\hat{x}_{2i}\}_{i=1}^{n_a}$ are both revealed to $\mathcal{A}_{(S_1, MU_1)}^{\text{SH}}$. Even though the adversary can know $\sum_{j=n_a+1}^{n} \hat{x}_{1j}\hat{x}_{2j}$, the value of $\{\hat{x}_{1j}\}_{j=n_a+1}^{n}$ and $\{\hat{x}_{2j}\}_{j=n_a+1}^{n}$ remains unknown. According to Theorem 3, it is secure to reveal gram matrix to S_1.

Step 3 Train SVM: Given the gram matrix K, we train SVM to get α. $\mathcal{A}_{S_1}^{\text{SH}}$ cannot infer anything about gene expression profiles based on α.

Step 4 Reconstruct β based on α: If gene expression profiling microarray is known to the public, then $\mathcal{A}_{S_1}^{\text{SH}}$ will know which genes are picked out as biomarkers corresponding to non-zero elements of β. We can remedy this vulnerability through permuting the gene expression profile before uploading.

References

1. Duffy, M.J., Crown, J.: A personalized approach to cancer treatment: how biomarkers can help. Clin. Chem. **54**(11), 1770–1779 (2008)
2. Barretina, J., Caponigro, G., Stransky, N., Venkatesan, K., Margolin, A.A., Kim, S., Wilson, C.J., Lehár, J., Kryukov, G.V., Sonkin, D., et al.: The cancer cell line encyclopedia enables predictive modelling of anticancer drug sensitivity. Nature **483**(7391), 603–607 (2012)
3. Garnett, M.J., Edelman, E.J., Heidorn, S.J., Greenman, C.D., Dastur, A., Lau, K.W., Patricia Greninger, I., Thompson, R., Luo, X., Soares, J., et al.: Systematic identification of genomic markers of drug sensitivity in cancer cells. Nature **483**(7391), 570–575 (2012)
4. Covell, D.G.: Data mining approaches for genomic biomarker development: applications using drug screening data from the cancer genome project and the cancer cell line encyclopedia. PloS One **10**(7), e0127433 (2015)
5. Jang, I.S., Neto, E.C., Guinney, J., Friend, S.H., Margolin, A.A.: Systematic assessment of analytical methods for drug sensitivity prediction from cancer cell line data. In: Pacific Symposium on Biocomputing, p. 63. NIH Public Access (2014)
6. Erlich, Y., Narayanan, A.: Routes for breaching and protecting genetic privacy. Nat. Rev. Genet. **15**(6), 409–421 (2014)
7. Schadt, E.E., Woo, S., Hao, K.: Bayesian method to predict individual SNP genotypes from gene expression data. Nat. Rev. Genet. **44**(5), 603–608 (2012)
8. Tibshirani, R.: Regression shrinkage and selection via the lasso. J. Roy. Stat. Soc. Ser. B (Methodol.) **58**, 267–288 (1996)
9. Friedman, J., Hastie, T., Tibshirani, R.: glmnet: lasso and elastic-net regularized generalized linear models. R Package Version, 1 (2009)
10. Zou, H., Hastie, T.: Regularization and variable selection via the elastic net. J. Roy. Stat. Soc. Ser. B (Stat. Methodol.) **67**(2), 301–320 (2005)
11. Liu, J., Ji, S., Ye, J., et al.: SLEP: sparse learning with efficient projections, vol. 6, p. 491. Arizona State University (2009)

12. Shalev-Shwartz, S., Zhang, T.: Accelerated proximal stochastic dual coordinate ascent for regularized loss minimization. In: ICML, pp. 64–72 (2014)

13. Zhou, Q., Chen, W., Song, S., Gardner, J.R., Weinberger, K.Q., Chen, Y.: A reduction of the elastic net to support vector machines with an application to GPU computing. In: Twenty-Ninth AAAI Conference on Artificial Intelligence (2015)

14. Lin, K.P., Chen, M.S.: Privacy-preserving outsourcing support vector machines with random transformation. In: Proceedings of the 16th ACM SIGKDD International Conference on Knowledge Discovery and Data Mining, pp. 363–372. ACM (2010)

15. Laur, S., Lipmaa, H., Mielikäinen, T.: Cryptographically private support vector machines. In: Proceedings of the 12th ACM SIGKDD International Conference on Knowledge Discovery and Data Mining, pp. 618–624. ACM (2006)

16. Tassa, T., Jarrous, A., Ben-Ya'akov, Y.: Oblivious evaluation of multivariate polynomials. J. Math. Crypt. 7(1), 1–29 (2013)

17. Liu, F., Ng, W.K., Zhang, W.: Encrypted SVM for outsourced data mining. In: 2015 IEEE 8th International Conference on Cloud Computing, pp. 1085–1092. IEEE (2015)

18. Yu, H., Jiang, X., Vaidya, J.: Privacy-preserving SVM using nonlinear kernels on horizontally partitioned data. In: Proceedings of the 2006 ACM Symposium on Applied Computing, pp. 603–610. ACM (2006)

19. Yu, H., Vaidya, J., Jiang, X.: Privacy-preserving SVM classification on vertically partitioned data. In: Ng, W.-K., Kitsuregawa, M., Li, J., Chang, K. (eds.) PAKDD 2006. LNCS, vol. 3918, pp. 647–656. Springer, Heidelberg (2006). doi:10.1007/11731139_74

20. Vaidya, J., Hwanjo, Y., Jiang, X.: Privacy-preserving svm classification. Knowl. Inf. Syst. 14(2), 161–178 (2008)

21. Van Dijk, M., Juels, A.: On the impossibility of cryptography alone for privacy-preserving cloud computing. HotSec 10, 1–8 (2010)

22. Chow, S.S., Lee, J.H., Subramanian, L.: Two-party computation model for privacy-preserving queries over distributed databases. In: NDSS (2009)

23. Erkin, Z., Veugen, T., Toft, T., Lagendijk, R.L.: Generating private recommendations efficiently using homomorphic encryption and data packing. IEEE Trans. Inf. Forensics Secur. 7(3), 1053–1066 (2012)

24. Demmler, D., Schneider, T., Zohner, M.: ABY-A framework for efficient mixed-protocol secure two-party computation. In: NDSS (2015)

25. Pedersen, T.B., Saygın, Y., Savaş, E.: Secret charing vs. encryption-based techniques for privacy preserving data mining (2007)

26. López-Alt, A., Tromer, E., Vaikuntanathan, V.: On-the-fly multiparty computation on the cloud via multikey fully homomorphic encryption. In: Proceedings of the Forty-Fourth Annual ACM Symposium on Theory of Computing, pp. 1219–1234. ACM (2012)

27. Peter, A., Tews, E., Katzenbeisser, S.: Efficiently outsourcing multiparty computation under multiple keys. IEEE Trans. Inf. Forensics Secur. 8(12), 2046–2058 (2013)

28. Wang, B., Li, M., Chow, S.S., Li, H.: Computing encrypted cloud data efficiently under multiple keys. In: 2013 IEEE Conference on Communications and Network Security (CNS), pp. 504–513. IEEE (2013)

29. Wang, B., Li, M., Chow, S.S., Li, H.: A tale of two clouds: computing on data encrypted under multiple keys. In: 2014 IEEE Conference on Communications and Network Security (CNS), pp. 337–345. IEEE (2014)

30. Blaze, M., Bleumer, G., Strauss, M.: Divertible protocols and atomic proxy cryptography. In: Nyberg, K. (ed.) EUROCRYPT 1998. LNCS, vol. 1403, pp. 127–144. Springer, Heidelberg (1998). doi:10.1007/BFb0054122

31. Catalano, D., Fiore, D.: Using linearly-homomorphic encryption to evaluate degree-2 functions on encrypted data. In: Proceedings of the 22nd ACM SIGSAC Conference on Computer and Communications Security, pp. 1518–1529. ACM (2015)

32. Bresson, E., Catalano, D., Pointcheval, D.: A simple public-key cryptosystem with a double trapdoor decryption mechanism and its applications. In: Laih, C.-S. (ed.) ASIACRYPT 2003. LNCS, vol. 2894, pp. 37–54. Springer, Heidelberg (2003). doi:10.1007/978-3-540-40061-5_3

33. Yang, W., Soares, J., Greninger, P., Edelman, E.J., Lightfoot, H., Forbes, S., Bindal, N., Beare, D., Smith, J.A., Richard Thompson, I., et al.: Genomics of drug sensitivity in cancer (GDSC): a resource for therapeutic biomarker discovery in cancer cells. Nucleic Acids Res. 41(D1), D955–D961 (2013)

34. Jaggi, M.: An equivalence between the Lasso and support vector machines. In: Argyriou, A., Signoretto, M., Suykens, J.A.K. (eds.) Regularization, Optimization, Kernels, and Support Vector Machines, pp. 1–26. Taylor & Francis, Boca Raton (2014)

35. Zhou, H., Wornell, G.: Efficient homomorphic encryption on integer vectors and its applications. In: Information Theory and Applications Workshop (ITA 2014), pp. 1–9. IEEE (2014)

36. Boneh, D., Goh, E.-J., Nissim, K.:.Evaluating 2-DNF formulas on ciphertexts. In: Kilian, J. (ed.) TCC 2005. LNCS, vol. 3378, pp. 325–341. Springer, Heidelberg (2005). doi:10.1007/978-3-540-30576-7_18

37. Fushiki, T.: Estimation of prediction error by using k-fold cross-validation. Stat. Comput. 21(2), 137–146 (2011)

38. Goldreich, O.: Foundations of Cryptography: Basic Applications, vol. 2. Cambridge University Press, Cambridge (2004)

39. Wong, W.K., Cheung, D.W.L., Kao, B., Mamoulis, N.: Secure kNN computation on encrypted databases. In: Proceedings of the 2009 ACM SIGMOD International Conference on Management of data, pp. 139–152. ACM (2009)

40. Halevi, S., Shoup, V.: An implementation of homomorphic encryption. https://github.com/shaih/HElib

41. Keerthi, S.S., Chapelle, O., DeCoste, D.: Building support vector machines with reduced classifier complexity. J. Mach. Learn. Res. **7**, 1493–1515 (2006)

42. Pedersen, T.P.: Non-interactive and information-theoretic secure verifiable secret sharing. In: Feigenbaum, J. (ed.) CRYPTO 1991. LNCS, vol. 576, pp. 129–140. Springer, Heidelberg (1992). doi:10.1007/3-540-46766-1_9

43. Hsieh, C.-J., Chang, K.-W., Lin, C.-J., Keerthi, S.S., Sundararajan, S.: A dual coordinate descent method for large-scale linear SVM. In: Proceedings of the 25th International Conference on Machine Learning, pp. 408–415. ACM (2008)

44. Tyree, S., Gardner, J.R., Weinberger, K.Q., Agrawal, K., Tran, J.: Parallel support vector machines in practice. arXiv preprint arXiv:1404.1066 (2014)

45. Chapelle, O.: Training a support vector machine in the primal. Neural Comput. **19**(5), 1155–1178 (2007)

Privacy-Preserving Community-Aware Trending Topic Detection in Online Social Media

Theodore Georgiou[✉], Amr El Abbadi, and Xifeng Yan

Department of Computer Science,
University of California, Santa Barbara, Santa Barbara, USA
{teogeorgiou,amr,xyan}@cs.ucsb.edu

Abstract. Trending Topic Detection has been one of the most popular methods to summarize what happens in the real world through the analysis and summarization of social media content. However, as trending topic extraction algorithms become more sophisticated and report additional information like the characteristics of users that participate in a trend, significant and novel privacy issues arise. We introduce a statistical attack to infer sensitive attributes of Online Social Networks users that utilizes such reported community-aware trending topics. Additionally, we provide an algorithmic methodology that alters an existing community-aware trending topic algorithm so that it can preserve the privacy of the involved users while still reporting topics with a satisfactory level of utility.

1 Introduction

With the explosive growth of Online Social Networks and the consequential unparalleled creation of an enormous amount of user generated content, algorithms that can extract meaningful insights and summarize this content have been widely studied and used. Specifically, the concept of *Trending Topics* has been popularly utilized in the detection of breaking news, hyper-local events, or memes, and also significantly contribute as marketing and advertising mechanisms. In its broader definition, a trending topic is a set of words or phrases that refer to a temporarily popular topic. Trending topics are used to understand and explain how information and memes diffuse through vast social networks with hundreds of millions of nodes. However, due to the open-access nature of Online Social Networks like Twitter, where everyone can see who says what, and depending on how much information a trending topic contains, novel notions of privacy emerge.

As a concrete example, Twitter reports Trending Topics by location, even at the city resolution. Their service also offers a search functionality which enables the discovery of all social postings (tweets) that contain certain keywords, and those tweets are always associated with a user of the social media service. When Twitter reports that a topic is trending in Athens, Greece, anyone can find the users that mentioned this topic through Search and may, therefore, assume that

Published by Springer International Publishing AG 2017. All Rights Reserved
G. Livraga and S. Zhu (Eds.): DBSec 2017, LNCS 10359, pp. 205–224, 2017.
DOI: 10.1007/978-3-319-61176-1_11

they live in Athens, Greece. The location of a user could be considered a sensitive attribute, if for example they post provocative political opinions and are afraid of physical repercussions. As we will show later, an attacker can easily infer the location of hundred of thousands of Twitter users through a simple crawling of Location-based trending topics using the official Twitter API. These users do not geocode their tweets neither publicly display their location on their profile. Thus, the correlation between trending topics and attributes like location can lead to privacy leaks. Building smarter trending topic extraction algorithms, which contain richer demographic information of the involved users can further increase the privacy risk of any reported topic. It is important that any algorithm that extracts multiple correlated user attributes takes privacy seriously into account.

In [8] we proposed an efficient method to identify trending topics on Twitter where the underlying user population (the users that mention the topic) share common attributes like age, location, gender, political affiliation, sports teams, etc. Through human-based evaluations we showed that topics correlated with surprising attribute values tend to be 27% more interesting and informative than trending topics that are extracted purely based on their raw frequency or burstiness. We call such an algorithm a *community-aware trending topic extraction algorithm* since the involved users in each topic form homogeneous groups (communities), even if they are not linked directly.

Due to the public nature of Online Social Networks like Twitter, apart from identifying the real identity of a user, an attacker will usually try to *infer* sensitive attribute values of certain users utilizing knowledge of the social network (who is a friend with whom, or who follows who). Furthermore, a sensitive attribute inference attack is also a significant risk in the context of community-aware trending topic reporting and to the best of our knowledge has not been studied before. At the same time, large Social Media websites like Facebook and Twitter already have proprietary methods for inferring social attributes of their users that are not explicitly provided by them. Recently, it was revealed that Facebook is able to learn a user's political preference between values like "Liberal", "Very Liberal", "Moderate", or "Conservative". This is a particularly interesting case since user content on Facebook is usually not accessible to anyone except the user's immediate social network. However, if sensitive attribute information, like political preference, is used in the context of enriching other features which are publicly known, like Facebook's Trending section, then this feature could start leaking sensitive information to virtually anyone.

To demonstrate how sensitive attribute inference could be applied as an attack in the context of trending topics, we provide a hypothetical example in Fig. 1 where users mention certain topics that were reported as trending from a community-aware algorithm (listed in the table at the top of the figure). The information in the table is public to everyone, similarly to the lists of Trending Topics that Facebook and Twitter already publish to their users in general. The main difference is that each topic is also linked with values for specific attributes like gender, age, location, political preference, etc. The association of an attribute value with a topic indicates that this specific attribute value is a

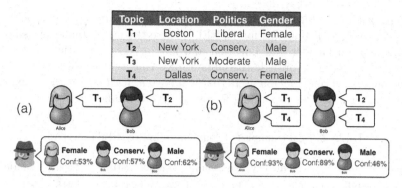

Fig. 1. Alice and Bob are two users who have discussed some topics. These topics were reported as trending and additionally, for each topic certain demographic information was extracted for 3 attributes: Location, Political Preference, and Gender. These values indicate that a significant *portion* of all the users that mentioned each topic, belong to the community defined by those values. An attacker can observe these values and can also find which topics Alice and Bob have discussed. Based on this knowledge, the attacker can infer certain attribute values of Alice and Bob with certain confidence. In case (a) (left), where Bob and Alice have only discussed a single topic, the attacker has low inference confidence. In case (b) (right), Bob and Alice have also discussed topic T_4 which increases the confidence of the attacker for Alice's gender and Bob's political preference but at the same time decreases the confidence for Bob's gender because T_2 and T_4 have mostly male and female communities correspondingly.

characteristic for the majority of the users that mentioned the topic (but not necessarily all of them). For an attacker, this means that they cannot be 100% confident that every user mentioning topic T_1 lives in Boston. However, when users discuss *several* topics, the attacker's confidence may increase. As shown in Fig. 1 Alice and Bob each mention some of the topics that happen to be listed in the table of trending topics. Since the attacker can obtain a list of the users that mentioned each topic (e.g., Twitter provides such search functionality), they can also *increase* their confidence (note the difference between cases (a) and (b)) in inferring Alice and Bob's sensitive attributes like political preference or gender without even accessing their posts or network.

In Table 1 we list some real examples of topics and their corresponding community characteristics (attribute values) that we extracted from Twitter data. The communities are characterized by values for several attributes including Location, Gender, Age, Political party (US only), or even Sports teams. Note that these attribute values are temporal and might change over time, even for the same topics. Each topic has a frequency (how many unique users mentioned it) and a community defined by the attributes that describe a significant part of the users that mentioned the topic. In practice, it is impossible to observe topics where the entirety of their population forms a homogeneous community on some attribute values, therefore, the reporting algorithm will only guarantee that at least some percentage of this user population shares the reported attribute values. Note, that a community is not necessary to have a value for every attribute, as it happens for "#NFL" where the user population is homogeneous only on Gender and Location and not in Age or Politics. In the last column of the table we provide the number of privacy violations for each topic, i.e. the number of social media users that will have *at least one attribute* exposed to an attacker if the corresponding trending topic is publicly reported.

Table 1. Real examples of community-aware trending topics

Topic	Frequency	Community characteristics	Size	Violations
#NavyYardShooting	5427	Location: USA, Age: 19–22	5218	2561
#NFL	1534	Gender: Male, Location: USA	1212	389
#FreeJustina	54	Location: Boston, Gender: Female, Political party: Democrats	51	13
#OscarTrial	1242	Location: Johannesburg: ZA, Gender: Female	1133	345
#ObamaCare	5090	Location: USA, Politics: Republicans	4818	1002
#ObamaIn3Words	246	Location: USA, Age: 19–22, Gender: Male, Politics: Republicans	224	76
#RedSox	528	Location: Boston, Age: 19–22, Gender: Male, Team: Red Sox	411	256

An attacker similar to the one in Fig. 1 can peruse the rows of Table 1 and attempt to infer sensitive attribute values for the involved users. If there is a user that mentioned both topics #ObamaCare and #ObamaInThreeWords then the attacker can be very confident that the user supports the Republican party, that they are located in the United States, and moderately confident that they are male and a young adult. This is becomes more important In the presence of even more sensitive attributes like sexual orientation, religion, or race. Note that this kind of attack is different from existing privacy scenarios where the attacker infers sensitive attributes through the user's local social graph (e.g., [26]). In the case of community-aware trending topics, membership to a community is implicit and happens just by mentioning certain topics. Therefore, even if a user is careful with which groups they subscribe to or become members of, sensitive information can still be exposed simply through the mention of a topic.

We tested how easily we can attack private attributes in existing Trending Topics reports. As mentioned earlier, Twitter provides Trending Topics by location (a total of 401 cities in the world). We crawled these topics through the Twitter API, and managed to infer the location of approximately 300k users that mentioned topics which were trending only in a single location just within a single day of crawling. 11.8% of these users had public location and from a sample we estimated that this location inference attack was 82.33% successful. This proved how easy an attacker can exploit existing Trending Topics to infer the location of thousands of users. Therefore, altering trending topic algorithms to protect the sensitive attributes of OSN users is an important area to study.

Main contributions: In this research we formally introduce a novel privacy model that captures the notion of sensitive attribute inference in the presence of community-aware trending topic reports where an attacker can increase their inference confidence by consuming these reports and the corresponding community characteristics of the involved users. We discuss a basic attack and provide an efficient algorithm that preserves the privacy of each individual user so that sensitive attributes can not be successfully inferred. To the best of our knowledge we are the first to address this notion of privacy and introduce an algorithm that uses the idea of attribute generalization in combination with Artificial Intelligence techniques to efficiently defend against such attacks.

In the next sections we provide related literature on the subject of sensitive attribute inference in Social Media (Sect. 2), discuss the data, attack, and privacy models (Sect. 3), and provide an analysis of the basic attack that is based on Naive Bayes inference (Sect. 4), which is commonly used in this line of research. We then present a novel approach to preserve privacy while maintaining topic reports with high utility (Sect. 5). Finally, we provide experimental results on the algorithm's performance and utility (Sect. 6).

2 Related Work

Data privacy is a thoroughly studied area and several families of algorithms have been proposed to deal with different kinds of attacks, mostly on published anonymized datasets. Most notably, the concepts of k-anonymity [16,17,20], l-diversity [11], t-closeness [10], and Differential Privacy [6] include methodologies to preserve data privacy and information anonymity. However, privacy in Online Social Networks follows a different data model where most of the information is publicly available: the Twitter social graph, the set of online postings by every user in Twitter, user membership in Facebook pages, etc. What is not accessible though, is information about sensitive characteristics that users might want to keep hidden from the general public. An attack to discover these characteristics is known as sensitive or private attribute inference.

There are studies and published algorithms for inferring user demographics based on the content posted by social media users or their social network. Schwartz et al. [18] developed language models to identify the gender and age of Facebook users. [5] describe a method to infer user demographics by utilizing external knowledge of website user demographics and correlating it with a social media service. Their approach mainly differs from Schwartz et al.'s in its ability to infer the user characteristics without analyzing the content of postings. Nazi et al. [12] proposed a methodology to discover hidden information from Social Media by exploiting publicly accessible interfaces like the search functionality.

While all the aforementioned work provides useful data mining tools and models, the privacy implications of the proposed methods are not examined.

Zheleva et al. [26] were the first to study the privacy of sensitive attributes in the context of Online Social Networks. They describe a variety of attack models to infer sensitive user attributes but the model most related to the current work, is the model that utilizes the membership of users in Facebook pages. This model is similar to the "membership" of a user to a trending topic's community. However, they do not provide any algorithmic solution since it is the choice of the user to subscribe to a page. A system called Privometer [21] measures how much privacy leaks from certain user actions (or from their friends' actions) and creates a set of suggestions that could reduce the risk of a sensitive attribute being successfully inferred, like "tell your friend X to hide their political affiliation". Similar to Privometer, in [3], and then in [14], a method is proposed for the prevention of information leakage by introducing noise, through the removal of edges or addition of fake edges, to the social graph. This idea was then extended to a finer-grained perturbation in [2] where edges are only added *partially*. Eunsu et al. [15] built a system called "curso" that identifies when a user's privacy is violated through the analysis of their local network. There are also studies that focused on the anonymization of network data where the attacker tries to statistically infer the relationship between members of the social network. Most prominent works in this area include [4,25]. Tassa et al. [22] also studied the same problem but specifically consider *distributed* social networks.

Dealing with privacy on a virtually infinite stream of data poses its own challenges and most of the aforementioned techniques focus on static datasets. Dwork et al. have studied privacy in streaming environments and proposed a family of algorithms called Pan-Private Streaming Algorithms [7]. The main focus of these algorithms is to deal with attackers with control of the machine(s) where the algorithm is running but no access to the stream, while in our case they have access to every social post.

3   Data and Attack Models

3.1   Data Model

The users of a Social Media service are represented as a set $U = \{u_1, u_2, ..., u_n\}$. Each user u is associated with a vector v of k *sensitive* attributes (e.g., location, age, etc.). The attribute a_i of a user u ($u.v.a_i$) can take on one of a set of possible values $\{a_{i1}, a_{i2}, ..., a_{im_i}\}$, where m_i is the corresponding attribute's total number of unique values. The values of an attribute form a *hierarchy* which for some attributes can have a significant depth (e.g., for location: cities, to regions, to countries, to continents, to wordwide) or be trivial (e.g. for gender: from male and female to any gender). An attribute value can be *generalized* by being replaced with an ancestor value from the hierarchy. A user can mark a set of attributes as sensitive and keep them private. Or depending on the nature of an attribute, e.g., race, which the social media service might infer using its own proprietary inference algorithm, it could be considered as sensitive for everyone.

The content of the Social Media service is represented as an infinite stream P of posts. Every post $p \in P$ has a unique author (user) $p.u$ and contains an arbitrary number of topic keywords $p.T = \{t_1, t_2, ...\}$. We define a publicly available search function $SEARCH$ that returns all the users mentioning a given topic keyword t: $SEARCHt = \{p.u | t \in p.T\}$. The number of users mentioning t is referred to as *topic population* and its size is equal to $|SEARCHt|$ and referred to as *topic frequency* (second column in Table 1). We can assume that each user that mentions topic t is counted only *once* to avoid bias from spamming. The search function $SEARCH$ is defined for multiple topics as well, and returns the *intersection* of the users that mention all the given topics.

We define a *homogeneous community* as a group of users with identical values in some of their attributes, but not necessarily connected in the social graph. More formally, a *homogeneous community* contains users that share the same values for a combination of attributes $C \in \mathcal{P}\{a_1, a_2, ..., a_k\}$ where $\mathcal{P}$ is the power-set symbol and a_i is a user attribute (e.g., location, age, etc.). Users that live in San Francisco, are 25 years old, and are male, form a homogeneous community that contains all the users identified by these values for the attribute combination {location, age, gender}. Users in New York form another homogeneous community defined by the singleton attribute combination {location}.

A *community-aware* trending topic algorithm (referred to as $CATT$ [8]) identifies topic keywords mentioned by a *homogeneous community* that has at least size ξ of the total topic population ($0 < \xi \leq 1$). For example, if $\xi = .7$, a topic with frequency 1000 will have *at least* 700 users forming a homogeneous community. The CATT algorithm reports records in the form of a stream of tuples: t_i, C_i, where C_i is the set of attribute values that define the homogeneous community CATT identified for topic t_i. If a topic t has no homogeneous community of size $\xi |SEARCHt|$ or larger associated with it then it isl not reported by CATT. We will refer to homogeneous communities simply as *communities* and to topics extracted via a community-aware algorithm as *community-aware topics*.

CATT extracts trending topics using a *batch-based* sliding window on the stream of social postings of the service. At the end of each window, CATT reports a set of pairs t_i, C_i) which includes all the extracted topics from the current window. We refer to the output of CATT for each window of social postings as a *batch*. Table 1 shows an example of such a batch that contains 8 pairs. Through the definition of community-aware trending topics, the users of the social media service inherit an implicit membership to communities just by mentioning certain topics. Using a single reported pair t_i, C_i one can infer that at least $\xi\%$ of the users in $SEARCHt_i$ are characterized by the values of C_i. This constantly increasing knowledge enables an attacker to gradually improve their inference confidence for a given user's sensitive attribute(s).

Note that execution of CATT requires the knowledge of community attributes for the involved users. Realistically, CATT is executed by the Social Media service itself which has access to private user information or even its own proprietary method to extract attributes. Attackers lack access to the necessary information to execute CATT themselves.

3.2   Attack Model

A CATT algorithm reports a stream of batches of pairs t_i, C_i. The attacker knows CATT's threshold ξ, as it is public knowledge, has access to the output stream, and to the search function $SEARCH$ which returns the set of users that have mentioned the provided topic(s). It is also safe to assume that the attacker has general knowledge of each attribute's prior distribution. For example, such knowledge might include the location distribution based on a Census, the age distribution based on published statistics from the social media service, the gender distribution based on users that have this information public, etc. We can safely assume that the attacker is omnipotent and can indefinitely store the pairs (t_i, C_i) and the corresponding sets of users $SEARCHt_i$. The goal of the attacker is to infer a user's sensitive attribute by exploiting the knowledge of each topic's community C_i and the users associated with it. In the presence of an omnipotent attacker a privacy preserving algorithm must maintain all previous trending topics and communities to accurately calculate the probability distribution of the sensitive attribute values, of each user.

In related literature on sensitive attribute inference [14,21,26], an attacker would train a Naive Bayes Classifier to choose the value of a sensitive attribute L that maximizes the probability distribution $PL|u.T$. However, though Naive Bayes is known to be a decent classifier, it is also known to be a bad estimator [24]. For the inference process to be accurate, a high probability bound is necessary, so we consider that attack to be successful only when the inference probability of an attribute value is greater than a set threshold θ (e.g., $\theta = .75$ or .85). We will be using a global value for θ across all attributes and users, but the proposed model and algorithm support different values for each attribute and user.

4   Privacy Model

4.1   Sensitive Attribute Inference

Having established the models for the data (social stream) and the attacker (inference of sensitive attributes) we can now formally define the privacy model. For every user in the social network that discusses several topics in a streaming fashion, we want to protect against having their sensitive attribute values leaked through the continuous reporting of community-aware trending topics. Specifically, any attacker that has access to current and historical reports of community-aware trending topics should not be able to infer any user's sensitive attribute with confidence that is higher that a set value θ. At no point should an attacker be able to infer a *lower bound* for the distribution $PL|u.T$ (probability distribution of sensitive attribute L of a user u given the topics T of u), that is higher than θ.

Definition: If there is even a single case where a user's sensitive attribute can be inferred with confidence larger than θ, this comprises a *privacy violation*.

A community-aware trending topic algorithm that is capable of maintaining a record of zero privacy violations while it continuously reports new batches of topics is called θ-*private*.

Referring back to the example of Fig. 1, if θ is set to .75 then an algorithm that reports the topics in the table of the figure is *not* θ-private in case (b), since the attacker can infer the gender of Alice and the political preference of Bob with confidence that is higher than θ. To make the algorithm θ-private we would need to obfuscate the gender and political preference associated with topics T_1, T_2, and T_4. If Alice and Bob had only discussed topics T_1 and T_2, as in case (a), then the algorithm would be θ-private for this specific instance.

The inference of a sensitive attribute involves estimating the probability of a specific value given some background knowledge. As already discussed, the attacker has access to prior attribute probabilities and the output and settings of CATT. The Naive Bayes classifier is a powerful and simple technique to calculate the probability of a sensitive attribute value. Arguably, if the attacker has additional information of other sensitive attributes (e.g., already knows that Alice is a woman because she has her own photo in her profile) then they can get a better estimation of the probability of another sensitive attribute, like her location, than they would from Naive Bayes. In the following subsection we focus on the calculations necessary to get a lower bound of the probability $P(L|u.T)$ using Naive Bayes. The end goal is to anticipate what values the attacker can successfully infer so that they can be kept private. This is typically easy since the attacker's knowledge is generally based on publicly available information and the privacy model can incorporate it if necessary. To keep things simple, for the rest of the paper we assume that the attacker has no existing knowledge of sensitive attribute values and therefore the Naive Bayes Classifier can set a precise upper bound. The introduced privacy model is independent of how $P(L|u.T)$ is calculated by an attacker and the privacy preserving algorithm proposed later can be easily adjusted to calculate these distributions differently.

4.2   Naive Bayes Inference

Given a collection of topic and community tuples t_i, C_i (the output of CATT) and a search function $SEARCH$, an attacker may attempt to infer the sensitive attributes of users that mention at least one of the topics t_i. Let u be a user that has *mentioned* k topics $t_1, t_2, ..., t_k$ and let L be one of the user's sensitive attributes (e.g., location). The probability distribution of L, given that the user mentioned some topics $t_1, t_2, ..., t_k$ is:

$$PL|t_1, t_2, ..., t_k = \frac{Pt_1, t_2, ..., t_k|LPL}{Pt_1, t_2, ..., t_k} \tag{1}$$

by applying the Bayes Rule. $P(L)$ is the prior multinomial distribution of the attribute L and can be assumed to be known to an attacker based on their general knowledge on such information. The probability distribution of a user mentioning topics $t_1, t_2, ..., t_k$ given L, $Pt_1, t_2, ..., t_k|L$, is equal to the number of users u that mention all the k topics and have a specific value for L, over the total number of users with that value of L. For example, for $L = a$:

$$Pt_1, ..., t_k | L = a = \frac{|\{u | u.v.L = a, t_1 \in u.T, ..., t_k \in u.T\}|}{|\{u | u.v.L = a\}|} \tag{2}$$

where $u.v.L$ is the attribute L in the user's vector of attributes v. Similarly, the prior probability of topics $Pt_1, t_2, ..., t_k$ is equal to the number of users that mentioned these topics over the total number of users n: $|SEARCH t_1, t_2, ..., t_k|/n$.

While an attacker might have knowledge of the attribute's multinomial distribution and the ability to calculate the prior probability of any topic combination (using the search function $SEARCH$), they cannot compute the set of users that have a specific attribute value $L = a$: $\{u | u.v.L = a\}$. Instead, they can obtain an approximate value of the probability distribution $Pt_1, t_2, ..., t_k | L$ based on the reported tuples from CATT. The attacker can exploit the guarantees provided by CATT that a reported trending topic t_i has a population of size $|SEARCH t_i|$ with a homogeneous community C_i with size at least $\xi |SEARCH t_i|$.

More specifically, if the attribute L is not part of C_i, then the topic population of t_i follows the prior distribution of L: $Pt_i | L = PL$. If $L \in C_i$ and has a value $L = a$, then applying the Bayes Rule we get:

$$P_{approx} t_i | L = a = \frac{PL = a | t_i Pt_i}{PL = a} = \frac{\xi}{PL = a} Pt_i \tag{3}$$

Similarly, the probability that a user with value $L = b$ mentions topic t_i is:

$$P_{approx} t_i | L = b = \frac{PL = b | t_i Pt_i}{PL = b}$$
$$= \frac{1 - \xi PL = b | SEARCH t_i|}{PL = bn} = 1 - \xi Pt_i \tag{4}$$

The attacker can now approximate the probability distribution (2) by assuming topic independence given L: $P_{approx} t_1, t_2, ..., t_k | L =_{i=1}^{k} Pt_i | L$ where each factor of the product can be computed using the probability formulas from (3) and (4). Note that topic independence given L is an assumption that can be true when the number of topics k is large. An attacker can use the following formula to approximate $PL | u.T$:

$$P_{approx} L | u.T = \frac{n PL^{t_i \in u.T} Pt_i | L}{|SEARCH u.T|} \tag{5}$$

If for any value of $L = l$, the probability $PL = l | u.T$ becomes larger than the threshold θ then we assume that the privacy of this user for L is violated.

5   Privacy Preservation Methodology

A *community-aware* trending topic algorithm is also *θ-privacy-preserving* if its output does not enable the inference of sensitive user attributes with a confidence greater than a threshold θ, for any of the users involved. We will refer to this modification of the CATT algorithm as θ-$CATT$. At the same time, the goal is to keep reporting trending topics with maximum *utility*. Maximizing the utility of the results is a competing goal with preserving privacy since the algorithm could report an empty result set and the privacy leakage would be zero. Issues arise when the algorithm reports at least one trending topic t_i and its community C_i and for all users in $SEARCHt_i$ some statistical information is leaked. Especially challenging is the fact that users continuously discuss new topics which results in a constant stream of information that an attacker can use to increase their inference confidence of sensitive attribute values (as demonstrated in Fig. 1).

We now introduce a novel approach that utilizes the concept of generalization in combination with Artificial Intelligence to efficiently solve the exponentially expensive anonymization problem while preserving significant utility.

5.1 Utility of Trending Topics

The goal behind extracting trending topics that certain communities focus on is to provide additional insight into why certain topics end up trending, understand which user demographics are interested in an event, product, etc., and generally provide more interesting, surprising and personalized trending topics to the users of the social media service. Using the notion of Self-information from Information Theory [19] we provide a measure of the information content for community-aware trending topics. Self-information can capture how surprising an event is based on the probability of the event. The total utility of θ-CATT's results is equal to the self-information sum of every reported topic's community. The *self-information of a community* C_i is $IC_i = -log_2 PrC_i$. Intuitively, the less likely a community is to be observed, the higher its self-information. Since we are using the logarithm with base 2, self-information is measured in bits. This metric provides a systematic way to measure the utility of the reported topics and can be used to calculate the information/utility loss when anonymization is applied. We define a *utility function* $util()$ which returns the utility over a set of tuples (t_i, C_i). Other metrics can be used as well without alterations to θ-CATT.

5.2 Community Attribute Anonymization

θ-CATT needs to constantly monitor the maximum confidence of a hypothetical attacker to infer every sensitive attribute of every user in the service. When θ-CATT identifies a trending topic t_i with a homogeneous community that involves $|SEARCHt_i|$ users, it has to make sure that none of the users $u \in SEARCHt_i$ will have their sensitive attributes leaked by publishing (t_i, C_i). To ensure that, it calculates the probability of each sensitive attribute for every user u: $PL|u.T$ and checks if the value becomes greater than θ. If it does not, then the pair (t_i, C_i) is published. If it does, θ-CATT will anonymize the sensitive attribute of the community before publishing, while preserving as much utility as possible.

We utilize the method of *attribute generalization* to achieve anonymization similarly to k-anonymity [9,16,17,20]: if the city of a user can be inferred, θ-CATT reports location at the state level instead, which will alter the inference probability since a much larger population is described by this value. Generalization of categorical attributes is achieved by moving up a level in the attribute hierarchy (as described in earlier section). Depending on the depth of an attribute's hierarchy, a single generalization (moving up a single level in the attribute's value hierarchy) might lead to complete anonymization which also means zero utility for this attribute. For example, generalizing the value "male" will result to "any gender" (or "*").

The θ-CATT algorithm encapsulates the privacy-agnostic CATT algorithm which just extracts the community-aware trending topics by consuming the social stream. θ-CATT receives the batch of topics and attributes pairs (t_i, C_i) (as described in earlier section), and combined with the knowledge of every user's sensitive attributes and the topics they have previously mentioned $(u.T)$, calculates if any user's privacy would leak with the publication of the batch.

5.3   Finding the Best Anonymization Strategy

In order to output a list of trending topics that contains no privacy violations, a decision must be made that involves choosing which topic communities should be anonymized without sacrificing too much utility. There are many solutions to this problem, each with a different level of utility loss. To avoid solving this problem in exponential time by trying all possible combinations and choosing the one that minimizes the utility loss, we propose an algorithm that efficiently finds the best strategy for identifying a near optimal combination to anonymize. The θ-CATT algorithm is able to identify the privacy risk each new topic-community pair poses before publishing it, ideally in real time. To achieve this computation, θ-CATT needs to store: (1) the history of trending topics previously reported by the algorithm, that each user u has mentioned, and (2) the communities that were reported to be correlated with those topics. With this information θ-CATT can simulate an attacker and identify privacy violations before they even occur.

Batch-Based Anonymization. When a batch of pairs (t_i, C_i) is reported by CATT, θ-CATT will iterate through all pairs, apply necessary anonymizations and publish the altered set of pairs. A naive approach to identify which pairs require anonymization, is to iterate through them one by one, and if a pair violates the privacy of at least one user, appropriately anonymize the community's sensitive attribute(s) before moving to the next topic. However, the iteration order might lead to non-optimal results where more communities get anonymized than necessary to preserve privacy and utility loss is not minimal. For example, it might be better to anonymize a single community C_3 instead of anonymizing two communities C_1 and C_2 and achieve the same privacy gain. Occasionally, the combination of two topic communities can enable their publication without anonymization while if we each pair is individually considered, then neither of them would get reported. For this reason, θ-CATT considers the privacy and utility of the whole batch to identify the best anonymization strategy which minimizes the required attribute generalization and utility loss.

Assume for simplicity that there is a single sensitive attribute L and let S be a batch of k pairs (t_i, C_i) with communities that have a value for attribute L. Since the generalization of an attribute in a community C_i lowers the total utility of the batch, we want to generalize L in the least possible number of communities. An *anonymized batch* S' is a *modified* version of S with an arbitrary number of the communities in S anonymized (a community is anonymized when its attribute L is generalized at least once as described earlier). If a community does not contain a value for attribute L, it is ignored since it will not alter any user's inference probability for L. Therefore, there is a total of 2^k different anonymized batches S' ranging from the case where nothing is anonymized to the case where all k communities are anonymized and every possible combination in between.

The goal for θ-CATT is to find the batch S' that has greater utility than any other S'': $utilS' \geq utilS''$ while at the same time S' preserves the privacy of every user's sensitive attribute. For example, in Table 1, $k = 7$ and S contains the eight topic-community pairs listed in the table. If reporting these 7 pairs violates the privacy of any of the involved users, then θ-CATT will identify an anonymized version of the batch that does not leak sensitive attributes.

A* State Encoding. To find the best anonymized batch S', a naive approach would be to enumerate all 2^k possible batches and keep the batch with the maximum utility, which at the same time does not leak any sensitive user attributes. However, this approach has exponential complexity $O(2^k)$. Instead, we propose a customized version of the A* algorithm, which is an Informed Search method [13], to identify a good batch S' *efficiently*. A* is a search algorithm, hence, it requires a search tree with a starting node and a goal node to reach. Each node of the tree is called a *state* and corresponds to a batch S'. The starting state would be the non-anynomized batch S while the goal state would be the anonymized batch S' that preserves the privacy of all involved users. There are many acceptable goal states, so additionally a cost function is needed to indicate the amount of sacrificed utility to reach a specific state.

Each anonymized batch S' corresponds to a state and all possible states form the search tree. We encode S' as a k-digit binary number where the i-th digit corresponds to the pair $t_i, C_i \in S'$. A value of 0 as the i-th digit indicates that the sensitive community attribute L in t_i, C_i is generalized, while a value of 1 indicates that it is not. Ideally, we would like to report the batch S' that corresponds to the value 111...1 (no anonymization). A batch S' is an *ancestor* of batch S'' in the search tree if their encoding differs in exactly one digit, where this digit is 0 in S' and 1 in S''. Using this notion of ancestors a search tree can be defined where the encoding 111...1 is the root node and a node's children contain all descendant encodings. For example, for $k = 4$, the children of root node 1111 are: 1110, 1101, 1011, and 0111. The children of 1110 are: 1100, 1010, and 0110, etc. A visual example for $k = 3$ is shown in Fig. 2(a). All search tree branches will have 00...0 as the common leaf node which corresponds to a fully anonymized batch and is the least desirable result since its utility is minimal.

As the *starting state* of A* θ-CATT selects the batch S (original, non-anonymized output of the CATT algorithm) which has encoding 111...1. The *goal state* will be the first state that has no privacy leaks (all sensitive attribute inference probabilities are below θ). Given a random state S', the neighbors are generated by flipping a single digit with value 1. If there are no such digits left, the search tree has reached its end. Given that the algorithm is stable across batches (all probabilities are below θ before a new batch), an acceptable goal state will always exist. In the worst case this will be the state with encoding 00...0 at the bottom of the search tree (Fig. 2(a)).

A* Cost Function. A* requires a cost function that returns the cost of visiting each state. θ-CATT utilizes the following cost function f: $fS' = gS' + hS'$. Function gS' returns the total utility loss: $gS' = utilS - utilS'$, where S is the original non-anonymized set of topics and communities. Function $h(S')$ is the *heuristic* that estimates how close the current state is to the goal state and we use the following measure: $h(S') = \#$ users with a privacy violation. The number of users with a privacy violation is obtained by iterating through all the involved users in the batch and calculating the probability of inferring their sensitive attribute(s) with confidence higher than θ (Eq. 5). The function g measures the cumulative cost to reach a node in the search tree (how much utility has been sacrificed) and function h *estimates* the remaining distance of the goal state, where there is no privacy violation for any user. Note that this specific heuristic is not *admissible* (it might overestimate the cost to reach the goal state), which means that A* might not find the optimal path. Not finding the optimal path means that some additional utility might be sacrificed in order to greedily reach a goal state in less steps. Since the two functions g and h measure different units we normalize them with two weights α and β: $f(S') = \alpha g(S') + \beta h(S')$ where $\alpha + \beta = 1$. The exact values of α and β depend on the total number of users (for g) and the specific utility function used (for h).

Algorithmic Complexity. A* checks recursively if the current node is an acceptable goal state—number of privacy violations is equal to zero—and if it is not, it expands its children nodes and adds them in a priority queue to visit them next. Priority is calculated using the $f(.)$ function. This strategy enables θ-CATT to find a path to a batch S' that does not violate the privacy of any user, while reducing the number of necessary steps. The only trade-off is that the utility of the reached S' might not be optimal. For multiple sensitive attributes, the same process can be executed in parallel.

Let V be the set of sensitive attributes, k the size of the batch with pairs of topics and communities, T the set of all topics in the batch, and n the total number of users in the social network. The *time complexity* of the algorithm is:

$$O(|V| \cdot k \cdot |SEARCHT| + |SEARCHT| \cdot |u.T|)$$

The main bottleneck of the algorithm is the calculation of the inference probability (Eq. 5) for a specific attribute and every involved user. First, the whole process must be repeated for every sensitive attribute. This entails linear complexity to the number of sensitive attributes. Second, probability calculations must be repeated every time the cost of a state in the search tree is valuated. While there are 2^k states to explore, the customized A* with the proposed greedy heuristic can reach a local optimum in logarithmic complexity. $log_2 2^k = k$, thus, the algorithm scales linearly (amortized) with the number of topics in the batch. Finally, we need to calculate probabilities for every involved user, so the time complexity will also be proportional to $|SEARCHT|$. The inference probability formula (Eq. 5) contains the product of the empirical probabilities $Pt_i|L$ where t_i is an old topic the user has mentioned and L is a sensitive attribute. To avoid calculating this product every time the inference probability is measured, we can instead store in memory the products for all topics the user has mentioned so far. The prior probability of PL needs to be calculated only once per batch and n is a fixed number (at least in the context of a batch). The only "problematic" term is the denominator of the fraction, $|SEARCHu.T|$, which requires the calculation of the intersection of every set of users that mentioned the same topics with user u. However, this value needs to be calculated only once per user, per batch. Therefore, the time complexity of the inference probability calculation is constant.

The necessary *space complexity* to store the probability products for each user and sensitive attribute is: $On|V|$.

6 Experimental Results

For our experiments we used a real Twitter dataset that contains a uniform 10% sample of the complete Twitter Firehose stream from a 39 day period between April 16 and May 24, 2014. Each tweet also contains the information of its author (user). The extracted topics include unigrams, hashtags or capitalized entities from the tweets' raw text. The four extracted user demographics include location, gender, age, and US political party preference. Location extraction was done on (1) the tweet level using Twitter's geo-tagging mechanism, and to further improve the recall, on (2) the user level using a user-provided raw text field (similarly to [1,23]). To extract gender and age we applied existing language models extracted from [18] on social media data. The hierarchy for gender includes the leaf nodes "male"/"female" and the top level of "all genders" or "*". Similarly, the hierarchy for age includes the leaf nodes "13–18"/"19–22"/"23–29"/"30+" and the top level "*". Finally, for political party affiliation we gathered the official Twitter accounts associated with the three most popular US political parties: Democratics, Republicans, and Libertarians. Then, a user's political affiliation was determined based on the simple majority of interactions (@-replies) with these accounts. More extensive details can be found in [8].

We consider all four attributes to be *sensitive* for every user. Then we ran two versions of our algorithms (simple CATT and θ-CATT) and compared the results. The algorithm settings are: $\theta = .7$ (attacker's inference confidence), $\xi = .5$ (community size as a ratio of the topic population), utility $util\{t_i, C_i\} =_{i=1}^{k} IC_i$ (self-information sum), $\alpha = .999$, and $\beta = .001$. The selected values were empirically chosen to reflect a realistic scenario with a plethora of violations.

The average number of extracted trending topics and community pairs in the dataset is 112 per window (a window of data corresponds to a single batch of trending topics as described in earlier section). We focus on the topics that have a specific city-level location, or age, or gender, or political party preference values, which on average is $k = 21.57$ topics per batch. The per-batch average number of unique location values is 15.2, number of unique gender and political party values is 2, and number of unique age values is 2.8. The average number of involved users is 8162. The average utility without any anonymization (simple CATT) is 43.1 bits but also contains an average of 213.2 privacy violations. Privacy violations were counted by identifying users that have inference probabilities (Eq. 5) for either location, age, gender, or political party preference, that is higher than θ. To preserve the privacy of the location attribute, θ-CATT anonymized on average 4.3 communities to bring the number of privacy violations to 0. The average utility of the anonymized results published by θ-CATT is 38.37 bits, so there is a total utility loss of 4.73 bits.

Examples that demonstrate cases where a community got anonymized to preserve the involved users' privacy are listed in Table 2. The 4th column lists how many privacy violations would occur if the original community was published. The 5th column shows how the proposed algorithm decided to anonymize the community by generalizing at least one attribute. *After anonymization, θ-CATT managed to bring all privacy violations to 0 so that the reported results are θ-private.* For the topic #OscarTrial the location attribute was generalized to hide the location of 345 users. For the topic #ObamaInThreeWords both age and party preference are generalized to preserve the privacy of 76 users.

Table 2. Examples of communities and the corresponding anonymized versions.

Topic	Original community	Size	Viol/ns	Anonymized community
#OscarTrial	Location: Johannesburg, ZA, Gender: Female	1133	345	**Location: ZA,** Gender: Female
#FreeJustina	Location: Boston, Gender: Female, Politics: Democrat	51	13	Location: Boston, **Gender: *,** Politics: Democrat
Bruins	Location: Boston, Gender: Male, Age: 19–22	196	58	**Location: *,** Gender: Male, Age: 19–22
#ObamaIn3Words	Location: USA, Age: 19–22, Gender: Male, Politics: Republican	224	76	Location: USA, **Age: *,** Gender: Male, **Politics: ***

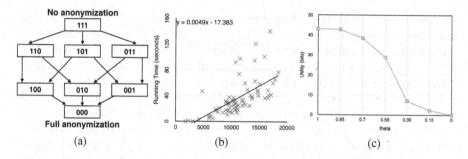

Fig. 2. (a) Full search tree ($k = 3$). "No anonymization" is the starting state of A*. (b) Running time for $k = 21.57$. (c) Utility loss for different values of θ.

In Fig. 2(c) it can be seen how the utility loss scales for different values of θ. As expected, when $\theta = 1$, an attacker must be 100% confident when inferring a sensitive attribute which in reality is practically impossible and results in maintaining the *full utility* of the results (equal to the utility of CATT's output). On the other end, for $\theta = 0$, no information leakage is permitted at all, therefore, full anonymization of the communities is necessary and utility becomes equal to 0. These two extremes are equally not practical for a meaningful and realistic combination of trending topics with utility and preserved privacy. Based on the values in Fig. 2 we observe that choosing a value of θ above .6 can maintain at least 73% of CATT's original utility of community-aware trending topics. This curve is a useful guide for choosing the desired privacy-utility trade-off.

Figure 2(b) shows the running time of our privacy preservation algorithm. All running times are recorded on a personal laptop with a 2.6 GHz Intel Core i5 processor and 16 Gb of RAM. There were 70 datapoints each corresponding to randomly sampled batches of topics. Since the complexity of the algorithm is mainly affected by the number of *involved users* (users mentioning one of the topics in the batch) the plot demonstrate how the running time is affected by this number. Each datapoint is an execution time (y-axis) of a single batch and corresponds a certain number of involved users (x-axis). The number of topics with sensitive attributes (batch size) was quite stable throughout our experiments with a mean of $k = 21.57$ and a standard deviation of 3.35. The plot also contains the corresponding least-square linear trendline and its equation. All reported running times are within the range of 0 s (no anonymizations were necessary for these batches so A* immediately found the goal state to be the starting state) and 160 s. Note that the time necessary to stream-in the data of a single batch takes around 3–4 min based on the rate of new tweets being created on Twitter, therefore, an average running time of 39.56 s is more than sufficient to produce results before the new batch is even ready for processing. This means that the algorithm can be used in a real-time fashion, a strong requirement for any streaming algorithm.

To examine if the running time is affected by the size of a batch k we also performed an experiment where we forced the number of topics to be always equal to 15—an arbitrarily selected value that is less than 21.57—by randomly dropping some topics. We observed that the running time is also increasing *linearly* with the number of users, as expected. Altering k had no apparent effect on how the running time scales with the number of users, similar to the slope of the trendline in Fig. 2(b), which proves that the greedy heuristic of A* has sublinear amortized complexity. Based on the projected trendlines in Fig. 2(b), we estimate that the running time for $100K$ users, which is a number that can be observed for trending topics on the Twitter web-page, would be approximately 490 s which is again acceptable based on the rate of generated tweets. Therefore, our algorithm satisfies the efficiency requirement of a practical real-world setting.

7   Conclusions

With the introduction of algorithms that extract trending topics that correlate with user demographics (community-aware topics), novel ways emerge to attack sensitive user information through attribute inference. We are the first to address privacy concerns in this context, by demonstrating how an attacker can statistically infer sensitive attribute values and introducing a privacy model for the preservation of these sensitive values of each individual user that discusses trending topics in a social network. Towards this end, we propose a new algorithmic approach that utilizes Artificial Intelligence methods in a novel way to efficiently identify when a privacy violation may occur and remedy all violations by efficiently extracting an optimal anonymization strategy which maximizes the utility of the reported trending topics and corresponding community characteristics.

Acknowledgments. This work is supported by NSF grant CNS 1649469.

References

1. Achrekar, H., Gandhe, A., Lazarus, R., Yu, S.H., Liu, B.: Predicting flu trends using Twitter data. In: Computer Communications Workshops, pp. 702–707 (2011)
2. Boldi, P., Bonchi, F., Gionis, A., Tassa, T.: Injecting uncertainty in graphs for identity obfuscation. Proc. VLDB Endow. **5**(11), 1376–1387 (2012). http://dx.doi.org/10.14778/2350229.2350254
3. Bonchi, F., Gionis, A., Tassa, T.: Identity obfuscation in graphs through the information theoretic lens. In: Proceedings of the International Conference on Data Engineering, pp. 924–935. ICDE, Washington, DC (2011). http://dx.doi.org/10.1109/ICDE.2011.5767905
4. Campan, A., Truta, T.M.: Data and structural k-anonymity in social networks. In: PinKDD 2008, pp. 33–54 (2009). http://dx.doi.org/10.1007/978-3-642-01718-6_4
5. Culotta, A., Ravi, N.K., Cutler, J.: Predicting the demographics of twitter users from website traffic data. In: Proceedings of the Conference on Artificial Intelligence, pp. 72–78 (2015). http://dl.acm.org/citation.cfm?id=2887007.2887018
6. Dwork, C.: Differential privacy. In: Bugliesi, M., Preneel, B., Sassone, V., Wegener, I. (eds.) ICALP 2006. LNCS, vol. 4052, pp. 1–12. Springer, Heidelberg (2006). doi:10.1007/11787006_1

7. Dwork, C., Naor, M., Pitassi, T., Rothblum, G.N., Yekhanin, S.: Pan-private streaming algorithms. In: Proceedings of the Innovations in Computer Science - ICS 2010, Tsinghua University, Beijing, China, pp. 66–80, 5–7 January 2010. http://conference.itcs.tsinghua.edu.cn/ICS2010/content/papers/6.html

8. Georgiou, T., El Abbadi, A., Yan, X.: Extracting topics with focused communities for social content recommendation. In: Proceedings of the 2017 ACM Conference on Computer Supported Cooperative Work and Social Computing, CSCW 2017, Portland, OR, USA, pp. 1432–1443, 25 February–1 March 2017. http://dl.acm.org/citation.cfm?id=2998259

9. LeFevre, K., DeWitt, D.J., Ramakrishnan, R.: Incognito: efficient full-domain k-anonymity. In: Proceedings of the 2005 ACM SIGMOD International Conference on Management of Data, pp. 49–60. ACM (2005)

10. Li, N., Li, T., Venkatasubramanian, S.: t-closeness: privacy beyond k-anonymity and l-diversity. In: ICDE 2007, pp. 106–115 (2007)

11. Machanavajjhala, A., Kifer, D., Gehrke, J., Venkitasubramaniam, M.: L-diversity: privacy beyond k-anonymity. ACM Trans. Knowl. Discov. Data 1(1) (2007). http://doi.acm.org/10.1145/1217299.1217302

12. Nazi, A., Thirumuruganathan, S., Hristidis, V., Zhang, N., Shaban, K., Das, G.: Query hidden attributes in social networks. In: 2014 IEEE International Conference on Data Mining Workshop, pp. 886–891, December 2014

13. Nilsson, N.J.: Problem-Solving Methods in Artificial Intelligence. McGraw-Hill Pub. Co., New York (1971)

14. Raymond, H., Murat, K., Bhavani, T.: Preventing private information inference attacks on social networks. IEEE Trans. Knowl. Data Eng. 25(8), 1849–1862 (2013). http://dx.doi.org/10.1109/TKDE.2012.120

15. Ryu, E., Rong, Y., Li, J., Machanavajjhala, A.: Curso: protect yourself from curse of attribute inference: a social network privacy-analyzer. In: Proceedings of the ACM SIGMOD Workshop on Databases and Social Networks, DBSocial 2013, New York, NY, USA, pp. 13–18 (2013). http://doi.acm.org/10.1145/2484702.2484706

16. Samarati, P.: Protecting respondents' identities in microdata release. IEEE Trans. Knowl. Data Eng. 13(6), 1010–1027 (2001). http://dx.doi.org/10.1109/69.971193

17. Samarati, P., Sweeney, L.: Generalizing data to provide anonymity when disclosing information. In: PODS, vol. 98, p. 188 (1998)

18. Schwartz, H., Eichstaedt, J., Kern, M., Dziurzynsk, L., Ramones, S.: Personality, gender, and age in the language of social media: the open-vocabulary approach. PLoS ONE 8(9), e73791 (2013). https://doi.org/10.1371/journal.pone.0073791

19. Shannon, C.E.: A mathematical theory of communication. SIGMOBILE Mob. Comput. Commun. Rev. 5(1), 3–55 (2001). http://doi.acm.org/10.1145/584091.584093

20. Sweeney, L.: Achieving k-anonymity privacy protection using generalization and suppression. Int. J. Uncertain. Fuzziness Knowl. Based Syst. 10(05), 571–588 (2002)

21. Talukder, N., Ouzzani, M., Elmagarmid, A.K., Elmeleegy, H., Yakout, M.: Privometer: privacy protection in social networks. In: Workshops Proceedings of the International Conference on Data Engineering, ICDE, pp. 266–269 (2010). http://dx.doi.org/10.1109/ICDEW.2010.5452715

22. Tassa, T., Cohen, D.J.: Anonymization of centralized and distributed social networks by sequential clustering. IEEE Trans. Knowled. Data Eng. 25(2), 311–324 (2013)

23. Vieweg, S., Hughes, A.L., Starbird, K., Palen, L.: Microblogging during two natural hazards events: what Twitter may contribute to situational awareness. In: Proceedings of the SIGCHI conference on human factors in computing systems, pp. 1079–1088. ACM (2010)
24. Zhang, H.: The optimality of Naive Bayes. AA **1**(2), 3 (2004)
25. Zheleva, E., Getoor, L.: Preserving the privacy of sensitive relationships in graph data. In: International Conference on Privacy, Security, and Trust in KDD, pp. 153–171 (2008). http://dl.acm.org/citation.cfm?id=1793474.1793485
26. Zheleva, E., Getoor, L.: To join or not to join: the illusion of privacy in social networks with mixed public and private user profiles. In: Proceedings of the International Conference on World Wide Web, pp. 531–540 (2009). http://doi.acm.org/10.1145/1526709.1526781

Privacy-Preserving Outlier Detection
for Data Streams

Jonas Böhler[1](✉), Daniel Bernau[1], and Florian Kerschbaum[2]

[1] SAP Research, Karlsruhe, Germany
{jonas.boehler,daniel.bernau}@sap.com
[2] University of Waterloo, Waterloo, Canada
florian.kerschbaum@uwaterloo.ca

Abstract. In cyber-physical systems sensors data should be anonymized at the source. Local data perturbation with differential privacy guarantees can be used, but the resulting utility is often (too) low. In this paper we contribute an algorithm that combines local, differentially private data perturbation of sensor streams with highly accurate outlier detection. We evaluate our algorithm on synthetic data. In our experiments we obtain an accuracy of 80% with a differential privacy value of $\epsilon = 0.1$ for well separated outliers.

1 Introduction

In cyber-physical systems, e.g. smart metering, connected cars or the Internet of Things, sensors stream data to a sink, e.g. a database in the cloud, which is commonly controlled by a different entity. The subjects observed by the sensors have a vested interest in preserving their privacy towards the other entity. A technical means to preserve privacy is to anonymize the data (at the source). However, the data itself may be personally identifiable information as was, e.g., shown for smart meter readings [10]. Local data perturbation with differential privacy guarantees [5] can be used to protect against such exploitation and can be applied by the sensor. However, the resulting utility in this non-interactive model is often much lower than in the interactive, trusted curator model of differential privacy. So far, successful, differentially private outlier detection was only achieved in the interactive model [8,17,18].

Our algorithm contributed in this paper shows that local data perturbation of sensor streams combined with highly accurate outlier detection is feasible. We achieve this by using a relaxed version of differential privacy and a privacy-preserving correction method. The relaxation is to adapt the sensitivity to the set of data excluding the outliers [4]. We assume a scenario where outliers are subject to subsequent investigation which requires precise data, e.g. a broken power line or water pipe. Our privacy-preserving correction method uses distribution of trust between a correction server and an analyst server (the database). The correction server never learns the real measurements, but only the random

G. Livraga and S. Zhu (Eds.): DBSec 2017, LNCS 10359, pp. 225–238, 2017.
DOI: 10.1007/978-3-319-61176-1_12

noise added by the data perturbation (with indices of data values). The analyst server never learns the random noise, but only indices of data values whose outlier status – false positives and false negatives – the correction server has adjusted[1]. The result provides an improved outlier detection and preserves differential privacy towards the data analyst, since data perturbation is applied at the source (independent of the algorithm). Furthermore, the correction server never learns enough information to reconstruct any of the data.

Our non-interactive data perturbation is applied once for all subsequent analyses and does not require a privacy budget distributed over a series of queries which is critical in many applications [6,20]. We evaluate our algorithm on synthetic data. In our experiments we detect 80% of outliers in a subset of 10% of all points with a differential privacy value of $\epsilon = 0.1$ on data sets with well separated outliers. Our error correction method has an average runtime of less than 40 ms on 100,000 data points.

2   Related Work

We perform outlier detection on sensor data perturbed with relaxed differential privacy at the source and correct the detection errors due to perturbation. We are not aware of any related work on this specific problem, however, there has been extensive work in related areas: releasing differentially private topographical information, relaxations of differential privacy and separation of outliers and non-outliers.

In the area of releasing topological proximities under differential privacy a foundation for privately deriving cluster centers is provided in [1,16,21]. Their approaches have two drawbacks due to the use of the interactive model: The complex determination of ϵ for an assumed number of iterations until convergence and the limitation to aggregated cluster centers. An approach towards non-interactive differential privacy in clustering through a hybrid approach of non-interactive and interactive computations is formulated in [22]. A foundation for increasing differential privacy utility by sensitivity optimizations is introduced in [17]. Furthermore, the authors formulate a differentially private approach to release near optimal k-means cluster centers with their *sample-and-aggregate* framework. In [18] the sample-and-aggregate approach is extended to detect the minimal ball B enclosing approximately 90% of the points; everything outside B is presumed to be an outlier. Their approach is formulated in the interactive model and requires to apply the calculated ball B to the original data for outlier identification. The work in [8] is similar in the desire to produce a sanitized data set representation allowing an unlimited series of adaptive queries. Their non-interactive approach for producing private coresets (a weighted subset of points capturing geometric properties of its superset) suitable for k-means and k-median queries is proven theoretically efficient, but does not allow to identify

---

[1] The analyst server learns also parameters about the data set which are however computable from the output of the algorithm.

individual outliers. Chances for great accuracy improvements in differentially private analysis are identified in [18] if outliers were identified and removed before analysis.

Several relaxations for differential privacy have been suggested. Either by adapting the sensitivity [17], additional privacy loss [3], by distinguishing between groups with different privacy guarantees [11,14], or by relaxing the adversary [19,23]. In [3] (ϵ,δ)-differential privacy is presented where the privacy loss does not exceed ϵ with probability at most $1-\delta$ where δ is negligible. For our scenario in which outliers should be treated as a separate group, δ would become very large. Instead, we argue for a noise distribution with different ϵ guarantees for outliers and non-outliers. By relaxing the assumed adversary knowledge about the data the work [23] shows that utility gains in Genome-Wide Association Studies are achievable. This relaxation is not discriminating between different groups found in the data set as in our case. Additionally, we avoid relaxing the adversary and instead decrease the privacy guarantee for outliers.

The discussion on separation of outliers and non-outliers has been addressed in [11] by questioning the equal right for privacy for all (i.e. citizens vs. terrorists). Their work is close to ours in enforcing privacy guarantees to differentiate between a protected and a target subpopulation. However, in [11] the original data is maintained and query answers are perturbed interactively with a trusted curator. We avoid giving access to original data and enforce perturbation at the source. It is concluded in [15] that sparse domains incorporate a high risk of producing outliers in the perturbed data and thus argue for the need of outlier identification and removal in the unperturbed data set. In contrast, we preserve outliers and enable the detection of outliers in the perturbed data. Tailored differential privacy is defined in [14] and aims to provide *stronger* ϵ-differential privacy guarantees to outliers. We decided to evaluate the opposite by granting them *less* protection since we see outliers as faulty systems or sensors one needs to detect.

3   Preliminaries

We model a database (or data set) D as a collection of records from $\mathcal{D}^n$, i.e. $D \in \mathcal{D}^n$, where each entry D_i of D represents one participant's information. The *Hamming distance* $d_H(\cdot,\cdot)$ between two databases $x, y \in \mathcal{D}^n$ is $d_H(x,y) = |\{i : x_i \neq y_i\}|$, i.e. the number of entries in which they differ. Databases x, y are called *neighbors* or *neighboring* if $d_H(x,y) = 1$.

Definition 1 (Differential Privacy). *A perturbation mechanism $\mathcal{M}$ provides ϵ-differential privacy if for all neighboring databases D_1 and D_2, and all $S \subseteq Range(\mathcal{M})$,*

$$Pr[\mathcal{M}(D_1) \in S] \leq \exp(\epsilon) \cdot Pr[\mathcal{M}(D_2) \in S].$$

The protection for an individual in the database is measured by the privacy level ϵ. While a small ϵ offers higher protection for individuals involved in the computation of a statistical function f, a larger ϵ offers higher accuracy on f. In case an individual is involved in a series of n statistical functions perturbed by a corresponding mechanism $\mathcal{M}_i$, where each function is requiring ϵ_i, her protection is defined as $\epsilon = \sum_{i=1}^{n} \epsilon_i$ by the basic sequential composition theorem of Dwork et al. [3,16]. A data owner can limit the privacy loss by specifying a maximum for ϵ called *privacy budget* [4]. Depending on the mutual agreement the exhaustion of the privacy budget can require the original data to be destroyed as mentioned in [20] since the privacy guarantee no longer holds.

The noise level of $\mathcal{M}$ in differential privacy is dependent on the *sensitivity* of f. For an overview of different notions of sensitivity with respect to the l_1-metric see Table 1. The global sensitivity GS_f of a function $f : \mathcal{D}^n \to \mathbb{R}^k$ determines the worst-case change that the omission or inclusion of a single individual's data can have on f. For example, if f is a counting query the removal of an individual can only change the result by 1. GS_f has to cover *all neighboring* databases, whereas local sensitivity $LS_f(D)$ covers *one fixed* database instance $D \in \mathcal{D}^n$ and all its neighbors. In certain cases databases with low local sensitivity can have neighbors with high local sensitivity, thereby allowing an attacker to distinguish them by the sensitivity-dependent noise magnitude alone. In contrast, smooth sensitivity $SS_f(D)$ compares a fixed database instance D with *all other database instances* but with respect to the distance between them and a privacy parameter β^2. Using the notation from Table 1, the parameters that differ in the various notions are: allowed distance between neighboring databases D_1, D_2 (1 for GS_f and LS_f, unrestricted for SS_f) and choice of databases D_1 (a single fixed database instance for LS_f and SS_f, unrestricted for GS_f). In Sect. 4 we introduce a new notion of sensitivity, *relaxed sensitivity*, where the choice of databases is generalized to allow the selection of a subgroup of all possible databases.

Table 1. Comparison of different sensitivity notions

Global [4]	$GS_f = \displaystyle\max_{D_1, D_2 : d_H(D_1, D_2)=1;\ D_1, D_2 \in \mathcal{D}^n} \|f(D_1) - f(D_2)\|_1$
Local	$LS_f(D_1) = \displaystyle\max_{D_2 : d_H(D_1, D_2)=1;\ D_2 \in \mathcal{D}^n} \|f(D_1) - f(D_2)\|_1$
Smooth [17]	$SS_{f,\beta}(D_1) = \displaystyle\max_{D_2 \in \mathcal{D}^n} \left(LS_f(D_2) e^{-\beta d_H(D_1, D_2)} \right)$

Two models for computation of a mechanism $\mathcal{M}$ have been suggested by [5]. In the *interactive* model a data analyst receives noisy answers to functions evaluated on unperturbed data D as long as the privacy budget is not exhausted. In contrast, the original data D can be discarded in the *non-interactive* model by producing a sanitized version $D' = \mathcal{M}(D, f)$ of D and results are calculated with

---

[2] SS_f needs to be a *smooth upper bound* S as defined in [17], i.e. $\forall D \in \mathcal{D}^n : S(D) \geq LS_f(D)$ and $\forall D_1, D_2 \in \mathcal{D}^n, d_H(D_1, D_2) = 1 : S(D_1) \leq e^{\beta} \cdot S(D_2)$. These requirements can be fulfilled by $S(D) = GS_f$ with $\beta = 0$. SS_f, however, is the *smallest* function to satisfy these requirements with $\beta > 0$.

D'. While [17] assumes that the majority of mechanisms utilize the interactive model the findings of [2] suggest that D' is inefficient but potentially useful for many classes of queries if computational constraints are ignored. However, the non-interactive model also has its benefits: First, there is no need for a curator who requires access to the sensitive D, analyzes and permits queries and adjusts the privacy budget. Second, storage constrained sensors do not need to retain D and instead release a locally sanitized D'. Third, the data owner is not left with the administrative decision on how to handle exhausted privacy budgets (e.g. destroy D or refresh budget periodically as discussed in [16,21]).

4   Relaxed Differential Privacy

Differential privacy is a strong privacy guarantee due to its two worst-case assumptions: the adversary is assumed to have complete knowledge about the data set except for a single record and all possible data sets are covered by the guarantee. To relax differential privacy one has to relax these assumptions. The first assumption was relaxed in [19] by using a weaker but more realistic adversary and bounding the adversary's prior and posterior belief. In [17] the second assumption is relaxed by their notion of smooth sensitivity. We focused on the latter approach due to the fact that we are concerned with the discovery of outliers: We do not need the guarantee to hold for all records equally.

4.1   Relaxed Sensitivity

Our following new notion of *relaxed sensitivity* allows for different privacy guarantees for groups $\mathcal{N}$ (non-outliers), $\mathcal{O}$ (outliers) within a single dataset $\mathcal{D}^n = \mathcal{N} \cup \mathcal{O}$.

Definition 2 (Relaxed sensitivity). *Let $\mathcal{D}^n = \mathcal{N} \cup \mathcal{O}$ then the relaxed sensitivity of a function $f : \mathcal{D}^n \to \mathbb{R}^k$ is*

$$RS_f^{\mathcal{N},\mathcal{D}^n} = \max_{\substack{D_1,D_2 \in \mathcal{N} \\ D_2 : d_H(D_1,D_2)=1}} \|f(D_1) - f(D_2)\|_1.$$

In the following we abuse notation slightly and say that in the case that $\mathcal{D}$ consists of multiple, independent columns the sensitivity and perturbation are calculated per column and not the entire database at once. While local sensitivity only holds for *one fixed* database instance the relaxed sensitivity covers *all* databases from the subset $\mathcal{N}$. Let LS_f^X, GS_f^X denote local and global sensitivity respectively over a database set X.

Theorem 1. *Relaxed sensitivity compares to local and global sensitivity as follows:*

$$LS_f^{\mathcal{N}}(D) \leq RS_f^{\mathcal{N},\mathcal{D}^n} = GS_f^{\mathcal{N}} \leq GS_f^{\mathcal{D}^n}$$

where $D \in \mathcal{N} \subseteq \mathcal{D}^n$.

The proof is omitted due to space constraints. We will omit the dataset in the sensitivity notation in the following when it is not explicitly needed. The privacy guarantee is enforced by noise whose magnitude is controlled by privacy parameter ϵ and the sensitivity. We adapt the popular Laplace mechanism [5] to allow its invocation with different sensitivity notions.

Definition 3 Laplace mechanism. *Given any function $f : \mathcal{D}^n \to \mathbb{R}^k$, the Laplace mechanism is defined as*

$$\mathcal{M}_L(x, f(\cdot), GS_f/\epsilon) = f(x) + (Y_1, \ldots, Y_k)$$

where Y_i are independent and identically distributed random variables drawn from the Laplace distribution $Laplace(GS_f/\epsilon)$.

To relax differential privacy, we will adapt the scaling parameter to RS_f/ϵ, thus sampling noise from $Laplace(RS_f/\epsilon)$. We view a database as consisting of multiple columns and the perturbation is performed *per column*: The Laplace mechanism receives a column, i.e. a vector, as input and outputs a perturbed vector.

Theorem 2. *Let $\mathcal{D}^n = \mathcal{N} \cup \mathcal{O}$ and f be a function $f : \mathcal{D}^n \to \mathbb{R}^k$. The Laplace mechanism $\mathcal{M}_L(x, f, RS_f/\epsilon)$ preserves ϵ'-differential privacy for $x \in \mathcal{D}^n$ and preserves ϵ-differential privacy for $x \in \mathcal{N}$, where $\epsilon' = \epsilon \cdot GS_f/RS_f \geq \epsilon$.*

The proof is omitted due to space limitations.

4.2   Approximation of Relaxed Sensitivity

We do not want to restrict the queries an analyst can perform in the non-interactive model. Therefore, we choose to evaluate the *identity function* $f_{id}(x) = x$ for sensitivity determination. The sensitivity for f_{id} can be unbounded depending on the input domain. In the following we assume elements in $\mathcal{N}$ to be bounded real numbers. With f_{id} as our function and bounded $\mathcal{N}$, we can express the relaxed sensitivity as $RS_{f_{id}} = \max(\mathcal{N}) - \min(\mathcal{N})$, i.e. the gap between possible databases, that we seek to close. We see sensors measurements as point coordinates and use the terms interchangeably.

With historical data and domain knowledge the approximation can be tailored more precisely to individual data sets. With knowledge about what measurements can be considered physically possible one can approximate a bound for $\mathcal{N}$. We approximate $RS_{f_{id}}$ in the following with q-th percentiles ρ_q, $q \in [0, 100]$ of $\mathcal{D}$. We denote with p_o the percentage of outliers in the data set – alternatively, it can be seen as a bound on $\mathcal{N}$. We set $q_{max} = 100 - p_o/2$, $q_{min} = 100 - q_{max}$, and approximate $RS_{f_{id}}$ with

$$\widehat{RS_{f_{id}}} = \rho_{q_{max}}(\mathcal{D}) - \rho_{q_{min}}(\mathcal{D}). \tag{1}$$

For this approximation we assume the following characteristics regarding our datasets:

1. We define outliers as points on an outer layer surrounding non-outliers,
2. the percentage of outliers or a bound for $\mathcal{N}$ can be approximated,
3. the data set contains only one cluster.

Assumption 1 is also used in depth-based outlier detection algorithms. Regarding Assumption 2, one can learn the outlier percentage or bounds via historical data and the range of plausible (i.e. non-faulty, physically possible) measurements. If multiple clusters exists the data can be split in cluster groups thus fulfilling Assumption 3. The split is either performed by the data owner or a third party in a privacy-preserving manner (see Sect. 2 for clustering approaches consuming a portion of ϵ).

$\widehat{RS_f}$ implicitly defines $\mathcal{N}'$, which is an estimation of $\mathcal{N}$. An estimation $\widehat{RS_f} > RS_f$ leads to $\mathcal{N} \subset \mathcal{N}'$, i.e. more elements than necessary are protected which does not decrease privacy for elements from $\mathcal{N}$. However, $\widehat{RS_f} < RS_f$ implicitly defines $\mathcal{N}' \subset \mathcal{N}$, i.e. elements in $\mathcal{N}\backslash\mathcal{N}'$ could suffer a privacy loss since they receive less noise than needed. We want to stress that even for an inaccurate approximation, the non-outliers are still protected and receive a privacy level of $\epsilon' = \epsilon \cdot RS_{f_{\mathrm{id}}}/\widehat{RS_{f_{\mathrm{id}}}} \geq \epsilon$ (see Theorem 2). Furthermore, we correct errors introduced by the perturbation (or estimation $\widehat{RS_f} > RS_f$). For this we classify and detect the errors based on their change in distance to the center after and before perturbation as described in the following sections.

5 Outliers and False Negative Types

Let f_{outlier} be an outlier detection function $f_{\mathrm{outlier}} : \mathcal{T} \to \{1, \ldots, |\mathcal{T}|\}$ which returns the indices (i.e. row numbers) of $\mathcal{T}$ which are outliers. We will refer to outliers detected in the unperturbed data set as *outliers* or $\mathcal{O} = f_{\mathrm{outlier}}(\mathcal{T})$. When referring to the perturbed version of $\mathcal{T}$, denoted as $\mathcal{T}'$, we will use *presumed outliers* or $\mathcal{O}' = f_{\mathrm{outlier}}(\mathcal{T}')$. We define outliers as points on an outer layer surrounding a (denser) core similar to [18]. Our goal is to find a small subset containing $\mathcal{O}$ on the perturbed data without having access to the original data.

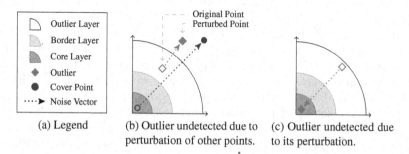

(a) Legend   (b) Outlier undetected due to   (c) Outlier undetected due
                    perturbation of other points.   to its perturbation.

Fig. 1. Types of false negatives after perturbation. Layers correspond to unperturbed data points.

We perturb the data set with the adapted Laplace mechanism using approximated relaxed sensitivity per column. For well-separated outliers and non-outliers and with our relaxed sensitivity notion $\mathcal{O}$ and $\mathcal{O}'$ can be equal. However, this is not necessarily the case and therefore we present a correction algorithm in Sect. 6 to find the *false negatives* i.e. missing outliers from $\mathcal{O}$ that are not in $\mathcal{O}'$. The presumed outliers in $\mathcal{O}'$ can be separated in two sets: *false positives*, i.e. presumed outliers in $\mathcal{O}'$ that are not in $\mathcal{O}$ and *true positives*, i.e. outliers in $\mathcal{O}$ that are also in $\mathcal{T}'$.

We distinguish between two different types of false negatives visualized in Fig. 1. Non-outliers lie in the core layer, outliers in the outlier layer and the empty border layer separates the two. The layers for the unperturbed data differ from the perturbed layers which are omitted in Fig. 1. The two types of false negatives can occur as follows: First, a non-outlier can become a *cover point* after perturbation, i.e. "cover" a real outlier to produce a false negative as shown in Fig. 1b. Second, an outlier can also become a false negative on its own when it lands in a non-outlier region after perturbation, e.g. a dense core as in Fig. 1c, where it will not be detected in the perturbed data.

6   Relaxed Differentially Private Outlier Detection and Correction

Given the data $\mathcal{T}'$, a relaxed differentially private version of $\mathcal{T}$, we want to find the outliers corresponding to the unperturbed $\mathcal{T}$. We use the *semi-honest model* introduced in [9] where corrupted protocol participants do not deviate from the protocol but gather everything created during the run of the protocol. (e.g. message transcripts, temporary memory). Furthermore, we assume that only one participant can be corrupted; an alternative assumption is that participants do not share their knowledge. The assumption that parties do not share their knowledge is similar to the interactive model of differential privacy, i.e. different analysts do not collaborate by combining their privacy budgets.

In the following we view data sets as consisting of two columns, one column per measured attribute. Let $\mathcal{T}$, $\mathcal{T}'$ be the unperturbed resp. perturbed data set. For the perturbation each column receives independently drawn noise from the Laplace mechanism with approximated relaxed sensitivity. We use sets of *indices* corresponding to rows in $\mathcal{T}$ and $\mathcal{T}'$. This is convenient since an index identifies the same point before and after perturbation. Let I be the set of indices from $\mathcal{T}$ (and thereby also $\mathcal{T}'$). We denote with $\mathcal{T}[i]$ the row at index $i \in I$ and with $\mathcal{T}[,j]$ the selection of column j. Recall that we denote with $\mathcal{O}$ the set of all indices corresponding to rows with outliers in $\mathcal{T}$ detected by f_{outlier}, i.e. $\mathcal{O} = f_{\text{outlier}}(\mathcal{T})$, and with $\mathcal{O}'$ the set of all presumed outlier indices from $\mathcal{T}'$. As before, *false negatives* are missing outliers from $\mathcal{O}$ that are not found in $\mathcal{O}'$, *false positives* are presumed outliers in $\mathcal{O}'$ that are not in $\mathcal{O}$, and *true positives* are outliers in $\mathcal{O}$ that are also in $\mathcal{O}'$. We denote the Euclidean distance of two points c, x as $d(c, x) = d_c(x)$ where c is the center of $\mathcal{T}$, determined by averaging each column.

Definition 4 (Distance Difference d_{diff}). *The distance difference between a point in T and T' at index i and the center c after and before perturbation respectively is*

$$d_{diff}[i] = d_c(T'[i]) - d_c(T[i]).$$

We denote with $w_\mathcal{O}$ the width of the outlier layer (visualized in Fig. 1) in the unperturbed data T. We denote with $\mathcal{FN}_{Lj}$ a set of indices for different layers $j \in \{1, 2, 3\}$ of presumed false negatives. These are spatial layers based on the false negative types of Fig. 1 used in the correction phase.

6.1   Correction Algorithm

Our goal is to detect presumed outliers on relaxed differentially private data T', find the undetected false negatives and remove additionally detected false positives. The algorithm presented in Fig. 2 operates as follows: Each sensor S has as input the data set T, the privacy parameter ϵ and approximated relaxed sensitivities $\widehat{RS_{f_{id},j}}$ for each data column j. The correction server CS has as input the outlier layer width $w_\mathcal{O}$. The values for $\widehat{RS_{f_{id},j}}$ and $w_\mathcal{O}$ are determined with historical data, i.e. knowledge about past outliers, and bounds for the non-outliers, e.g. normal, non-faulty measurement values for sensors. S scales the data in line 1a and generates the perturbed data set T' via perturbation of each column j with the adapted Laplace mechanism parameterized with $\widehat{RS_{f_{id},j}}/\epsilon$. Then it sends T' to the analyst A and the distance differences d_{diff} to the correction server CS. In line 3 the server CS filters $\mathcal{O}'$ in two sets $\mathcal{FP}$ and $\mathcal{TP}$ for presumed false positives and true positives respectively. The filtering is based on comparison of d_{diff} against a threshold – the distance difference with the biggest change between sorted distance differences. We use the fact that false positives, i.e. non-outliers that were detected as outliers in the perturbed set, have a higher distance difference, i.e. receive more noise, than true positives when non-outliers and outliers are well-separated. We do not want to remove true positives (actual outliers) under any circumstances. Thus, we err on the side of removing not enough false positives if the separation between outliers and non-outliers is low. In line 4 we reduce the set of indices to check for potential false negatives. Without the reduction true negatives can land in $\mathcal{FN}_{Lj}$, since they have the same distance difference (d_{diff}) but not the same core distance (d_c) as false negative candidates. The server CS detects false negatives in line 5, i.e. outliers not contained in $\mathcal{O}'$, in three spatial layers $\mathcal{FN}_{L1}$, $\mathcal{FN}_{L2}$, and $\mathcal{FN}_{L3}$, where the first corresponds to the false negative type from Fig. 1c and the latter to 1b. The use of two layers $\mathcal{FN}_{L2}$ and $\mathcal{FN}_{L3}$ for one false negative type is due to the row reduction and a simplification for $\mathcal{FN}_{L2}$ explained in the following.

The inequalities are based on the unperturbed position of outliers in respect to non-outliers (i.e. on an outer, less dense layer) and the distance difference after perturbation. The reasoning for $\mathcal{FN}_{L1}$ is that non-outliers, who are by definition closer to the center c, are distanced further away from c after perturbation due to the noise magnitude. Whereas outliers, who are already further away, can

Input: Each sensor $\mathcal{S}_{id}$ has data $\mathcal{T}$, privacy level ϵ and approximated relaxed sensitivities $\widehat{RS}_{f_{id},j}$ per column j. Correction server $\mathcal{CS}$ has outlier layer width $w_\mathcal{O}$ of sensor domain.

1. **Each sensor $\mathcal{S}_{id}$**
   (a) Scales each column $\mathcal{T}[,j]$: subtraction of mean and division of standard deviation.
   (b) Perturbs each column to $\mathcal{T}'[,j] = \mathcal{M}_L(\mathcal{T}[,j], f_{id}, \widehat{RS}_{f_{id},j}/\epsilon)$ and sends its id and perturbed data, i.e. $(id, \mathcal{T}')$, to analyst $\mathcal{A}$.
   (c) Calculates the data center c by averaging every dimension, calculates the distance differences $d_{\text{diff}} = d_c(\mathcal{T}') - d_c(\mathcal{T})$ and sends (id, d_{diff}) to correction server $\mathcal{CS}$.
2. **Analyst $\mathcal{A}$** performs standard outlier detection on $\mathcal{T}'$ to get the list $\mathcal{O}'$ of presumed outliers indices and sends $(id, \mathcal{O}')$ to $\mathcal{CS}$.
3. **Correction Server $\mathcal{CS}$**
   (a) Calculates the threshold index t for the biggest change between ascending d_{diff} values of presumed outliers: $t = \arg\max_{j_k \in J} d_{\text{diff}}[j_{k+1}] - d_{\text{diff}}[j_k]$ where $J = \{j_1, j_2, \dots\}$ are the indices from $\mathcal{O}'$ sorted according to ascending d_{diff} values. (For convenience we define $j_{k+1} = j_k$ for $k+1 > |J|$.)
   (b) Separates indices via t into false positives $\mathcal{FP} = \{i \in \mathcal{O}' \mid d_{\text{diff}}[i] > d_{\text{diff}}[t]\}$ and true positives $\mathcal{TP} = \mathcal{O}' \backslash \mathcal{FP}$ and calculates $d_{\mathcal{TP}} = \min_{i \in \mathcal{TP}} d_{\text{diff}}[i]$.
   (c) Sends $(id, d_{\mathcal{TP}}, d_{\mathcal{TP}} + w_\mathcal{O})$ to $\mathcal{A}$.
4. **Analyst $\mathcal{A}$** creates

$$I_2 = \{i \in I \backslash \mathcal{O}' \mid d_c(\mathcal{T}'[i]) \geq d_{\mathcal{TP}}\},$$
$$I_3 = \{i \in I \backslash \mathcal{O}' \mid d_c(\mathcal{T}'[i]) \geq d_{\mathcal{TP}} + w_\mathcal{O}\},$$

and sends (id, I_2, I_3) to $\mathcal{CS}$.
5. **Correction Server $\mathcal{CS}$** creates false negatives layer sets

$$\mathcal{FN}_{L1} = \{i \in I \backslash \mathcal{O}' \mid d_{\text{diff}}[i] < 0\},$$
$$\mathcal{FN}_{L2} = \{i \in I_2 \mid 0 \leq d_{\text{diff}}[i] \leq d_{\mathcal{TP}}\},$$
$$\mathcal{FN}_{L3} = \{i \in I_3 \mid d_{\mathcal{TP}} \leq d_{\text{diff}}[i] \leq d_{\mathcal{TP}} + w_\mathcal{O}\}.$$

Output: $\mathcal{CS}$ outputs $(id, \mathcal{TP}, \mathcal{FN}_{L1}, \mathcal{FN}_{L2}, \mathcal{FN}_{L3})$ to $\mathcal{DO}$.

Fig. 2. Algorithm for correction of outlier detection.

reduce their distance to the center as seen in Fig. 1c. Hence, we look for indices i fulfilling $d_{\text{diff}}[i] < 0$. The idea behind $\mathcal{FN}_{L2}$ is a simplification that all outliers lie on the same "orbit" around the center (same center distance). In this case the minimal distance difference d_{diff} to become a true positive is the minimal d_{diff} one can find from the set of presumed true positives $\mathcal{TP}$, i.e. $d_{\mathcal{TP}} = \min_{t \in \mathcal{TP}}(d_{\text{diff}}[t])$. Only unperturbed outliers with a d_{diff} greater than $d_{\mathcal{TP}}$ could be detected as an outlier after perturbation. Hence, the remaining undetected have d_{diff} larger than 0 (otherwise they land in $\mathcal{FN}_{L1}$) but smaller than $d_{\mathcal{TP}}$. However, not all outliers do lie on the same orbit. Therefore, we collect in $\mathcal{FN}_{L3}$ indices with distance difference greater than $d_{\mathcal{TP}}$. We also know that no undetected outlier's distance difference can be greater than the distance difference of false positives in $\mathcal{FP}$, i.e. $d_{\mathcal{FP}} = \min\limits_{f \in \mathcal{FP}}(d_{\mathrm{diff}}[f])$. More precise is the outlier layer width $w_{\mathcal{O}}$ (line 4).

6.2 Privacy of the Correction Algorithm

The server $\mathcal{CS}$ knows the distance difference for points at a given index i, i.e. $d_{\mathrm{diff}}[i]$, but $\mathcal{CS}$ does not know which perturbed point is identified by i since this knowledge remains at the analyst $\mathcal{A}$. The only information $\mathcal{CS}$ sends to $\mathcal{A}$ is $d_{\mathcal{TP}}, w_{\mathcal{O}}$, i.e. information regarding outliers which we like to detect. Even if $\mathcal{A}$ had access to all outlier information, including the noise used to perturb them, it would not lessen the protection of non-outliers. In exchange $\mathcal{A}$ sends $\mathcal{CS}$ I_{L2}, I_{L3}, i.e. sets of indices j of perturbed points with $d_c(\mathcal{T}'[j]) \geq d_{\mathcal{TP}}$ and $d_c(\mathcal{T}'[j]) \geq d_{\mathcal{TP}} + w_{\mathcal{O}}$ respectively. However, $\mathcal{CS}$ does not know which points correspond to these indices. If $\mathcal{A}$ and $\mathcal{CS}$ were to collaborate (i.e. not semi-honest) they could only narrow down the possible origin of a perturbed point. The information collaborating servers could learn can be reduced by using *frequency-hiding order-preserving encryption* as presented in [12] for the distance differences. Secure computation is not practical for our scenario (e.g. computation constrained IoT sensors) due to the bidirectional communication – although the communication complexity can be seen as almost independent of the network performance as shown in [13].

7 Outlier Detection Evaluation

We compared detected outliers on the original data with presumed outliers found on the perturbed data. Our algorithm was implemented in R 3.3.1 and run on a Apple MacBook Pro (Intel Core i7-4870HQ CPU, 16 GB main memory). We selected DBSCAN [7] to realize f_{outlier} and used the fpc package implementation. DBSCAN utilizes density and proximity of points, thus matching our spatial definition of the outlier topology where the outliers lie on an outer layer surrounding a denser core. DBSCAN is parameterized via *eps* for neighborhood reachability (point proximity and connections) and *minPts* (threshold density within the reachable neighborhood). We used the same DBSCAN parameters for the unperturbed and perturbed data. Our error correction logic only requires between 30 and 40 ms for 100,000 points. Four synthetic datasets were created to examine the impact of varying separation between outliers and non-outliers with the following characteristics: 100,000 points in $\mathbb{R}^2$ where each dimension is sampled independently from a normal distribution with standard deviation 3 and mean 0. Outlier percentage in the unperturbed data is 10%. After sampling the distances between outliers and non-outliers were increased to $\approx 50, 120, 220, 400$; we denote these data sets with *Separation 50*, etc. These separation distances were chosen based on decreasing probabilities that Laplace distributed noise preserves the outlier topology.

With *accuracy* we denote the percentage of all false negatives and true positives, i.e. all outliers, found with our approach. Furthermore, *subset size* is the

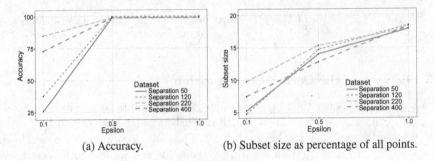

(a) Accuracy.   (b) Subset size as percentage of all points.

Fig. 3. Accuracy and subset size for synthetic data sets; mean of 5 runs with $\epsilon \in \{0.1, 0.5, 1\}$.

size of the output of algorithm in Fig. 2, i.e. $\{\mathcal{TP}, \mathcal{FN}_{L1}, \mathcal{FN}_{L2}, \mathcal{FN}_{L3}\}$. Accuracy and subset size for $\epsilon \in \{0.1, 0.5, 1\}$ is shown in Fig. 3a and b respectively. The results for $\epsilon = 0.1$ indicate that our correction algorithm achieves meaningful accuracy of $\approx 75\%$ for separation ≥ 220. However, $\epsilon = 0.1$ is a strong privacy guarantee and $\epsilon \in \{0.5, 1\}$ still offers meaningful protection. For $\epsilon = 0.5$ the found outlier percentage increases to 95–100% for all data sets. With a privacy guarantee of $\epsilon = 1$ our correction algorithm is not always needed since the separation between outliers and non-outliers is preserved even after perturbation. The subset size is always below 20% as is evident from Fig. 3b.

8 Conclusion

We implemented and evaluated an algorithm for detection of individual outliers on data perturbed in the local, non-interactive model of differential privacy, which is especially useful for IoT scenarios. We introduced a new notion of sensitivity, *relaxed sensitivity*, to provide different differential privacy guarantees for outliers in comparison to non-outliers. Furthermore, we presented a correction algorithm to detect false negatives and false positives. In our experiments we detect 80% of outliers in a subset of 10% of all points with a differential privacy value of $\epsilon = 0.1$ for data with well separated outliers.

Acknowledgments. This work has received funding from the European Union's Horizon 2020 research and innovation programme under grant agreement No. 700294 (C3ISP) and 653497 (PANORAMIX).

References

1. Blum, A., Dwork, C., McSherry, F., Nissim, K.: Practical privacy: the SuLQ framework. In: Proceedings of the ACM Symposium on Principles of Database Systems (PODS) (2005)
2. Blum, A., Ligett, K., Roth, A.: A learning theory approach to noninteractive database privacy. J. ACM (JACM) **60**(2), 12 (2013)

3. Dwork, C., Kenthapadi, K., McSherry, F., Mironov, I., Naor, M.: Our data, ourselves: privacy via distributed noise generation. In: Vaudenay, S. (ed.) EUROCRYPT 2006. LNCS, vol. 4004, pp. 486–503. Springer, Heidelberg (2006). doi:10.1007/11761679_29

4. Dwork, C., McSherry, F., Nissim, K., Smith, A.: Calibrating noise to sensitivity in private data analysis. In: Halevi, S., Rabin, T. (eds.) TCC 2006. LNCS, vol. 3876, pp. 265–284. Springer, Heidelberg (2006). doi:10.1007/11681878_14

5. Dwork, C., Roth, A.: The algorithmic foundations of differential privacy. Found. Trends Theor. Comput. Sci. 9(3–4), 211–407 (2014)

6. Erlingsson, Ú., Pihur, V., Korolova, A.: RAPPOR: randomized aggregatable privacy-preserving ordinal response. In: Proceedings of the Conference on Computer and Communications Security (CCS) (2014)

7. Ester, M., Kriegel, H.P., Sander, J., Xu, X.: A density-based algorithm for discovering clusters in large spatial databases with noise. In: Proceedings of the International Conference on Knowledge Discovery and Data Mining (KDD) (1996)

8. Feldman, D., Fiat, A., Kaplan, H., Nissim, K.: Private coresets. In: Proceedings of the ACM symposium on Theory of computing (STOC) (2009)

9. Goldreich, O.: Foundations of Cryptography. Basic Applications, vol. 2. Cambridge University Press, Cambridge (2009)

10. Jawurek, M., Johns, M., Kerschbaum, F.: Plug-in privacy for smart metering billing. In: International Symposium on Privacy Enhancing Technologies Symposium, pp. 192–210. Springer (2011)

11. Kearns, M., Roth, A., Wu, Z.S., Yaroslavtsev, G.: Privacy for the protected (only). ArXiv e-prints, May 2015

12. Kerschbaum, F.: Frequency-hiding order-preserving encryption. In: Proceedings of the 22nd ACM SIGSAC Conference on Computer and Communications Security, pp. 656–667. ACM (2015)

13. Kerschbaum, F., Dahlmeier, D., Schröpfer, A., Biswas, D.: On the practical importance of communication complexity for secure multi-party computation protocols. In: Proceedings of the 2009 ACM Symposium on Applied Computing, pp. 2008–2015. ACM (2009)

14. Lui, E., Pass, R.: Outlier privacy. In: Dodis, Y., Nielsen, J.B. (eds.) TCC 2015. LNCS, vol. 9015, pp. 277–305. Springer, Heidelberg (2015). doi:10.1007/978-3-662-46497-7_11

15. Machanavajjhala, A., Kifer, D., Abowd, J., Gehrke, J., Vilhuber, L.: Privacy: theory meets practice on the map. In: Proceedings of the International Conference on Data Engineering (ICDE) (2008)

16. McSherry, F.: Privacy integrated queries: an extensible platform for privacy-preserving data analysis. In: Proceedings of the ACM International Conference on Management of Data (SIGMOD) (2009)

17. Nissim, K., Raskhodnikova, S., Smith, A.: Smooth sensitivity and sampling in private data analysis. In: Proceedings of the ACM Symposium on Theory of Computing (STOC) (2007)

18. Nissim, K., Stemmer, U., Vadhan, S.: Locating a small cluster privately. In: Proceedings of the ACM Symposium on Principles of Database Systems (PODS) (2016)

19. Rastogi, V., Hay, M., Miklau, G., Suciu, D.: Relationship privacy: output perturbation for queries with joins. In: Proceedings of the ACM Symposium on Principles of Database Systems (PODS) (2009)

20. Roth, A.: New Algorithms for Preserving Differential Privacy. Ph.D. thesis, Carnegie Mellon University (2010)

21. Roy, I., Setty, S.T., Kilzer, A., Shmatikov, V., Witchel, E.: Airavat: security and privacy for mapreduce. In: Proceedings of the USENIX Conference on Networked Systems Design and Implementation (NSDI) (2010)
22. Su, D., Cao, J., Li, N., Bertino, E., Jin, H.: Differentially private k-means clustering. In: Proceedings of the ACM Conference on Data and Applications Security and Privacy (CODASPY) (2016)
23. Tramèr, F., Huang, Z., Hubaux, J.P., Ayday, E.: Differential privacy with bounded priors: reconciling utility and privacy in genome-wide association studies. In: Proceedings of the ACM Conference on Computer and Communications Security (CCS) (2015)

Undoing of Privacy Policies on Facebook

Vishwas T. Patil$^{(\boxtimes)}$ and R.K. Shyamasundar

Department of Computer Science and Engineering,
Information Security R&D Center, Indian Institute of Technology Bombay,
Mumbai 400076, India
ivishwas@gmail.com, shyamasundar@gmail.com

Abstract. Facebook has a very flexible privacy and security policy specification that is based on intensional and extensional categories of user relationships. The former is fixed by Facebook but controlled by users whereas the latter is facilitated by Facebook with limited control to users. Relations and flows among categories is through a well-defined set of protocols and is subjected to the topology of underlying social graph that continuously evolves by consuming user interactions. In this paper, we analyze how far the specified privacy policies of the users in Facebook preserve the standard interpretation of the policies. That is, we investigate whether Facebook users really preserve their privacy *as they understand it* or certain of their innocuous actions leak information contrary to their privacy settings. We demonstrate the kind of possible breaches and discuss how plausibly they could be set right without compromising performance. The breaches are validated through experiments on the Facebook.

1 Introduction

Social networks help individuals, groups of individuals, organizations, and governments to establish their digital identity online and allow other digital identities to interact with it. Establishment of connections and subsequent interactions allow the digital identities to engage with their audience. The platform also facilitates search for new audience. It keeps track of interactions among identities and categorizes them according to their individual likes and dislikes, which helps the platform in presenting relevant content and advertisement to its audience.

Protection of one's digital identity and associated information is desired and expected. Social network platforms deploy access control systems to ensure that security and privacy of their users is maintained. However, by definition, social networks are formed by dynamic connections among stakeholders that independently control their privacy specifications. It is challenging to ensure the privacy of a conservative individual who is connected with a liberal individual. In this paper, we shall study a set of such scenarios and analyze the impact of actions of independent stakeholders' on privacy.

The navigability over the Facebook has been succinctly captured by Boyd and Ellison [3] who characterize social network systems like Facebook by three

© IFIP International Federation for Information Processing 2017
Published by Springer International Publishing AG 2017. All Rights Reserved
G. Livraga and S. Zhu (Eds.): DBSec 2017, LNCS 10359, pp. 239–255, 2017.
DOI: 10.1007/978-3-319-61176-1_13

functions: *(i) identity representation:* allow users to create a profile by populating a pre-defined template of fields with personal information, *(ii) distributed relationship articulation:* facilitate relationship with other identities and organize the relationships into categories like Friend, Family, etc., *(iii) traversal-driven access:* allow users to traverse the social space and grant access based on the access policy specified on the reachable node.

Abstraction of relationships into categories like Family, Friends, Friends of Friends, etc., help individuals to specify access control in a natural and understandable way. On the other hand, it makes access control enforcement very challenging when there are billions of nodes in the social graph and edges, which are added/deleted spontaneously as users interact. It is indeed an achievement by Facebook that they achieve the underlying access control in quite a performance-centric manner [4,7]. Thus, it is interesting and important to understand the access control mechanism of such a dynamic system so that one can attempt to analyze or reason about its properties and compliances with settings permissible in a user's profile.

The paper is organized as follows: Sect. 2 highlights user representation in Facebook and overall global model to enable analysis of access control and policy preservation mechanisms. In Sect. 3, we discuss how the categories, as defined and facilitated by Facebook, can be used to provide/restrict access to users. Section 4 illustrates the gaps between user interpretation of privacy-preservation and user actions that undo privacy. In Sect. 5, we discuss possible approaches to mitigate various privacy anomalies discovered. Section 6 presents current related work and other works on Facebook privacy evaluations that have become irrelevant as over the years Facebook has updated its architecture, features and mechanisms for access control.

2   Access Control in Facebook: User Representation, Social Graph

Facebook organizes its content retrieval and distribution around 3 pillars: *(i) Newsfeed:* responsible to present content/updates to users about their friends, *(ii) Timeline:* where users curate their own content, and *(iii) Graph search:* also known as *social graph* that consumes all of the actions of Facebook users and simultaneously allows them to query the graph. Queries to the graph are resolved in context of the requester and the content being requested. Access to content is governed by an access policy specified by its owner. To study access control in Facebook we briefly explain its social graph and some of the relevant query functions.

2.1   Social Graph of Facebook

Social graph in Facebook is a representation of user information on Facebook. Each and every action or event created by Facebook's users is consumed by the social graph. The graph evolves reflecting user actions. Facebook's social graph is composed of [8]:

- **nodes** - basically "things" such as a *user*, a *photo*, a *page*, a *comment*
- **edges** - the connections between those "things", such as relationships (*friends*)
- **fields** - info about those "things", such as a *user*'s *birthday*, or the *name* of a *page*

Each user of Facebook is represented by a node (identified uniquely by a 64-bit number) on Facebook's social graph and a user's relationship with other nodes is captured through labelled edges. For example, $(A) \xleftrightarrow{\text{friends}} (B)$ is a representation of *friendship* relationship between user A and B. When user Z *follows* user B on Facebook, that relationship is absorbed by the social graph and the graph evolves to: $(A) \xleftrightarrow{\text{friends}} (B) \xleftarrow{\text{follower}} (Z)$.

Updates to social graph happen by adding/deleting nodes (or updating fields of nodes), and adding/deleting/updating labelled edges. Social graph allows its nodes to be queried. A user is allowed to compose a query by specifying a particular node (of type *root* [8]) about which the requester needs information. It is very likely that different sets of information about a node are presented based on who the requester is. For example, in the graph shown above, assume user B posts a photo with access set to his "Friends"; the access control mechanism of Facebook will allow user A to reach the post, whereas user Z will be denied access. If user B changes the access setting to "Public"; both user A and user Z will be allowed to access the post. In the following subsections we shall briefly introduce the reader to Facebook's social graph and its access control mechanism based on lists (information classification labels/categories).

2.2   Representation of User Events and Interpretation of Privacy Policies

Consider a typical user event on Facebook as depicted in Fig. 1(a). The event is created by Alice and its interaction with Bob, Cathy, and David is depicted in Fig. 1(b) through the social graph. Upon careful observation on nodes in Fig. 1(b), one can notice the node *ids* are in chronological order, signifying sequence of events and user actions. As per Facebook's current (05/2017) working, we can reason the following about Fig. 1(b):

- Alice has "checked-in" to a "location" and has *tagged* Bob in that event, therefore Bob has visibility (access) to the event. This is because Facebook's default policy allows a tagged user to access the event in which the user is tagged.
- Cathy could comment on Alice's event because Cathy has visibility of the event. This could happen in two circumstances: either Alice has set her access policy on this event to "Friends" or "Public".
- David could like the comment made by Cathy because David had visibility to the event. This implies Alice's access policy for this event must be "Public" since David and Alice are not friends.

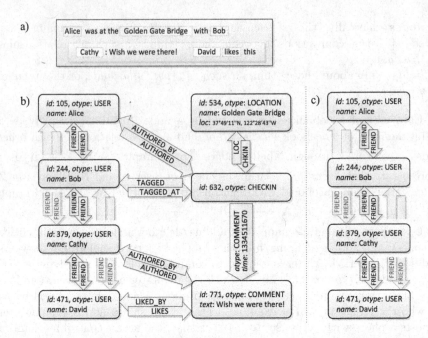

Fig. 1. (a) Creation of event, (b) its representation on social graph [4], (c) deletion of event

Similarly, reasoning about Fig. 1(c), we can conclude that the change in social graph is not because of Alice setting her event's access policy to "Only Me" but because of Alice deleting the event altogether. Owner of an event can delete the event at will and all dependent nodes of that event node also get deleted. If Alice sets the access policy of her event to "Only Me", then the event node, its dependent nodes, and their associations will be part of the social graph but visibility of the event gets restricted to Bob alone. To understand the scenario in which Alice changes the access policy of the event to "Only Me" we need to understand the access control model of Facebook. Social graph provides traversal paths from a requester to the node being requested. Existence of path between a requester and the node being requested is not a sufficient condition to access the node but the requester has to also satisfy the access policy specified on that node by node's owner. To analyze Facebook's access control model, let us first understand the types of nodes in its social graph, and set of queries that can be run on them, followed by an understanding of categories of access/privacy policies provided by Facebook.

Querying the Social Graph

1. Let $\mathbb{G}(\mathbb{V}, \mathbb{E})$ denote the social graph of Facebook. Let $\mathbb{V}$ denote the set of vertices or nodes and $\mathbb{E}$ as the set of edges, where $\mathbb{E} \subseteq \mathbb{V} \times \mathbb{V}$.

2. Let $T_V : \mathbb{V} \longrightarrow \mathbb{S}_V$ provide type of a node (e.g., *user, photo, event, ...*)
   The members of this set can further be divided into two sets: *subjects* and *objects*. Users are subjects and the content generated by the users is treated as objects. The fields populated by a user in her profile are also treated as objects.

3. Let $T_E : \mathbb{E} \longrightarrow \mathbb{S}_E$ provide type of an edge (e.g., *friend, like, comment, ...*)
   The members of this set are associations between subjects and objects.
   Let $\mathcal{E} : \mathbb{S}_E \longrightarrow 2^{\mathbb{E}}$ be a generic function that returns the set of all edges of same type. Let $\mathcal{E}_t \; \forall \; t \in \mathbb{S}_E$ be a function on edge of type t.

(E_1) $\mathcal{E}_t(x) := \{y \mid (x, y) \in \mathcal{E}_t\}$

(E_2) $\mathcal{E}_{friends}(user) = \{subject \mid (user, subject) \in \mathcal{E}_{friends}\}$

(E_3) $\mathcal{E}_{authored}(user) = \{object \mid (user, object) \in \mathcal{E}_{authored}\}$

(E_4) $\mathcal{E}_{authored\_by}(object) = \{user \mid (object, user) \in \mathcal{E}_{authored\_by}\}$

(E_5) $\mathcal{E}_{likes}(user) = \{object \mid (user, object) \in \mathcal{E}_{likes}\}$

(E_6) $\mathcal{E}_{liked\_by}(object) = \{user \mid (object, user) \in \mathcal{E}_{liked\_by}\}$

(E_7) $\mathcal{E}_{tagged\_at}(user) = \{object \mid (user, object) \in \mathcal{E}_{tagged\_at}\}$

(E_8) $\mathcal{E}_{tagged}(object) = \{user \mid (object, user) \in \mathcal{E}_{tagged}\}$

While the above set of query functions is useful for a requester to traverse the social graph, it also raises questions as how Facebook preserves the access of users/resources in the system. When a requester traverses towards a node of its interest on the social graph, underlying topology of the graph determines existence of a path from the requester to the node in question. However, existence of traversal path is not a sufficient condition for access. Each non-*root* type of node in the social graph is access controlled by a policy specified by its owner. In the following, we shall discuss the various types of policies Facebook provides its users.

3   Policy Specification for Access over Users in Facebook

In this section, we shall discuss usage of relationship categories (i.e., labels) as policy handles and its efficacy enforcing user stated privacy policies.

3.1   Lists as Policy: Extensional vs Intensional Information Classification

Each user on Facebook is provided with pre-defined relationship categories, called lists, along which users can organize their relationships with others. "Friends" is the basic relationship category to which every user-to-user relationship (friendship) is added. A user is allowed to organize friendship relations into other pre-defined categories like "Family", "Close Friends", "Acquaintances" so that a distinct affinity level could be imposed on relations. This is how people, in real-world, intend to organize their relationships. Aforementioned list of relationship categories is common across all user profiles. This notion of categorizing (or listing of) friends into affinity levels help users to specify who can have access to their information. Users are also provided with an option to create "Private Lists". When a user posts something, the post is labelled with one of these list names. Any requester who satisfies membership to the label assigned with the post can access the post. Since members of these aforementioned lists are finite and known to the list owner, this category of labels is called *extensional* labels. And the information published using this type of labels is classified as extensional information. In other words, extensional labels are labels in user's local namespace.

There is another set of labels for information classification that is *intensional*. The labels that fall under this category are: "Friends of friends", "Public", and the set of "Social Lists". Social Lists[1] are automatically added to user's account at the time of profile creation/update, when a user provides her school, university, affiliation information. A post labelled with intensional label is available to a requester who satisfies membership to it. Owners of such posts do not have *complete* control over who will be accessing the intensional information because membership to intensional labels is not directly under their control. In other words, intensional labels are labels in user's extended namespace.

Labels are used as access control policies over a user's information. A typical access policy consists of a label available to the user. Facebook also provides *Custom* policies that involves a combination of labels. Custom policy's interpretation involves *set algebraic union and intersection* operations over labels used to compose it. We shall study the exact evaluation sequence of labels, at the time of access, later. The whole gamut of information labelling in Facebook provides a very rich and flexible access (thus privacy) policy specification over a user's information. Users are allowed to change labels of their objects as per their discretion. However, this flexibility in policy specification is not well-understood by majority of the users and users end up in a state where their policy specification may look innocuous, whereas it may not. We shall study such scenarios in next section. Given below is a typical set of labels (information classification categories) provided to express access control policies:

- Only Me: is a label/list in which user herself is the only member
- Public: is a label, when used, the associated object is accessible publicly
- Friends: is the primary list under which all friendship relations are enlisted
- Restricted: is a list of friends to whom only Public labelled information is allowed
- Family: is a list of friends who are assigned as family members
- Close Friends: is a list of friends who are assigned as close friends
- Acquaintances: is a list of friends who are assigned as acquaintances
- Friends of friends: is an *intensional* list consisting of users who have friendship relation with some member in "Friends" list, therefore;

---

[1] Facebook categorizes this type of list as Smart List because they are created based on common affiliations across the Facebook users. Facebook treats Close Friends, Family also as Smart Lists because, based on interactions among users, smart membership suggestions to these categories are provided by Facebook. To disambiguate, we use the term Social List.

(E_9) $\mathcal{L}_{ffriends}(user) = \{subject \mid \exists\ y\ s.t.\ (user, y) \in \mathcal{L}_{friends}$ and $(y, subject) \in \mathcal{L}_{friends}\}$

- *University*: is a social list of friends who are also members of Smart List *University*
- *School*: is a social list of friends who are also members of Smart List *School*
- *Cycling*: is a Private List to which user has assigned a set of friends

Let us understand the usage of labels for access control with an example: assuming current state of social graph is represented by $\text{(A)} \xleftarrow{\text{friends}} \text{(B)} \xleftarrow{\text{follower}} \text{(Z)}$ Let P_1 be a post by user B. Access to object P_1 is determined by the access policy associated with P_1. User B can change access policy of his objects it owns at will. Let us discuss access implications of different policies on P_1

1. $\text{B} \xrightarrow[\text{Only Me}]{\text{authored}} P_1$: since B himself is the only member of list named "Only Me", no one else except B will have access to P_1

2. $\text{B} \xrightarrow[\text{Friends}]{\text{authored}} P_1$: since $\mathcal{L}_{friends}(B) = \{A\}$ [cf. (E_2)], user A can access P_1

3. $\text{B} \xrightarrow[\text{Public}]{\text{authored}} P_1$: since the policy is "allow all", both A and Z can access object P_1

4. $\text{B} \xrightarrow[\text{Only Me}]{\text{authored}} P_2 \xrightarrow{\text{tagged}} \text{A}$: since $\mathcal{L}_{tagged}(P_2) = \{A\}$ [cf. (E_8)], even though the access policy on P_2 is "Only Me", user A will be able to access object P_2. Tagging can be seen as adding an exception to the default access policy of an object.

Access Control of objects in Facebook is a simple check on associated list's membership. If a requester of an object is a member of the list with which the object is protected, the requester gets access. Tagging is a positive exception to the membership check. There are two negative exceptions to the membership check: "Restricted" list and "Blocked" list. If a requester of an object is member of one of these lists, access is denied even when the requester is member of the list with which the object is protected.

When a *Custom* policy is used for access control, sequence of evaluation of lists used to build *Custom* policy becomes important. In the following we explain how policies are evaluated and enforced at the time of access.

3.2   Policy Evaluation and End-to-End Enforcement

Each object in Facebook is labelled with an access policy, which is a combination of labels with positive/negative exceptions. At the time of access, policies are evaluated in context of object's owner. In other words, evaluation of labels (i.e., Lists) is done in the namespace of object's owner. Every object has an access policy. Users are allowed to change access policies of own objects. Any combination of labels, from a user's namespace, is permitted to express access policies.

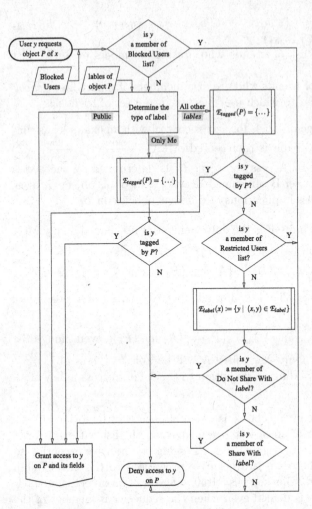

Fig. 2. Sequence of policy evaluation [cf. (E_1), (E_8)]

Figure 2 outlines the sequence of policy evaluation and enforcement when user y requests an access to object P. All labels, except the labels "Public" and "Only Me", are treated as Custom policy internally. Since a Custom policy is nothing but a policy composed of a user's extensional and intensional labels and exceptions.

In other words, a Custom policy is a combination of two fields: "Share with" and "Do not share with". The "Share with" field can accept any user label, except "Public". The "Do not share with" field accepts any user label, except "Public", "Friends", and "Friends of friends". Both the fields accept user's individual friends. Every user profile is provisioned with two lists: Blocked Users and Restricted Users. Blocked Users consists of any user on Facebook with whom the user does not want to interact. If a user blocks a current friend, the relationship between them gets unfriended. Restricted Users consists of friends of the user who will be restricted to have access to the user's public posts only, unless tagged. These two lists are not allowed to be used to express access policy but they are internally handled at the time of each access enforcement.

Table 1. Snapshots of associations formed in a social graph

$$\xrightarrow{time}$$

t_1	t_2	t_3	t_4	t_5
A $\xrightarrow[\text{OnlyMe}]{\text{authored}} P_{A_1}$	A $\xrightarrow{\text{authored}} P_{B_1}$	A $\xrightarrow[\text{Public}]{\text{authored}} P_{A_1}$	A $\xrightarrow{\text{likes}} P_{F_2}$	A $\xrightarrow{\text{likes}} P_{F_3}$
B $\xrightarrow[\text{Family}]{\text{authored}} P_{B_1}$	B $\xrightarrow[\text{Public}]{\text{authored}} P_{D_1}$	B $\xrightarrow[\text{Friends}]{\text{authored}} P_{B_1}$		
C $\xrightarrow[\text{Friends}]{\text{authored}} P_{C_1}$	C $\xrightarrow{\text{authored}} P_{D_1}$	C $\xrightarrow{\text{authored}} P_{C_1}$	C $\xrightarrow{\text{authored}} P_{B_1}$	C $\xrightarrow{\text{likes}} (A \xrightarrow{\text{authored}} P_{B_1})$
D $\xrightarrow[\text{FFriends}]{\text{authored}} P_{D_1}$	D $\xrightarrow{\text{likes}} P_{E_1}$			
E $\xrightarrow[\text{Public}]{\text{authored}} P_{E_1}$	E $\xrightarrow{\text{likes}} P_{F_1}$	E $\xrightarrow{\text{likes}} P_{D_1}$	E $\xrightarrow[\text{FFriends}]{\text{authored}} P_{F_3}$	E $\xrightarrow[\text{OnlyMe}]{\text{authored}} P_{E_1}$
F $\xrightarrow[\text{School}]{\text{authored}} P_{F_1}$	F $\xrightarrow[\text{School-E}]{\text{authored}} P_{F_2}$	F $\xrightarrow[\text{University-School}]{\text{authored}} P_{F_3}$	F $\xrightarrow{\text{likes}} P_{E_1}$	F $\xrightarrow{\text{likes}} P_{A_1}$

Assumptions:

at time t_0 (following is the state of different sets)
$users = \{A, B, C, D, E, F\}$ $objects = \{P_{A_1}, P_{D_1}, P_{C_1}, P_{D_1}, P_{E_1}, P_{F_1}, P_{F_2}, P_{F_3}\}$
friendship edges $= \{(A,B), (B,C), (C,D), (D,E), (E,F), (F,A)\}$ $Family_B = \{A\}$
$University = \{E, F\}$ $School = \{A, E, F\}$

at time t_4 (user E has disassociated from list School) $School = \{A, F\}$

at time t_5 $University = \{A, E, F\}$ $School = \{F\}$

3.3 Reasoning About Access Control in Facebook w.r.t. Social Graph

To analyze the access control on Facebook, two broad categories of information represented in Facebook becomes very handy: (i) The utility of social graph in Facebook to record relationships and interactions among users and their objects (content). (ii) The way Facebook allows its users to categorize their relationships through intensional, extensional classes and, in turn, using these class names as labels on their respective objects to express access policies.

Table 2. Node reachability in social graph at t_1

At time t_1

- P_{A_1} is accessible to its owner alone
- P_{B_1} is accessible to A, as B has added A to list Family
- P_{C_1} is accessible to B and D [cf. (E_2)]
- P_{D_1} is accessible to all except A [cf. (E_9)]
- P_{E_1} is accessible to all
- P_{F_1} is accessible to A and E

Now let us reason about access control in Facebook with respect to an evolving social graph. Various possible associations that are formed between users and objects are captured through five snapshots depicted in Table 1. (cf. Appendix A)

Table 2 describes accessibility of objects to users at time t_1. Object can be operated upon (e.g., to like/comment) by a requester only when the object is accessible. Tables 3, 4, 5 and 6 describe the actions taken or events created by users between time t_2 and t_5.

Table 3. Events and actions at t_2

At time t_2
• A comments on object P_{B_1}
• B shares object P_{D_1} as Public
• C comments on object P_{D_1}
• D likes object P_{E_1}
• E likes object P_{F_1}
• F creates object P_{F_2}, accessible to A

Table 4. Events and actions at t_3

At time t_3
• A changed access policy of P_{A_1} from Only Me to Public, so it becomes accessible to all
• B changed access policy of P_{B_1} from Family to Friends, so it becomes accessible to A and C
• C comments on her own post
• E likes P_{D_1}
• F creates object P_{F_3} with with custom policy in which "Share with = University" and "Do not share with = School". The object is not accessible to anyone. University and School are social lists and they will be evaluated in the context of user F. Therefore University$_F$ = {E} and School$_F$ = {A,E}. Though user E is part of "Share with" field of this custom policy, he will not get access to P_{F_3} because "Do not share with" is evaluated before "Share with" (cf. Fig. 2)

Table 5. Events and actions at t_4

At time t_4
• A likes object P_{F_2}
• C comments on object P_{B_1}. This object was not accessible to C until t_2
• E disassociates from list School
• E shares P_{F_3} with FFriends. Though user A is a member of user E's FFriends, she will not get access to P_{F_3} because it is governed by its owner's base policy University-School.
• F likes object P_{E_1}

Table 6. Events and actions at t_5

At time t_5
• A disassociates from list School
• A likes object P_{F_3} as it is now accessible, since she has associated herself with list University and is no more part of list School
• C likes the comment made by A on P_{B_1}
• E changes the access policy of P_{E_1} from Public to Only Me
• F likes object P_{A_1}

4   Analysis of Privacy-Preservation in Facebook Through User Specified Policies/Actions

Issue of privacy comes to fore as soon as an unintended observer observes an information and learns something more, which later could be associated with that subject under observation. The observed information need not necessarily be "personally identifiable information" as defined in prevalent definitions of privacy [13,21]. We believe that users trust Facebook as they voluntarily divulge [19] personal information, during profile creation and thereupon, to Facebook. In this section we analyze some of the instances of privacy violations in Facebook using the analysis done in Sects. 2 and 3. All of these breaches have been validated using Facebook as of May 2017.

Our analysis of Facebook policies using its access control is elaborated using a hypothetical scenario captured in Table 1. In this table, each row represents a user's actions in chronological fashion. Therefore, we use $C.t_4$ to denote an action of user C at time t_4. All other actions, of every user, up to time t_4 is the environment/status of social graph w.r.t. $C.t_4$. Thus, an action should be analyzed in context with its current environment. Note that, *since social graph is a co-creation by its users, an individual has little or no control over the environment in which he/she is operating. An action/setting that seems privacy-preserving can later be compromised by a change in environment.* This will become clear as we navigate through the scenarios.

Nonrestrictive Change in Policy of an Object Risks Privacy of Others. Consider user action $A.t_2$ in context with environment trace $B.t_1$, $B.t_3$. User B has changed policy of his object P_{B_1} from Family to Friends. Since user A is member of B's Family, through action $A.t_2$ user A has authored a comment on P_{B_1}. The environment at time t_2 ensured that only Family members of B had access to object P_{B_1}. Nonrestrictive change in policy from Family to Friends on P_{B_1} gets enforced on all of its dependent nodes (*comment, reply*) and fields (*like*). *User A's comment is exposed to friends of B without A's consent.*

Restrictive Change in Policy of an Object Suspends Others' Privileges. Consider user action $E.t_5$ in context with two other events in environment $D.t_2$, $F.t_4$. User E has changed policy of her object P_{E_1} from Public to Only Me. Prior to policy change, users D and F have liked the object. When a query on "likes of a user" is made to social graph at time t_4, object P_{E_1} will be listed in the reply. Restrictive change in policy from Public to Only Me on P_{E_1} gets enforced on all of its dependent nodes (*comment, reply*) and fields (*like*). A "likes of user D" query will not list object P_{E_1} at t_5, neither user D/F will be able to disassociate themselves from the object under control of user E. The like edges to P_{E_1} from D and F will only be accessible again on social graph when user E makes a nonrestrictive policy change on P_{E_1}. Assume P_{E_1} is a sensitive post to which some users have liked/commented. *A restrictive change in policy over P_{E_1} locks out users from updating/retracting their own comments or likes.*

At a later point in time, user E can divulge list of users associated with her post from the past.

Share Operation is Privacy-Preserving. Consider user action $B.t_2$ in context with environment $D.t_1$. Since user B is member of D's "Friends of friends" list, object P_{D_1} appears on his *Newsfeed*, which he shares by action $B.t_2$. Sharing an object of others creates a local node and it is allowed to specify access policy on this new node under the control of sharer. This shared node can have comments, likes as any other object node created by the sharer. However the sharer cannot increase or decrease the accessibility sphere of the original object (on social graph) which it has shared. Original access policy of the base object continues to be carried along with the object in its shared form. Thus, the intended reach of the object by its original author is always enforced. *Restrictive or non-restrictive change in policy on base object reflects wherever the object exists on social graph in a shared form. This is similar to the notion of capability lists* [15] *except that instead of user carrying the capability list, the object is carrying it.*

Policy Composition Using Intensional Labels Is Not Privacy-Preserving. Consider user actions $F.t_3$, $E.t_4$, $A.t_5$ in context with social list "School" at t_4 and t_5. Through action in $F.t_3$, user F has created an object P_{F_3} with a custom policy University-School. Here the intention of the user is to make the object available to his friends from University but not from his School. According to the state of social graph at time t_3 nobody gets access to object P_{F_3} because University $\subseteq$ School. At time t_4, user E disassociates herself from social list School and thus could get access to P_{F_3}. Through action in $E.t_4$, user E shares P_{F_3} with access policy as School. As shared objects carry their access policy wherever they exist on social graph, user A will not receive access to P_{F_3} due to $E.t_4$. Whereas by disassociating herself from social list School, user A will have access to P_{F_3} at time t_5 as shown in $A.t_5$. *Disassociation from a social list allows users to bypass the privacy/access intention of a custom policy when an intensional label is part of "Do not share with" field. Similarly, association with a social list allows users to bypass the privacy/access intention of a custom policy when an intensional label is part of "Share with" field. Note that "Friends of Friends" is also an intensional label.*

Like, Comment Operations are Not Privacy-Preserving. Consider user actions $D.t_2$ and $F.t_4$ in context with environment event $E.t_1$. On Facebook, *List of Friends* is an object of user profile. In its privacy settings, Facebook allows to choose intended audience for this object. We assume all users in this scenario have set their audience to "Only Me" for this object. The intention behind such a setting is not to let the profile visitors know who their friends are, except their mutual friends. However, the way Facebook works, *Newsfeed* of a user is supplied with relevant content from user's social circle. With a high probability friends' posts appear in *Newsfeed* to which a user may interact by making a comment or like. These interactions get recorded on social graph. *When a user interacts with objects with access policy set to Public, those interactions also become public. Social graph allows queries to public content.* For example, https://fb.com/search/FBID-Alice/photos-commentedon returns all the "photo" type of objects on which Alice has commented. Similarly, /photos-liked returns all photos liked by Alice. *For a typical user, these queries return objects from their friends. Any user of Facebook can make these queries to social graph for any other user of Facebook.* For a complete list of search attributes please refer Facebook API page [8].

5   Is There a Way to Preserve the Intentions of Policies?

Having seen scenarios of breach of policies, the question is: is it possible to adhere to the intended privacy in the policies without compromising too much on the performance or is it possible to consider policy streamlining to avoid such breaches? Due to lack of space (cf. [20]), we shall briefly discuss some of the possible approaches below.

Intercept and Resolve Intensional Labels at the Time of Object Creation. Association and disassociation with intensional label of type social list is unverified, unsupervised and easy. Whereas associating with Friends of Friends label of a user is relatively tough but possible. To restrict users from bypassing access control when intensional label is part of access policy, one may think of resolving intensional labels into its prevalent member set and use list of member IDs as explicit custom policy. To automate this process we are experimenting with a browser extension that intercepts, resolves intensional labels and writes equivalent custom policy on-the-fly.

Anonymize Unsafe Operations via a *Proxy* node. Information leakage on *List of Friends* object has serious security implications. By knowing friends of a user an attacker creates profiles similar to friend of the user and launch a spear phishing attack. Therefore, it is important to prevent this leakage *at least* when the object's policy is set to Only Me. Collecting likes on objects is one of the important input Facebook relies on for *Newsfeed* content and targeted advertising. Without forgoing these objectives, Facebook can introduce a special node to its social graph called *Proxy* node. All like and comment operations of users with privacy setting as Only Me for *List of Friends* object should be rerouted through this *Proxy* node on social graph.

Treat Comment Nodes Similar to Nodes with Access Policy. A comment is a node on social graph of Facebook. However comment type of nodes have shared ownership as long as the post on which comment is made is accessible to the commenter. Once the owner of the post changes post policy to a restrictive one, commenter's ownership is suspended until the node policy is reverted. Facebook can extend the treatment of post type of nodes to comment type of nodes.

6   Related Work

The access control mechanism of Facebook is ad-hoc and hybrid therefore it is different from standard, prevalent lattice-based access control models [22] like Mandatory Access Control (MAC), Discretionary Access Control (DAC), Role-Based Access Control (RBAC), Capability Systems [15] etc. Facebook's access control mechanism is based on the topology of the underlying volatile social graph. Content and identities are represented by nodes in the graph and edges represent their relationships with other nodes. An access is granted to a requester (a node on the social graph) only if there exists a path to the node being accessed and the requester fulfils the access control specified on the node by its owner. Access control in Facebook involves a subtle element of delegation similar in RBAC [1,6] in the midst of discretionary access control [11,16]. Access control mechanism of Facebook is a function of communication history among users (e.g., existence of friendship is necessary for certain policies), which has parallels with characterizations presented in [23] and works presented in [18] has a security and privacy conformance model based on labels controlled by independent domains.

Owing to the difficulty for users to fully comprehend the privacy consequences of adjusting their privacy settings or re-adjusting their relationships; Fong et al. [10], presented a paradigm for access control in Facebook-style SNSs and also discussed the possibility of overcoming Sybil attacks [9] using the conditions (prevalent on Facebook in year 2009) needed for satisfying Denning's principle of privileged attenuation (POPA); it easily follows from our illustrated privacy breaches that POPA cannot be preserved in the current setup of Facebook. In [5], a rule-based access control mechanism for Facebook-style SNSs is presented. The mechanism relies on relationship inputs (node's type, trust level of relationships, etc.) from underlying social graph. Social graph of Facebook, as of today, does not provide a normal user the attributes required to write rules that are presented in it.

In works on web transparency and accountability [17], it has been argued that vast amounts of profile tracking leads to various breaches of policies and hence there is a need for regulations on privacy while profile tracking. In [2], the authors present an inference, from social graph queries about friendship, that eight friendship links are enough to construct a "public view" of a user. Works highlighting the importance of privacy to friendship links are presented in [12,14]. In future, privacy will be *the most* important parameter for any application's acceptability, especially so for applications dealing with social networks [19].

7   Conclusion

Through our experiments on Facebook, we have discovered certain user actions and configurations that undo the privacy guarantees set by the user. We presented our findings with the help of a hand-crafted scenario and proposed plausible mitigation techniques for privacy-preservation. Our mitigation techniques in the existing access control model would pave way for a general purpose access control model for global-scale social applications. It is of interest to note that there is a serious impact on privacy due to App ecosystem of Facebook. While the impact of cross-domain access can be analyzed through several techniques like the one explored in [18], the aspect just pointed out need further exploration. Assuming there will be a regulation for privacy [19], one would need to understand how compliance of regulations on the policy settings can be ensured.

Acknowledgement. The work was carried out as part of research at ISRDC (Information Security Research and Development Center), supported by 15DEITY004, Ministry of Electronics and Information Technology, Govt of India.

A   Appendix

Figure 3 shows instance of the graph at time t_1. Each *user* node of the graph has authored an object. Their access policies are marked on respective (*user*,*object*) edges. Therefore, at time t_1, following are the objects introduced to the social graph. Accessibility to these objects is evaluated according to their respective labels as described in Table 2.

Figures 4, 5 and 6 show the state of social graph at time t_2, t_4, and t_5, respectively.

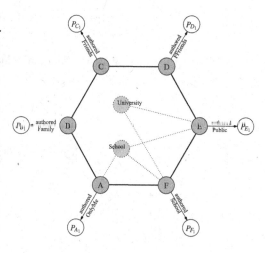

Fig. 3. Social graph at time t_1

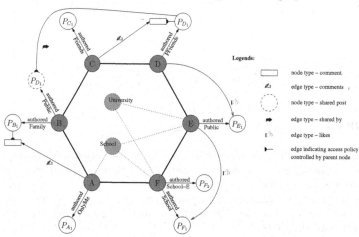

Fig. 4. Social graph at time t_2

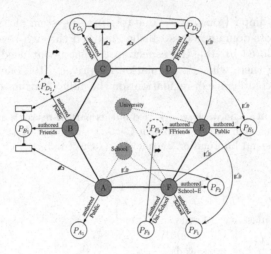

Fig. 5. Social graph at time t_4

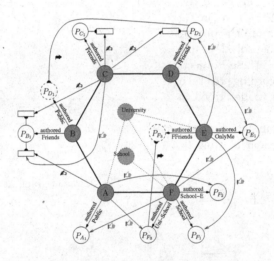

Fig. 6. Social graph at time t_5

References

1. Barka, E., Sandhu, R.: Framework for role-based delegation models. In: Proceedings of the 16th Annual Computer Security Applications Conference, p. 168. IEEE Computer Society (2000)
2. Bonneau, J., et al.: Eight friends are enough: Social graph approximation via public listings. In: 2nd EuroSys Workshop on Social Network Systems, SNS 2009, pp. 13–18. ACM (2009)
3. Boyd, D.M., Ellison, N.B.: Social network sites: Definition, history, and scholarship. J. Comput.-Mediated Commun. **13**(1), 210–230 (2007)

4. Bronson, N., Amsden, Z., et al.: TAO: facebook's distributed data store for the social graph. In: USENIX ATC 2013, pp. 49–60 (2013)

5. Carminati, B., Ferrari, E., Perego, A.: Enforcing access control in web-based social networks. ACM TISSEC **13**(1), 6:1–6:38 (2009)

6. Crampton, J., Khambhammettu, H.: Delegation in role-based access control. Int. J. Inf. Secur. **7**(2), 123–136 (2008)

7. Curtiss, M., Becker, I., Bosman, T., et al.: Unicorn: a system for searching the social graph. Proc. VLDB Endow. **6**(11), 1150–1161 (2013)

8. Facebook: Graph API Overview (2017). https://developers.facebook.com/docs/graph-api/overview/

9. Fong, P.W.L.: Preventing sybil attacks by privilege attenuation: a design principle for social network systems. In: IEEE Symposium on Security and Privacy, pp. 263–278 (2011)

10. Fong, P.W.L., Anwar, M., Zhao, Z.: A privacy preservation model for facebook-style social network systems. In: Backes, M., Ning, P. (eds.) ESORICS 2009. LNCS, vol. 5789, pp. 303–320. Springer, Heidelberg (2009). doi:10.1007/978-3-642-04444-1_19

11. Graham, G.S., Denning, P.J.: Protection: principles and practice. In: Proceedings of the Spring Joint Computer Conference, AFIPS 1972, 16–18 May 1972, pp. 417–429. ACM (1972)

12. Hangal, S., Maclean, D., Lam, M.S., Heer, J.: All friends are not equal: using weights in social graphs to improve search. In: 4th SNA-KDD Workshop. ACM (2010)

13. International Association of Privacy Professionals: What is privacy? (2017). https://iapp.org/about/what-is-privacy/

14. Jernigan, C., Mistree, B.: Project 'Gaydar' Computes Orientation. In: CACM (2009). http://cacm.acm.org/news/42191-project-gaydar-computes-orientation

15. Levy, H.M.: Capability-Based Computer Systems. Digital Press, Bedford (1984)

16. Li, N., Tripunitara, M.V.: On safety in discretionary access control. In: 2005 IEEE Symposium on Security and Privacy (SP 2005), pp. 96–109, May 2005

17. Narayanan, A., Reisman, D.: The princeton web transparency and accountability project. In: Cerquitelli, T., Quercia, D., Pasquale, F. (eds.) Transparent Data Mining for Big and Small Data. Studies in Big Data, vol. 11, 45–57. Springer, Cham (2017)

18. Narendra Kumar, N.V., Shyamasundar, R.K.: Dynamic labelling to enforce conformance of cross domain security/privacy policies. In: Krishnan, P., Radha Krishna, P., Parida, L. (eds.) ICDCIT 2017. LNCS, vol. 10109, pp. 183–195. Springer, Cham (2017). doi:10.1007/978-3-319-50472-8_15

19. Patil, V.T., Shyamasundar, R.K.: Privacy as a currency: un-regulated? In: 14th International Conference on Security and Cryptography, SECRYPT 2017 (2017, to appear)

20. Patil, V.T., Shyamasundar, R.K.: Social networks and collective unravelling of privacy. Technical report, ISRDC, IIT Bombay (2017). http://isrdc.iitb.ac.in/reports/isrdc-tr-2017-rks-vtp-sns-privacy.pdf

21. Renaud, K.G.D.: Privacy: Aspects, definitions and a multi-faceted privacy preservation approach. In: 2010 Information Security for South Africa, pp. 1–8, August 2010

22. Sandhu, R.S.: Lattice-based access control models. Computer **26**(11), 9–19 (1993)

23. Schneider, F.B.: Enforceable security policies. ACM TISSEC **3**(1), 30–50 (2000)

Cloud Security

Towards Actionable Mission Impact Assessment in the Context of Cloud Computing

Xiaoyan Sun[1(✉)], Anoop Singhal[2], and Peng Liu[3]

[1] California State University, Sacramento, CA 95819, USA
xiaoyan.sun@csus.edu
[2] National Institute of Standards and Technology, Gaithersburg, MD 20899, USA
anoop.singhal@nist.gov
[3] Pennsylvania State University, University Park, PA 16802, USA
pliu@ist.psu.edu

Abstract. Today's cyber-attacks towards enterprise networks often undermine and even fail the mission assurance of victim networks. Mission cyber resilience (or active cyber defense) is critical to prevent or minimize negative consequences towards missions. Without effective mission impact assessment, mission cyber resilience cannot be really achieved. However, there is an overlooked gap between mission impact assessment and cyber resilience due to the non-mission-centric nature of current research. This gap is even widened in the context of cloud computing. The gap essentially accounts for the weakest link between missions and attack-resilient systems, and also explains why the existing impact analysis is not really actionable. This paper initiates efforts to bridge this gap, by developing a novel graphical model that interconnects the mission dependency graphs and cloud-level attack graphs. Our case study shows that the new cloud-applicable model is able to bridge the gap between mission impact assessment and cyber resilience. As a result, it can significantly improve the effectiveness of cyber resilience analysis of mission critical systems.

1 Introduction

Due to the increasing severity of cyber-attacks, mission assurance entails critical demands of active cyber defense and cyber resilience more than ever. Especially in the public cloud where a large number of enterprise networks reside, failure of effective cyber defense would generate huge negative impact towards enterprises' missions. Mission cyber resilience or active cyber defense means capabilities to make prioritized, proactive and resource-constraint-aware recommendations on taking cyber defense actions, including network and host hardening actions, quarantine actions, adaptive MTD (Moving Target Defense) actions, roll-back actions, repair and regeneration actions. Due to the fundamental necessity and importance of situational awareness to decision making, cyber situational awareness plays a critical role in achieving mission cyber resilience. Specifically, mission

G. Livraga and S. Zhu (Eds.): DBSec 2017, LNCS 10359, pp. 259–274, 2017.
DOI: 10.1007/978-3-319-61176-1_14

cyber resilience cannot be really achieved without impact assessment. That is, knowing which mission and how a mission is impacted by an attack is the key for making correct resilience decisions.

However, there is actually a largely overlooked gap between mission impact assessment and cyber resilience, though both mission impact assessment and attack-resilient systems have been extensively researched in the literature:

(1) Despite extensive research on attack-resilient survivable systems and networks [1], most if not all existing cyber resilience techniques are unfortunately not mission-centric. The cyber resilience analysis is usually constrained to the level of assets, without consideration in regards to the mission impact. For example, the recommendation of having a backup server would be made by performing asset-level resilience analysis, but the mission-level impact is rarely analyzed. Lack of mission models and mission dependency analysis is a common limitation of existing attack resilience techniques. Without mission dependency analysis, existing cyber resilience techniques cannot quantify the effectiveness of the recommended cyber response actions in terms of mission goals, and hence cannot convincingly justify the superiority of the recommended response actions.

(2) From another aspect, despite extensive research on mission impact assessment, mission impact assessment results cannot be automatically used to make mission-centric recommendations on taking cyber response actions. For instance, even if the dependency relationship between a server and a mission is definite, it's still difficult to make any resilience recommendations regarding this server due to the absence of asset-level cyber resilience analysis. This gap is even widened in the context of cloud computing. In public cloud, each enterprise network has its own missions. These missions are usually expected to be independent and isolated from each other. However, multi-step attacks may penetrate the boundaries of individual enterprise networks from the same cloud, and thus impact missions of multiple enterprise networks. That is, attacks that happen in one enterprise network may be able to affect missions of another enterprise network in the same cloud. Therefore, mission impact should be re-assessed in cloud environment.

(3) In addition, most mission impact assessment techniques are generally one-dimensional, without explicitly considering the dimension of service dependency. For example, a mission dependency graph usually specifies the dependency relations among a task and its supporting services, but may not include dependencies among services. However, service dependency is also indispensable for accurate mission impact analysis. For example, a web service may depend on the authentication service to verify users, while the authentication service may further depend on the database service to query users' credentials. If the database service is down, missions related to the web service and authentication service may also be impacted, although they do not directly depend on the database service. As a result, the service dependency relations should be explicitly included for accurate mission cyber resilience analysis.

Hence, lack of automation tools in associating missions with attack-resilient systems is a weakest link in achieving cyber resilience. Without such association, existing mission impact analysis results are not really actionable: it's difficult to find out why and how a mission has been impacted. Since bridging this gap may significantly boost the cyber resilience of mission critical systems, how to bridge this gap is a very important problem.

Therefore, the primary objective of this paper is to take the first steps towards systematically bridging the critical gap between mission impact assessment and cyber resilience in the context of cloud computing. We aim to model the mission impact process and enable the automatic reasoning of this process. To achieve this goal, we identified and addressed the following challenges:

First, it is very challenging to envision a never-seen-before graphical model that can integrate mission dependency graphs and cloud-level attack graphs in such a way that can effectively bridge the gap between existing mission impact analysis results and attack-graph-based active cyber defense. No graphical model has yet been proposed to bridge this gap, though two schools of thoughts have been respectively developed on mission impact analysis and attack-graph-based active cyber defense.

Second, a cloud environment gives rise to new challenges in bridging the gap. Cloud services such as Infrastructure as a Service (IaaS), make attack graphs more complicated and harder to get analyzed. Conventional attack graphs cannot capture the stealthy information flows introduced by certain cloud features. Attackers could leverage the hidden security vulnerabilities caused by inappropriate cloud management to launch zero-day attacks.

The significance of this paper's contributions is two-fold: (1) We have developed a novel graphical model, the mission impact graph model, to systematically bridges the critical gap between impact assessment and cyber resilience. Bridging this gap significantly boosts the cyber resilience of mission critical systems; (2) To the best of our knowledge, this is the first work that investigates the mission impact assessment problem by considering the special features in cloud computing environment; (3) We have extended the attack graph generation tool MulVAL [9] to enable logical reasoning of mission impact assessment and automatic generation of mission impact graphs.

2   Our Approach

In this paper, we aim to bridge the gap between mission impact assessment and cyber resilience.

On the side of mission impact assessment, different types of mission dependency graphs have been developed to associate missions with component tasks and assets. As shown in Fig. 1, the status of assets (hosts, virtual machines, etc.) will generate direct impact towards missions through dependency relations. In current literature, such dependency relations among assets, tasks, and missions are usually very loose and not well defined. As a result, the corresponding mission impact assessment is also inaccurate. In addition, without considering the

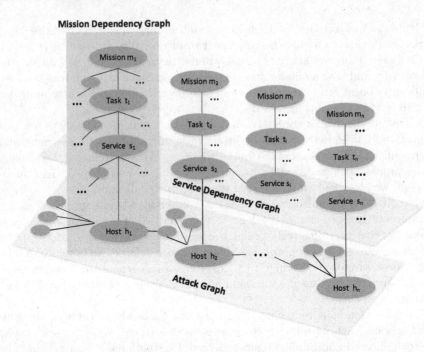

Fig. 1. The Mission dependency graph, service dependency graph, and attack graph.

possibility of multi-step attacks caused by combinations of vulnerabilities, the mission impact assessment is usually not sufficiently comprehensive. Missions that seem irrelevant to some compromised assets may also be impacted due to the existence of multi-step attacks. For example, in Fig. 1, let's assume mission m_1 and mission m_n depends on host h_1 and host h_n respectively. Analysis of the individual mission dependency graphs (vertical graphs in Fig. 1) could lead to a conclusion: if h_1 is compromised, then m_1 will be impacted but m_n will not. This is because the individual mission dependency graphs do not show any direct relations between mission m_n and host h_1. However, if the possibility of a multi-step attack is considered, host h_1 can be used as a stepping-stone to compromise h_n. In this case mission m_n has the possibility of being impacted as well. Therefore, with only mission dependency graphs, it is not sufficient to perform accurate and comprehensive mission impact assessment.

On the side of cyber resilience, attack graphs have become mature techniques for analyzing the causality relationships between vulnerabilities and exploitations. As in Fig. 1, by analyzing the vulnerabilities existing in the network, attack graphs are able to generate potential attack paths that show a sequence of attack steps (e.g., from host h_1 to h_n). This capability enables security admins to proactively analyze the influence of some security operations towards the potential attack paths. For example, security admins could check how potential attack paths would be changed if a vulnerability is patched. However, the traditional attack graph has two limitations. First, it is not mission-centric. The attack graph is able to generate potential attack paths through logical reasoning, but it lacks the capability of reasoning potential impacts towards missions. Second, traditional attack graphs do not consider potential attacks enabled by some special features of public cloud environment, such as virtual machine image sharing and virtual machine co-residency. New attack types may arise in cloud due to these features, but are not captured in traditional attack graphs.

In addition, service dependency discovery has been studied in a number of research works [20–22]. In a network, the normal functioning of an application or service may depend on the normal functioning of other services. Service dependency graphs are able to capture such dependency relations. The service dependencies are important for correct mission impact assessment in that missions may be indirectly impacted by other seem-to-be irrelevant services. For example, mission m_i in Fig. 1 does not directly depend on service s_2 in the vertical mission dependency graph. Without considering the service dependencies, m_i seems to be irrelevant with s_2 and thus will not be impacted by its status. However, by taking service dependencies into account, the status of s_2 will affect service s_i, which will eventually impact m_i. Hence, service dependency graphs should also be incorporated when performing mission impact assessment for the purpose of cyber resilience.

Therefore, considering the respective capabilities and disadvantages of mission dependency graphs and attack graphs, this paper proposes to develop a logical graphical model, called *attack graph based mission impact graph* (referred as mission impact graph in the rest of this paper), to integrate mission dependency graphs, service dependency graphs, and cloud-level attack graphs. Our approach contains three steps. First, there exist essential semantic gaps between mission dependency graphs and attack graphs. We identify the semantic gaps and unify the representation of nodes and edges. This makes interconnecting mission dependency graphs and attack graphs feasible. Services are already components of mission dependency graphs, so service dependency graphs are align with mission dependency graphs in terms of semantics. Second, to bridge the gap inside a cloud environment, we extend traditional attack graphs into cloud-level attack graphs. The cloud-level attack graphs are incorporated into new mission impact graphs. Third, we implement a set of interaction rules in MulVAL [8,9] to enable automatic generation of logical mission impact graph.

3 The Semantic Gap Between the Attack Graph and the Mission Dependency Graph

Generally speaking, a mission dependency graph is a mathematical abstraction of assets, services, mission steps (also known as tasks) and missions, and all of their dependencies [6]. A mission dependency graph has five types of nodes, including assets, services, tasks, missions and logical dependency nodes. The logical dependency nodes are basically AND-nodes and OR-nodes that represent logical dependencies among other nodes. The AND-node represents that a parent nodes depends on all of its children nodes. The OR-node denotes that a parent node depends on at least one of its children nodes. For example, a successful task may depend on all of the supporting services being functional, while a complete mission could require only one of its tasks being fulfilled. Edges in a mission dependency graph represent the interdependencies existing among nodes.

As for the attack graph, it usually shows the potential attack steps leading to an attack goal. Several different types of attack graphs have been developed, such as state enumeration attack graphs [10–12] and dependency attack graphs [13–15]. This paper uses the dependency attack graph for analysis. Figure 2 is part of a simplified attack graph. A traditional attack graph generated by MulVAL is composed of two types of nodes, fact nodes (including primitive fact nodes and derived fact nodes) and derivation nodes (also known as rule nodes). Primitive fact nodes (denoted with rectangles in Fig. 2) present objective conditions of the network, such as the network, host, and vulnerability information. Derived fact nodes (denoted with diamonds) are the facts inferred by applying the derivation rule. Each derivation node (denoted with ellipse) represents the application of a derivation rule. The derivation rules are implemented as interaction rules in MulVAL. Simply put, one or more fact nodes could be the preconditions of a derivation node, while the derived fact node is the post-condition of the derivation node. For example, in Fig. 2, if node 4 "the attacker has access to the server", node 5 "the server provides a service with an application" and node 6 "the application has a vulnerability" are all satisfied, then the rule in node 7 will take effect and make node 8 become true. That is, attacker is able to execute arbitrary code on the server. In this example, node 4, 5 and 6 are the fact nodes; node 7 is the derivation node; node 8 is the derived fact node.

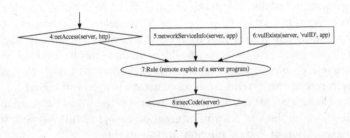

Fig. 2. Part of a simplified attack graph.

Mission dependency graphs and traditional attacks graphs have the following semantic gaps:

(1) The meaning of nodes differs. In a mission dependency graph, a node denotes an entity, such as an asset, a service, a task, or a mission. The node does not specify the status of the entity. In a traditional attack graph, a node represents a statement, be it a rule or a fact. For example, a primitive fact node could be "the web server provides OpenSSL service" or "the openssl

program has a vulnerability called CVE-2008-0166". A rule node could be "the remote exploit of a server program could happen".

(2) The meaning of edges differs. In a mission dependency graph, the edges represent general interdependencies among nodes, and do not specify concrete dependency types. The logical relations are specially denoted with AND and OR nodes. In a traditional attack graph, directed edges represent the causality relationship among nodes. One or more fact nodes could cause a derivation node to take effect, which further enables a derived fact node.

(3) The representation of logical relations among nodes differs. In a mission dependency graph, the logical relations are represented specifically with AND and OR nodes. In traditional attack graph, the logical relations are not provided explicitly, but are implied in the graph structure: derivation nodes (rule nodes) imply AND relations and derived fact nodes imply OR relations. That is, fact nodes that serve as preconditions of a derivation node have AND relations, while derivation nodes leading to a derived fact node have OR relations. The underlying principle is that all of the preconditions have to be satisfied to enable a derivation rule, while a derived fact node can become true as long as one rule is satisfied.

Hence, in the proposed mission impact graph, we will bridge the semantic gaps between mission dependency graphs and attack graphs.

4   Incorporating Cloud-Level Attack Graphs

In the public cloud, each enterprise network can generate its own individual attack graph by scanning hosts and virtual machines in the network. These individual graphs may not be complete because new attack paths enabled by the cloud environment could be missed. Therefore, a cloud-level attack graph is needed to capture potential missing attacks by taking some features of public cloud into consideration, such as virtual machine image sharing and virtual machine co-residency. Hence, [16] proposed the construction of cloud-level attack graphs. A cloud-level attack graph contains three levels: virtual machine level, virtual machine image level, and host level. The virtual machine level mainly captures the causality relationship between vulnerabilities and potential exploits inside the virtual machines. The virtual machine image level focuses on attacks related to virtual machine images. For example, a virtual machine image may be instantiated by different enterprise networks. As a result, its security holes are also inherited by all the instance virtual machines. The virtual machine image level is able to reflect such inheritance relationship. The host level mainly captures potential attacks to hosts, including exploits leveraging the virtual machine co-residency relationship.

Therefore, the mission impact graph needs to be extended to incorporate cloud-level attack graphs. The semantics of mission impact graphs remain the same because cloud-level attack graphs have the same semantics as traditional attack graphs. However, the mission impact graph is now composed of two parts:

cloud-level attack graph part, and the cloud-applicable mission dependency part. New nodes should be added as derivation nodes and fact nodes to incorporate special features of cloud.

To achieve this goal, we crafted a set of Datalog clauses in MulVAL as the primitive facts, derived facts and interaction rules. For the cloud-level attack graph part, new facts and rules are crafted to model virtual machine image vulnerability existence, vulnerability inheritance, backdoor problem, and virtual machine co-residency problem, and so on. For mission dependency part, new rules are added to model the residency dependencies among virtual machines and hosts, service dependencies among virtual machines and services, etc. For example, the residency dependency relationship between a host and the dependent virtual machines can be modeled with the following interaction rule:

```
interaction rule(
   (hostImpact(VM):-
      residencyDepend(Vm, Host),
      HostImpact(Host)),
   rule_desc('An compromised host will impact the dependent
virtual machines')).
```

5   Mission Impact Graph and Graph Generation

The new graphical model, which is referred to as mission impact graph, is formally defined as follows: (1) It is a directed graph that is composed of three parts: attack graph part, service dependency part and mission-task-service-host dependency part. (2) It contains two kinds of nodes: derivation nodes and fact nodes. Each fact node represents a logical statement. Each derivation node represents an interaction rule that is applied for derivation. There are two types of fact nodes, primitive fact nodes and derived fact nodes. A primitive fact node represents a piece of given information, such as host configuration, vulnerability information, network connectivity, service information, progress status of a mission (e.g. which mission steps are already completed and which are not), and so on. Derived fact nodes are computing results of applying interaction rules iteratively on input facts. (3) The edges in the mission impact graph represent the causality relations among nodes. A derived fact node depends on one or more derivation nodes (which have OR relations); a derivation node depends on one or more fact nodes (which have AND relations).

In mission impact graphs, we need to combine attack graphs and mission dependency graphs by unifying their representation of nodes and edges. It is composed of four steps:

Step 1, the entity nodes in mission dependency graphs become either primitive fact nodes or derived fact nodes in mission impact graphs. A fact node represent a statement about an entity. Primitive fact nodes usually represent already known information provided by network scanners or human administrators, such as host configuration, network configuration, and vulnerability information, etc. Derived fact nodes are computed information by applying interaction rules towards the primitive fact nodes. One entity in the mission dependency

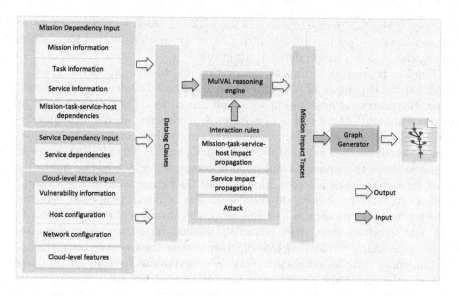

Fig. 3. Logical mission impact graph generation.

graph may become a number of fact nodes depending on its related known information and computed information. For example, a service node s in the mission dependency graph may be related to two pieces of information: the known information that showing "s is disabled" through a port scanning, and the computed information that "s is impacted" by applying interaction rules. In this case, the s node in the mission dependency graph could become two fact nodes in the mission impact graph: a primitive fact node with the predicate *serviceDisabled*; and a derived fact node with the predicate *serviceImpacted*. Similarly, an asset node h may in mission dependency graph could become a derived fact node with predicate *execCode*, meaning "attackers can execute arbitrary code on the host h".

Step 2, derivation nodes are added into the mission impact graph to model the causality relationships among fact nodes. The interdependencies among entities such as assets, services, tasks, and missions in the mission dependency graph can be interpreted into specific impact causality rules, which become derivation nodes in mission impact graph. For example, the dependency between a task t and a service s could be interpreted into a rule R: "t will be compromised if s is disabled and t is not completed yet". When node "s is disabled" and node "t is not completed yet" are both satisfied, the derivation node stating rule R will take effect.

Step 3, logical relation nodes in mission dependency graphs are removed, including AND and OR nodes. The logical relations among nodes are implied with graph structure as in the mission impact graph: derivation nodes imply AND, and derived fact nodes imply OR.

Step 4, the fact nodes and derivation nodes in mission impact graphs are connected with edges to represent direct causality relations, rather than general dependencies as in mission dependency graphs.

Finally, to enable automatic generation of mission impact graphs, we extended the capability of MulVAL by creating new Datalog clauses. Figure 3 shows the process of mission impact graph generation. Three sets of input are provided to MulVAL in terms of mission dependencies, service dependencies, and cloud-level attacks. The input sets are converted to corresponding Datalog clauses, which are then fed to MulVAL reasoning engine. The MulVAL reasoning engine analyzes the input sets and perform computation by applying interaction rules. Interaction rules are very important for the logical reasoning. For mission impact analysis, we created three different sets of interaction rules, including rules for mission-task-service-host impact propagation, service impact propagation, and attack analysis. After reasoning, mission impact traces are generated. Graph generators such as Graphviz [23] can analyze the traces to build the final mission impact graphs. Originally MulVAL only contains the input set and integration rules for attack graph generation. We added two more input sets and two more integration rule sets, and also extended the attack-related input set and interaction rule set by including cloud-level features.

In the implementation, three sets of Datalog clauses are added as primitive facts, derived facts, and interaction rules for the function of mission impact analysis. For primitive facts, we crafted clauses that describe mission-task dependencies, the service types, task service dependencies, and mission progress status, and so on. Some information can be provided by system administrators. For derived facts, we added clauses for the status of missions, tasks, services and assets. To enable the logical reasoning, we created interaction rules to model the causality relationships between pre-conditions and post-conditions. For example, attacks towards servers will impact services that are provided by these servers. The interaction rule describing this causality relationship could be represented with Datalog clauses as follows:

```
interaction rule(
    (serviceImpacted(Service, H, Perm):-
       hostProvideService(H, Service),
       execCode(H, Perm)),
    rule_desc('An compromised server will impact the dependent
service')).
```

6 Case Study

As shown in Fig. 4, our scenario contains three enterprise networks in cloud: A is a start-up company, B is a medical group, and C is a vaccine supplier. In addition to providing existing vaccines, C is also developing a new type of vaccine together with its collaborators. The formula of the new vaccine is still very confidential. For security purposes, C only accepts client requests from trusted IPs. The relationships between A, B and C are: (1) they are on the same cloud; (2) A's webserver and B's database server are two virtual machines

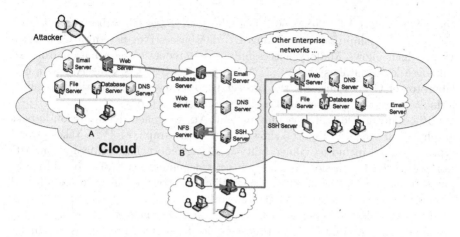

Fig. 4. The attack and mission scenario.

that co-reside on the same physical host; This is not uncommon in that cloud providers generally host virtual machines on arbitrary hosts. This co-residency relationship can be leveraged by attackers. (3) B is a trusted client to C.

The mission for medical group B, *Bm1*, is to provide medical services to all of its patients. Sample tasks include: *Bt1*, patients make appointments; *Bt2*, access medical records; *Bt3*, order shots or medicine; *Bt4*, administer shots; *Bt5*, update medical records, and so on. The mission for vaccine supplier C, *Cm1*, is to supply vaccines to authorized medical groups, and develop the new type of vaccine with collaborators. The sample tasks include: *Ct1*, ask for login ID and password; *Ct2*, check the ID and password. If the user is medical group, go to *Ct3*. If the user is collaboration partner, go to *Ct4*; *Ct3*, order vaccine; *Ct4*, check and update new vaccine information. *Ct3* and *Ct4* are then composed of a number of subtasks.

In our attack scenario, attacker Mallory (could be a competitor of victim companies) is very interested in the new vaccine, and wants to steal its formula from supplier C. To break into the supplier network, Mallory performs the following attack steps: (1) Mallory compromises A's webserver by exploiting vulnerability CVE-2007-5423 in tikiwiki 1.9.8; (2) Mallory leverages the co-residency relationship to take over B's database server, based on a side channel attack in cloud. Extensive research has been done on side channel attacks [17–19]. (3) B's NFS server has a directory (*/exports*) that is shared by all the servers and workstations inside the company. Normally B's web server should not have write permission to this shared directory. However, due to a configuration error of the NFS export table, the web server is given write permission. Therefore, Mallory uploads a Trojan horse to the shared directory, which is crafted as a management software named *tool.deb*. (4) The innocent Workstation user from B downloads *tool.deb* from NFS server and installs it. This creates an unsolicited connection back to Mallory. (5) The Workstation has access to C's webserver as a trusted client. Mallory then managed to take over it via a brute-force key guessing attack (CVE-2008-0166); (6) Mallory leverages C's webserver as a stepping stone to compromise C's MongoDB database server based on CVE-2013-1892, which allows Mallory successfully steal credential information from an employee login database table; (7) Mallory logins into C's webserver as a collaborator of C, and accesses the project proprietary documentation to collect formula-related vaccine research and development records.

By performing logical reasoning in MulVAL, we generated a mission impact graph for our scenario. Figure 5 shows a part of the graph. MulVAL takes several types of inputs, including vulnerability-scanning report, host configuration, network connectivity, mission-task-service-asset dependencies, and so on. The output is a mission impact graph showing which missions are likely to be affected by considering current status of the networks.

The result cloud-level mission impact graph is very helpful for understanding potential threats to missions in this scenario. First, individual mission impact graph may miss important attacks leveraging some features of cloud, and thus generate incorrect evaluation about possible threats to missions. For example, without considering the co-residency relationship between A's webserver and B's database server, B seems to be very safe as the database has no exploitable vulnerability. As a result, mission *Bm1* is viewed as safe. However, our mission impact graph shows that *Bm1* has the possibility of being impacted because the virtual machine co-residency can be leveraged for attack. Second, the result

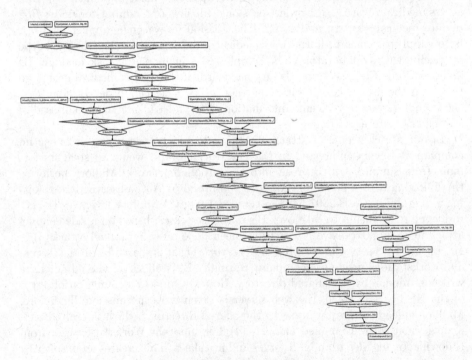

Fig. 5. Mission impact graph.

cloud-level mission impact graph is mission-centric. Although attack paths for the scenario network can be generated in the traditional attack graph, the potential impact towards mission cannot be assessed without analyzing the dependency relationships among missions, tasks, services and assets. For example, the attack graph is able to show the attack path from A's web server to C's database server, but the impact of attacks to missions is unclear. Our mission impact graph is able to show such impact towards missions by considering both attacks and missions.

One function of our mission impact graph is to perform automated "taint" propagation through logical reasoning. Given a "taint", be it a vulnerability, a compromised machine, or a disabled service, the impact of the "taint" can be analyzed through logical reasoning. The mission impact graph is able to reflect affected entities such as assets, services, tasks, and missions. For example, in our case study, if C's web server is compromised, the mission impact graph will show that task $Ct1$ and mission $Cm1$ are impacted.

The generated mission impact graph enables effective mission-centric cyber resilience analysis. Propagating the attack-graph-based active cyber defense from the attack graph side to the mission impact side is helpful for performing advanced proactive "what-if" mission impact assessment. Through logical reasoning, impact analysis can be performed all the way from inside a machine to a mission. Given the input information, attack graphs can predict the potential attack paths and identify possibly to-be-affected assets. Mission impact graph extends attack graphs in a way that the prediction of potential attack paths directly enables the prediction of potential mission impact. Therefore, we can perform proactive "what-if" mission impact analysis by changing the input conditions. For example, what if we remove a server? What if we patch a vulnerability on a host? Which tasks or missions will be affected? In our case study, if we break the co-residency relationship between A's webserver and B's database server by moving one of the virtual machines, attacks towards B and C will prevented. As a result, missions in B and C won't be affected. Similarly, if the vulnerability on C's webserver is patched, attacks towards C can be stopped and mission $Cm2$ will be safe. In addition, we can also analyze the potential mission impact by assuming vulnerability existence on other servers. For example, what if an unknown security hole exists on a host? Which tasks or missions will be affected in this case? In addition, as the situation knowledge regarding a network is continuously collected, such knowledge can be interpreted into input files to the automated tool for iterative "what-if" mission impact analysis based on the current situation. Therefore, performing such "what-if" analysis enables interactive mission impact analysis, and thus helps security admins make correct decisions for cyber resilience.

7   Related Work

A literature review is performed to disclose the mismatch between mission impact assessment and cyber resilience: (1) the formal models used by the existing mission impact assessment techniques cannot be directly used by the existing attack-resilient system and network designs; (2) lack of mission models and mission dependency analysis is a common limitation of existing cyber resilience techniques.

Mission impact assessment. In the past decade or so, extensive research has been conducted on modeling the mission dependencies to help facilitate computer-assisted analysis of current missions. The existing mission-oriented impact assessment techniques can be classified into four categories: (1) mission impact assessment through use of ontology based data collection. The basic idea is to create the ontology of mission dependencies. For example, the Cyber Assets to Mission and Users (CAMUs) approach [2] assumes that a cyber asset provides a cyber capability that in turn supports a mission. Their approaches mine existing logs and configurations, such as those from LDAP, NetFlow, FTP, and UNIX to create these mission-asset mappings; (2) mission impact assessment through use of dependency graphs [4,5]. The basic idea is the use of mission dependency graphs for cyber impact assessment and a hierarchical (time-based) approach to mission modeling and assessment; (3) mission impact assessment through use of mission thread modeling [6]. The basic idea is to leverage mission metrics supported by resource model and value model; (4) mission impact assessment through use of Yager's aggregators [3]. The basic idea is to utilize a tree-based approach to calculate the impact of missions. The mission tree is a tree-structure that utilizes Yager's aggregators [7] to intelligently aggregate the damage of assets to calculate the impact on each individual mission.

Cyber resilience and active cyber defense. Since 2000, a tremendous amount of research has been conducted on how to make systems and networks resilient to cyber-attacks. For example, the two volumes of DARPA Information Survivability Conference and Exposition proceedings described the design, implementation and evaluation of the first set of survivable and attack-resilient systems and networks [1]. The cyber response actions adopted in these systems include replication actions, honeypot actions, software diversification actions, dynamic quarantine actions, adaptive defense actions, roll-back actions, proactive and reactive recovery actions. Since then, a variety of cyber response actions have emerged, including migration actions, regeneration actions, MTD actions, decoy actions, CFI (control flow integrity) actions, ASLR (Address Space Layout Randomization) actions, IP randomization actions, N-variant defense actions, and software-defined network virtualization actions.

8   Conclusion

This paper makes the first efforts to close a gap between mission impact assessment and cyber resilience. In the cloud environment it is even more difficult to analyze the impact of vulnerabilities and security events on missions. To fill the gap and associate missions with current attack-resilient systems, this paper develops a novel graphical model that interconnects mission dependency graphs and cloud-level attack graphs. Our case study shows that the mission impact graph model successfully bridges the gap and can significantly boost the cyber resilience analysis of mission critical systems.

Acknowledgement. We thank the anonymous reviewers for their valuable comments. This work was supported by ARO W911NF-15-1-0576, ARO W911NF-13-1-0421 (MURI), CNS-1422594, NIETP CAE Cybersecurity Grant, and NIST 60NANB16D241.

Disclaimer

References

1. Proceedings of DARPA Information Survivability Conference and Exposition, Anaheim, California, 12–14 June 2001, Volume I & Volume II (2001)
2. D'Amico, A., Buchanan, L., Goodall, J.: Mission impact of cyber events: scenarios and ontology to express the relationships between cyber assets, missions, and users. In: Proceedings of the 5th International Conference on Information Warfare and Security (2010)
3. Holsopple, J., Yang, S.J., Sudit, M.: Mission impact assessment for cyber warfare. Intelligent Methods for Cyber Warfare, vol. 563, pp. 239–266. Springer, Cham (2015)
4. Jakobson, G.: Mission cyber security situation assessment using impact dependency graphs. In: Information Fusion (FUSION) (2011)
5. Sawilla, R.E., Ou, X.: Identifying critical attack assets in dependency attack graphs. In: Jajodia, S., Lopez, J. (eds.) ESORICS 2008. LNCS, vol. 5283, pp. 18–34. Springer, Heidelberg (2008). doi:10.1007/978-3-540-88313-5_2
6. Musman, S., Temin, A., Tanner, M., Fox, D., Pridemore, B.: Evaluating the Impact of Cyber Attacks on Missions. MITRE Corporation, Bedford (2009)
7. Yager, R.R.: On ordered weighted averaging aggregation operation in multicriteria decision making. IEEE Trans. Syst. Man Cybern. **18**, 183–190 (1988)
8. Ou, X., Boyer, W.F., McQueen, M.A.: A scalable approach to attack graph generation. In: ACM CCS (2006)
9. Ou, X., Govindavajhala, S., Appel, A.W.: MulVAL: a logic-based network security analyzer. In: USENIX Security (2005)
10. Sheyner, O., Haines, J., Jha, S., Lippmann, R., Wing, J.M.: Automated generation and analysis of attack graphs. In: Security and Privacy (S&P) (2002)
11. Ramakrishnan, C.R., Sekar, R.: Model-based analysis of configuration vulnerabilities. J. Comput. Secur. **10**, 189–209 (2002)
12. Phillips, C., Swiler, L.P.: A graph-based system for network-vulnerability analysis. In: Proceedings of Workshop on New Security Paradigms (1998)
13. Jajodia, S., Noel, S., O'Berry, B.: Topological analysis of network attack vulnerability. In: Kumar, V., Srivastava, J., Lazarevic, A. (eds.) Managing Cyber Threats. Massive Computing, vol. 5. Springer, USA (2005)
14. Ammann, P., Wijesekera, D., Kaushik, S.: Scalable graph-based network vulnerability analysis. In: ACM CCS (2002)

15. Ingols, K., Lippmann, R., Piwowarski, K.: Practical attack graph generation for network defense. In: ACSAC (2006)

16. Sun, X., Dai, J., Singhal, A., Liu, P.: Inferring the stealthy bridges between enterprise network islands in cloud using cross-layer Bayesian networks. In: Tian, J., Jing, J., Srivatsa, M. (eds.) International Conference on Security and Privacy in Communication Networks. Lecture Notes of the Institute for Computer Sciences, Social Informatics and Telecommunications Engineering, vol. 152. Springer, Cham (2015)

17. Zhang, Y., et al.: Homealone: co-residency detection in the cloud via side-channel analysis. In: 2011 IEEE Symposium on Security and Privacy. IEEE (2011)

18. Ristenpart, T., et al.: Hey, you, get off of my cloud: exploring information leakage in third-party compute clouds. In: Proceedings of the 16th ACM Conference on Computer and Communications Security. ACM (2009)

19. Younis, Y., Kifayat, K., Merabti, M.: Cache side-channel attacks in cloud computing. In: International Conference on Cloud Security Management (ICCSM) (2014)

20. Chen, X., Zhang, M., Mao, Z.M., Bahl, P.: Automating network application dependency discovery: experiences, limitations, and new solutions. In: Proceedings of the 8th USENIX Conference on Operating Systems Design and Implementation (OSDI) (2008)

21. Natarajan, A., Ning, P., Liu, Y., Jajodia, S., Hutchinson, S.E.: NSDMiner: automated discovery of network service dependencies. In: Proceeding of IEEE International Conference on Computer Communications (2012)

22. Peddycord III, B., Ning, P., Jajodia, S.: On the accurate identification of network service dependencies in distributed systems. In: USENIX Association Proceedings of the 26th International Conference on Large Installation System Administration: Strategies, Tools, and Techniques (2012)

23. http://www.graphviz.org/

Reducing Security Risks of Clouds Through Virtual Machine Placement

Jin Han[1], Wanyu Zang[2], Songqing Chen[3], and Meng Yu[1(✉)]

[1] University of Texas at San Antonio, San Antonio, USA
meng.Yu@utsa.edu
[2] Texas A&M University at San Antonio, San Antonio, USA
[3] George Mason University, Fairfax, USA

Abstract. Cloud computing is providing many services in our daily life. When deploying virtual machines in a cloud environment, virtual machine placement strategies can significantly affect the overall security risks of the entire cloud. In recent years, some attacks are specifically designed to collocate with target virtual machines in the cloud. In this paper, we present a Security-aware Multi-Objective Optimization based virtual machine Placement algorithm (SMOOP) to seek a Pareto-optimal solution to reduce overall security risks of a cloud. SMPPO also considers resource utilization on CPU, memory, disk, and network traffic using several placement strategies. Our evaluation results show that security of clouds can be effectively improved through virtual machine placement with affordable overheads.

1 Introduction

Cloud computing is the basis of many services in our daily life, such as email services, services of smart Internet of Things (IoT) devices and file sharing services. In an Infrastructure as a Service (IaaS) cloud like Amazon EC2 [4], many virtual machines (VMs) share a physical server. The placement of virtual machines can have different strategies, leading to different computing performance, energy consumption, and resource utilization. Therefore, given different resource constraints, how to achieve multiple objectives is a very important problem in cloud computing. Such a problem has attracted extensive attention recently [6,14,16].

With resource and other constraints, the virtual machine placement (VMP) is essentially a multiple-objective optimization problem. Phan et al. [16] used an Evolutionary Multi-Objective Optimization (EMOA) algorithm to build Green Clouds when considering energy consumption, cooling energy consumption and user-to-service distance in VMP optimization. Xu and Fortes [22] proposed a genetic algorithm with fuzzy multi-objective evaluation to minimize the total resource wastage, power consumption and thermal dissipation costs in VMs placement. Shigeta et al. [20] suggested to assign different weights to multi-objectives on cost and performance and built a cost evaluation plug-in module

G. Livraga and S. Zhu (Eds.): DBSec 2017, LNCS 10359, pp. 275–292, 2017.
DOI: 10.1007/978-3-319-61176-1_15

to search for the optimal VMs placement. Some other research focus on minimizing the overall network cost while considering large communication requirements [3,15], or applying the constraint programming (CP) engine to optimize VMP [2,7]. While above multi-objective optimization placement schemes greatly improve the overall performance of the cloud, the security risk of the entire cloud environment was not considered as an objective or at most considered as one constraint in the initialization phase.

At the same time, there are new types of attacks targeting at the cloud infrastructure. For example, some attacks, such as those discussed in [8,9,12,18] , exploit the vulnerabilities of hypervisor (or Virtual Machine Monitor, VMM), e.g., Xen [5] or KVM [11]. Once the attacker compromises the hypervisor, he or she can take over all the VMs running on it. In [17] (HYG attack), the initial stage of the attack is to locate a target VM. Upon success, the attacker will try to launch a VM on the same physical server. It is a placement based attack and the success of the attack depends on the placement strategies of the cloud, or the configuration policy of the cloud. Apparently, collocating with vulnerable virtual machines, or "bad neighbors", on the same physical server does increase the security risks to cloud users. Thus, the security risk exposed to the user depends not only on how secure the VM itself is, such as the operating system and applications running inside, but also the Virtual Machine Monitor (VMM or Hypervisor), running underlying the VMs, and other VMs coresident on the same node.

We believe that security should be considered as one key element, the same as the energy and performance, in VM placement. In the previous work [14], we proposed a VMP scheme based on the security risk of each VM. However, the security analysis of our previous work mainly focused on dependency relations. Yuchi and Shetty [26] extended our previous work to the VM placement initialization. Yu et al. [25] proposed isolation rules to formulate the VMP behavior based on the Chinese wall policy. Unfortunately, this work mainly focuses on improving security and overlooks other objectives, such as energy saving and resource utilization. Besides, the security measurements in this work mainly consider the vulnerabilities of VMs or hypervisor, or security regulations, without considering the security assessment of a VMP.

When comparing different VMP schemes, the security metrics can only be evaluated after a placement is specified. For example, a specific placement scheme has an unique attack path exposed by co-residence that may disappear in a different placement. Therefore, there is no generic function to map a placement scheme into a security assessment value. We cannot simply apply any existing evolutionary multi-objective optimization algorithm (EMOA) to solve our problem directly. Furthermore, the low efficiency and the complicated security assessment require us to design our own crossover and mutation procedures in the the EMOA algorithm.

In this paper, we propose a VM placement specific security measurement of the cloud, and a new VMP approach to provide better intrusion resilience, resource utilization, and network performance. In the proposed VM placement specific security assessment, we consider the vulnerabilities not only on VMs and hypervisor themselves, but also the host co-resident and network connections that will change with the VM placement. Based on the proposed security measurement scheme, we propose an evolutionary multi-objective optimization algorithm, named as Security-aware Multi-Objective Optimization based virtual machine Placement algorithm (SMOOP), to seek a Pareto-optimal solution balancing the multiple objectives on security, resource utilization, and network traffic.

Our proposed scheme features an innovative combination of the following contributions.

- We conduct security assessment of the cloud from four aspects: networking, co-residence, hypervisor vulnerabilities, and VM vulnerabilities. The proposed security risk assessment is placement specific and crosses multiple dimensions. We provide detailed metrics and approach to measure the security of the cloud in the case study and experiments.
- We consider security as one objective in VMP strategies, with other objectives and constraints at the same time. To the best of our knowledge, this is the first work that includes a placement specific security assessment in the context of multi-objective optimization based VMP.
- We propose a high-scalable approach, SMOOP with five placement strategies, to achieve Pareto-optimal placement given multiple objectives. Each objective can have different weight according to the application context of our approach. The experimental results show the effectiveness of our strategies. SMOOP can provide improved security of the cloud with affordable overheads.

The rest of the paper is organized as follows. In Sect. 2, we compare our contribution with related work. Section 3 describes the formulation of VMP optimization problem. Section 4 describes the design and implementation of SMOOP. The evaluation results are discussed in Sect. 5 and Sect. 6 summarizes our work.

2   Related Work

As cloud computing become more popular, VMP has become one of the most critical security problems of cloud. Recently, a lot of research on cloud computing have set the goal to improve the security level of data center [1]. Existing research on the co-residence based attacks, e.g., side channel attacks, demonstrates the real threat to the normal users if they are collocated with a vulnerable or malicious VM [17,23,24,28]. Thus, security aware VMP has been investigated as a practical solution to mitigate such attacks [2,14,19].

Saeed et al. [2] presented a security-aware approach for resource allocation in clouds which allows for effective enforcement of defense-in-depth for cloud VMs. They tried to enhance the security level by modeling the cloud provider's constraints and customer's requirements as a constraint satisfaction problem (CSP). However, the placement generated by this method can only satisfy the input constraints, rather than being an optimal placement to meet multiple objectives.

Some other research utilizes isolation rules in the VMP. Afoulki et al. [1] proposed a VMP algorithm which improves the security of cloud computing by performing isolation between users. Each user can submit a list of adversary users with whom it does not want to share a physical machine. Yu et al. [25] also proposed isolation rules to formulate the VMs placement behavior based on Chinese wall policies.

Our previous work [14] proposed a VM placement scheme based on security risk of each VM, and Yuchi and Shettey [26] extended it to the VM placement initialization. Both of them mainly focused on the dependency relations. Yuchi and Shettey's method also over simplified the problem and did not reflect the potential risk caused by co-resident VMs [21,27]. Previously, we have investigated to periodically migrate VMs based on the game theory, making it much harder for the adversaries to locate the target VMs in terms of survivability measurement [29]. But we did not consider the risk caused by the co-resident VMs in the same physical machine.

Our work in this paper differs from the aforementioned work mainly in two aspects. First, existing work simplifies the security consideration in the placement. They mainly consider the security constraints or regulations, or vulnerabilities of VMs or hypervisor in the placement. They often overlook co-residency attacks, which is a key factor in VM placement. In our security-aware VMP, we comprehensively consider security assessment associated with placement, including the security risks in the network connection, co-residence, VMs and hypervisor. Second, existing work often emphasizes on security while overlooking other performance factors. We propose an optimal solution satisfying multiple objectives on security, resource utilization, and network traffic.

3   Problem Formulation

In this section, we describe our system and metrics to model the objectives, and constraints of virtual machine placement in a cloud.

3.1   Threat Model and Security Assumptions

In this paper, we mainly consider co-residency based attacks, such as cross-VM side channel attacks. Also, we assume that the attackers are capable of utilizing vulnerabilities in both VMs and virtual machine monitors (VMMs, or hypervisor) of the clouds.

We have the following assumptions for the cloud: ① the cloud management, placement related software components, and the migration process are all secure; ② for simplicity, each migration of a VM will result in affordable cost in terms of service interruption and consume the same amount of resources; ③ the cloud provider has enough CPU, network bandwidth, and other resources to perform arbitrary migration of VMs; and ④ the cloud provider has sufficient resources as the reward, e.g., extra memory or CPUs, to motivate VM migration. The above assumptions ensure that change of VM placement is both acceptable and affordable for cloud provider and clients.

3.2 Security Assessment

In a cloud, an attacker can compromise a VM through different attack paths. They can compromise a VM through the vulnerabilities (in the operating system, or applications) carried by the VM, the co-resident VMs, the host VMM, or VMs on different physical machines having network connections. Therefore, we cannot simply use the vulnerabilities of VMs, or the vulnerabilities of the hypervisor to evaluate the security risk of an entire cloud. We need a comprehensive approach to measure the security risks of a specific placement scheme.

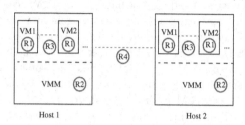

Fig. 1. Security Risk Metrics

For this purpose, we propose a four dimensional security risk evaluation model, as shown in Fig. 1, to assess the security risk of a cloud. The new evaluation model covers all possible attack paths in a cloud. Four different types of security risks are described as follows.

- VM risk (R_1 in the figure): the risk/vulnerability carried by a VM itself. If a VM has more vulnerabilities than others, it is more likely to be compromised first. Vulnerable VMs can be used as stepping stones to attack co-resident VMs and underlying hypervisor to gain more privileges.
- VMM/hypervisor risk (R_2): the risk/vulnerability carried by a VMM/Hypervisor. An adversary may gain the administrative privileges via the vulnerabilities in a hypervisor or the control VM. Such vulnerabilities will enable the adversary to compromise all guest VMs on the hypervisor.
- Co-residency risk (R_3): the risk caused by the VMs co-resident on the same hypervisor. Assume that, in the figure, VM1 (an attacker VM) and VM2 (a normal user VM) share the same CPU core or are located on the same physical machine, the attacker will be able to steal the user's private information, such as the cryptographic key, via side-channel attacks.
- Network risk (R_4): the security risk of a VM caused by the network connections. For example, VM1, located in host 1, provides web services, and the VM2, located in host 2, contains a database server. The attacker may compromise the database server through accessing the web server, e.g., SQL injection.

3.3   An Example Using Our Model and Metrics

Using the proposed security risk assessment model, we can assign or calculate the values of each type of the risks based on specific hardware, software, and network configuration. In this section, we provide an example to show how to quantify the values of each types of security risks, and also how to calculate the overall security risk of the entire cloud. In the example, we assume we have N VMs and M physical machines.

- R_1: CVSS (Common Vulnerability Scoring System) is a popular tool to measure the vulnerabilities of software or hardware [14]. We can use vulnerability scanner tools, such as Nessus and Qualys, to generate the vulnerability list for every VM. We can score each VM's risk based on the list. For example, we can use CVSS Base Score as a VM's risk value, with the assumption that the vulnerable level of a VM is not higher than the worst vulnerability of that VM. The CVSS score uses an interval scale of $(0, 10)$ to measure the severity of vulnerabilities. For a VM v_i, its VM risk R_1 is $VM_{R_1}^i = SCORE_i/10$, so its range would be limited in a scale of $(0, 1)$. Note that R_1 is not affected by a specific placement.

- R_2: The risk level of a hypervisor is determined by two factors: its own vulnerability and the VMs running on it. For hypervisor's own vulnerability, we can use scanner tools to generate the vulnerability list and also use the most severe one to obtain the $SCORE_{hypervisor}$ from CVSS. We use $Risk_{hypervisor} = SCORE_{hypervisor}/10$ to indicate the vulnerability of the hypervisor so that the value is fit in 0 to 1, similar to R_1. There are different ways to calculate how the guest VMs can affect the security of hypervisors. In this paper, we mainly consider the VM with the highest risk since this may be the most vulnerable attacking surface to the hypervisor. Assume VM v_i is on the host K, its hypervisor risk R_2 is calculated as: $R_2^i = Risk_{hypervisor}(1 + \max(R_1^j p_{jK}))$, where $j = 1$ to N, and $p_{jK} = 1$ if VM v_j is placed on host K as well. Different from R_1, R_2 is affected by different VM placements.

- R_3: A malicious VM may compromise a normal VM if they are collocated on the same physical machine. If an attacker compromises a VM, he can compromise, with enough time, other co-resident VMs eventually. So a VM can survive only if all other co-resident VMs can survive. For a VM v_i on the host K, its co-residency risk R_3 is calculated as: $R_3^i = 1 - \prod_{j=1}^{N}(1 - R_1^j p_{jK})$, where $p_{jK} = 1$ if VM v_j is placed on the physical machine K. Similar to R_2, R_3 will change if VM placement changes.

- R_4: If an attacker compromises a VM, he is able to compromise (with enough time) all other VMs with network connections to the compromised VM. Considering the cascades [13] in the network, giving a VM, a depth first algorithm can be applied to build all possible attack paths. In our previous work, we also discussed how to evaluate risks of being attacked based on Markov Chains [14]. Thus, there are different ways to evaluate the risk levels based on network connections. In this paper, we simply consider the risk caused by only direct network connections for simplicity, while other approaches can also be applied. Thus, for a VM v_i, its network risk R_4 is $R_4^i = 1 - \prod_{j=1}^{N}(1 - R_1^j)$, where v_j is a VM sending packets to v_i directly, and v_i and v_j are not on the same host. R_4 changes with different VM placements as well.

With all types of risks defined as above, we define the security risk, R^i of a VM v_i as the following.

$$R^i = 1 - (1 - R_1^i)(1 - R_2^i)(1 - R_3^i)(1 - R_4^i) \tag{1}$$

Based on our discussions, our metrics show the risk levels of different VMs. For example, a VM with risk level 70% is safer than a VM with risk level 80%.

3.4 Objectives in VM Placement

Assume that we have N VMs and M physical machines. There are three values to optimize: security risk (SR), resource wastage (RW) and network traffic (NT). Our goal is to find solutions to minimize these values.

Security Risk. Minimizing the security risk of the entire cloud is our first objective. The security risk of a VM v_i is $R^i = 1-(1-R_1^i)(1-R_2^i)(1-R_3^i)(1-R_4^i)$. To evaluate the security risk of the entire cloud, we need to consider the security risk of all VMs in the cloud. In security assessment, we use the median value of all VMs' risk values as the risk level of the cloud. Thus, security risk is calculated as follows.

$$f_{SR} = \text{median}(\mathrm{R}^i) \tag{2}$$

where $i = 1$ to N. The reason to choose the median value is twofold. First, the median value is more robust than the mean value. Second, in our placement generation, a dangerous VM will be isolated from other VMs. Thus, VMs with high risk values are outliers in our system. It is the same for VMs with low risk values.

Resource Wastage. Minimizing resource wastage, while complying with the constraints, is the second objective in VMP optimization. In this paper, we consider the wastage of multiple resources, including CPU, memory, and disk. In stead of using one value to measure the resource wastage, we use a vector to represent the resources wastage.

Assume the CPU, memory and disk capacity for a host J as $\langle CPU^J, MEM^J, DISK^J \rangle$. A VM v_i requests resources as $\langle cpu_i, mem_i, disk_i \rangle$, therefore, the CPU wastage of the host J is $W_J^c = (CPU^J - \sum_{i=1}^{N} cpu_i p_{iJ})/CPU^j$, where $p_{iJ} = 1$ if VM v_i is placed on host J, otherwise it is 0. The memory wastage of host J is $W_J^m = (MEM^J - \sum_{i=1}^{N} mem_i p_{iJ})/MEM^J$. The disk wastage of host J is $W_J^d = (DISK^J - \sum_{i=1}^{N} disk_i p_{iJ})/DISK^J$.

For a physical machine J, we choose the maximum value from $\{W_J^c, W_J^m, W_J^d\}$ to represent the resource wastage of host J. We would like to minimize the total amount of the resource wastage of the entire cloud.

$$f_{RW} = \sum_{J=1}^{M} \max(W_J^c, W_J^m, W_J^d) \tag{3}$$

while subject to the following capacity constraints in each host J:

$$\sum_{i=1}^{M} cpu_i \times p_{iJ} < CPU^J \tag{4}$$

$$\sum_{i=1}^{M} mem_i \times p_{iJ} < MEM^J \tag{5}$$

$$\sum_{i=1}^{M} disk_i \times p_{iJ} < DISK^J \tag{6}$$

Network Traffic. The third optimization objective is to minimize the network traffic in cloud. One way to reduce the network traffic is to identify correlated VMs that exchange high volume of data with each other, and then put them on the same physical machine if possible. We use the following equation to measure the network traffic from VM v_i to VM v_j:

$$T_{ij} = P_{ij}/t \tag{7}$$

where P_{ij} is the number of packets sent from VM v_i to VM v_j in time period of t. Therefore, the total network traffic in the time period of t is:

$$f_{NT} = \sum_{j=1}^{N} \sum_{i=1}^{N} T_{ij} g_{ij} \tag{8}$$

where $g_{ij} = 0$ if VM v_i and VM v_j are placed on the same host, otherwise it is 1.

Note that our system does not limit the number of objectives or constraints. The users can add more objectives or constraints, such as energy or migration cost, based on their preferences.

4   SMOOP Design

With the proposed security metric of VMs, we can quantify the risk level of a cloud. As a typical multi-objective optimization problem, the objectives may conflict with each other. For example, if we place more VMs on a physical server, it will be less secure due to co-residency problem. However, it can reduce the resource wastage and network traffic. It is impractical to always find the optimal solution minimizing all objectives. The evolutionary multi-objective algorithms (EMOA), such as NSGAII [10], are popular solutions to such multi-objective optimization problems. Using EMOA, we can obtain Pareto-optimal solutions balancing on the objectives of security, network traffic and resource utilization.

Challenges. The VMP can be considered as a bin-packing problem, where each VM needs to be placed on a physical server once and only once, so it is an $\mathcal{NP}$ hard problem. The challenge is that the security metrics can only be evaluated after the placement is specified. For example, in a specific placement scheme, a unique attack path exposed by co-residence may disappear in a different placement. Therefore, there is no generic function mapping a placement scheme into a security assessment value. As a result, we cannot use any existing multi-objective programming solutions to solve our problem. Furthermore, we have complicated security strategies in each placement generation, we have to design a new crossover and a new mutation procedure in the EMOA algorithm.

4.1 Security-Aware Multi-objective Optimization Based VMP

In this section, we present our Security-aware Multi-Objective Optimization based virtual machine Placement (SMOOP). The algorithm is shown in Algorithm 1. Table 1 describes the variables used in the algorithm.

Table 1. Variable definition

Variable	Description
V_N	Number of virtual machines
P_N	Number of physical machines
N_G	Number of iteration
N_IS	Number of placement in candidate pool
N_Elite	Number of elite would be passed to next iteration
N_C	Times of crossover operation in one iteration
N_M	Times of mutation operation in one iteration

In practice, FFD (First-Fit with the possible fullest node) has been widely used in VMP. It can quickly provide a placement with consideration on resource utilization. Thus, we use it to generate a baseline for future comparison in the algorithm. As shown in Algorithm 1, SMOOP generates hundreds of placements and passes those with high fitness value to the next iteration. In each iteration, randomly chosen parents are applied to crossover and mutation operations. An elite choosing function is designed to improve efficiency. For each generated temporary placement in an iteration, we apply a multi-objective evaluation function to assign ranking values. The highly ranked placements are put into a candidate pool, and used as the parents for next iteration. The preference of multi-objective evaluation can be adjusted (described in Sect. 4.3) in our algorithm.

In an *initialization phase*, the consideration for migration cost can be avoided. Our goal is to search for the best possible placement plan based on the multi-objectives requirement. The crossover operation is used to improve the overall efficiency. In the *re-optimization phase*, which is triggered by adding VMs or removing VMs, the migration cost is considered as an important factor in the mutation operation to limit the number of migrating VMs (in switch() function of Algorithm 3). Note that our goal is to improve the survivability of entire cloud, so we do not optimize our solution for a specific VM. Therefore, our approach

---

Algorithm 1. SMOOP

Ensure: Canditate = init() by Strategies
  for $G = 1 \rightarrow N\_G$ **do**
    for $i = 1 \rightarrow N\_E$ **do**
      Elite[i] = Elite_choosing(Canditate)
    end for
    for $j = 1 \rightarrow N\_C$ **do**
      (X, Y) = Random_select(Canditate)
      Off_C[j] = Crossover(X, Y)
    end for
    for $k = 1 \rightarrow N\_M$ **do**
      X = Random_select(Canditate)
      Off_M[k] = Mutation(X)
    end for
    Temp = fitness_sorting(Elite, Off_C, OFF_M)
    for $i = 1 \rightarrow N\_G$ **do**
      Candidate[i] = temp[i]
    end for
  end for

---

may lower the security level of a specific VM, while the overall security level of whole cloud can still be improved.

4.2   Crossover and Mutation Operation

The crossover operation, shown in Algorithm 2, is one of the key elements in our algorithm. The main purpose of the crossover operation is to guarantee that there is always a chance to generate new improved placement based on the existing placement in the current iteration.

Isolated Zones. Since the security is a key factor in the placement generation, we introduce isolated zones in our algorithm to accommodate different security demand. Physical machines with the highest hypervisor risk levels are put into isolated zones. The most dangerous VMs and VMs connected to them are placed into the isolated zones by priorities. The purpose of the isolated zones is to isolate the most dangerous VMs first and reduce the number of attack paths through network connections.

If all physical machines use the same copy of hypervisor, the vulnerabilities of all the hypervisors will also be the same. In such a situation, we have the following assumptions. ① The possibility to compromise any physical machine through the hypervisor attack surface for a specific VM is the same. ② If the communication bandwidth between two VMs is larger than zero, the possibility to compromise one VM through another VM will be non-zero.

We propose five security related strategies to reduce security risk during each placement generation. *Placement strategy I: Put a VM into a physical machine*

Algorithm 2. Crossover(X, Y)

Temp = $Blank\_Placement\_Object$
Rank_A = Rank(X)
Rank_B = Rank(Y)
for $i = 1 \rightarrow P\_N$ **do**
    if Rank_A[i] > Preset_value **then**
        temp ← Rank_A[i]
    end if
    if Rank_B[i] > Preset_value **then**
        temp ← Rank_B[i]
    end if
end for
Remove_duplicate_VM(temp)
Ran = Gen_Random_list(VM)
for $i = 1 \rightarrow V\_N$ **do**
    if Ran[i] is not in temp **then**
        temp ← Ran[i] by Strategies
    end if
end for
Return Temp

which has network connections with it. The purpose is to reduce R_4 caused by network connections.

Assume that physical machine PM_I already has a set of VMs, $S_i = (v_a, \ldots, v_i)$ and PM_J has a set of VMs, $S_j = (v_b, \ldots, v_j)$. There is no network connections between S_i and S_j. When a new VM v_n is to be placed and it has network connections with at least one VM in PM_I. If v_n is placed into PM_I, S_j on PM_J will not be affected by attacks through network connections. It will only affect S_i on physical machine PM_I. For any VM $v_i \in S_i$, R_3^i will be updated by adding the new VM v_n. Assume that the current co-residency risk of v_i is $Old(R_3^i)$, then we have

$$New(R_3^i) = Old(R_3^i) + (1 - Old(R_3^i))R_1^n \qquad (9)$$

where R_1^n is the risk level of v_n.

If v_n is placed into a physical machine PM_J, not only R_3 will be updated, R_4 introduced by network connections will also be increased by $\sum_{v=c}^{K} T_{nv}$ (All traffic through VM v_n to connected VMs on physical machine PM_I).

According to the security metrics defined earlier, in this case, the co-residency risk of each VM $v_j \in S_j$ will be

$$New(R_3^j) = Old(R_3^j) + (1 - Old(R_3^j))R_1^n \qquad (10)$$

Also, R_4 of VM v_j increases as well.

$$New(R_4^j) = Old(R_4^j) + (1 - Old(R_4^j))R_1^n \qquad (11)$$

Therefore, following strategy 1, assign VM v_n on to PM_I rather than PM_J can reduce security risk.

We have more strategies applied in the placement generation. *Placement strategy II: high risk VMs should be put into the isolated zones. Placement strategy III: low risk VM without any connection with VMs in isolated zones should be put into low risk physical machines.* Strategy II and III generate physical machines that contain only low risk VMs and have no network connections with high risk VMs in isolated zones. *Placement strategy IV: marked lowest and highest hypervisor risk physical machines should have a higher probability to be kept during crossover operation.* This is based on our strategy II and III. *Placement strategy V: If a VM on one physical machine has connection with a VM on a different physical machine, we should migrate one of them on to the same physical machine.*

Mutation Operation. Mutation operations, shown in Algorithm 3, operate on a random-chosen temporary placement, trying to obtain an improved result. Its purpose is to keep evolving the existing placement with limited migration cost.

Algorithm 3. Mutation(X)

Temp = *Blank_Placement_Object*
temp ← X
for $i = 1 \rightarrow$ Preset_Maximum_number **do**
    temp ← Switch(temp) by Strategy of Switching VM
end for
Return Temp

When a Pareto-optimal solution is generated, our algorithm double checks the workload balance in every physical machine, migrating marked VMs among physical machines. In the `switch()` function of the algorithm, a VM can only be switched (migrated) to another physical machine to which it has network connections. The strategy is to guarantee that random evolving will not jeopardize the isolation of dangerous VMs and reduce unnecessary switching operation.

4.3 Prioritize the Objectives

In our current fitness function, we have three objectives, including minimizing the security risk, minimizing the resource wastage, and minimizing network traffic. Our algorithm tries to provide a Pareto-optimal solution which can be as good as possible in every degree based on the three objectives. To enable users to prioritize the objectives according to their business preference, we can add weight factors into the fitness function.

$$f = w_1 f_{SR} + w_2 f_{RW} + w_3 f_{NT} \tag{12}$$

where w_i represent different weights and $\sum_{i=1}^{3} w_i = 1$. If we consider that the security is more important than the other two objectives, we can assign higher weight on the security risk. Currently, our algorithm can optimize and balance security, the utilization of CPU, memory and disk, and the network traffic. Our algorithm can be easily extended to support more objectives and constraints, such as energy.

5   Evaluation

We implemented our solution in Java. All input data are provided through configuration files. Multiple threads are used to improve the performance. We randomly generate a large number of VMs with different parameters to evaluate SMOOP. In our evaluation, for each VM, we randomly assign the requirement of CPU, memory, and disk. The vulnerability score is assigned based on the uniform distribution. Following the same method, we configure the physical machine.

5.1   Computing Complexity

Assume that there are M physical servers and each server has N VMs. Our algorithm iterates for k times. Assume that the fitness value of a VM can be calculated in constant time with a fitness function, which is $O(N)$. We sort the candidate pool with complexity of $O(N log N)$. The elite choosing operation will take constant time. The crossover and mutation operation in our algorithm is bounded by $O(MN^2)$. The overall combined complexity of our algorithm is $O(k(MN^2 + N \log N + N))$, thus $O(kMN^2)$. Multi-threading is used in our implementation. We used 8 threads to conduct elite selection, crossover and mutation operation simultaneously.

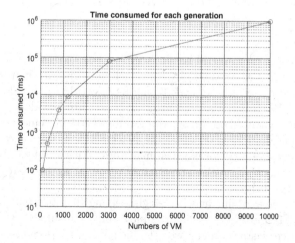

Fig. 2. Scalability

We test our implementation in a 8 core processor with 16GB memory. The overall performance of our algorithm is affected by the number of VMs, the number of physical machines, and the number of candidate placement generated in each generation. Figure 2 shows the computing time for each generation under the following setting: ① 100 different placements are generated for each generation. ② 270 operations are done in each generation. With 10000 VMs and 500 physical machines, each generation takes about 15–20 min. If we reduce the number of VMs to 3000, each generation takes around 2 min.

5.2   Effectiveness in Risk Reduction

Security risk is a key consideration in VMP. To evaluate if our strategies can improve the security level of the entire cloud, we conduct the experiments considering risk level as the only objective in the placement. Figure 3 shows the security risk with 800 VMs and 60 physical machines. At the beginning of each simulation, we always generate 100 placement with the random-FFD algorithm and use the lowest risk level as the baseline reference. We collect the placement with the lowest risk level in each generation. Within 20 generations, the risk level of whole cloud can be reduced by 25% to 30%.

Figure 4 shows the security risk with different number of VMs and physical machines in each generation. Despite the increased number of the VMs, the median value of the risk level of VMs is stable within the range of 0.82 to 0.84. If we check placement with the lowest risk level in the first generation, our algorithm improves with the increased number of the VMs. We repeat our experiment 20 times with different numbers of VMs and physical machines. The reduced risk level is from 5% (400 VMs and 20 physical machines) to 15% (6400 VMs and 400 physical machines) just in the first generation.

5.3   Effectiveness of Multi-objective Optimization

In this evaluation, we consider multi-objectives on risk level, resource wastage, and network traffic.

Figure 5 shows experimental results with weight setting (0.8, 0.1, 0.1) in an environment of 800 VMs and 60 physical machines. The risk level has weight of 80%, resource wastage and network traffic have weight of 10% for each in the fitness function. We collect the placement with best fitness value. The baseline is still the best placement chosen from 100 random-FFD placements. If a physical machine can hold hundreds of VMs, the placement generated by FFD will be using the minimum number of physical machines. With setting of (0.8, 0.1, 0.1), the active number of physical machines and resource wastage are limited, with much improved security.

We also run the experiment with weight setting (0.4, 0.3, 0.3) in an environment of 3000 VMs and 200 physical machines, and the results are shown in Fig. 6. Since the resource wastage and network traffic have higher weights, the allowance of resource wastage was controlled and it also affects the security improvement

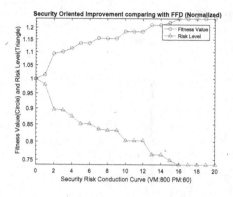

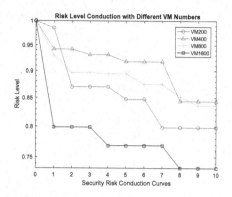

Fig. 3. Comparing with random-FFD

Fig. 4. Security improvement with different number of VMs.

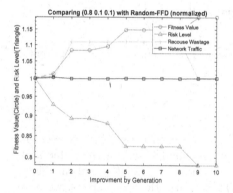

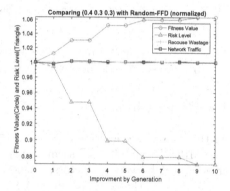

Fig. 5. Multi-objective optimization

Fig. 6. Multi-objective optimization 2

we can achieve. A cloud provider can always change the optimization preferences by changing the weights of different objectives.

5.4   Comparison with Random-FFD Algorithm

In the experiment, we use with 1600 VMs and 120 physical machines, we generate 100 placements with the random-FFD algorithm. We choose the placement with the lowest median value of risk level. After running our algorithm to reduce the risk level, we choose the best placement. As shown in Fig. 7, we can see that the risk level of the entire VM set has been effectively reduced. In the figure, the X-axis is the risk level of VMs. For example, 10% means that the risk level is between 10% and 20%. With 1600 VMs, the risk level under 50% is improved by 15% to 35%. The risk level above 80% dropped from 54% to

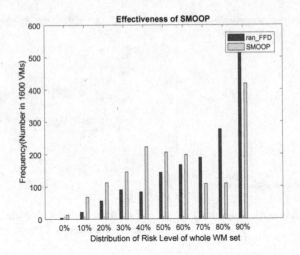

Fig. 7. Comparison with distribution in 1600 VMs and 120 PMs

33%. The experimental results demonstrate our placement strategies can greatly improve the security level of the entire cloud.

6 Conclusion

In this paper, we describe an approach for comprehensive security assessment of VMP. We quantify the security risks of the cloud based on the vulnerabilities caused by various factors, including the network, the physical machines, the VMs, and the co-residency of VMs. To optimize these objectives, we have designed a new scheme to generate VMP based on multiple objectives optimization with the given resource and other constraints. Our proposed strategy seeks the Pareto-optimal placement while considering multiple optimization objectives and constraints. The experimental results demonstrate the effectiveness of our approach and the improvement compared with existing solutions.

Acknowledgment. This project is partially supported by ARO grant W911NF-15-1-026 and NSF CNS-1634441.

References

1. Afoulki, Z., Bousquet, A., Rouzaud-Cornabas, J.: A security-aware scheduler for virtual machines on iaas clouds. Report 2011 (2011)
2. Al-Haj, S., Al-Shaer, E., Ramasamy, H.V.: Security-aware resource allocation in clouds. In: 2013 IEEE International Conference on Services Computing, pp. 400–407, (June 2013)
3. Alicherry, M., Lakshman, T.V.: Optimizing data access latencies in cloud systems by intelligent virtual machine placement. In: 2013 Proceedings IEEE INFOCOM, pp. 647–655, (April 2013)

4. Amazon: Amazon web services. http://aws.amazon.com
5. Barham, P., Dragovic, B., Fraser, K., Hand, S., Harris, T., Ho, A., Neugebauer, R., Pratt, I., Warfield, A.: Xen and the art of virtualization. SIGOPS Oper. Syst. Rev. **37**(5), 164–177 (2003). http://doi.acm.org/10.1145/1165389.945462
6. Bin, E., Biran, O., Boni, O., Hadad, E., Kolodner, E.K., Moatti, Y., Lorenz, D.H.: Guaranteeing high availability goals for virtual machine placement. In: 2011 31st International Conference on Distributed Computing Systems, pp. 700–709, (June 2011)
7. Caron, E., Le, A.D., Lefray, A., Toinard, C.: Definition of security metrics for the cloud computing and security-aware virtual machine placement algorithms. In: 2013 International Conference on Cyber-Enabled Distributed Computing and Knowledge Discovery, pp. 125–131, (October 2013)
8. CVE-2007-4993: Xen guest root can escape to domain 0 through pygrub. http:// cve.mitre.org/cgibin/cvename.cgi?name=CVE-2007-4993
9. CVE-2007-5497: Vulnerability in xenserver could result in privilege escalation and arbitrary code execution. http://suuport.citrix.com/article/CTX118766
10. Deb, K., Pratap, A., Agarwal, S., Meyarivan, T.: A fast and elitist multiobjective genetic algorithm: Nsga-ii. IEEE Trans. Evol. Comput. **6**(2), 182–197 (2002)
11. default, M.: Kernel based virtual machine. http://www.linux-kvm.org
12. Hacker, A.: Xbox 360 hypervisor privilege escalation vulnerability. http://www. securityfocus.com/archive/1/461489
13. Horton, J.D., Cooper, R., Hyslop, W., Nickerson, B.G., Ward, O., Harland, R., Ashby, E., Stewart, W.: The cascade vulnerability problem. J. Comput. Secur. **2**(4), 279–290 (1993)
14. Li, M., Zhang, Y., Bai, K., Zang, W., Yu, M., He, X.: Improving cloud survivability through dependency based virtual machine placement, (2012)
15. Maziku, H., Shetty, S.: Network aware vm migration in cloud data centers. In: 2014 3rd GENI Research and Educational Experiment Workshop, pp. 25–28, (March 2014)
16. Phan, D.H., Suzuki, J., Carroll, R., Balasubramaniam, S., Donnelly, W., Botvich, D.: Evolutionary multiobjective optimization for green clouds. In: Proceedings of the 14th Annual Conference Companion on Genetic and Evolutionary Computation, GECCO 2012, pp. 19–26. ACM, New York (2012)
17. Ristenpart, T., Tromer, E., Shacham, H., Savage, S.: Hey, you, get off of my cloud: Exploring information leakage in third-party compute clouds. In: Proceedings of the 16th ACM Conference on Computer and Communications Security, CCS 2009, pp. 199–212. (2009). http://doi.acm.org/10.1145/1653662.1653687
18. Rutkowska, J., Wojtczuk, R.: Xen owning trilogy. Talk at Black Hat (2008)
19. Shetty, S., Yuchi, X., Song, M.: Security-aware virtual machine placement in cloud data center. In: Moving Target Defense for Distributed Systems, WN, pp. 13–24. Springer, Cham (2016). doi:10.1007/978-3-319-31032-9_2
20. Shigeta, S., Yamashima, H., Doi, T., Kawai, T., Fukui, K.: Design and implementation of a multi-objective optimization mechanism for virtual machine placement in cloud computing data center. In: Yousif, M., Schubert, L. (eds.) Cloud-Comp 2012. LNICSSITE, vol. 112, pp. 21–31. Springer, Cham (2013). doi:10.1007/ 978-3-319-03874-2_3
21. Varadarajan, V., Zhang, Y., Ristenpart, T., Swift, M.M.: A placement vulnerability study in multi-tenant public clouds. In: USENIX Security, pp. 913–928 (2015)

22. Xu, J., Fortes, J.A.B.: Multi-objective virtual machine placement in virtualized data center environments. In: Green Computing and Communications (GreenCom), 2010 IEEE/ACM International Conference on Cyber, Physical and Social Computing (CPSCom), pp. 179–188, (December 2010)

23. Xu, Y., Cui, W., Peinado, M.: Controlled-channel attacks: Deterministic side channels for untrusted operating systems. In: 2015 IEEE Symposium on Security and Privacy, pp. 640–656, (May 2015)

24. Xu, Z., Wang, H., Wu, Z.: A measurement study on co-residence threat inside the cloud. In: 24th USENIX Security Symposium (USENIX Security 15), pp. 929–944, USENIX Association, Washington, D.C. https://www.usenix.org/conference/usenixsecurity15/technical-sessions/presentation/xu

25. Yu, S., Gui, X., Tian, F., Yang, P., Zhao, J.: A security-awareness virtual machine placement scheme in the cloud. In: 2013 IEEE 10th International Conference on High Performance Computing and Communications 2013, IEEE International Conference on Embedded and Ubiquitous Computing, pp. 1078–1083, (November 2013)

26. Yuchi, X., Shetty, S.: Enabling security-aware virtual machine placement in iaas clouds. In: 2015 IEEE Military Communications Conference, MILCOM 2015, pp. 1554–1559, (October 2015)

27. Zhang, W., Jia, X., Wang, C., Zhang, S., Huang, Q., Wang, M., Liu, P.: A comprehensive study of co-residence threat in multi-tenant public paas clouds. In: Lam, K.-Y., Chi, C.-H., Qing, S. (eds.) ICICS 2016. LNCS, vol. 9977, pp. 361–375. Springer, Cham (2016). doi:10.1007/978-3-319-50011-9_28

28. Zhang, Y., Juels, A., Reiter, M.K., Ristenpart, T.: Cross-tenant side-channel attacks in paas clouds. In: Proceedings of the 2014 ACM SIGSAC Conference on Computer and Communications Security, CCS 2014, pp. 990–1003. ACM, New York (2014)

29. Zhang, Y., Li, M., Bai, K., Yu, M., Zang, W.: Incentive compatible moving target defense against VM-colocation attacks in clouds. In: Gritzalis, D., Furnell, S., Theoharidou, M. (eds.) SEC 2012. IAICT, vol. 376, pp. 388–399. Springer, Heidelberg (2012). doi:10.1007/978-3-642-30436-1_32

Firewall Policies Provisioning Through SDN in the Cloud

Nora Cuppens[1], Salaheddine Zerkane[1,2,3]([✉]), Yanhuang Li[1], David Espes[2,3], Philippe Le Parc[2,3], and Frédéric Cuppens[1,2]

[1] IMT Atlantique, 2 Rue de la Chataigneraie, 35510 Cesson-sevigne, France
zerkanesalaheddine@gmail.com
[2] BCOM, 1219 Avenue des Champs Blancs, 35510 Cesson-sevigne, France
[3] Lab-STICC, Université de Bretagne Occidentale,
20 Avenue Le Gorgeu, 29200 Brest, France

Abstract. The evolution of the digital world drives cloud computing to be a key infrastructure for data and services. This breakthrough is transforming Software Defined Networking into the cloud infrastructure backbone because of its advantages such as programmability, abstraction and flexibility. As a result, many cloud providers select SDN as a cloud network service and offer it to their customers. However, due to the rising number of network cloud providers and their security offers, network cloud customers strive to find the best provider candidate who satisfies their security requirements. In this context, we propose a negotiation and an enforcement framework for SDN firewall policies provisioning. Our solution enables customers and SDN providers to express their firewall policies and to negotiate them via an orchestrator. Then, it reinforces these security requirements using the holistic view of the SDN controllers and it deploys the generated firewall rules into the network elements. We evaluate the performance of the solution and demonstrate its advantages.

Keywords: Security policies · Software Defined Networking · Cloud computing · Orchestration · Firewall · OpenFlow · Service providers · ABAC

We implement our solution into an existing SDN Firewall solution by enhancing its orchestration layer. Then, we deploy a use case for our proposition in an SDN infrastructure. The scenario compounds a NSC and 3 different NSPs. Each provider delivers a type of SDN firewall whether it is a stateful proactive SDN firewall [36], a stateful reactive SDN firewall [37] or a stateless SDN firewall [16,31]. All the peers express their firewall policies using our language. The Orchestrator mediates between them, selects the best provider and runs a negotiation process in order to reach a mutual agreement with the customer. Afterwards, it sends the contract to the chosen firewall service. The latter interprets it into OpenFlow [1,12,25] rules and installs them in the network elements.

© IFIP International Federation for Information Processing 2017
Published by Springer International Publishing AG 2017. All Rights Reserved
G. Livraga and S. Zhu (Eds.): DBSec 2017, LNCS 10359, pp. 293–310, 2017.
DOI: 10.1007/978-3-319-61176-1_16

We evaluate the performance of our framework by setting a test-bed for the aforementioned scenario.

The rest of the paper is organized as follows: Sect. 1 reviews existing proposals on policy-related solutions. Section 2 describes the formalism and all the processes from policy expression till the interpretation process of the firewall policies into OpenFlow rules and their deployment. Section 3 presents the integration of our solution into an existing SDN firewall environment and its performance experimentation results. Section 4 concludes the paper and outlines future work.

1 Related Work

The literature lacks of propositions that integrate to SDN security applications (especially SDN firewalls) firewall policy provisioning in the orchestration layer. There is an open research field on the subject in terms of SDN as a service and firewall policies orchestration in the cloud. To the extent of our studies the majority of SDN security solutions do not support negotiation between firewall policies, neither propose mutual agreement processes between providers nor reinforcement function of the client-provider agreement. CloudWatcher [32], FRESCO [13,33] and OpenSec [17] are three famous SDN propositions that rely on specific policy script languages. They lack interoperability and openness since they are plateform specific. In addition, they do not integrate a policy management process in order to interact with the cloud level. Other solutions focus on policy expression and enforcement. Tang et al. [34] develop a service oriented high level policy language to specify network service provisioning between end nodes. Batista et al. [2] propose the PonderFlow, an extension of Ponder [9] language to OpenFlow network policy specification. EnforSDN [3] proposes a management process that exploits SDN principles in order to separate the policy resolution layer from the policy enforcement layer in network service appliances. The concept improve the enforcement management, network utilization and communication latency, without compromising network policies. However, it can not handle stateful security applications like stateful firewalls.

Many solutions have been proposed for selecting NSP in the cloud. There are two trends in the literature. The major one focusses on NSC's security requirements without taking into consideration NSP's constraints. [18,23] define the selection strategy exclusively on NSC's capacity. Bernsmed et al. [4] present a security SLA (Service Level Agreement) framework for cloud computing to help potential NCSs to identify necessary protection mechanisms and facilitate automatic service composition. In [6], different virtual resource orchestration constraints are resumed and expressed by Attribute-Based Paradigm from NSC perspective. The other trend which is part of our work takes into consideration NSP capacities and offers in order to perform the selection. In [20], both NSP and NSC can express security requirements in SLA contract then these security requirements are transformed to OrBAC [15] policies. Li et al. [21] proposes a method to measure the similarity between security policies and suggest using the solution in SP selection process.

Most of the work in the literature define security policies negotiation based on access control negotiation. The literature is classified in 3 types of negotiations: (1) negotiation with no constraints, (2) negotiation with global constraints, (3) negotiation with local constraints [11]. For example, [5] examines the problem of negotiating a shared access state, assuming all negotiators use the RBAC [30] policy model. Based on a mathematical framework, negotiation is modeled as a Semiring-based Constraint Satisfaction Problem (SCSP) [7]. In [35], authors argue that the guidance provided by constraints is not enough to bring practical solutions for automatic negotiation. Thus, they define an access control policy language which is based on Datalog [14] with constraints and the language can be used to define formal semantics of XACML [28]. Towards the need for human consent in organizational settings, Mehregan et al. [22] develop an extension of Relationship-Based Access Control (ReBAC) model [10] to support multiple ownership, in which a policy negotiation protocol is in place for co-owners to come up with and give consent to an access control policy. Some autors consider security policy negotiation as a process of contract establishment. For example, Li et al. [19] propose to integrate policy negotiation in contract negotiation by introducing bargaining process. An extension of the negotiation model is proposed in [29] and the model is designed for privacy policy negotiation in mobile health-Cloud environments. In our work we combine the 3 negotiation types and adapt them to negotiate SDN firewall policies in the cloud.

Our solution fills the aforementioned gaps. It meets key-functional requirements for user-centric clouds as (1) it addresses the firewall service configuration at the management layer. (2) it offers a language for firewall policies expression. (3) It supports multiple policy models in order to translate attribute-based security expression to concrete policies. (4) It selects the best SDN Firewall NSP. (5) It provides a negotiation protocol for NSC and NSP based on 3 types of negotiations. (6) it establishes a service level agreement between NSC and NSP. (7) It reinforces the contract according to SDN infrastructure configuration. (8) Then, it interprets it into OpenFlow rules and deploys them inside the network. To the best of our knowledge, there is no method in the literature that considers all these points together.

2   SDN Firewall Policy Provisioning Model

Both NSC and NSPs specify their firewall policies using our proposed expression Language. Then the Orchestrator assesses the expressions by comparing NSP's service templates with NSC's policies after receiving them. It ranks the NSPs and selects the best one which fulfills the most NSC's requirements. It starts a negotiation process with NSC in the case it is necessary. A successful negotiation generates a firewall policy agreement. This contract is derived from high-level firewall policies and sent to the chosen SDN firewall service. The latter reinforces the received policies according to its view (topology, previous security policies and other network configurations) and translates them into OpenFlow rules. Afterwards, it sends the OpenFlow rules to the SDN controllers. Each one of them deploys the received OpenFlow rules on its network elements.

2.1 Scenario Description

We introduce a use case to experiment our concept. The subjects involved in the scenario are NSC, SDN orchestrator and 3 NSPs. NSC requires an SDN Firewall service that meets its firewall policies (Requirements). Each NSP provides a type of SDN firewalls and a set of firewall policies (Obligations). The three SDN firewall services are as follows:

1. **NSP1: SDN Reactive Stateful Firewall** [37]. It forwards systematically all the packets to the stateful firewall Application over the SDN controller. The application verifies each packet using its access control table and reacts to these network events by installing the proper stateful firewall OpenFlow rules in the Network Elements. It relies on the Finite State Machine of all the connections. This service spares Network elements resources. However, it shifts the computing and memory loads on the controller. As a result, the latter became vulnerable to some DDoS attacks.
2. **NSP2: SDN Proactive Stateful Firewall** [36]. The service is based on a white list approach. It closes all the accesses and opens only the routes to the authorized connections while tracking their states. The service pre-installs all the Stateful firewall OpenFlow rules in the Network Elements. The latter sends each time a copy of their events to the Firewall service. The proactive service protects against DDoS attacks. It delegates also the access control to the Network Elements.
3. **NSP3: Stateless SDN Firewall** [16,31]. The service does not track the connections states and it is vulnerable to DDoS attacks. However it consumes fewer resources in the Network Elements and in the Controller comparing with the above services.

2.2 Expression of Firewall Policies

We propose an SDN firewall policy language to homogenize NSP's obligations and NSC's requirements. The proposed language is inspired from the Attribute-Based Access Control Model (ABAC) [8]. It allows expressing firewall policies based on a common template. These unification guarantees the interoperability between the Obligations and the Requirements. The grammar of our language is as follows:

Π is the set of all the firewall policies. It describes the access controls within the dynamic environment of the allocated cloud resources: $\Pi = \{\pi_1, \pi_2, ..., \pi_m\}$ where $\pi_{i=1..m}$ are firewall policies.

Θ is a set of Obligations. It encompasses all the firewall policies of Π expressed by NSPs. $\Theta = \{\theta_1, \theta_2, ..., \theta_k\}$

Φ is a set of Requirements. It encompasses all the firewall policies of Π expressed by NSC. $\Phi = \{\phi_1, \phi_2, ..., \phi_j\}$

Where $\Pi = \Theta \cup \Phi$ and $\pi_{i=1..m} \equiv \theta_{i=1..k} \vee \phi_{i=1..j}$

Each firewall policy π_i is formed by many atomic elements $\varepsilon_{i=1..n}$:

$$\pi_{i=1..m} \equiv \varepsilon_1 \wedge \varepsilon_2 \wedge ... \wedge \varepsilon_n$$

$\varepsilon_{i=1..n}$ is defined by a preposition of predicates. Each predicates defines a propriety of the *element*.

Theorem 1. $A(\varepsilon_i)$ and $B(\varepsilon_i)$ are two predicates defining ε_i proprieties. Predicates equivalence is determined by the preposition : $A(\varepsilon_i) \in \Omega, B(\varepsilon_i) \in \Psi \mid (\Psi = \Omega) \rightarrow (A(\varepsilon_i) \equiv B(\varepsilon_i))$.

The atomic rule element $\varepsilon_{i=1..n}$ is formed by the following predicates :

1. **Type:** $type(\varepsilon_i) \equiv subject(\varepsilon_i) \vee action(\varepsilon_i) \vee object(\varepsilon_i) \vee context(\varepsilon_i)$
2. **Domain:** $domain(type(\varepsilon_i)) \in \{protocol, time...\}$. Domain restricts the unit of an element.
3. **Value:** $value(type(\varepsilon_i)) \equiv variable(type(\varepsilon_i)) \vee non\text{-}variable(type(\varepsilon_i))$.
   variable has not an assigned value while *non-variable* has an already assigned value. Both *variable* and *non-variable* can be assigned by three kinds of data types:
   (a) **constant:** numeric or semantic value, ex. $value(type(\varepsilon_i)) = TCP$.
   (b) **interval:** numeric interval, ex. $value(type(\varepsilon_i)) = [8:00,\ 20:00]$
   (c) **set:** a collection of values, ex. $value(type(\varepsilon_i)) = \{15:00, 16:00\}$

   For simplification, we use x_i to present a *variable*, $x_i \equiv variable(type(\varepsilon_i))$

4. **Scope:** it defines the access to the values of a *variable*. It can be:
   (a) **Public preference:** $\mathbf{pub_{pre}}(x_i)$ a public preference *variable* is accessible as public information.
   (b) **Private preference:** $\mathbf{pri_{pre}}(x_i)$ a private preference *variable* is a local configuration that can not be disclosed.

If *context* is not specified in a policy we add a universal context element $\top$. It indicates that all the obligations for the context are acceptable.

$$(context(\varepsilon_i) \equiv \top) \rightarrow ((domain(context(\varepsilon_i)) \equiv \top) \wedge (value(context(\varepsilon_i)) \equiv \top))$$

Finally we write:
$\varepsilon_i \equiv type(\varepsilon_i) \wedge domain(type(\varepsilon_i)) \wedge value(type(\varepsilon_i)) \wedge (\mathbf{pub_{pre}}(x_i) \vee \mathbf{pri_{pre}}(x_i))$
When the scope is not defined: $\varepsilon_i \equiv type(\varepsilon_i) \wedge domain(type(\varepsilon_i)) \wedge value(type(\varepsilon_i))$
The firewall policies given in Sect. 2.1 using our language are defined in Table 1.

2.3   Assessment of Firewall Policies

The assessment of firewall policies is based on matching the Obligations with the Requirements in order to determine which NSPs' policies satisfies NSC's requests. This process depends on two level of relationships. Element-Element relation which relies on corresponding the predicates of the firewall policies elements. The second level (Policy-Policy relation) focuses on finding the relationships between the matched elements.

Table 1. Firewall policy expression for NSC, NSP1, NSP2 and NSP3

			ϕ_1		
	element	ε_1	ε_2	ε_3	ε_4
	type domain value	subject organization {NSC, NSP}	action firewall operation pass	object protocol {HTTP, TCP, ICMP}	context time [0:00,24:00]
			ϕ_2		
	element	ε_1	ε_2	ε_3	ε_4
NSC	type domain value Scope	subject organization {NSC, NSP} -	action firewall operation x_2 $\mathbf{pri_{pre}}$ ({quarantine, block, alert})	object protocol TCP -	context connection_exc TCP_failed_Time >30 -
			ϕ_3		
	Element	ε_1	ε_2	ε_3	ε_4
	type domain value	subject organization {NSC, NSP}	action firewall operation block	object protocol ICMP	context attack_detection DoS_detection
			θ_1		
	element	ε_1	ε_2	ε_3	ε_4
	type domain value Scope	subject organization NSP -	action firewall operation x_2 $\mathbf{pub_{pre}}$({pass, block})	object protocol x_3 $\mathbf{pub_{pre}}$({HTTP, TCP, ICMP})	context time x_4 ⊤
NSP1			θ_2		
	element	ε_1	ε_2	ε_3	ε_4
	type domain value Scope	subject organization NSP -	action firewall operation x_2 $\mathbf{pri_{pre}}$({block, alert})	object protocol x_3 $\mathbf{pub_{pre}}$({TCP, HTTP, SSH, ICMP})	context connection_exc x_4 ⊤
			θ_1 (**same policy as NSP1**)		
			θ_2 (**same policy as NSP1**)		
			θ_3		
NSP2	Element	ε_1	ε_2	ε_3	ε_4
	type domain value Scope	subject organization NSP -	action firewall operation block -	object protocol x_3 $\mathbf{pub_{pre}}$({HTTP, TCP,SSH, ICMP})	context attack_detection x_4 $\mathbf{pub_{pre}}$({Poisoning, DoS_detection})
			θ_1 (**same policy as NSP1**)		
			θ_4		
NSP3	element	ε_1	ε_2	ε_3	ε_4
	type domain value Scope	subject organization NSP -	action firewall operation x_2 $\mathbf{pub_{pre}}$({pass, block})	object IP_address x_3 ⊤	context ⊤ x_4 ⊤

Element-Element Relation. There are five relations between the elements:

1. **inconsistent:** $(type(\varepsilon_i) \not\equiv type(\varepsilon_j)) \rightarrow (\varepsilon_i \nvdash \varepsilon_j)$. If two rule elements ε_i and ε_j have not equivalent *type* predicates then they are in *inconsistent* relation denoted: $\varepsilon_i \nvdash \varepsilon_j$. For example, in Table 1, $\phi_1.\varepsilon_1 \nvdash \theta_1.\varepsilon_2$ because $subject(\phi_1.\varepsilon_1) \not\equiv action(\theta_1.\varepsilon_2)$

Theorem 2. *Not equivalence of type is defined as follows:*
$type(\varepsilon_i) \in \Omega, type(\varepsilon_j) \in \Psi \mid ((\Omega \not\subseteq \Psi) \wedge (\Psi \not\subseteq \Omega)) \rightarrow (type(\varepsilon_i) \not\equiv type(\varepsilon_j))$

2. **comparable:** $((type(\varepsilon_i) \equiv type(\varepsilon_j)) \wedge (domain(\varepsilon_i) \cong domain(\varepsilon_j))) \rightarrow (\varepsilon_i \sim \varepsilon_j)$. If two rule elements ε_i and ε_j have equivalent *type* predicates and their domain predicates are in congruence, then they are in *comparable* relation. It is denoted with $\varepsilon_i \sim \varepsilon_j$. For example, in Table 1, $\phi_1.\varepsilon_2 \sim \theta_1.\varepsilon_2$ because their *subject* predicates are equivalent and their *domain* predicates are congruent.

Theorem 3. *Domain congruence is defined as follows:*
$domain(\varepsilon_i) \in \Omega, domain(\varepsilon_j) \in \Psi \mid ((\Omega \subseteq \Psi) \vee (\Psi \subseteq \Omega)) \rightarrow (domain(\varepsilon_i) \cong domain(\varepsilon_j))$

3. **equal:** $((\varepsilon_i \sim \varepsilon_j) \wedge (value(type(\varepsilon_i)) \cong value(type(\varepsilon_j)))) \rightarrow (\varepsilon_i = \varepsilon_j)$. If two rule elements ε_i and ε_j are *comparable* and their values predicates are in congruence, then they are in *equal* relation denoted with $\varepsilon_i = \varepsilon_j$. For example, in Table 1, $\phi_2.\varepsilon_3 = \theta_3.\varepsilon_3$ because both elements are *comparable* and their *value* predicates are congruent ($\{TCP\} \subseteq \{HTTP, TCP, SSH, ICMP\}$).

Theorem 4. *Value congruence is defined as follows:*
$value(type(\varepsilon_i)) \in \Omega, value(type(\varepsilon_j)) \in \Psi \mid ((\Omega \subseteq \Psi) \vee (\Psi \subseteq \Omega)) \rightarrow (value(type(\varepsilon_i)) \cong value(type(\varepsilon_j)))$

4. **unequal:** $((\varepsilon_i \sim \varepsilon_j) \wedge (value(type(\varepsilon_i)) \not\equiv value(type(\varepsilon_j)))) \rightarrow (\varepsilon_i \neq \varepsilon_j)$. If two rule elements ε_i and ε_j are *comparable* but do not have equivalent value, they are in *unequal* relation denoted with $\varepsilon_i \neq \varepsilon_j$. For example, in Table 1, $\phi_1.\varepsilon_2 \neq \theta_3.\varepsilon_2$ because both are comparable however they have not equivalent values (*pass* $\not\equiv$ *block*).

Theorem 5. *Not equivalence of value is defined as follows:*
$type(\varepsilon_i) \in \Omega, type(\varepsilon_j) \in \Psi \mid ((\Omega \not\subseteq \Psi) \wedge (\Psi \not\subseteq \Omega)) \rightarrow (value(type(\varepsilon_i)) \not\equiv value(type(\varepsilon_j)))$

5. **incomparable:** $((type(\varepsilon_i) \cong type(\varepsilon_j)) \wedge (domain(\varepsilon_i) \not\equiv domain(\varepsilon_j))) \rightarrow (\varepsilon_i \nsim \varepsilon_j)$. If two rule elements ε_i and ε_j have equivalent *type* predicates and not equivalent *domain* predicate, then they have *incomparable* relation denoted with $\varepsilon_i \nsim \varepsilon_j$. For example, in Table 1, $\phi_1.\varepsilon_3 \nsim \theta_4.\varepsilon_3$ because they have congruent *type* predicates but their domains are not equivalent (*protocol* $\not\equiv$ *IP_address*).

Theorem 6. *Congruence of type is defined as follows:*
$type(\varepsilon_i) \in \Omega, type(\varepsilon_j) \in \Psi \mid ((\Omega \subseteq \Psi) \vee (\Psi \subseteq \Omega)) \rightarrow (type(\varepsilon_i) \cong type(\varepsilon_j))$

Policy-Policy Relations. We derive from Element-Element relations, three relations between policies. These relations are as follows:

1. **match:** $(P_1 \wedge P_2 \wedge ... \wedge P_n) \rightarrow (\pi_\alpha \bowtie \pi_\beta)$
   $P_i \equiv ((\pi_\alpha.\varepsilon_i = \pi_\beta.\varepsilon_1) \vee ... \vee (\pi_\alpha.\varepsilon_i = \pi_\beta.\varepsilon_n)) \mid \pi_\alpha, \pi_\beta \in \Pi, i = 1..n$
   If any *element* of a policy α is in equal relation with another element of a policy β, then the two policies are in *match* relation denoted with $\pi_\alpha \bowtie \pi_\beta$. For example, in Table 1, $\phi_3 \bowtie \theta_3$ because $(\phi_3.\varepsilon_1 = \theta_3.\varepsilon_1) \wedge (\phi_3.\varepsilon_2 = \theta_3.\varepsilon_2) \wedge (\phi_3.\varepsilon_3 = \theta_3.\varepsilon_3) \wedge (\phi_3.\varepsilon_4 = \theta_3.\varepsilon_4)$.
2. **mismatch:** $\exists \varepsilon_i \exists \varepsilon_j (\pi_\alpha.\varepsilon_i \nsim \pi_\beta.\varepsilon_j) \rightarrow (\pi_\alpha \asymp \pi_\beta) \mid \pi_\alpha, \pi_\beta \in \Pi$. If there are at least *incomparable* elements ε_i and ε_j from two policies π_α and π_β, then the two policies have *mismatch* relation denoted with $\pi_\alpha \asymp \pi_\beta$. For example, in Table 1, $\phi_3 \asymp \theta_2$ because $\phi_3.\varepsilon_4 \nsim \theta_2.\varepsilon_4$.
3. **potential match:** $((\forall \varepsilon_i \forall \varepsilon_j (\pi_\alpha.\varepsilon_i \sim \pi_\beta.\varepsilon_j)) \wedge (\exists \varepsilon_k \exists \varepsilon_l (\pi_\alpha.\varepsilon_k \neq \pi_\beta.\varepsilon_l))) \rightarrow (\pi_\alpha \propto \pi_\beta) \mid \pi_\alpha, \pi_\beta \in \Pi$. If all the elements of the policies π_α and π_β are *comparable* but it exists at least an *unequal* relation between two of their respective elements then the two policies are in a *potential match* relation denoted $\pi_\alpha \propto \pi_\beta$. For example, in Table 1, $\phi_1 \propto \theta_1$ because $((\phi_1.\varepsilon_1 \sim \theta_1.\varepsilon_1) \wedge (\phi_1.\varepsilon_2 \sim \theta_1.\varepsilon_2) \wedge (\phi_1.\varepsilon_3 \sim \theta_1.\varepsilon_3) \wedge (\phi_1.\varepsilon_4 \sim \theta_1.\varepsilon_4)) \wedge (\phi_1.\varepsilon_1 \neq \theta_1.\varepsilon_1)$.

NSP Ranking. The orchestrator ranks each NSP based on the relations between the *Requirements* and the *Obligations* (see Sect. 2.3). This process enables selecting the most compliant NSP according to the Algorithm 1.

2.4   Establishment of Contract

Negotiation Protocol. When an agreement is not reached between the peers, the orchestrator negotiates the offers of the chosen NSP with NSC. We propose the Rule-Element Based Negotiation Protocol (RENP) in order to manage the negotiation process. Our protocol specifies for each element the next action regarding the proposed values (v_{rec}) of NSP and the local configuration (v_{loc}) of NSC. The protocol contains three types of actions:

1. **accept:** it indicates that the proposed value is agreed.
2. **refuse:** it indicate that the proposed value is aborted.
3. **propose:** It generates a counter-offer .

Table 4 in the Appendix presents the detail of RENP protocol. (vp) is the proposed variable upon negotiation. The results of the assessment and negotiation processes for the example defined in Sect. 2.1 are as follows. The orchestrator finds *potential match* relations between the pairs: (ϕ_1, θ_1), (ϕ_2, θ_2) and (ϕ_3, θ_3). NSP1 does not meet the firewall requirement of ϕ_3 and NSP3 does not fulfill ϕ_2 and ϕ_3. As a consequence, the resulting relations are *mismatch*. The orchestrator puts NSP2 into the ranking list.

The orchestrator conducts then policy negotiation with NSC on behalf of NSP2. It accepts the obligations θ_1 for ϕ_1 and θ_3 for ϕ_3. However, it proposes a

Algorithm 1. NSP Ranking

```
 1: rank_list is Empty {Initial ranking list is Empty}
 2: for All NSPs do
 3:     rank ← 0 {Default ranking value}
 4:     matching ← True {To make sure that there are only matches for Gold ranking}

 5:     for i=0, i≤Length(Requirement), i++ do
 6:         for j=0, j≤Length(Obligation), j++ do
 7:             if Match(Requirement[i], Obligation[j]) = True then
 8:                 rank ← 2
 9:                 Break
10:             else if Potential_Match(Requirement[i], Obligation[j]) = True then
11:                 rank ← 1
12:                 matching ← False
13:                 Break
14:             else if Mismatch(Requirement[i], Obligation[j]) = True then
15:                 matching ← False
16:             end if
17:         end for
18:         if rank = 0 then
19:             Break
20:         end if
21:     end for
22:     if (rank = 2) and (matching = True) then
23:         Add (rank_list, (NSP, NSP_gold)){Tag NSP as NSP gold and add it to the
             ranking list}
24:     else if rank = 1 then
25:         Add (rank_list, (NSP, NSP_silver)){Tag NSP as NSP silver and add it to the
             ranking list}
26:     else
27:         print NSP is not compliant with NSC, it will not be add to the ranking list
28:     end if
29: end for
30: return rank_list
```

counter offer for ϕ_2. This case corresponds to column 5 in row 4 of Table 4 because the received value (in ϕ_2) $vp_{rec} : quarantine$ has no intersection with $pri_{pre_{loc}} :$ {$block, alert$} (in θ_2). Thus, the orchestrator chooses another value $= block$ in $pri_{pre_{loc}}$ as a new proposition. After receiving the proposition, NSC accepts the new value because $block$ belongs to the local private configuration of ϕ_2. Then the orchestrator establishes the contract between NSC and NSP2 (Table 2).

General Agreement. Algorithm 2 illustrates the contract building process conducted by the orchestrator. It chooses the NSP_{top} in the top of rank_list. The orchestrator accepts directly without negotiation NSP_{gold} and establishes contract with NSC. While for NSP_{silver}, it starts a negotiation process with NSC by executing the proposed negotiation protocol (see Table 4). It transforms potential match relations into match relations. If the negotiation fails NSP is deleted from rank_list and the negotiation process is re-conducted.

Table 2. Final agreement between NSP and SDN orchestrator

NSC	Policy 1				
	Element	ϵ_1	ϵ_2	ϵ_3	ϵ_4
	Type	Subject	Action	Object	Context
	Domain	Organization	Firewall operation	Protocol	Time
	Value	NSP	Pass	{HTTP, TCP, ICMP}	[0:00,24:00]
	Policy 2				
	Element	ϵ_1	ϵ_2	ϵ_3	ϵ_4
	Type	Subject	Action	Object	Context
	Domain	Organization	Firewall operation	Protocol	connection_exc
	Value	NSP	Block	TCP	TCP_failed_Time>30
	Policy 3				
	Element	ϵ_1	ϵ_2	ϵ_3	ϵ_4
	Type	Subject	Action	Object	Context
	Domain	Organization	Firewall operation	Protocol	attack_detection
	Value	NSP	Block	ICMP	DoS_detection

2.5   Enforcement of Security Policy

Policy Transformation. SDN NSP contains two levels of policy abstraction:
(1) a service level abstraction which defines the business logic. This high level is
expressed by administrators and tenants. (2) An OpenFlow level which interprets
the high-level abstraction into infrastructure specific rules. The abstraction at the
service level hides the details of the network configuration and service deployment.
It simplifies the expression of the service policies. While the OpenFlow level ensures
deploying the policies into the network elements according to the network state.

The orchestrator sends the high level policies to SDN Firewall Applications.
Each one interprets the high level policies into OpenFlow rules and sends them
to the controller.

The interpretation process is based on mapping the elements of the high level
policy model with OpenFlow elements. A high level policy can be interpreted to
more than one OpenFlow rule.

OpenFlow is based on flow rules. Mainly, it structures policies into 6 parts. (1)
OF.Type can be *Flow ADD rules, Flow MODIFY rules* and *Flow DELETE rules*.
(2) *Matching Fields* define the characteristics of the traffic. They describe the
header of a packet in order to identify network flows. (3) *Actions* specify the operations on the matched. These actions can be *Drop* traffic, *Forward to controller*,
Forward to Port. (4) *Timers* indicate the lifetime of the rule (*Hard_Timeout*)
or the ejection time if the rule is not matched for a time interval (*Idle_Timeout*).
(5) *Metadata* can be used to save any extra information. (6) *Counters* allow to
specify rules based on traffic statistics. Table 3 shows the interpretation of the
final agreement (Table 2) into OF rules. We applied the following mappings.

1. *Object* corresponds to *OF Matching Fields*. It is the first element which is
   mapped to its OF counterparts.
   The interpretation of the object element will generate at least an OF rules
   for each object value.

Algorithm 2. Establishment of contract

```
 1: rank_list← call (NSP Ranking) {Making NSP Ranking List}
 2: while NSP in rank_list do
 3:     Best_NSP = NSP_top{Choose NSP_top from rank_list}
 4:     if Best_NSP is NSP_gold then
 5:         Accept (Best_NSP.Obligation()) { Accept NSP_top offer }
 6:         Contract =Generate_Contract (Best_NSP, NSC) {Establish  contract
            with NSC}
 7:         return
 8:     else
 9:         Negociate (Best_NSP.Obligation()), NSC.Requirement()) {Start negotia-
            tion with NSC}
10:         negotiation_Result = RENP (potential match) {Execute negotiation proto-
            col between potential match rule pairs}
11:         if negotiation_Result = Accepted then
12:             Contract =Generate_Contract (NSP, NSC)  {Establish contract with
                NSC}
13:             return
14:         else
15:             Delete (NSP_top, rank_list) { Delete NSP_top from rank_list }
16:         end if
17:     end if
18: end while
```

2. *Action* of the high level policy corresponds to OF *Action Field*. OpenFlow offers also the possibility to express many actions (*ACTION_List*) and to associate them to the same OF rule. The orchestrator verifies that there is no contradiction between actions (for example: *block* and *allow*) in the same policy. For example, *block* corresponds to the OF action: *DROP*.

3. *Context* element is mapped to OF components such as $TIME\_OUTs$ and OF *Counters* but also to firewall specific functions. The interpretation to firewall function triggers a condition in order to execute the OF rule of the *Context*. For example, the context: $TCP\_Failed\_Time > 30$ in *Policy2* triggers TCP connection counting function and when it exceeds 30 connections it installs the corresponding rule for *policy2*.

4. NSP is mapped to the topology of the service provider. For each link between the nodes, the interpretation module generates the corresponding OF rules taking the 3 aforementioned mappings. The OF Matching fields that correspond to the topology are at least $IP_{src}, IP_{dst}, PORT_{src}$, and $PORT_{dst}$. If the topology is not provided, the Firewall Application installs the OF rules without specifying the topology matching fields.

5. The default type of OF rule is ADD. The firewall application verifies firstly that the rule is not a duplicate of a previous interpreted rule by comparing both matching parts. If only the contexts are different, then the OF rule type is set to $MODIFY$. If the firewall application receives from the Controller an error upon sending a $MODIFY$ rule, the firewall application changes the $MODIFY$ rule to ADD rule and re-sends it to the controller.

Table 3. Interpretation of the final agreement into OpenFlow rules

	OF type	Matching field	Action	Timer	Firewall function
Policy 1	ADD	ETH_Type=2048 IP_Proto=6 DST_P=80	Forward controller	Idle_Timeout=0 Hard _Timeout=0	
Policy 1	ADD	ETH_Type=2048 IP_Proto=6	Forward controller	Idle_Timeout=0 Hard _Timeout=0	
Policy 1	ADD	ETH_Type=2048 IP_Proto=1	Forward controller	Idle_Timeout=0 Hard _Timeout=0	
Policy 2	MODIFY	ETH_Type=2048 IP_Proto=6	DROP		TCP_Count(30)
Policy 3	MODIFY	ETH_Type=2048 IP_Proto=1	DROP		SNORT.ALERT = DDOS

Policy Deployment. The controller opens a secure channel with the data plane devices (network elements) and communicates by exchanging OpenFlow messages. Upon receiving OF rules from SDN firewall, the controller sends them to the corresponding data plane devices which then install the rules. The rules are parsed and their elements are saved in the Flow tables of the data plane devices. At this level, the OF rules become Flow entries in the data plane devices. When the data plane devices receive network packets, they parse their headers and match the contents with all the matching fields of each flow entry. Once a matching is found, the corresponding action is executed on the packet. If a match is not found, the data plane device drops the packet or sends it to the controller.

3   Evaluation

We implement the proposed solution using Python programming language. The Framework has an orchestrator layer which integrates the firewall provisioning model. In addition, it has a stateful SDN firewall application that executes the security policies and the finite state machines of stateful network protocols.

We deploy our solution in the B-Secure platform which is a cloud environment to test the performance of SDN security solutions. It consists of a central machine (16 GB of RAM and Intel i7 processors), a data plane device machine (16 GB of RAM, Intel i7 processors, 6 physical network interfaces of 1 GB/s) connected to the central machine, and two machines (4 GB of RAM and Intel i3 processors) connected to the data plane device.

We install the orchestrator, the firewall application and the SDN controller RYU [24] in the central machine. In the data plane device machine, we run the OpenVswitch (OVS) [26, 27]. It is a virtual switch framework widely used in both industry and research. The physical network link between the controller and the OVS is 1 GB/s.

In the central machine, we deploy NSC and NSP2 of the use case (see Sect. 2.1). The orchestrator generates the contract and sends the high level policies to the SDN firewall Application. Then the latter interprets the policies to OpenFlow rules and asks RYU controller to install them on OVS. We vary the number of NSC's *Requirements* from 1 to 2500 policies. We measure the following performance metrics:

1. *Policy Processing Time (PPT)* is the time that the orchestrator needs to process the policy expression.
2. *Orchestrator Processing Time (OPT)* is the total time taken by the orchestrator from the first policy expression to sending the last policy.
3. *Firewall Application Processing Time (FAPT)* is the total time taken by the firewall application to process the policies.
4. *Controller Processing Time (CPT)* is the total time taken by the controller to send all the OpenFlow rules.
5. *Infrastructure Processing Time (IPT)* is the total time that the infrastructure (Controller-OVS link and OVS) needs to install all the OF rules.
6. *Policy Processing Total Time (PPTT)* is the total policy provisioning time.
7. *Orchestrator Setup Rate (OSR)* is the speed of the orchestrator:

$$OSR = OPT/Number\ of\ Policies \qquad (1)$$

8. *Firewall Setup Rate (FSR)* is the speed of the firewall Application:

$$FSR = FAPT/Number\ of\ Policies \qquad (2)$$

9. *Controller Setup Rate (CSR)* is the speed of the Controller:

$$CSR = CPT/Number\ of\ Openflow\ Rules \qquad (3)$$

10. *Infrastructure Setup Rate (ISR)* is the speed of the Infrastructure:

$$ISR = IPT/Number\ of\ Openflow\ Rules \qquad (4)$$

Figures 1 and 2 display the different measured processing times during the experiment. The processing times in all the figures increase with the rise of rule number. In Fig. 1, we observe that PPT increases slowly with a starting value of 0.00236 s for 10 policies to a maximal value of 0.6421 s for 2500 policies. In addition, PPT's values are the lowest among all processing times. $FAPT$ is slightly higher than CPT. Moreover, OPT is slower than both $FAPT$ and CPT. For example, the values for 2500 policies are: 2.700 s, 2.254 s and 2.141 s. However, the largest processing times are those of IPT as shown in Fig. 2 ((10,0.0034 s) and (2500, 11.025 s)).

The reasons are the amount and nature of the processing that each layer performs. The orchestrator runs many processes in order to generates the final agreement. The firewall application performs the interpretation to OF rules and the controller deploys the generated OF rules in the network element. The infrastructure is impacted by the performance of OVS and the data link with

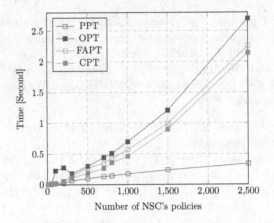

Fig. 1. Policy processing times

the controller. Our solution accumulates a processing time of 7.96 s with 2500 rules, while the infrastructure processing time is 1.5 times higher (2500, 11.02 s). Around 50% (in Average) of $PPTT$ is taken by the infrastructure to deploy the rules. For example, for 2500 rules, it takes 18.12 s from the time of releasing the initial policy request in the orchestrator to the time of deployment of the final OF rule in the network element. 60% of this time is taken by the infrastructure alone (see Fig. 2).

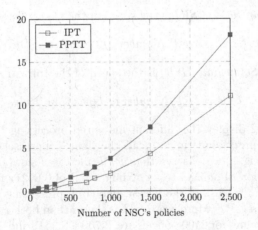

Fig. 2. Policy deployment times

Figure 3 displays the different policy setup rates in the infrastructure. We observe 3 different states. In the first state, CSR, FSR, OSR and ISR increase to reach their top values respectively (100, 5186), (100, 4838), (50, 3150), (10, 2941). In the second stage, all the rates decrease rapidly. The diminution is linear for CSR, FSR and OSR while fluctuating for ISR. In the third stage, we observe that all the rates reduce with the increase of the number of policies. The rates of our solution reach a value around 1000 Policies/s at 2500 while ISR continues to hold lower values. This observation comforts the previous results. ISR low rates are caused by the load on the link Controller-OVS. We observe that the orchestrator has lower performance than the firewall application. This observation consolidates the explanations provided previously. Our solution has good performances with around 1000 policies/s. In practice, the number of firewall rules depends on the size of the topology and the granularity of each rule. Furthermore, policies changes do not need the repetition of all the process because OpenFlow enables to update directly the installed rules.

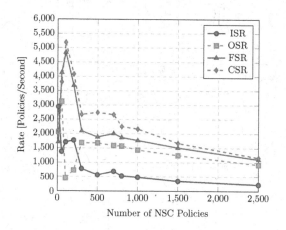

Fig. 3. Solution's policy rates

4 Conclusion and Perspectives

We propose a solution to express firewall policies, assess NSC's requirements with NSPs' obligations, select the best NSP candidate, negotiate and agree on a common policy contract then deploy the agreement in a SDN cloud platform. We integrate and deploy the solution in an existing SDN firewall framework. Moreover, we evaluate its performance and scalability. The evaluation shows promising results with a rate of 1000 deployed policies/s.

Our framework brings many advantages. It offers interoperability between different NSCs and NSPs through a unified language that simplifies administrator's tasks. It abstracts the complexity of the network by hiding the infrastructure details. In addition, it automatizes firewall policies orchestration.

In order to improve our solution, we plan to include in the process new elements such as quality of service, pricing and NSP's reputation. This improvement will resolve specific cases in our ranking algorithm. For example, the empty rank_list or multiple matching and potential matching NSPs.

Acknowledgement. The work of Nora Cuppens and Frédéric Cuppens reported in this paper has been partially carried out in the SUPERCLOUD project, funded by the European Unions Horizon 2020 research and innovation programme under grant N643964.

A   RENP Protocol

Table 4. RENP protocol

v_{loc} v_{rec}	non variable	variable $\mathbf{pub_{pre}}$	variable $\mathbf{pri_{pre}}$	proposed value (vp)
non variable	$(v_{rec} = v_{loc})$ $\to accept(v_{rec})$ $(v_{rec} \neq v_{loc})$ $\to refuse$	$(\{v_{loc}\} \subseteq \{pub_{pre_{rec}}\})$ $\to propose(v_{loc})$ $(\{v_{loc}\} \not\subseteq \{pub_{pre_{rec}}\})$ $\to refuse$	$propose(v_{loc})$	-
variable $\mathbf{pub_{pre}}$	$(\{v_{rec}\} \subseteq \{pub_{pre_{loc}}\})$ $\to accept(v_{rec})$ $(\{v_{rec}\} \not\subseteq \{pub_{pre_{loc}}\})$ $\to refuse$	$((\{pub_{pre_{loc}}\} \cap \{pub_{pre_{rec}}\}) \neq \emptyset)$ $\to propose(x)$ $x = (\{pub_{pre_{loc}}\} \cap \{pub_{pre_{rec}}\})$ $((\{pub_{pre_{loc}}\} \cap \{pub_{pre_{rec}}\}) = \emptyset)$ $\to refuse$	$propose(x)$ $x = pub_{pre_{loc}}$	$(\{vp_{rec}\} \subseteq \{pub_{pre_{loc}}\})$ $\to accept(vp_{rec})$ $(\{vp_{rec}\} \not\subseteq \{pub_{pre_{loc}}\})$ $\to refuse$
variable $\mathbf{pri_{pre}}$	$(\{v_{rec}\} \subseteq \{pri_{pre_{loc}}\})$ $\to accept(v_{rec})$ $(\{v_{rec}\} \not\subseteq \{pri_{pre_{loc}}\})$ $\to refuse$	$((\{pri_{pre_{loc}}\} \cap \{pub_{pre_{rec}}\}) \neq \emptyset)$ $\to propose(x)$ $x = (\{pri_{pre_{loc}}\} \cap \{pub_{pre_{rec}}\})$ $((\{pri_{pre_{loc}}\} \cap \{pub_{pre_{rec}}\}) = \emptyset)$ $\to refuse$	$propose(x)$ $x \in \{pri_{pre_{loc}}\}$	$(\{vp_{rec}\} \subseteq \{pri_{pre_{loc}}\})$ $\to accept(vp_{rec})$ $((\{vp_{rec}\} \cap \{pri_{pre_{loc}}\}) = \emptyset)$ $\wedge \neg negotiate)$ $\to refuse$ $((\{vp_{rec}\} \cap \{pri_{pre_{loc}}\}) = \emptyset)$ $\wedge negotiate)$ $\to propose(x)$ $x \in \{pri_{pre_{loc}}\}$ $((\{vp_{rec}\} \cap \{pri_{pre_{loc}}\}) \neq \emptyset)$ $\wedge(\{vp_{rec}\} \not\subseteq \{pri_{pre_{loc}}\})$ $\wedge negotiate)$ $\to propose(x)$ $x \in (\{vp_{rec}\} \cap \{pri_{pre_{loc}}\})$

References

1. Adrian Lara, A.K., Ramamurthy, B.: Network innovation using openflow: a survey. IEEE Commun. Surv. Tutorials **16**(1), 493–511 (2014)
2. Batista, B., Fernandez, M.: Ponderflow: a policy specification language for openflow networks. In: The Thirteenth International Conference on Networks, pp. 204–209 (2014)
3. Ben-Itzhak, Y., Barabash, K., Cohen, R., Levin, A., Raichstein, E.: EnforSDN: network policies enforcement with SDN. In: 2015 IFIP/IEEE International Symposium on Integrated Network Management (IM), pp. 80–88. IEEE (2015)

4. Bernsmed, K., Jaatun, M.G., Undheim, A.: Security in service level agreements for cloud computing. In: CLOSER, pp. 636–642 (2011)
5. Bharadwaj, V.G., Baras, J.S.: Towards automated negotiation of access control policies. In: Policy, pp. 111–119 (2003)
6. Bijon, K., Krishnan, R., Sandhu, R.: Virtual resource orchestration constraints in cloud infrastructure as a service. In: Proceedings of the 5th ACM Conference on Data and Application Security and Privacy, pp. 183–194. ACM (2015)
7. Bistarelli, S., Montanari, U., Rossi, F., Schiex, T., Verfaillie, G., Fargier, H.: Semiring-based CSPs and valued CSPs: frameworks, properties, and comparison. Constraints 4(3), 199–240 (1999)
8. Chernov, D.V.: Attribute based access control models. Prikladnaya Diskretnaya Matematika, Suppl., 79–82 (2012)
9. Damianou, N., Dulay, N., Lupu, E., Sloman, M.: The ponder policy specification language. In: Sloman, M., Lupu, E.C., Lobo, J. (eds.) POLICY 2001. LNCS, vol. 1995, pp. 18–38. Springer, Heidelberg (2001). doi:10.1007/3-540-44569-2_2
10. Fong, P.W.: Relationship-based access control: protection model and policy language. In: Proceedings of the First ACM Conference on Data and Application Security and Privacy, pp. 191–202. ACM (2011)
11. Gligor, V.D.: Negotiation of access control policies. In: Christianson, B., Malcolm, J.A., Crispo, B., Roe, M. (eds.) Security Protocols 2001. LNCS, vol. 2467, pp. 202–212. Springer, Heidelberg (2002). doi:10.1007/3-540-45807-7_29
12. Hegr, T., Bohac, L., Uhlir, V., Chlumsky, P.: Openflow deployment and concept analysis. Adv. Electr. Electron. Eng. 11(5), 327 (2013)
13. Hu, H., Han, W., Ahn, G.J., Zhao, Z.: Flowguard: building robust firewalls for software-defined networks. In: Proceedings of the Third Workshop on Hot Topics in Software Defined Networking, pp. 97–102. ACM (2014)
14. Huang, S.S., Green, T.J., Loo, B.T.: Datalog and emerging applications: an interactive tutorial. In: Proceedings of the 2011 ACM SIGMOD International Conference on Management of Data, pp. 1213–1216. ACM (2011)
15. Kalam, A.A.E., Baida, R., Balbiani, P., Benferhat, S., Cuppens, F., Deswarte, Y., Miege, A., Saurel, C., Trouessin, G.: Proceedings of the IEEE 4th International Workshop on Organization based access control. In: Policies for Distributed Systems and Networks, POLICY 2003, pp. 120–131. IEEE (2003)
16. Kaur, K., Kaur, S., Gupta, V.: Software defined networking based routing firewall. In: 2016 International Conference on Computational Techniques in Information and Communication Technologies (ICCTICT), pp. 267–269. IEEE (2016)
17. Lara, A., Ramamurthy, B.: Opensec: policy-based security using software-defined networking. IEEE Trans. Netw. Serv. Manag. 13(1), 30–42 (2016)
18. Leite, A.F., Alves, V., Rodrigues, G.N., Tadonki, C., Eisenbeis, C., de Melo, A.: Automating resource selection and configuration in inter-clouds through a software product line method. In: 2015 IEEE 8th International Conference on Cloud Computing, pp. 726–733. IEEE (2015)
19. Li, Y., Cuppens-Boulahia, N., Crom, J.M., Cuppens, F., Frey, V.: Reaching agreement in security policy negotiation. In: 2014 IEEE 13th International Conference on Trust, Security and Privacy in Computing and Communications, pp. 98–105. IEEE (2014)
20. Li, Y., Cuppens-Boulahia, N., Crom, J.-M., Cuppens, F., Frey, V.: Expression and enforcement of security policy for virtual resource allocation in IaaS cloud. In: Hoepman, J.-H., Katzenbeisser, S. (eds.) SEC 2016. IFIP AICT, vol. 471, pp. 105–118. Springer, Cham (2016). doi:10.1007/978-3-319-33630-5_8

21. Li, Y., Cuppens-Boulahia, N., Crom, J.-M., Cuppens, F., Frey, V., Ji, X.: Similarity measure for security policies in service provider selection. In: Jajodia, S., Mazumdar, C. (eds.) ICISS 2015. LNCS, vol. 9478, pp. 227–242. Springer, Cham (2015). doi:10.1007/978-3-319-26961-0_14

22. Mehregan, P., Fong, P.W.: Policy negotiation for co-owned resources in relationship-based access control. In: Proceedings of the 21st ACM on Symposium on Access Control Models and Technologies, pp. 125–136. ACM (2016)

23. Nathani, A., Chaudhary, S., Somani, G.: Policy based resource allocation in iaas cloud. Future Gener. Comput. Syst. 28(1), 94–103 (2012)

24. NTT: Component-based software defined networking framework (2017). www.osrg.github.io/ryu/

25. ONF: Openflow switch specification, December 2014

26. Pfaff, B., Pettit, J., Amidon, K., Casado, M., Koponen, T., Shenker, S.: Extending networking into the virtualization layer. In: Hotnets (2009)

27. Pfaff, B., Pettit, J., Koponen, T., Jackson, E.J., Zhou, A., Rajahalme, J., Gross, J., Wang, A., Stringer, J., Shelar, P., et al.: The design and implementation of open vSwitch. In: NSDI, pp. 117–130 (2015)

28. Rissanen, E.: extensible access control markup language (XACML) version 3.0 (committe specification 01). Technical report, OASIS (2010). http://docs.oasisopen.org/xacml/3.0/xacml-3.0-core-spec-cd-03-en.pdf

29. Sadki, S., El Bakkali, H.: An approach for privacy policies negotiation in mobile health-cloud environments. In: 2015 International Conference on Cloud Technologies and Applications (CloudTech), pp. 1–6. IEEE (2015)

30. Sandhu, R.S., Coynek, E.J., Feinsteink, H.L., Youmank, C.E.: Role-based access control models yz. IEEE Comput. 29(2), 38–47 (1996)

31. Satasiya, D., et al.: Analysis of software defined network firewall (sdf). In: International Conference on Wireless Communications, Signal Processing and Networking (WiSPNET), pp. 228–231. IEEE (2016)

32. Shin, S., Gu, G.: Cloudwatcher: network security monitoring using openflow in dynamic cloud networks (or: How to provide security monitoring as a service in clouds?). In: 2012 20th IEEE International Conference on Network Protocols (ICNP), pp. 1–6. IEEE (2012)

33. Shin, S., Porras, P.A., Yegneswaran, V., Fong, M.W., Gu, G., Tyson, M.: Fresco: modular composable security services for software-defined networks. In: NDSS (2013)

34. Tang, Y., Cheng, G., Xu, Z., Chen, F., Elmansor, K., Wu, Y.: Automatic belief network modeling via policy inference for SDN fault localization. J. Internet Serv. Appl. 7(1), 1 (2016)

35. Xue, W., Huai, J., Liu, Y.: Access control policy negotiation for remote hot-deployed grid services. In: First International Conference on e-Science and Grid Computing (e-Science 2005), 9 p. IEEE (2005)

36. Zerkane, S., Espes, D., Le Parc, P., Cuppens, F.: A proactive stateful firewall for software defined networking. In: Risks and Security of Internet and Systems - 11th International Conference, CRiSIS 2016, Roscoff, France, 5–7 September 2016, Revised Selected Papers, pp. 123–138 (2016)

37. Zerkane, S., Espes, D., Le Parc, P., Cuppens, F.: Software defined networking reactive stateful firewall. In: Hoepman, J.-H., Katzenbeisser, S. (eds.) SEC 2016. IFIP AICT, vol. 471, pp. 119–132. Springer, Cham (2016). doi:10.1007/978-3-319-33630-5_9

Budget-Constrained Result Integrity Verification of Outsourced Data Mining Computations

Bo Zhang[1], Boxiang Dong[2], and Wendy Wang[1(✉)]

[1] Stevens Institute of Technology, Hoboken, NJ, USA
{bzhang41,Hui.Wang}@stevens.edu
[2] Montclair State University, Montclair, NJ, USA
dongb@montclair.edu

Abstract. When outsourcing data mining needs to an untrusted service provider in the Data-Mining-as-a-Service (DMaS) paradigm, it is important to verify whether the service provider (server) returns correct mining results (in the format of data mining objects). We consider the setting in which each data mining object is associated with a weight for its importance. Given a client who is equipped with limited verification budget, the server selects a subset of mining results whose total verification cost does not exceed the given budget, while the total weight of the selected results is maximized. This maps to the well-known *budgeted maximum coverage* (BMC) problem, which is NP-hard. Therefore, the server may execute a heuristic algorithm to select a subset of mining results for verification. The server has financial incentives to cheat on the heuristic output, so that the client has to pay more for verification of the mining results that are less important. Our aim is to verify that the mining results selected by the server indeed satisfy the budgeted maximization requirement. It is challenging to verify the result integrity of the heuristic algorithms as the results are non-deterministic. We design a probabilistic verification method by including *negative candidates* (NCs) that are guaranteed to be excluded from the budgeted maximization result of the ratio-based BMC solutions. We perform extensive experiments on real-world datasets, and show that the NC-based verification approach can achieve high guarantee with small overhead.

Keywords: Data-Mining-as-a-Service (DMaS) · Cloud computing · Result integrity · Budgeted maximization

1 Introduction

Due to the fast increase of data volume, many organizations (clients) with limited computational resources and/or data mining expertise have outsourced their data and data mining needs to a third-party service provider (server) that is computationally powerful. This emerges the Data-Mining-as-a-Service (DMaS) paradigm [8,21]. Although DMaS offers a cost-effective solution for the client,

© IFIP International Federation for Information Processing 2017
Published by Springer International Publishing AG 2017. All Rights Reserved
G. Livraga and S. Zhu (Eds.): DBSec 2017, LNCS 10359, pp. 311–324, 2017.
DOI: 10.1007/978-3-319-61176-1_17

it raises several security issues. One of the issues is *result integrity verification*, i.e., how to verify that the untrusted server returns the correct mining results [9,18]. There are many incentives for the server to cheat on the mining results. As data mining is critical for decision making and knowledge discovery, it is essential for the client to verify the result integrity of the outsourced mining computations.

The key idea of the existing result integrity verification solutions (e.g., [7]) is to associate the mining results with a *proof*, which enables the client to verify the result correctness with a deterministic guarantee. It has been shown that the construction of integrity proofs can be very costly [7]. We assume that the client has to pay for the cost of proof generation. If the client has a tight budget for proof construction, she only can afford the proof construction of a subset of mining results. We consider that the mining results that involve different data items bring different benefits to the client (e.g., the shopping patterns of luxury goods are more valuable than that of bread and milk). Therefore, it is desirable that the mining results that bring the maximum benefits to the client are picked for the proof construction. This problem can be mapped to the well-known budgeted maximum coverage problem (BMC) [13].

Formally, given a set of objects $O = \{o_1, \ldots, o_n\}$, each of which associated with a weight (value) w_i and a cost c_i, the *budgeted maximization problem* (BMC) is to find a subset of objects $O' \subseteq O$ whose total cost does not exceed the given budget value B, while the total weight is maximized. A variant of BMC problem considers the case when the cost of overlapping objects is amortized, i.e., $c(\{o_i, o_j\}) < c_i + c_j$. This motivates the *budgeted maximization with overlapping costs (BMOC)* problem [6]. It has been proven that both BMC and BMOC problems are NP-hard [13].

Given the complexity of BMC/BMOC problems, the server may execute a heuristic BMC/BMOC algorithm (e.g., [6,19]) to pick a subset of mining results for proof construction. The server is incentivized to cheat on proof selection by picking the results with cheaper computations (e.g., by randomly picking a subset of mining results), and claims that those picked results are returned by the BMC/BMOC heuristic algorithm. To catch such cheating of proof construction, it is important to design efficient methods for verification of the result integrity of heuristic BMC/BMOC computations.

In general, it is difficult to verify the result correctness of heuristic algorithms, as the output of these algorithms has much uncertainty. Prior work (e.g., metamorphic testing [4,5]) mainly use software testing techniques that execute the algorithm multiple times with different inputs, and verify if multiple outputs satisfy some required conditions. These techniques cannot be applied to the DMaS paradigm, as the client may not be able to afford to execute the (expensive) data mining computations multiple times. Existing works [7,21] on integrity verification of outsourced data mining computations mainly focus on the authentication of *soundness* and *completeness* of the data mining results, but not the result correctness of budgeted maximization algorithm.

We aim to design efficient verification methods that can catch any incorrect result of heuristic BMC/BMOC algorithms with high *probabilistic* integrity guarantee and small computational overhead. To our best knowledge, this is the first work that focuses on verifying the result integrity of the budgeted maximization algorithm. We focus on the type of *ratio-based* BMC/BMOC heuristic algorithms that rely on the weight/cost ratio to pick the objects, and make the following main contributions. First, we design an efficient result correctness verification method for the ratio-based BMC/BMOC problem. Following the intuition of [7,14–16] to construct evidence patterns for the purpose of verification, our key idea is to construct a set of *negative candidates (NCs)* that will *not* be picked by any ratio-based heuristic BMC/BMOC algorithm. A nice property is that NCs do not impact the original output of BMC/BMOC algorithm (i.e., all original mining results selected by the BMC/BMOC algorithm are still picked under the presence of NCs). The verification mainly checks if the server picks any NC for proof construction. If there is, then the server's cheating on proof construction is caught with 100% certainty. Otherwise, the mining results that the server picked for proof construction are trusted with a probabilistic guarantee. We formally analyze the probabilistic guarantee of the NC-based method. Second, the basic NC-based approach is weak against the attacks that utilize the knowledge of how NCs are constructed. Therefore, we improve the basic NC-based approach to be robust against these attacks. Last but not least, we take *frequent itemset mining* [11] as a case study, and launch a rich set of experiments on three real-world datasets to evaluate the efficiency and robustness of the NC-based verification. The experiment results demonstrate that the NC-based verification method can achieve high guarantee with small overhead (e.g.,it takes at most 0.001 s for proof verification of frequent itemset results from 1 million transactions).

The remaining of the paper is organized as following. Section 2 introduces the preliminaries. Section 3 presents the NC-based verification method. Section 4 reports the experiment results. Section 5 discusses the related work. Section 6 concludes the paper.

2   Preliminaries

2.1   Budgeted Maximization Coverage (BMC) Problem

Given a set of objects $O = \{o_1, \ldots, o_n\}$, each associated with a cost c_i and a weight (i.e., value) w_i, the budgeted maximization coverage (BMC) problem is to find the set $O' \subseteq O$ s.t. (1) $\sum_{o_i \in O'} c_i \leq B$, for any specified budget value B, and (2) $\sum_{o_i \in O'} w_i$ is maximized. When the overlapping objects share the cost, i.e., $c(o_i, o_j) < c_i + c_j$, the problem becomes the budgeted maximization coverage with overlapping cost (BMOC) problem. Both BMC and BMOC problems are NP-hard [6,13]. Various heuristic algorithms have been proposed [6,19]. Most of the heuristic BMC/BMOC algorithms [3,6,13,19] follow the same principle: pick the objects repeatedly by their $\frac{weight}{cost}$ ratio (denoted as *WC-ratio*) in descending order, until the total cost exceeds the given budget. The major difference between these algorithms is how the cost is computed for a given set of objects; BMC algorithms simply sum up the cost of these objects, while BMOC algorithms compute the total cost of these objects *with sharing*. We call these *ratio-based* heuristic algorithms the GREEDY algorithms. We assume that the server uses a ratio-based greedy algorithm to pick the subset of mining results for proof construction.

2.2 Budget-Constrained Verification

In this paper, we consider the scenario where the data owner (client) outsources her data D as well as her mining needs to a third-party service provider (server). The server returns the mining results R of D to the client. We represent R as a set of *mining objects* $\{o_1, \ldots, o_n\}$, where each mining object o_i is a pattern that the outsourced data mining computations aim to find. Examples of the mining objects include outliers, clusters, and association rules. Different mining objects can be either *non-overlapping* (e.g., outliers) or *overlapping* (e.g., association rules that share common items).

Since the server is potentially untrusted, it has to provide an *integrity proof* of its mining results, where the proof can be used to verify the correctness of the mining results [7]. In general, each mining object is associated with a single proof [2,17]. We use c_i to denote the cost of constructing the proof of the mining object o_i. According to [7], the overlapping mining objects share the proofs (and its construction cost) of the common data items. The total proof cost of R is $c_R = \sum_{\forall o_i \in (\cup_{\forall o_j \in R} o_j)} c_i$.

We assume that the client has to pay for the cost of proof construction. The proof construction cost is decided by the number of mining objects that are associated with the proofs. Intuitively, the more the client pays for the proof construction, the more mining objects that she can verify the correctness. We assume that different mining objects bring different benefits to the client. We use w_i to denote the *weight* of the mining object o_i. We follow the same assumption of the BMC/BMOC problem that the weights of different mining objects never share, regardless of the overlaps between the mining objects. Thus given a set of mining objects R, the total weight of R is computed as: $w_R = \sum_{\forall o_i \in R} w_i$.

In this paper, we consider the client who has the limited budget for proof construction. Given the budget B and the mining results R from the outsourced data D, the client asks the server to pick a subset of mining results $R' \subseteq R$ for proof construction where $c_{R'} \leq B$, while $w_{R'}$ is maximized. Apparently, this problem can be mapped to either BMC (for share-free proof construction scenario) or BMOC (for shared proof construction scenario). Given the complexity of BMC/BMOC algorithms, the server runs the GREEDY algorithm to pick a subset of mining objects for proof construction.

2.3 Verification Goal

Due to various reasons, the server may cheat on proof construction. For instance, in order to save the computational cost, it may randomly select a set of data mining objects, instead of executing the GREEDY algorithm. Therefore, after the client receives the mining objects R from the server, in which $R' \subseteq R$ is associated with the result integrity proof, she would like to verify whether R' indeed satisfies *budgeted maximization*. Formally,

Definition 1 (Verification Goal). *Given a set of mining objects $R = \{o_1, \ldots, o_n\}$, let R' denote the mining objects picked by the server for proof construction. A verification algorithm should verify whether R' satisfies the following two budgeted maximization requirements: (1) The total cost of R' does not exceed the budget B; and (2) The total weight of R' is maximized.*

The verification of Goal 1 is trivial, as the client can simply calculate the total cost of R' and compare it with B. The verification of Goal 2 is extremely challenging due to two reasons: (1) due to the NP-hardness complexity of the BMC/BMOC problem, it is impossible that the client re-computes the budgeted maximization solutions; and (2) as the server uses a heuristic method to pick R', it is expected that R' may not be the optimal solution. A naive solution is to pick a set of objects $R'' \subseteq R$ (either randomly or deliberately), and compare the total weight of R'' with that of R'. However, this naive solution cannot deliver any verification guarantee, even if the total weight of R' is smaller than that of R'', as the output of the heuristic algorithms is not expected to be optimal. Our goal is to design efficient verification method that can deliver *probabilistic* guarantee of the results of heuristic BMC/BMOC algorithms. *We must note that the verification of correctness and completeness of the mining results is not the focus of our paper.*

3 NC-based Verification Approach

3.1 Basic Approach

The key idea of our verification method is to use the *negative candidates* (NCs) of budgeted maximization. The NCs are guaranteed to be *excluded* from the output of any GREEDY algorithm. Therefore, if the server's results include any NC, the client is 100% sure that the server fails the verification of budgeted maximization. Otherwise, the client has a probabilistic guarantee of budgeted maximization.

Intuitively, if the WC-ratio of any NC is smaller than the WC-ratio of any real object in D, the GREEDY algorithm never picks any NC if it is executed faithfully. Besides this, we also require that the presence of NCs should not impact the original results (i.e., all original mining objects selected by GREEDY should still be selected under the presence of NCs). Following this, we define the two requirements of NCs:

- Requirement 1: There is no overlap between any NC and any real object or NC;
- Requirement 2: The WC-ratio of any NC o_i is lower than the ratio of any real object $o_j \in R$, i.e., $\frac{w_i}{c_i} < \frac{w_j}{c_j}$, where R is the mining results of the original dataset D.

Based on this, we have the following theorem:

Theorem 1. *Given a set of objects R and the GREEDY algorithm F, for any NC $o_i \in R$ that satisfies the two requirements above, it is guaranteed that $o_i \notin F(R)$.*

The correctness of Theorem 1 is straightforward. Requirement (1) ensures that NCs do not share cost with any other object, thus WC-ratio is its exact ratio used in the GREEDY algorithm. It also ensures that the presence of NCs do not change the WC-ratio of any other objects, and thus do not change the output of the GREEDY algorithm. Requirement (2) ensures that NCs always have the smallest WC-ratio, and thus are never picked by the GREEDY algorithm.

We formally define (ϵ, θ)-budgeted maximization as our verification goal.

Definition 2 ((ϵ, θ)-Budgeted Maximization). *Given a set of objects R, let $R' \subseteq R$ be the set of objects picked by any GREEDY algorithm. Assume the server cheats on θ percent of R'. We say a verification method $\mathcal{M}$ can verify (ϵ, θ)-budgeted maximization if the probability p to catch the cheating on R' satisfies that $p \geq \epsilon$.*

Suppose we insert K NCs into the original n objects. Suppose the GREEDY algorithm picks $m < n$ objects. Also suppose that the attacker picks $\ell \geq 1$ budget-maximized objects from the m objects (i.e., $\ell = m\theta$, where θ is the parameter for (ϵ, θ)-budgeted maximization) output by the GREEDY algorithm, and replaces them with other objects with lower weight/cost ratios. Note that if such replacement involves any NC, then the verification method can catch the cheating. Now let us compute the probability that the attacker can escape by picking no NCs when replacing budget maximized objects. If an attacker replaces an object o in the original budgeted-maximization result with another object o', the probability that o' is not a NC is $\frac{n-m}{n-m+K}$. Thus, with probability $\frac{n-m}{n-m+K}$, the attacker's wrong result is not caught. Now, given $\ell \geq 1$ budgeted-maximized objects that are not returned by the server, the probability p that the server can be caught (i.e., it picks at least one NC):

$$p = 1 - \prod_{i=0}^{\ell-1} \frac{n-m-i}{n-m+K-\ell}. \tag{1}$$

Since $\frac{n-i}{n+K-i} \geq \frac{n-i-1}{n+K-i-1}$, it follows that:

$$1 - \left(\frac{n-m}{n-m+K}\right)^\ell \leq p \leq 1 - \left(\frac{n-m-\ell}{n-m+K-\ell}\right)^\ell. \tag{2}$$

Based on this reasoning, we have the following theorem:

Theorem 2. *Given n original objects, among which $m < n$ itemsets are picked by GREEDY, the number K of NCs to verify (ϵ, θ)-budgeted maximization satisfies that: $K \geq \left(\frac{1}{\sqrt[m\theta]{1-\epsilon}} - 1\right)(n-m)$.*

We calculate the value of K with regard to various θ and ϵ values, and observe that our NC-based verification method does not require large number NCs when the budget is large, even though θ is a small fraction. For example, when $\theta = 0.1$ and $m = 400$, our method only requires 16 and 25 NCs in order to achieve p of at least 95% and 99%, respectively. A detailed analysis of the relationship between K and the verification parameters (i.e., θ and ϵ) can be found in the full paper [22].

A challenge to calculate the exact K value is that the client may not possess the knowledge of the real m value (i.e., the number of original objects picked by GREEDY). Next, we discuss how to estimate m. Intuitively, for any $m' \leq m$, $(\frac{1}{m'\sqrt[\theta]{1-\epsilon}} - 1)n) \geq (\frac{1}{m\sqrt[\theta]{1-\epsilon}} - 1)n)$. Therefore, we can simply calculate the lowerbound $\hat{m}$ of m, and estimate $K = (\frac{1}{\hat{m}\sqrt[\theta]{1-\epsilon}} - 1)\hat{n}$. To calculate $\hat{m}$, we can calculate the upperbound cost c_{max} of any object o_j in O, then we compute $\hat{m} = \lceil \frac{B}{c_{max}} \rceil$.

3.2   A More Robust Approach

If the server possesses the distribution information of the original dataset D, as well as the details of the NC-based verification, it may try to escape from the verification by distinguishing NCs from the original objects based on their characteristics. In particular, there are two possible attacks that can be launched: (1) the *overlap-based* attack: the server may identify those objects that are disjoint with other objects as NCs, since NCs do not overlap with any other object; and (2) the *ratio-based* attack: the server can sort all objects (including real ones and NCs) by their WC-ratio, and pick K' objects of the *lowest* ratio as NCs. Next, we discuss the strategy to mitigate these attacks.

In order to defend against the overlap-based attack, we introduce the overlap between NCs. In particular, we require the overlap between NCs should be similar to the overlap between the original objects. We define the *overlapping degree* of the object o_i as $od_i = \frac{d_i}{n-1}$, where d_i is the number of objects in D that o_i overlaps with. We assume the overlapping degree follows a normal distribution, i.e., $od \sim \mathcal{N}(u_o, \sigma_o^2)$. We estimate $\hat{u}_o$ and $\hat{\sigma}_o$ from the overlapping degrees of original objects. We require that the overlapping degree of NCs should follow the same distribution $\mathcal{N}(\hat{u}_o, \hat{\sigma}_o^2)$. To make sure that ratio of NCs remains smaller than that of any original object, we introduce the overlap by increasing the cost of NCs, while keeping the weight unchanged. Note that we only introduce overlapping between NCs; NCs and real mining objects do not overlap. Thus NCs still satisfy Requirement 1 of NCs.

Regarding to the ratio-based attack, identifying the NCs of the lowest WC-ratio leads to insufficient number of NCs to satisfy (ϵ, θ)-budgeted maximization. In particular, suppose that the server uniformly picks $K' \in [0, K]$ by random guess, and takes K' objects with the lowest WC-ratio as the NCs. Then $K - K'$ NCs remain to be unidentified by the server. Next, we analyze the the probabilistic integrity guarantee that can be provided if the server launches the ratio-based attack. We have the following theorem:

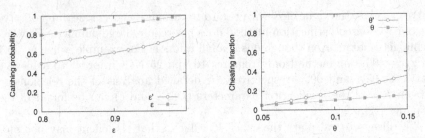

(a) Comparison between ϵ and ϵ' ($\theta = 0.1$) (b) Comparison between θ and θ' ($\epsilon = 0.99$)

Fig. 1. Evaluation of ϵ' and θ'

Theorem 3. *Given a set of K NCs that satisfies (ϵ, θ)-budgeted maximization, our NC-based approach is able to verify (ϵ', θ')-budgeted maximization, where $\epsilon' = 1 - \frac{\epsilon(n-m)-K(1-\epsilon)}{K(m\theta-1)}$, and $\theta' = \frac{(n-m)\ln(1-\epsilon)}{Km}(\ln\frac{K}{n-m} + \frac{K}{n-m} + \frac{K^2}{8(n-m)^2})$, where n is the number of original objects, and m is the number of objects picked by GREEDY.*

Due to the space limit, we omit the proof of Theorem 3.

We plot ϵ and ϵ' (defined in Theorem 3) in Fig. 1 (a) (θ and θ' in Fig. 1 (b) respectively). The results show that the ratio-based attack only degrades (ϵ, θ)-budgeted maximization slightly. For example, consider $n = 600$, and $m = 400$. When $\epsilon = 0.99$, and $\theta = 0.1$, it requires $K = 25$ NCs to verify $(0.99, 0.1)$-budgeted maximization (Theorem 2). If the server exploits the ratio-based attack, the NC-based approach needs 495 NCs to provide the same security guarantee.

4   Experiments

4.1   Setup

Hardware. We execute the experiments on a testbed with Intel i5 CPU and 8GB RAM, running Mac OS X 10. We implement all the algorithms in C++.

Datasets. We take *frequent itemset mining* [11] as the mining task. Given a dataset D and a threshold value min_{sup}, it aims at finding all the itemsets whose number of occurrences is at least min_{sup}. We use three datasets, namely *Retail*, *T10* and *Kosarak* datasets, available from the *Frequent Itemset Mining Data Repository*[1]. We use various support threshold values min_{sup} for mining. The data description and the mining setting can be found in our full paper [22].

Budget setting. We vary the budget to observe its impact on the performance of the NC-based approach. We define *budget ratio* $br = \frac{B}{Tb}$, where B is the given

---

[1] http://fimi.ua.ac.be/data/.

budget, and Tb is the total proof cost for all the frequent itemsets. Intuitively, a higher budget ratio leads to the higher budget for proof construction.

Cost and weight model. For each item x, its cost is computed as $c_x = \frac{supp_D(\{x\})}{max_{\forall x \in \mathcal{X}} supp_D(\{x\})}$, where $supp_D(\{x\})$ denotes the frequency of $\{x\}$ in D. We use four different strategies to assign the item weights. We consider that all items have the same weight. Given any itemset $X = \{x_1, \ldots, x_t\}$, the cost and weight of X is the sum of the individual items. In other words, $c_X = \sum c_{x_i}$, while $w_X = \sum w_{x_i}$.

4.2 Robustness of Probabilistic Verification

We measure the robustness of our NC-based approach. In particular, we simulate the cheating behavior by the attacker on budgeted maximization, and measure the probability p that our approach can catch the attacker. We compare p with the pre-defined desired catching probability ϵ. The experiment results on the datasets demonstrate that the detection probability is always larger than ϵ, which verifies the robustness of our verification approach. The detailed result can be found in our full paper [22].

4.3 Verification Performance

Verification preparation time. We measure the time of creating artificial transactions that produce NCs for verification of budgeted maximization. Figure 2 shows the NC creation time for the datasets. In general, the NC creation procedure is very fast. In all the experiments, it takes at most 0.9 s to generate the NCs, even for small $\theta = 0.02$ and large $\epsilon = 0.95$. Furthermore, we observe that larger ϵ and smaller θ lead to longer time for generating NCs. This is because the number of NCs is positively correlated to ϵ but negatively correlated to θ. In other words, we need more NCs to verify smaller errors of the budgeted maximization results with higher guarantee.

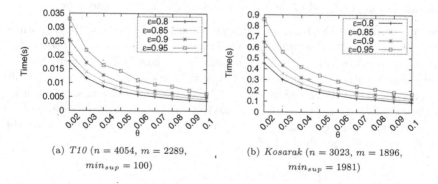

(a) *T10* ($n = 4054$, $m = 2289$,   (b) *Kosarak* ($n = 3023$, $m = 1896$,
$min_{sup} = 100$) $min_{sup} = 1981$)

Fig. 2. NC construction time

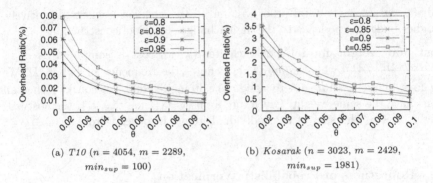

(a) $T10$ ($n = 4054$, $m = 2289$, $min_{sup} = 100$)

(b) $Kosarak$ ($n = 3023$, $m = 2429$, $min_{sup} = 1981$)

Fig. 3. Mining overhead (various θ setting for (ϵ, θ)-budget maximization)

Verification time. We measure the verification time on the datasets for various settings of θ and ϵ. We observe that the verification time is negligible; it never exceeds 0.039 s. The minimal verification time is 0.001 s, even for the $Kosarak$ dataset, which consists of nearly 1 m transactions. We omit the results due to the space limits.

Mining overhead by verification. We measure the mining overhead by adding artificial transactions for NC creation. We measure the *overhead ratio* as $\frac{(T_{D'} - T_D)}{T_D}$, where T_D and $T_{D'}$ are the mining time of the original dataset D and D' after adding artificial transactions. Intuitively the lower overhead ratio is, the smaller additional mining cost incurred by our NC-based verification.

Various θ. Figure 3 reports the overhead ratio for the datasets when changing θ setting for (ϵ, θ)-budgeted maximization. First, similar to the observation of NC creation time, the mining overhead ratio increases when θ becomes smaller and ϵ rises, as it requires more NCs, which results in larger mining overhead. Second, for all datasets, the overhead ratio is small (no more than 4%). The mining overhead is especially small for the $T10$ dataset. It never exceeds 0.08% even though $\epsilon = 0.95$. The reason is that the mining of the original $T10$ dataset consumes a long time, which leads to the small mining overhead ratio.

Various budgets. In Fig. 4, we display the effect of the budgets (in the format of the budget ratio) on the mining overhead. On both datasets, we observe that the mining overhead drops with the increase of budgets. Given a large budget, the GREEDY algorithm picks more frequent itemsets for proof construction. This results in the decreased number of NCs. Therefore, the mining overhead on the artificial transactions reduces, as the number of artificial transactions drops. In particular, the mining overhead can be very small, especially on the $T10$ dataset (no larger than 0.25%).

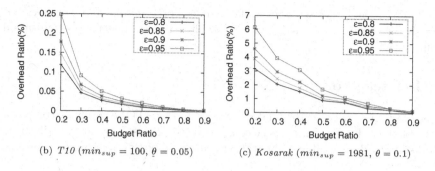

(b) *T10* ($min_{sup} = 100$, $\theta = 0.05$)   (c) *Kosarak* ($min_{sup} = 1981$, $\theta = 0.1$)

Fig. 4. Mining overhead (various budgets)

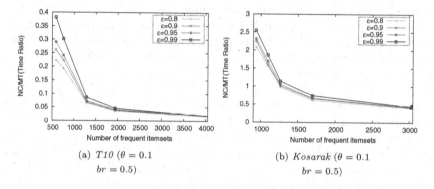

(a) *T10* ($\theta = 0.1$ (b) *Kosarak* ($\theta = 0.1$
     $br = 0.5$) $br = 0.5$)

Fig. 5. NC-based approach vs. metamorphic testing (MT)

4.4   Comparison with Metamorphic Testing (MT)

We compare our NC-based approach with the metamorphic testing (MT) approach [4,5]. For the implementation of the MT approach, we treat the GREEDY algorithm as a black box M, and run it on the discovered frequent itemsets $\mathcal{I}$ *twice*. The output of the first execution $M(\mathcal{I})$ is used as the input of the second execution of M. Apparently, if the server is honest, the output of the second execution of M should be the same as the output of the first execution, i.e., $M(\mathcal{I}) = M(M(\mathcal{I}))$. The overhead of MT is measured as the time of the second execution of the GREEDY algorithm. We measure the total time of our NC-based approach for verification, which includes the time of both verification preparation and verification. In Fig. 5, we show the ratio of the time performance of our NC-based approach to that of MT approach. Overall, our NC-based approach shows great advantage over MT approach, especially when the number of frequent itemsets is large. For example, on the *T10* dataset, our verification approach is at least 10 times faster than MT when the number of frequent itemset is larger than 3,000. We also observe that when the number of frequent itemsets increases, the ratio decreases (i.e., the NC-based approach is much faster than MT). This is because MT takes more time to pick the budget-maximized objects from the larger set of candidates. It is worth noting that MT does not provide any formal guarantee that the result satisfies budgeted maximization requirement.

5   Related Work

In this section, we discuss the related work, including: (1) the verifiable computation techniques for general-purpose computations; (2) result integrity verification of outsourced data mining computations; and (3) software testing for heuristic algorithms.

Gennaro et al. [9] define the notation of *verifiable computation* (VC) that allows a computationally limited client to be able to verify the result correctness of some expensive general-purpose computations that are outsourced to a computationally powerful server. Babai and Goldwasser et al. [1,10] design the probabilistic checkable proofs. However, the incurred proofs might be too long for the verifier to process. This is not ideal for the DMaS paradigm where commonly the client outsources the data mining computations with a single input dataset.

Verification of result integrity of outsourced data mining computations have caught much attention recently. The existing methods have covered various types of data mining computations, including clustering [15], outlier mining [14], frequent itemset mining [7,8], Bayesian networks [16], and collaborative filtering [20]. All these works focus on correctness and completeness verification of mining results, while ours aims at the problem of verification of budgeted maximization.

In software testing, it has been shown that the verification of heuristic algorithms is very challenging, as the heuristic methods may not give exact solutions for the computed problems. This makes it difficult to verify outputs of the corresponding software by a *test oracle* - a mechanism to construct the test cases. This is known as the test oracle problem [12] in software testing. A popular technique that is used to test programs without oracles is metamorphic testing (MT) [4,5], which generates test cases by making reference to metamorphic relations, that is, relations among multiple executions of the target program. Though effective, MT is not suitable to the budget-constrained DMaS paradigm, as it involves multiple executions of the same function.

6   Conclusion

In this paper, we investigate the problem of result verification of budgeted maximization problem in the setting of DMaS paradigm. We present our probabilistic verification method that authenticates whether the server's results indeed reach budgeted maximization. We analyze the formal guarantee of the budgeted maximization of our method. Our experiments demonstrate the efficiency and effectiveness of our methods. For the future work, one interesting direction is to investigate the deterministic verification methods that can authenticate budgeted maximization with 100% certainty.

References

1. Babai, L.: Trading group theory for randomness. In: Symposium on Theory of Computing (1985)
2. Benabbas, S., Gennaro, R., Vahlis, Y.: Verifiable delegation of computation over large datasets. In: Rogaway, P. (ed.) CRYPTO 2011. LNCS, vol. 6841, pp. 111–131. Springer, Heidelberg (2011). doi:10.1007/978-3-642-22792-9_7
3. Bleiholder, J., Khuller, S., Naumann, F., Raschid, L., Wu, Y.: Query planning in the presence of overlapping sources. In: Ioannidis, Y., Scholl, M.H., Schmidt, J.W., Matthes, F., Hatzopoulos, M., Boehm, K., Kemper, A., Grust, T., Boehm, C. (eds.) EDBT 2006. LNCS, vol. 3896, pp. 811–828. Springer, Heidelberg (2006). doi:10.1007/11687238_48
4. Chen, T.Y., et al.: Metamorphic testing: a new approach for generating next test cases. Technical report, Hong Kong University of Science and Technology (1998)
5. Chen, T.Y., et al.: Fault-based testing in the absence of an oracle. In: International Conference on Computer Software and Applications (2001)
6. Curtis, D.E., et al.: Budgeted maximum coverage with overlapping costs: monitoring the emerging infections network. In: Algorithm Engineering & Expermiments (2010)
7. Dong, B., et al.: Integrity verification of outsourced frequent itemset mining with deterministic guarantee. In: ICDM (2013)
8. Dong, B., Liu, R., Wang, H.W.: Result integrity verification of outsourced frequent itemset mining. In: Wang, L., Shafiq, B. (eds.) DBSec 2013. LNCS, vol. 7964, pp. 258–265. Springer, Heidelberg (2013). doi:10.1007/978-3-642-39256-6_17
9. Gennaro, R., Gentry, C., Parno, B.: Non-interactive verifiable computing: outsourcing computation to untrusted workers. In: Rabin, T. (ed.) CRYPTO 2010. LNCS, vol. 6223, pp. 465–482. Springer, Heidelberg (2010). doi:10.1007/978-3-642-14623-7_25
10. Goldwasser, S., et al.: The knowledge complexity of interactive proof systems. SIAM J. Comput. 18(1), 186–208 (1989)
11. Han, J., et al.: Mining frequent patterns without candidate generation. In: ACM Sigmod Record (2000)
12. Kanewala, U., et al.: Techniques for testing scientific programs without an oracle. In: International Workshop on Software Engineering for Computational Science and Engineering (2013)
13. Khuller, S., et al.: The budgeted maximum coverage problem. Inf. Process Lett. 70(1), 39–45 (1999)
14. Liu, R., Wang, H.W., Monreale, A., Pedreschi, D., Giannotti, F., Guo, W.: AUDIO: An integrity auditing framework of outlier-mining-as-a-service systems. In: Flach, P.A., Bie, T., Cristianini, N. (eds.) ECML PKDD 2012. LNCS, vol. 7524, pp. 1–18. Springer, Heidelberg (2012). doi:10.1007/978-3-642-33486-3_1
15. Liu, R., et al.: Integrity verification of k-means clustering outsourced to infrastructure as a service (IAAS) providers. In: SDM (2013)
16. Liu, R., et al.: Result integrity verification of outsourced Bayesian network structure learning. In: SDM (2014)
17. Papamanthou, C., Tamassia, R., Triandopoulos, N.: Optimal verification of operations on dynamic sets. In: Rogaway, P. (ed.) CRYPTO 2011. LNCS, vol. 6841, pp. 91–110. Springer, Heidelberg (2011). doi:10.1007/978-3-642-22792-9_6

18. Parno, B., Raykova, M., Vaikuntanathan, V.: How to delegate and verify in public: verifiable computation from attribute-based encryption. In: Cramer, R. (ed.) TCC 2012. LNCS, vol. 7194, pp. 422–439. Springer, Heidelberg (2012). doi:10.1007/978-3-642-28914-9_24
19. Sindelar, M., et al.: Sharing-aware algorithms for virtual machine colocation. In: Symposium on Parallelism in Algorithms and Architectures (2011)
20. Vaidya, J., et al.: Efficient integrity verification for outsourced collaborative filtering. In: ICDM (2014)
21. Wong, W.K., et al.: Security in outsourcing of association rule mining. In: VLDB (2007)
22. Zhang, B., et al.: Budget-constrained result integrity verification of outsourced data mining computations (2017). http://www.cs.stevens.edu/hwang4/papers/dbsec2017full.pdf

Searchable Encryption to Reduce Encryption Degradation in Adjustably Encrypted Databases

Florian Kerschbaum[1]([✉]) and Martin Härterich[2]

[1] University of Waterloo, Waterloo, Canada
florian.kerschbaum@uwaterloo.ca
[2] SAP, Karlsruhe, Germany
martin.haerterich@sap.com

Abstract. Processing queries on encrypted data protects sensitive data stored in cloud databases. CryptDB has introduced the approach of adjustable encryption for such processing. A database column is adjusted to the necessary level of encryption, e.g. order-preserving, for the set of executed queries, but never reversed. This has the drawback that long running cloud databases will eventually transform into only order-preserving encrypted databases. In this paper we propose searchable encryption as an alternative in order to reduce this encryption degradation. It maintains security while only marginally impacting performance when applied only to infrequently used queries for searching. We present a budget-based encryption selection algorithm as part of query planning for making the appropriate choice between searchable and deterministic or order-preserving encryption. We evaluate our algorithm on a long-tail distributed TPC-C benchmark on an experimental implementation of encrypted queries in an in-memory database. In one choice of parameters our algorithm incurs only a 1.5% performance penalty, but one of 15 columns is not decrypted to order-preserving or deterministic encryption. Our selection algorithm is configurable, such that higher security gains are possible at the cost of performance.

1 Introduction

In order to protect cloud databases data can be processed in encrypted form [1, 9, 10, 29, 30]. A common way to enable processing of encrypted data is order-preserving encryption [1, 2, 19, 23, 28]. Order-preserving encryption allows processing many SQL queries without modification. However, order-preserving encryption is susceptible to simple attacks on the static data [27].

In order to increase security CryptDB has introduced adjustable encryption [29]. The idea is to layer encryption in onions. Queries are analyzed and the encryption layer is adjusted before their execution.

This has the positive effect that only the layers necessary for the query execution, e.g. deterministic encryption instead of order-preserving encryption, are revealed and thus security is increased. A database starts in a completely secure

© IFIP International Federation for Information Processing 2017
Published by Springer International Publishing AG 2017. All Rights Reserved
G. Livraga and S. Zhu (Eds.): DBSec 2017, LNCS 10359, pp. 325–336, 2017.
DOI: 10.1007/978-3-319-61176-1_18

(cold) mode and transforms into a (hot) mode which is very efficient, since no more decryption operations are necessary, but all queries can be processed on the data as is. This transformation is also never reversed. Since it is not possible to determine when a cloud database has been compromised, there is no reason to encrypt data which has once been revealed to the cloud service provider.

This lack of reversion has the negative consequence that many databases may ultimately reach a state that has only order-preserving encrypted columns. Hence, in a long running database system adjustable encryption may be no better than pure order-preserving encryption. The set of all queries determines the encryption level, even if those queries contribute little to the overall load of the database. Particularly, the long tail of the query distribution may have a severe negative security effect. These queries are infrequently executed, e.g. only once, presumably using columns that are infrequently used for searching, but have the same impact on the security as the most frequently reoccurring ones. This paper proposes dealing with these infrequent queries differently in order to confine their impact.

This raises two research questions: First, how to detect infrequently used columns and second, how to handle them. For the second problem we propose to use searchable encryption [3,5,11,12]. Searchable encryption is a randomized, strongly secure encryption scheme where the key holder can issue tokens for equality or range searches. The search algorithm is different than in a regular table scan and also significantly slower. Yet, all data for which no token has been issued remains semantically secure. Hence, searchable encryption is particularly suited for infrequently searched columns, since the search pattern is sparse.

For the first problem we propose a more intelligent encryption selection algorithm. It now has two choices: searchable encryption and order-preserving or deterministic encryption. It will first try searchable encryption until a certain threshold has been reached and only then decrypt. This increases the time for transforming from a cold to a hot database, but handles infrequently used columns with searchable encryption.

We perform a set of experiments using the TPC-C benchmark on our algorithms in an implementation of encrypted queries in an in-memory, column-store database. Our algorithm is configurable in order to allow different trade-offs between security and performance according to the preference of the database administrator. However, we were particularly interested in a set of parameters where the gain of security is clearly higher than the cost of security. In one particular choice of parameters our algorithm may incur only a small 1.5% performance penalty, but the most infrequently used column is not decrypted. Note that the economic value of a single non-decrypted column can be very high for sensitive data, such as salaries, outstanding sales prices or health care data. Due to the difficulty of scientifically assessing sensitivity values, we only report the percentage of still randomly encrypted data; in our case a 6.7% security improvement which is still more than 4 times the performance penalty. We also show in detail the different trade-offs between security and performance in query planning for different parameters of the algorithm (Sect. 5).

2   Related Work

2.1   Queries on Encrypted Data

Hacigümüs et al. introduce the first database for processing SQL queries on encrypted data [10]. They use deterministic encryption and binning for range queries. Binning requires the client to post-process the result and filter non-matching entries. Agrawal et al. improve on this by order-preserving encryption [1]. In order-preserving encryption plaintexts are mapped to ciphertext in the same order. This removes the necessity for post-processing (and query rewriting) and range queries can be processed with the same relational operator as on plaintexts. Later, Popa et al. extend this concept using adjustable encryption which adjusts the ciphertext to the query [29]. Hang et al. also show how to implement this over multiple keys [13].

Boldyreva et al. formalize order-preserving encryption [2]. They provide a proof that their scheme is the best possible stateless encryption scheme [2]. Recently Naveed et al. have shown that this security definition is rather weak and simple attacks can exploit the static leakage of order-preserving encryption [27]. A stronger notions of security – indistinguishability under (frequency-analysing) order-preserving chosen plaintext attack – are achieved by the scheme by Popa et al. [28] and Kerschbaum [19], respectively. Yet, their schemes are not efficiently compatible with adjustable encryption. Hence, system builders have to make a choice between the two. Our experiments indicate that adjustable encryption has a high security improvement, e.g. in the TPC-C benchmark only 15 columns need to use deterministic encryption. We hence believe that adjustable encryption is preferable to a stronger order-preserving encryption.

We build on adjustable encryption providing a further enhancement of security. Particularly, we introduce a query planning algorithm for searchable encryption in order to handle infrequently used columns in adjustable encryption, such that even not all 15 columns need to be deterministic.

2.2   Searchable Encryption

Searchable encryption offers stronger security than order-preserving or deterministic encryption. Using a token generated by the secret key one can search for values or within ranges. Without the token the ciphertext is as secure as common standard encryption.

The first sub-linear search time, (inverted) index-based searchable encryption scheme was introduced by Curtmola et al. [5]. Its idea is to provide an index of deterministically encrypted keywords and an encrypted list of documents. Since each deterministic ciphertext is unique, these schemes are not as susceptible to frequency analysis, but still more efficient to search. Hahn and Kerschbaum showed how the index can be built from the access pattern [11].

Searchable encryption also supports complex queries. The fastest method has been proposed by Demertzis et al. in [6] where all subranges in the disjunctions are indexed (and leaked on a match).

There exist also a large number of applications which have been developed on top of encrypted, in-memory, column-store database with specific protocols, e.g., benchmarking [4,15,16,20,22,26], RFID tracking [18,21,25], smart metering [14], supply chain planning [7,24], web applications [8] or reputation systems [17].

3   Searchable Encryption

3.1   Definitions

We propose to use searchable encryption as an alternative to deterministic and order-preserving encryption in adjustable encryption. Searchable encryption allows the (private) key holder to issue a search token for a query string. Using this search token the ciphertext holder can compare a ciphertext to the query string. The result of this comparison (match/no match) is immediately revealed in plaintext.

We employ the symmetric key variant due to better performance and the lack of need for a public key in our scenario. A searchable encryption in our scenario consists of the following algorithms:

- $sk \leftarrow KeyGen(\lambda)$: Generates a secret key sk for a security parameter λ.
- $c \leftarrow Enc(sk, x)$: Encrypts a plaintext x into a ciphertext c using secret key sk.
- $t \leftarrow TrapDoor(sk, x)$: Generates a trapdoor search token t for plaintext x using secret key sk.
- $\top/\bot \leftarrow Test(t, c)$: Returns $\top$ if the search token matches and $\bot$ if not.

Note that the ability to decrypt is optional in our scenario and hence not implemented, since we can use another encrypted column of the same data for decryption.

3.2   Performance Calibration

We compare the execution time of SQL queries on searchable encryption to that of on deterministic encryption or order preserving encryption. These results are used to calibrate our query planning algorithm. This algorithm compares the runtime of a query using searchable encryption with its equivalent using deterministic or order-preserving encryption, respectively. Both executions – on searchable and on deterministic or order-preserving encryption – share some common effort which includes query pre-processing and decryption of the result set on the client and data transfer between the client and the server. We omit this time in the following evaluation, because it does not contribute to the advantage of one execution strategy over the other.

For searchable encryption we must consider the generation of the trapdoor and the runtime of the UDF. Our tests show that the UDF scales very well, so that we have the following linear cost model. Let N denote the number of entries in the searched database column, $t_{\text{UDF scan}}$ denote the scan time per database row and $t_{\text{trapdoor generation}}$ includes both the actual execution of the cryptographic algorithm and the row-independent execution time for query processing.

$$t_{\text{searchable}} = N t_{\text{UDF scan}} + t_{\text{trapdoor generation}}$$

Table 1. Constants used for calibration

$t_{\text{UDF scan}}$	$4.5\,\mu s$
$t_{\text{trapdoor generation}}$	$65\,ms$
t_{scan}	$\sim 0\,\mu s\ (< 1ns)$
$t_{\text{encryption}}$	$20\,ms$

When executing equality or range searches on deterministic or order-preserving encryption, respectively, we use unmodified relational operators that compare the values stored in the database to the search values. In this case the main execution time stems from encrypting these literal values to deterministic or order-preserving ciphertexts. Hence in our linear model

$$t_{\text{deterministic}} = N t_{\text{scan}} + t_{\text{encryption}}$$

the slope t_{scan} is very small. Note that in particular for in-memory databases we find that $t_{\text{scan}} \ll t_{\text{UDF scan}}$. In fact, for table sizes up to 100 million entries there is a total runtime of less than 50 ms.

Our measurements lead to the following values for $t_{\text{UDF scan}}$, $t_{\text{trapdoor generation}}$, t_{scan} and $t_{\text{encryption}}$ which we use for the calibration of our query planning algorithm (Table 1).

4   Detecting and Handling Infrequently Used Columns

Searchable encryption in our UDF can handle selection similar to deterministic or order-preserving encryption but at higher security and lower performance. We now aim to identify infrequently used columns, such that we can decide to handle them by searchable encryption keeping the performance impact low, but maximizing the relative security gain.

4.1   Problem

Consider an adjustably encrypted database and the following two sequences A and B of queries:

Sequence A:

```
SELECT x FROM T WHERE y > 10
SELECT x FROM T WHERE y > 10
SELECT x FROM T WHERE y > 10
SELECT x FROM T WHERE y > 10
SELECT x FROM T WHERE y > 10
```

Sequence B:

```
SELECT x FROM T WHERE y = 10
SELECT x FROM T WHERE y > 10
SELECT x FROM T WHERE y = 10
SELECT x FROM T WHERE y = 10
SELECT x FROM T WHERE y = 10
```

Using the standard adjustment algorithm both sequences result in an order-preserving encryption of column y. Yet, if in sequence B the second query is handled using searchable encryption, then the security would remain at deterministic encryption and the performance impact would be small. Our problem is to identify and handle differently this specific (infrequent) query.

The problem is a typical scheduling problem where an optimizer has to make a decision based on future inputs (queries). The decision problem in case of the first query of sequence A and the second query of sequence B is almost identical: A query requiring a database adjustment appears for the first time. The optimizer has to decide whether to use searchable encryption or to decrypt.

The only sensible choice is to treat first-time appearing queries as infrequent until they reach a certain threshold and then decrypt. We present our algorithm in the next section.

4.2 Algorithm

We use a budget mechanism in order to determine infrequently used columns. For each column col we maintain a budget $budget[col]$. This budget describes the extra amount of time allowed for searchable encryption compared to deterministic or order-preserving encryption. It can be maintained in an arbitrary but fixed unit of time (say milliseconds). For each column we use searchable encryption until the budget is used up (i.e. reaches 0) and subsequently switch to the other encryption schemes.

We fix two parameters α and β for our algorithm. Whenever a query is executed the parameter α is added to the budget. The parameter β defines the upper bound of the budget, i.e. the budget is never increased beyond β. We call this process budget refilling.

When we choose searchable encryption we deduct the additional cost of the query from the budget. Let σ be the cost of searchable encryption for equality and τ be the cost of searchable encryption for ranges as determined in Sect. 3.2.

We can run our algorithm – in particular budget refilling – for several different sets of columns. A SQL query uses a number of columns, not only the ones in selection. We consider and later evaluate the following options for the budget update strategy, i.e. choosing the column col.

- S_1: Increase the budget for all columns used as selection parameters.
- S_2: Increase the budget for all columns occurring in the query in any role (e.g. also in the result list).
- S_3: Increase the budget for all columns of all tables occurring in the query.
- S_4: Increase the budget for all columns of the database scheme used.

4.3   Cost Estimation

Note that the costs σ and τ used in our algorithm depend on the number of rows to which the test function needs to be applied. For simple scans on complete database tables this number is readily available. However, as soon as there are other selection conditions which narrow the result set it is important that they are applied first. Hence the actual number of rows the function acts on has to be estimated. This is a difficult problem and lies at the heart of many query optimizations already for non-encrypted data. In our implementation we use a straight-forward approach and assume that the selection conditions occurring in our queries are independent and reduce the result set by a fixed factor.

5   Experimental Results

5.1   Security Measure

We define security as the encryption state of the database after one test run, i.e. a series of queries chosen according to the distribution described before. Each column that is decrypted to deterministic encryption lowers the security compared to columns encrypted using randomized and searchable encryption. Our security measure is hence simply the number of columns *not* decrypted to order-preserving or deterministic encryption.

This security measure is independent of the number of queries we have executed. We simply measure the state of the database. Note that without our encryption selection algorithm all 15 columns would be decrypted to deterministic encryption after the first query accessing them, i.e. after each test run. We hypothesize that using our algorithm the database will remain in a more secure state.

We report the percentage of columns not decrypted. The baseline in our set of queries from the TPC-C benchmark is 15 columns that may be decrypted without our encryption scheme selection algorithm. Hence, each not decrypted column is a 6.7% security improvement. We ignore any different sensitivity level of columns, since they are difficult to assess scientifically, but note that any non-decrypted column may already have high economic value.

We devise a simple theoretical test whether a column is likely to be decrypted or not. Given the distribution of the test queries and the strategy used for budget refilling one can calculate the expected value of the change per query execution of the budget for a given data base column col. Let Q_i be the query type of query i, p_j be the probability of picking a query type j ($1 \leq j \leq 11$), and $\chi(Q_i, col) = \chi^{\text{refill strategy}}(Q_i, col)$ be 1 if query type Q_i leads to a refill for column col and 0 else. The cost $\sigma(Q_i, col)$ (cf. Sect. 4.3) of query i depends on the query type and the column. Then, we have

$$E(\Delta budget[col]) = \sum_i p_{Q_i} \left(\chi(Q_i, col)\alpha - \sigma(Q_i, col) \right) \tag{1}$$

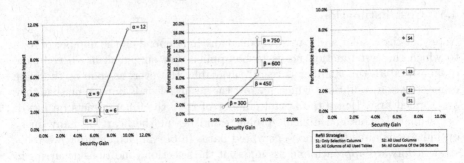

Fig. 1. (left) Security vs. Performance for $\alpha = 3, 6, 9, 12$, $\beta = 100\alpha$ and strategy S_1. (middle) Security vs. Performance for $\alpha = 3$, $\beta = 300, 450, 600, 750$ and strategy S_1. (right) Security vs. Performance for $\alpha = 3$, $\beta = 300$ and all strategies.

If the expected value $E(\Delta budget[col])$ is negative, then the budget will eventually reach 0. If, however, the expected value is positive then there is a good chance that no decryption of the column col to a weaker security level is ever necessary.

5.2   Performance Measure

We measure the wall clock time the database requires for performing the queries on encrypted data. Let t_i be time for the i-th query. We use the sum $s = \sum_i t_i$ of all queries as the measured performance.

In an encrypted "hot" database without our encryption selection algorithm the query would always be performed using deterministic or order-preserving encryption. Our encryption scheme selection algorithm improves security, but infrequent queries may be slower. In order to measure this performance penalty we first measured a baseline. We executed each query type of our set from the TPC-C benchmark 500 times on deterministic encryption. We use the median b_j as the baseline performance for this query type $(1 \leq j \leq 11)$.

Using the randomly chosen queries in a test run we compute a baseline for the entire test run. Let Q_i be the query type of query i $(1 \leq Q_i \leq 11)$. Then, our baseline B is $B = \sum_i b_{Q_i}$. We report the performance penalty of our encryption scheme selection algorithm as $\frac{s}{B} - 1$. A performance penalty of 5% means that an encrypted database using our algorithm (for the chosen set of parameters) executes 5% slower than encrypted database without our algorithm. Recall that our algorithm improves security, i.e. we trade performance for security.

Each test run consists of 500 queries $(0 < i \leq 500)$ chosen according to the geometric distribution described before. In order to have the database reach a "hot" state in our experiments, we disregard the first 100 queries in our performance measurement. Even when using our encryption scheme selection algorithm, several columns need to be decrypted to deterministic encryption. Including this time will skew our measurements, since it is not included in the baseline. We argue that the performance of a "hot" database is critical for practical use rather than the cost of reaching this state, since this can be included in an installation or setup phase. Hence, we focus our measurements on the "hot" phase.

5.3 Experimental Setup

We execute all experiments on an SAP HANA database (SP05 release) running on an HP Z820 workstation with 128 GB RAM and 16 dual cores (Intel Xeon CPU running at 2.60 GHz). There was no network access, connections were performed via the loopback interface. Our performance measurement is solely based on the database execution time and hence independent of network performance. Our security measurement is obviously independent of network performance.

Our client is implemented in Java 1.7 as a JDBC driver and running on the 64-bit JVM. The crypto routines are implemented in C++, compiled with GCC 4.3 and accessed via JNI. The UDFs use the same crypto routines accessible as linked libraries.

We execute series of test runs. Each test run consists of 500 queries chosen according to the geometric distribution described before. A series consists of 20 test runs and we report the median performance penalty and median security improvement as described before of those 20 test runs. Note that the median of an even number of values is the mean of the middle two values. Hence, we sometimes have "half" decrypted columns.

Each series of test runs has fixed parameters for the budget increment α, the budget upper limit β and the budget refilling strategy. The initial value for the budget is chosen to be $\frac{1}{2}\beta$. We conduct experiments varying each parameter individually. In the first experiment we vary α, in the second β and in the third the strategy. For each experiment we report the security improvement vs. performance penalty trade-off for the different parameter choices. We intend to give the database administrator guidance on configuring this trade-off in our algorithm.

5.4 Budget Increment α

We measure the impact of the budget increment α in our algorithm. As mentioned before α is linear in the expected value of the budget of a column and, hence, its probability of decryption. We choose four values for $\alpha = 3, 6, 9, 12$. We choose the budget upper limit $\beta = 100\alpha$. We choose the budget update strategy S_1, i.e. increase the budget for all columns used as selection parameters. We explain our choice of strategy using the results from the appropriate experiments in Sect. 5.6. Our random choices and experimental setup are as described before. Our results are depicted in Fig. 1.

Discussion. We can see in Fig. 1 that the performance impact of searchable encryption is high, since the slope is steep. This can be expected from our calibration in Sect. 3.2. Nevertheless, we also see that for the values of $\alpha = 3, 6, 9$

the security gain is higher than the performance penalty whereas for $\alpha = 12$ the performance penalty is higher than the security gain. Particularly, for $\alpha = 3$ and $\beta = 300$ we have performance penalty of 1.5% and one non-decrypted column, i.e. a security gain of 6.7%. Hence, we conclude that for truly infrequent queries searchable encryption is indeed a viable alternative. In our subsequent experiments we use $\alpha = 3$ in order to address this optimal set of infrequent queries. Note that in our test data, smaller values for $\alpha < 3$ make little sense, since there is only one non-decrypted column left.

5.5   Budget Upper Limit β

We measure the impact of the budget upper limit β in our algorithm. It is linear in the expected time to reach decryption (hot state), but it also counters the probabilistic nature of the query distribution. It has to be high enough in order to allow bursts of infrequent queries in an otherwise "stable" query set. We choose the four values for $\beta = 300, 450, 600, 750$. We chose $\alpha = 3$ from the results in the first experiment and the same budget update strategy S_1 as before. Our random choices and experimental setup are as described before. Our results are depicted in Fig. 1.

Discussion. We can see in Fig. 1 again a high performance impact of searchable encryption. Nevertheless, our choice of budget increment $\alpha = 3$ leads to higher security gains than performance penalties for $\beta = 300, 450, 600$. We conclude that $\alpha = 3$ is indeed addressing a promising set of infrequent queries. As expected the higher the budget upper limit β, the less columns are decrypted and the better the security, but the worse performance. Note that due to our budget update strategy S_1 of refilling only selection columns very few column budgets get increased and the budget limit (and initial budget) is more critical. For $\beta = 300$, we have the best security gain to performance penalty ratio. Hence, we use $\alpha = 3$ and $\beta = 300$ in our subsequent experiment for the budget update strategy.

5.6   Budget Update Strategy

We measure the impact of the budget update strategy S_1 to S_4. We expect less columns to be decrypted from S_1 to S_4, since budget increments occur more frequently. This implies more security gain and less performance penalty. We chose $\alpha = 3$ and $\beta = 300$ from the results in the first two experiments. Our random choices and experimental setup are as described before. Our results are depicted in Fig. 1.

Discussion. We can see in Fig. 1 somewhat surprising results. For our choices of $\alpha = 3$ and $\beta = 300$ the security is not impacted by the budget update strategy. All four strategies on average do not decrypt one column. Still, we can see the expected decreased performance penalty.

Different strategies may impact different columns and hence we believe that there may be a qualitative difference between the strategies. Still, even if the security differences are just too small to be measured, the strategy S_1 is clearly superior. Hence, we recommend and used in all our other experiments strategy S_1 of refilling only selection columns.

References

1. Agrawal, R., Kiernan, J., Srikant, R., Xu, Y.: Order preserving encryption for numeric data. In: Proceedings of the 2004 ACM International Conference on Management of Data, SIGMOD (2004)
2. Boldyreva, A., Chenette, N., Lee, Y., O'Neill, A.: Order-preserving symmetric encryption. In: Joux, A. (ed.) EUROCRYPT 2009. LNCS, vol. 5479, pp. 224–241. Springer, Heidelberg (2009). doi:10.1007/978-3-642-01001-9_13
3. Cash, D., Jarecki, S., Jutla, C., Krawczyk, H., Roşu, M.-C., Steiner, M.: Highly-scalable searchable symmetric encryption with support for boolean queries. In: Canetti, R., Garay, J.A. (eds.) CRYPTO 2013. LNCS, vol. 8042, pp. 353–373. Springer, Heidelberg (2013). doi:10.1007/978-3-642-40041-4_20
4. Catrina, O., Kerschbaum, F.: Fostering the uptake of secure multiparty computation in e-commerce. In: Proceedings of the 3rd International Conference on Availability, Reliability and Security, ARES (2008)
5. Curtmola, R., Garay, J., Kamara, S., Ostrovsky, R.: Searchable symmetric encryption: improved definitions and efficient constructions. J. Comput. Secur. 19(5), 895–934 (2011)
6. Demertzis, I., Papadopoulos, S., Papapetrou, O., Deligiannakis, A., Garofalakis, M.: Practical private range search revisited. In: Proceedings of the ACM International Conference on Management of Data, SIGMOD (2016)
7. Dreier, J., Kerschbaum, F.: Practical privacy-preserving multiparty linear programming based on problem transformation. In: Proceedings of the 3rd IEEE International Conference on Privacy, Security, Risk and Trust, PASSAT (2011)
8. Fuhry, B., Tighzert, W., Kerschbaum, F.: Encrypting analytical web applications. In: Proceedings of the 8th ACM Cloud Computing Security Workshop, CCSW (2016)
9. Hacigümüs, H., Iyer, B., Mehrotra, S.: Efficient execution of aggregation queries over encrypted relational databases. In: Proceedings of the 9th International Conference on Database Systems for Advances Applications, DASFAA (2004)
10. Hacigümüs, H., Iyer, B.R., Li, C., Mehrotra, S.: Executing SQL over encrypted data in the database-service-provider model. In: Proceedings of the 2002 ACM International Conference on Management of Data, SIGMOD (2002)
11. Hahn, F., Kerschbaum, F.: Searchable encryption with secure and efficient updates. In: Proceedings of the 21st ACM Conference on Computer and Communications Security, CCS (2014)
12. Hahn, F., Kerschbaum, F.: Poly-logarithmic range queries on encrypted data with small leakage. In: Proceedings of the 8th ACM Cloud Computing Security Workshop, CCSW (2016)
13. Hang, I., Kerschbaum, F., Damiani, E.: Enki: access control for encrypted query processing. In: Proceedings of the ACM International Conference on Management of Data, SIGMOD (2015)

14. Jawurek, M., Kerschbaum, F., Danezis, G.: SOK: privacy technologies for smart grids - a survey of options. Technical report MSR-TR-2012-119, Microsoft (2012)

15. Kerschbaum, F.: Building a privacy-preserving benchmarking enterprise system. Enterp. Inf. Syst. **2**(4), 421–441 (2008)

16. Kerschbaum, F.: Practical privacy-preserving benchmarking. In: Proceedings of the IFIP International Information Security Conference, SEC (2008)

17. Kerschbaum, F.: A verifiable, centralized, coercion-free reputation system. In: Proceedings of the 8th ACM Workshop on Privacy in the Electronic Society, WPES (2009)

18. Kerschbaum, F.: An access control model for mobile physical objects. In: Proceedings of the 15th ACM Symposium on Access Control Models and Technologies, SACMAT (2010)

19. Kerschbaum, F.: Frequency-hiding order-preserving encryption. In: Proceedings of the 22nd ACM Conference on Computer and Communications Security, CCS (2015)

20. Kerschbaum, F., Dahlmeier, D., Schröpfer, A., Biswas, D.: On the practical importance of communication complexity for secure multi-party computation protocols. In: Proceedings of the ACM Symposium on Applied Computing, SAC (2009)

21. Kerschbaum, F., Oertel, N.: Privacy-preserving pattern matching for anomaly detection in RFID anti-counterfeiting. In: Proceedings of the International Workshop on Radio Frequency Identification: Security and Privacy Issues, RFIDSec (2010)

22. Kerschbaum, F., Schneider, T., Schröpfer, A.: Automatic protocol selection in secure two-party computations. In: Proceedings of the 12th International Conference on Applied Cryptography and Network Security, ACNS (2014)

23. Kerschbaum, F., Schröpfer, A.: Optimal average-complexity ideal-security order-preserving encryption. In: Proceedings of the 21st ACM Conference on Computer and Communications Security, CCS (2014)

24. Kerschbaum, F., Schröpfer, A., Zilli, A., Pibernik, R., Catrina, O., de Hoogh, S., Schoenmakers, B., Cimato, S., Damiani, E.: Secure collaboratiue supply-chain management. IEEE Comput. **44**(9), 38–43 (2011)

25. Kerschbaum, F., Sorniotti, A.: RFID-based supply chain partner authentication and key agreement. In: Proceedings of the 2nd ACM Conference on Wireless Network Security, WISEC (2009)

26. Kerschbaum, F., Terzidis, O.: Filtering for private collaborative benchmarking. In: Proceedings of the International Conference on Emerging Trends in Information and Communication Security, ETRICS (2006)

27. Naveed, M., Kamara, S., Wright, C.: Inference attacks on property-preserving encrypted databases. In: Proceedings of the 21st ACM Conference on Computer and Communications Security, CCS (2014)

28. Popa, R.A., Li, F.H., Zeldovich, N.: An ideal-security protocol for order-preserving encoding. In: Proceedings of the 34th IEEE Symposium on Security and Privacy, S&P (2013)

29. Popa, R.A., Redfield, C.M.S., Zeldovich, N., Balakrishnan, H.: CryptDB: protecting confidentiality with encrypted query processing. In: Proceedings of the 23rd ACM Symposium on Operating Systems Principles, SOSP (2011)

30. Tu, S., Kaashoek, M.F., Madden, S., Zeldovich, N.: Processing analytical queries over encrypted data. In: Proceedings of the 39th International Conference on Very Large Data Bases, PVLDB (2013)

Efficient Protocols for Private Database Queries

Tushar Kanti Saha[1][(✉)], Mayank[2], and Takeshi Koshiba[3]

[1] Division of Mathematics, Electronics, and Informatics,
Graduate School of Science and Engineering, Saitama University, Saitama, Japan
saha.t.k.512@ms.saitama-u.ac.jp
[2] Department of Computer Science and Engineering,
Indian Institute of Technology (Banaras Hindu University), Varanasi, India
mayank.cse14@iitbhu.ac.in
[3] Faculty of Education and Integrated Arts and Sciences,
Waseda University, Tokyo, Japan
tkoshiba@waseda.jp

Abstract. We consider the problem of processing private database queries over encrypted data in the cloud. To do this, we propose a protocol for conjunctive query and another for disjunctive query processing using somewhat homomorphic encryption in the semi-honest model. In 2016, Kim et al. [IEEE Trans. on Dependable and Secure Comput.] showed an FHE-based query processing with equality conditions over encrypted data. We improve the performance of processing private conjunctive and disjunctive queries with the low-depth equality circuits than Kim et al.'s circuits. To get the low-depth circuits, we modify the packing methods of Saha and Koshiba [APWConCSE 2016] to support an efficient batch computation for our protocols with a few multiplications. Our implementation shows that our protocols work faster than Kim et al.'s protocols for both conjunctive and disjunctive query processing along with a better security level. We are also able to provide security to both attributes and values appeared in the predicate of the conjunctive and disjunctive queries whereas Kim et al. provided the security to the values only.

Keywords: Private Database Queries · Conjunctive · Disjunctive · Packing method · Homomorphic Encryption · Batch technique

1 Introduction

Private database queries (PDQ) plays an important role for accessing these data securely from any part of the world. In addition, users are not interested to disclose their queries and results to the database owners or any other parties. At the same time, database owners are not interested to disclose their whole database to their users. Besides, they do not like to keep their data in their personal computer or server because of high maintenance cost. They are now interested in storing their data to another party like the cloud so that database owners and

G. Livraga and S. Zhu (Eds.): DBSec 2017, LNCS 10359, pp. 337–348, 2017.
DOI: 10.1007/978-3-319-61176-1_19

their allowed users can access the data from anywhere in the world with a low cost. However, they also want to secure their data at the same time. Moreover, database owners want to secure their data using the encryption method of a cryptographic scheme. But the encrypted data needs to be decrypted by some trusted parties before utilizing it for some purposes which raises another security problem. In reality, it is hard to find such trusted parties. So it is desirable to execute some queries on encrypted data without decryption. On the contrary, homomorphic encryption (HE) is the encryption scheme which allows the meaningful operation like addition and multiplication on encrypted data without decryption. Therefore, we use homomorphic encryption scheme in case of private database queries. The concept of privacy homomorphism was coined by Rivest et al. in 1978 [10]. Also, the role of homomorphic encryption was limited to either addition or multiplication before introducing Gentry's revolutionary work in 2009 [6]. Moreover, the homomorphic encryption scheme can be classified into three types. Firstly, partial homomorphic encryption (PHE) allows either addition or multiplication but not both. Secondly, somewhat homomorphic encryption (SwHE) allows many additions and few multiplications. Finally, fully homomorphic encryption (FHE) allows any number of additions and multiplications. Here Gentry proposed the fully homomorphic encryption scheme by applying bootstrapping technique into somewhat homomorphic encryption scheme. But fully homomorphic encryption scheme is far behind from practical implementation due to its speed [11]. Therefore, we use somewhat homomorphic encryption scheme [9] which is faster than FHE due to supporting a limited number of multiplications. In this paper, we consider the security of the attributes and values appeared in the predicate of a conjunctive or disjunctive query with equality conditions. An example of the conjunctive query is that a managing staff of a hospital is trying to find out the patient's information from a hospital database who are suffering from 'Leukemia' and age is less than 30. In addition, a doctor is searching for the patients who are suffering from fever or cold, which is an example of the disjunctive query. In 2016, Kim et al. [8] showed an approach to address private database queries like conjunctive, disjunctive, and threshold queries using leveled FHE [3] with SIMD techniques. They also showed its implementation in [7] which took about 26.54 s to perform a query on 326 elements (0.08 sec./per record) including 11 attributes of 40-bit values with the 93-bit security level. But this speed of processing query is not satisfactory to process big data stored in the cloud. So there should be an efficient method to improve the performances of conjunctive and disjunctive queries with a better security level.

1.1   Reviews of Recent Works

In 2013, Boneh et al. [2] showed an efficient method of processing conjunctive queries only with somewhat homomorphic encryption. But the performance of their scheme is still far from practicality. In 2016, Cheon et al. [5] showed an approach to private query processing on encrypted databases using leveled FHE in [3] with a better security level. But their performances of query processing was highly time-consuming in a practical sense. They also declared performance improvement challenge for processing the private queries. At the same time, Kim et al. [8] also used BGV scheme in [3] to describe another protocol for processing conjunctive, disjunctive, and threshold queries over encrypted data in the cloud. They showed the practical implementation of the protocol in another paper [7]. It took about 0.11 s to access each record with 11 attributes of 40-bit values. In another paper, Kim et al. [7] showed a better security for processing these same types of queries. Here they provide security to both attributes and values in the predicate of a query. But it took about 0.12 s to access each record with 11 attributes of 40-bit values. Here they used equality circuits of depth $\lceil \log l \rceil$ to compare two l-bit integers which can be improved by the private batch equality protocol in [11]. Besides, none of the above protocols were able to achieve a remarkable efficiency regarding practicality. Recently, Saha and Koshiba [12] showed an efficient protocol than that in [5] for processing a conjunctive query. But their computation technique is only useful for processing a conjunctive query.

1.2 Our Contribution

In this paper, we consider the problem of processing private conjunctive and disjunctive queries with equality conditions over encrypted database. We also think the security both attributes α_i and values v_i with $1 \leq i \leq k$ appeared in the predicate of a conjunctive and disjunctive query. Here we follow the conventional approach of processing conjunctive and disjunctive queries. For example, let us consider the conjunctive and disjunctive queries with k equality conditions as *"select V from Record where $\alpha_1 = v_1$ and $\alpha_2 = v_2$ and ... and $\alpha_k = v_k$"* and *"select V from Record where $\alpha_1 = v_1$ or $\alpha_2 = v_2$ or ... or $\alpha_k = v_k$"* respectively. To process these queries with k equality conditions, the conventional solution is that client needs to send k queries firstly to the database server. After that, database server does the equality matching of attributes and the values of its Record table and sends back the results to the client. Then client needs to do the intersection and union of those k results from the database to get the actual result of conjunctive and disjunctive query respectively. In addition, most of the existing solutions with homomorphic encryption [4,5,7,8] used the equality circuits of depth $\lceil \log l \rceil$ for comparing two l-bit binary values. By developing new batch technique and packing methods, our equality circuit is reduced to a constant-depth circuit which includes many equality comparisons. Then we propose an efficient method to improve the performances of private conjunctive and disjunctive queries using ring learning with errors (RLWE) based SwHE of a better security level.

2 Our Protocols

In this section, we describe the protocols of processing private database queries for conjunctive and disjunctive cases. To address private database query (PDQ), we consider the security of both attributes and values in the predicate of a conjunctive or disjunctive query. Here we use the same protocol settings as in Kim et al. [7] with a different scenario.

2.1 Attribute Matching

Suppose a medical research institute (MRI) is maintaining its database of some patients in the cloud. Since patient's information are sensitive, MRI has uploaded its database using a public key encryption scheme. Here consider Bob has m encrypted records $\{\mathcal{R}_1, \ldots, \mathcal{R}_m\}$ in its Record table of the MRI's database with λ attributes where $\lambda \geq k$. Here we require only the k attributes and their values from the predicate of a query to process that query. Furthermore, we denote each attribute name with a δ-bit binary vector $\alpha_i = (g_{i,0}, \cdots, g_{i,\delta-1})$ where $g_{i,c}$ is the c-th bit of the i-th attribute with $1 \leq i \leq k$ and $0 \leq c \leq \delta - 1$. Also, we need to consider k attributes among λ attributes in each record required for our conjunctive and disjunctive query processing. We also denote each attribute name in the Record table using a δ-bit binary vector $\beta_j = (h_{j,0}, \cdots, h_{j,\delta-1})$ where $h_{j,e}$ is the e-th bit of the j-th attribute with $1 \leq j \leq \lambda$ and $0 \leq e \leq \delta - 1$. Since we consider the security of attributes, first the protocol matches encrypted attribute α_i with any attribute β_j in the Record table. Here Bob does the matching by computing the Hamming distance $\mathbb{H}_{i,j}$ between α_i and β_j as $\mathbb{H}_{i,j} = |\alpha_i - \beta_j|$. Here if $\mathbb{H}_{i,j} = 0$, then we can say that $\alpha_i = \beta_j$; otherwise $\alpha_i \neq \beta_j$. According to our protocol, α_i must be matched with any β_j for some $1 \leq i \leq k$ and $1 \leq j \leq \lambda$.

2.2 Batch Processing

For our Record table, each record is represented as $\mathcal{R}_\mu = \{w_{\mu,1} \ldots, w_{\mu,\lambda}\}$ where each value $w_{\mu,j} = (b_{\mu,j,0}, \ldots, b_{\mu,j,l-1})$ is considered as a binary vector of the same length l with $1 \leq \mu \leq m$ and $1 \leq j \leq \lambda$. We know that our Record table contains m records. If we want to compute the Hamming distance of each v_i from each $w_{\mu,j}$ one by one then it is more time-consuming. Here we use the batch technique of the private batch equality protocol in [11]. Actually, batch processing is the method of executing a single instruction on multiple data. The performance of our protocols can be increased by using the batch technique within the lattice dimension n. Generally, a big database consists of many tables where each table contains numerous records. For our conjunctive query processing with batch technique, if we compare all the values of a certain attribute of a particular table using a single computation then we will be required higher lattice dimension n which requires more memory to compute. This high requirement of memory may exceed the usual capacity of a machine in the cloud. So we divide all records of a table into blocks. For our given m records, we divide the total records m into p blocks as $p = \lceil m/\eta \rceil$. Here each block consists of η records with λ attributes from which we have to access k attributes. If we access each record of our Record table one after another then it requires $m \cdot k$ rounds communication between Alice and Bob in the cloud for accessing m records. On the contrary, the batch technique allows us to access all values of any attribute β_j at a time. By utilizing the batch technique, we reduce the communication complexity between Alice and Bob in the cloud from $m \cdot k$ to $\lceil (m \cdot k)/\eta \rceil$. Now we can pack the η values of the β_j-th attribute of each block in a single polynomial to support batch computation where $1 \leq j \leq \lambda$.

2.3   Protocol for Conjunctive Query

A conjunctive query is a query which contains multiple conditions in the predicate of the query connected by 'and'/'∧'. For instance, a research staff is trying to find the information of the patients who suffer from Leukemia and are 30 years old. This is a conjunctive query request to the cloud. In this scenario, consider Alice has a conjunctive query with k conditions in its predicate as "*select V from Record where* $\alpha_1 = v_1$ *and* $\alpha_2 = v_2$ *and* ... *and* $\alpha_k = v_k$". Here we follow the conventional approach of processing a conjunctive query. So it can be computed by intersecting 'IDs' from the result the k sub-queries as $\bigcap_{i=1}^{k} Q(\alpha_i = v_i)$ where $Q(\alpha_i = v_i) = \{$ID | the attribute α_i of ID takes v_i as the value.$\}$ Moreover, the values of k attributes $\{\alpha_1, \ldots, \alpha_k\}$ appeared in the predicate of the query is represented as a set $V = \{v_1, \ldots, v_k\}$ where $v_i = (a_{i,0}, \ldots, a_{i,l-1})$ is considered as a binary vector of length l. Here we consider the security of both attributes α_i and values v_i appeared in the predicate of the query. Firstly, Alice sends the encrypted attributes to find the required column in the Record table that are needed in the conjunctive query computation. Then she sends the encrypted values v_i to Bob to be matched with some $w_{\mu,j}$ using the multiple Hamming distance computation where $1 \leq \mu \leq m$ and $1 \leq j \leq \lambda$. To speed up the computation using batch processing as discussed in Sect. 2.2, let us form a query vector $\mathbf{A}_i = (a_{i,0}, \ldots, a_{i,l-1})$ from the values of the i-th condition of the query where $1 \leq i \leq k$. We also assume the i-th attribute of query condition matches with β_j-th attribute of the Record table where $1 \leq j \leq \lambda$. Again we form another record vector from η values of the attribute β_j of each block σ as $\mathbf{B}_{\sigma,j} = (w_{\sigma,j,1}, \ldots, w_{\sigma,j,\eta})$ where $w_{\sigma,j,d} = (b_{\sigma,j,d,0}, \ldots, b_{\sigma,j,d,l-1})$ with $1 \leq \sigma \leq p$ and $1 \leq d \leq \eta$. Here $|\mathbf{A}_i| = l$ and $|\mathbf{B}_{\sigma,j}| = \eta \cdot l$. Here we find the distance between two vectors $\mathbf{A}_i$ and $\mathbf{B}_{\sigma,j}$ by the multiple Hamming distance computation as

$$\mathbb{H}_{\sigma,d,i} = \sum_{r=1}^{l-1} |a_{i,r} - b_{\sigma,j,d,r}| = \sum_{r=1}^{l-1} (a_{i,r} + b_{\sigma,j,d,r} - 2a_{i,r}b_{\sigma,j,d,r}) \qquad (1)$$

where $1 \leq d \leq \eta$, $1 \leq \sigma \leq p$, and $j \in \{1, 2, \ldots, \lambda\}$. Moreover, if $\mathbb{H}_{\sigma,d,i}$ in Eq. (1) is 0 for some position d in the block σ then we can say that $\mathbf{A}_i = \mathbf{B}_{\sigma,j,d}$; otherwise $\mathbf{A}_i \neq \mathbf{B}_{\sigma,j,d}$ where $\mathbf{B}_{\sigma,j,d}$ is the d-th sub-vector of $\mathbf{B}_{\sigma,j,d}$. Here the multiple Hamming distance means the distances between the vector $\mathbf{A}_i$ and each sub-vector in $\mathbf{B}_{\sigma,j}$. So we need to define another packing method than that in [14]. In this way, Alice gets some IDs for each value v_i in the predicate of a query. Then she gets conjunctive query matched IDs after the intersection of all IDs for each v_i. She then sends the IDs to Bob in the cloud again and Bob returns the corresponding records to Alice. Now we explain our protocol for conjunctive query by the following steps.

1. Alice generates the public key and secret key by herself and encrypts each column of the database D' along with attributes. Then she uploads the database to the cloud.
2. Then she also parses both attributes and values from the predicate part of her query. She encrypts attributes α_i and v_i using her public key and sends it to Bob in the cloud.
3. For $1 \leq i \leq k$ and $1 \leq j \leq \lambda$, Bob tries to find out the β_j-th attribute of the Record table that matches α_i using the Hamming distance $\mathbb{H}_{i,j}$ with every attribute in the database. Here Alice helps Bob to find the β_j-th attribute after decryption of the Hamming distance result of Bob.
4. For $1 \leq \sigma \leq p$, Bob does secure computation of batch equality test as in Eq. (1) and sends the encrypted result $\mathbb{H}_{\sigma,d,i}$ to Alice to verify whether at least one of $\mathbb{H}_{\sigma,d,i}$'s is equal to 0.
5. For $1 \leq i \leq k$ and $1 \leq d \leq \eta$, Alice decrypts $\mathbb{H}_{\sigma,d,i}$ using her secret key and checks each value $\mathbb{H}_{\sigma,d,i}$ and gets the IDs for some position d where $\mathbb{H}_{\sigma,d,i} = 0$; In this way, she gets k sets of IDs for k conditions in the query.
6. Then Alice computes the intersection of k sets of IDs and sends the result to Bob to get her desired result.
7. Bob sends the encrypted data to Alice depending on those IDs given by Alice. Then Alice decrypts the data and gets her desired result.

2.4 Protocol for Disjunctive Query

A disjunctive query is a query which contains multiple conditions in the predicate of the query connected by 'or'/'∨'. As discussed in Sect. 2.3, we consider the same database settings for processing a disjunctive query. Let us look at a disjunctive query with k conditions in its predicate as "*select V from Record where* $\alpha_1 = v_1$ *or* $\alpha_2 = v_2$ *or* ... *or* $\alpha_k = v_k$". Here we also follow the conventional approach of processing a disjunctive query. Now we can compute by taking union 'IDs' from the result the k sub-queries as $\bigcup_{i=1}^{k} Q(\alpha_i = v_i)$. We process this query with the same multiple Hamming distance computation as in Eq. (1). Our protocol for processing the disjunctive query is same as discussed by 7 steps in Sect. 2.3 except step 6. In case of disjunctive query, Alice needs to compute the union of k sets of IDs instead of intersection (see step 6 in Sect. 2.3) required for conjunctive query protocol.

Remark 1. Here our protocols are secure under the assumption that Bob is semi-honest (also known as honest-but-curious), i.e., he always follows the protocols but tries to learn information from the protocols. Here we use somewhat homomorphic encryption scheme in [12] and skip its review due to page limitation.

3 Packing Method

In information theory, the method of encoding many bits in a single polynomial is called packing method. In 2011, Lauter et al. [9] used a packing method for an efficient encoding of an integer in a polynomial ring to facilitate arithmetic operations (see Sect. 4.1 in [9] for details). Here we need the packing methods for both attributes and value matching. Let us consider a binary vector $M = (11001101)$ with $l = 8$ which can be encoded as $Poly(M) = 1 + x^2 + x^3 + x^6 + x^7$ using the packing method in [9]. Here we review and modify the packing methods in Saha et al. [11] which was used in their private batch equality test protocol. Here we skip the discussion of our packing method for attribute matching due to page limitation which is a variant of the following packing method.

3.1 Our Packing Method for Value Matching

First, let us review some parameters in [12]. Let t (resp. q) defines the ring for a message space (resp. ciphertext space) as $R_t = \mathbb{Z}_t[x]/(x^n + 1)$ (resp. $R_q = \mathbb{Z}_q[x]/(x^n + 1)$) which is a ring of integer polynomials of degree less than n with coefficient modulo t (resp. q) (see [12] for details). To accelerate the processing of a conjunctive and disjunctive query, we need to compute the multiple Hamming distance $\mathbb{H}_{\sigma,d,i}$ in Eq. (1) with few polynomial multiplications. As discussed in Sect. 2.3, for $1 \le i \le k$ and $j \in \{1, \ldots, \lambda\}$, we consider two same integer vectors $\mathbf{A}_i = (a_{i,0}, \cdots, a_{i,l-1}) \in R_t$ of length l and $\mathbf{B}_{\sigma,j} = (w_{\sigma,j,1}, \ldots, w_{\sigma,j,s}) \in R_t$ where $w_{\sigma,j,d} = (b_{\sigma,j,d,0}, \ldots, b_{\sigma,j,d,l-1})$ of length $\eta \cdot l$ with $1 \le \sigma \le p$ and $1 \le d \le \eta$. Here we need to find the Hamming distances between $\mathbf{A}_i = (a_{i,0}, \ldots, a_{i,l-1})$ and $\mathbf{B}_{\sigma,j} = (b_{\sigma,j,1,0}, \ldots, b_{\sigma,j,1,l-1}, \ldots, b_{\sigma,j,\eta,0}, \ldots, b_{\sigma,j,\eta,l-1})$. Furthermore, we know from [14] that the secure inner product $\langle \mathbf{A}_i, \mathbf{B}_{\sigma,j} \rangle$ helps to compute the Hamming distance between $\mathbf{A}_i$ and $\mathbf{B}_{\sigma,j}$. Here we pack these integer vectors by some polynomials with the highest $\text{degree}(x) = n$ in such a way so that inner product $\langle \mathbf{A}_i, \mathbf{B}_{\sigma,j} \rangle$ does not wrap-around a coefficient of x with any degrees. For the integer vectors $\mathbf{A}_i$ and $\mathbf{B}_{\sigma,j}$ with $n \ge \eta \cdot l$ and $1 \le d \le \eta$, the packing method of [11] in the same ring $R = \mathbb{Z}[x]/(x^n + 1)$ can be rewritten as

$$Poly_1(\mathbf{A}_i) = \sum_{c=0}^{l-1} a_{i,c} x^c, \quad Poly_2(\mathbf{B}_{\sigma,j}) = \sum_{d=1}^{s} \sum_{e=0}^{l-1} b_{\sigma,j,d,e} x^{l \cdot d - (e+1)}. \tag{2}$$

Here if we multiply the above two polynomials, it will help us to find the inner product $\langle \mathbf{A}_i, \mathbf{B}_{\sigma,j} \rangle$ which in turn helps the multiple Hamming distances computation between the vectors $\mathbf{A}_i$ and $\mathbf{B}_{\sigma,j}$. Here each Hamming distance can be found as a coefficient of x with different degrees. Now the polynomial multiplications of $Poly_1(\mathbf{A}_i)$ and $Poly_2(\mathbf{B}_{\sigma,j})$ in the same base ring R can be represented as follows.

$$\left(\sum_{c=0}^{l-1} a_{i,c} x^c \right) \times \left(\sum_{d=1}^{s} \sum_{e=0}^{l-1} b_{\sigma,j,d,e} x^{l \cdot d - (e+1)} \right) = \sum_{d=1}^{s} \sum_{c=0}^{l-1} \sum_{e=0}^{l-1} a_{i,c} b_{\sigma,j,d,e} x^{c + l \cdot d - (e+1)}$$

$$= \sum_{d=1}^{s} \sum_{c=0}^{l-1} a_{i,c} b_{\sigma,j,d,c} x^{l \cdot d - 1} + \text{ToHD} + \text{ToLD} = \sum_{d=1}^{s} \langle \mathbf{A}_i, \mathbf{B}_{\sigma,j,d} \rangle x^{l \cdot d - 1} + \cdots \tag{3}$$

Here, $\mathbf{A}_i$ is the i-th vector of length l that appeared in the predicate of a conjunctive or disjunctive query where $1 \le i \le k$. Also, $\mathbf{B}_{\sigma,j,d}$ is the d-th sub-vector of $\mathbf{B}_{\sigma,j}$ of the block σ and β_j attribute of the Record table with $1 \leq \sigma \leq p$, $1 \leq d \leq \eta$ and $j \in \{1, \ldots, \lambda\}$. Moreover, the ToHD (terms of higher degree) means $deg(x) > l \cdot d - 1$ and the ToLD (terms of lower degrees) means $deg(x) < l \cdot d - 1$. The result in Eq. (3) shows that one polynomial multiplication includes the multiple inner products of $\langle \mathbf{A}_i, \mathbf{B}_{\sigma,j,d} \rangle$. According to the SwHE in Sect. 2 of [12], the packed ciphertexts for $Poly_\tau(A) \in R$ are defined for some $\tau \in \{1, 2\}$ as

$$ct_\tau(A) = \text{Enc}(Poly_\tau(A), pk) \in (R_q)^2. \tag{4}$$

Proposition 1. *Let $\mathbf{A}_i = (a_{i,0}, \cdots, a_{i,l-1})$ be an integer vector where $|\mathbf{A}_i| = l$ and $\mathbf{B}_{\sigma,j} = (b_{\sigma,j,1,0}, \ldots, b_{\sigma,j,1,l-1}, \ldots, b_{\sigma,j,\eta,0}, \ldots, b_{\sigma,j,\eta,l-1})$ be another integer vector of length $\eta \cdot l$. For $1 \leq d \leq \eta$, the vector $\mathbf{B}_{\sigma,j}$ includes η sub-vectors where the length of each sub-vector is l. If the ciphertext of $\mathbf{A}_i$ and $\mathbf{B}_{\sigma,j}$ can be represented as $ct_1(\mathbf{A}_i)$ and $ct_2(\mathbf{B}_{\sigma,j})$ respectively by Eq. (4) then under the condition of Lemma 1 (see Sect. 2.3 in [12] for details), decryption of homomorphic multiplication $ct_1(\mathbf{A}_i) \boxtimes ct_2(\mathbf{B}_{\sigma,j}) \in (R_q)^2$ will produce a polynomial of R_t with $x^{l \cdot d - 1}$ including coefficient $\langle \mathbf{A}_i, \mathbf{B}_{\sigma,j,d} \rangle = \sum_{d=1}^{s} \sum_{c=0}^{l-1} a_{i,c} b_{\sigma,j,d,e} x^{l \cdot d - 1} \mod t$. Alternatively, we can say that homomorphic multiplication of $ct_1(\mathbf{A}_i)$ and $ct_2(\mathbf{B}_{\sigma,j})$ simultaneously computes the multiple inner products for $1 \leq i \leq k$, $1 \leq \sigma \leq p$, $1 \leq d \leq \eta$, $0 \leq c \leq (l-1)$, and $j \in \{1, \ldots, \lambda\}$.*

4   Secure Computation Procedure

We need to securely compute both attribute and value matching as discussed in our protocol in Sect. 2.3. Now we present the matching technique of both attributes and values of a conjunctive and disjunctive query with the Record table in the following sub-sections. Due to page limitation, we skip the discussion of secure computation procedure of attribute matching (similar to the following secure computation of value matching).

4.1   Matching the Values in the Record

Now we compute our protocol using the SwHE scheme in [12] and the packing method in Sect. 3.1 for matching the records. In addition, according to Eq. (1), we need to find out the values of the multiple Hamming distance $\mathbb{H}_{\sigma,d,i}$. As discussed in Sect. 3.1, we consider two same integer vectors $\mathbf{A}_i = (a_{i,0}, \cdots, a_{i,l-1}) \in R_t$ and $\mathbf{B}_{\sigma,j} = (b_{\sigma,j,1,0}, \ldots, b_{\sigma,j,1,l-1}, \ldots, b_{\sigma,j,\eta,0}, \ldots, b_{\sigma,j,\eta,l-1}) \in R_t$ from which $\mathbb{H}_{\sigma,d,i}$ can be computed. Here, for $1 \leq d \leq \eta$, $\mathbb{H}_{\sigma,d,i}$ is computed by the multiple Hamming distance between $\mathbf{A}_i$ and $\mathbf{B}_{\sigma,j}$ using Eq. (1). For these two integer vectors $\mathbf{A}_i$ and $\mathbf{B}_{\sigma,j}$, the multiple Hamming distance $\mathbb{H}_{\sigma,d,i}$ in Eq. (1) can be computed by the packing method in Eq. (2) and inner product property in Eq. (3). Moreover, the packed ciphertext of the vectors $\mathbf{A}_i$ and $\mathbf{B}_{\sigma,j}$ is computed by the Eq. (4). So $\mathbb{H}_{\sigma,d,i}$ is computed from Proposition 1 and the packed ciphertext vector $ct_1(\mathbf{A}_i)$ and $ct_2(\mathbf{B}_{\sigma,j})$ in three homomorphic multiplications and two homomorphic additions as $ct(\mathbb{H}_{\sigma,d,i})$ equals

$$ct_1(\mathbf{A}_i) \boxtimes ct_2(V_2) \boxplus ct_2(\mathbf{B}_{\sigma,j}) \boxtimes ct_1(V_1) \boxplus (-2ct_1(\mathbf{A}_i) \boxtimes ct_2(\mathbf{B}_{\sigma,j})) \tag{5}$$

where V_1 denotes an integer vector like $(1, \ldots, 1)$ of length l and V_2 denotes another integer vector like $(1, \ldots, 1)$ of length $\eta \cdot l$. The above encrypted polynomial $ct(\mathbb{H}_{\sigma,d,i})$ includes many Hamming distances between the sub-vectors of $\mathbf{A}_i$ and sub-vectors of $\mathbf{B}_{\sigma,j}$. Here we need the Hamming distance $\mathbb{H}_{\sigma,d,i}$ in Eq. (1). Bob sends $ct(\mathbb{H}_{\sigma,d,i})$ to Alice for decryption. According to Proposition 1 and our protocols, Alice decrypts $ct(\mathbb{H}_{\sigma,d,i})$ in the ring R_q using her secret key and extracts $\mathbb{H}_{\sigma,d,i}$ as a coefficient of $x^{l \cdot d - 1}$ from the plaintext of $ct(\mathbb{H}_{\sigma,d,i})$. Then Alice checks whether at least one of the $\mathbb{H}_{\sigma,d,i}$ contains 0 or not to decide whether $\mathbf{A}_i = \mathbf{B}_{\sigma,j,d}$ or $\mathbf{A}_i \neq \mathbf{B}_{\sigma,j,d}$.

4.2 Secure Computation of Our Protocols

For the secure computation of conjunctive query protocol, Alice sends both encrypted attributes and values from the predicate to Bob in the cloud. Bob first securely matches attributes. Then Bob matches each v_i with the j-th column of Record table to find the equalities according to Eq. (5) and sends result $ct(\mathbb{H}_{\sigma,d,i})$ to Alice. Then Alice decrypts the results $ct(\mathbb{H}_{\sigma,d,i})$ and gets some IDs where she gets $\mathbb{H}_{\sigma,d,i}=0$ for some d of the σ-th block. In this way, Alice gets k sets of IDs for k values in the predicate of the query. After that, she does the intersection of those sets of IDs to support conjunctive query computation and sends IDs to Bob. Later, Bob sends the corresponding encrypted records from the Record table depending on those IDs. Finally, Alice decrypts the encrypted records using her secret key to get her desired result. On the contrary, to support disjunctive query computation, Alice and Bob do the same thing as required for conjunctive query except that Alice does the union of those sets of IDs and sends those IDs to Bob. In this way, we process secure computation for both of our protocols.

4.3 Hiding Additional Information from Leakage

During decryption, Alice can know some additional information from the computation of the Hamming distance $\mathbb{H}_{\sigma,d,i}$ in Eq. (1) than she needs due to sending encrypted polynomials $ct(\mathbb{H}_{\sigma,d,i})$ to her. But Alice needs to know only those coefficients which has degree $x^{l \cdot d - 1}$. We solve the problems by adding a random polynomial at the cloud (Bob) ends separately. For securing the polynomial $\mathbb{H}_{\sigma,d,i}$, Bob also adds another random polynomial r_b to $ct(\mathbb{H}_{\sigma,d,i})$ for masking extra information. Since Alice needs to check only the coefficient of $x^{l \cdot d - 1}$ from the large polynomial $ct(\mathbb{H}_{\sigma,d,i})$ produced by Bob, then random polynomial in the ring R can be represented by $r_b = \sum_{d=0}^{n/l} \sum_{i=0}^{l-2} r_{b,l \cdot d+i} x^{l \cdot d+i}$. Here $ct(\mathbb{H}_{\sigma,d,i})$ consists of three ciphertext components as $ct(\mathbb{H}_{\sigma,d,i}) = (c_0, c_1, c_2)$. So Bob adds r_b to the ciphertext as $ct(\mathbb{H}'_{\sigma,d,i}) = ct(\mathbb{H}_{\sigma,d,i}) \boxplus r_b = (c_0 \boxplus r_b, c_1, c_2)$. Here the ciphertext $ct(\mathbb{H}'_{\sigma,d,i})$ contains all required information as a coefficient of $x^{l \cdot d - 1}$ and hide all other coefficients using the randomization. In this way, we hide $ct(\mathbb{H}'_{\sigma,d,i})$ to disclose any information to Alice except the coefficient of $x^{l \cdot d - 1}$.

5   Performance Analysis

In this section, we present the both theoretical and practical performance of our protocols in comparison to Kim et al. [8] protocol. Here, we experimented our two protocols and compared their performances with conjunctive and disjunctive query results in [8]. Here we use the same scenario as Kim et al. [8] protocol along with the database and queries.

5.1   Theoretical Evaluation

In this section, we figure out the multiplicative depth of equality circuits for Kim et al. [8] and our protocol. To measure the equalities of attributes and values as discussed in Sect. 4, we required the Hamming distance computation in Eq. (1). In addition, the encrypted computation of these Hamming distances required only three polynomial multiplications as in Eq. (5). Furthermore, Kim et al. needed a multiplicative depth of $\lceil \log l \rceil + \lceil \log(1 + \rho) \rceil$ for their equality circuits comparing two l-bit message with ρ attributes. On the contrary, our method required only $\log 3$ due to using our packing method. Also, the communication complexity of our protocols is $\mathcal{O}(k \cdot m \cdot l \log q)$.

5.2   Parameter Settings and Security Level

Here, we used the same database settings as shown in [7]. So we consider a database where each record includes 11 attributes with $l = 40$ bits data. Besides, we also consider two cases of 100 and 1000 records in the Record table for our conjunctive and disjunctive query processing with $k = 10$ conditions. Moreover, we encoded the name of each attribute with $\delta = 8$ bits integer. Furthermore, we also set the values of some other security parameters required for the SwHE in the experiments. We also considered the equality as a comparison operator. Moreover, we took the block size $\eta = 100$. We also considered appropriate values for the parameters (n, q, t, ω) of our security scheme as discussed in Sect. 2 of [12] for successful decryption and achieving a certain security level. As mentioned in Sect. 3 of our protocols, we need the lattice dimension $n \geq (\lambda \cdot \delta)$ for attributes comparison and $n \geq (\eta \cdot l)$ for values comparison. For this reason, we set $n = 100 \cdot 40 = 4000$ for values matching. In addition, we set $n = 2048$ for attribute matching to provide better security in the computation. Furthermore, we set $t = 2048$ for our plaintext space R_t. According to the work in [9], we choose the standard deviation $\omega = 8$ and $q \geq 16n^2t^2\omega^4 = 2^4 \cdot 2^{22} \cdot 2^{22} \cdot 2^{12} = 2^{60}$ for the ciphertext space R_q during attribute matching. Therefore, we fix our parameters as $(n, q, t, \omega) = (2048, 61\text{-bits}, 2048, 8)$. Similarly, $q \geq 16n^2t^2\omega^4 = 2^4 \cdot 2^{24} \cdot 2^{22} \cdot 2^{12} = 2^{62}$ for values matching. So we set $(n, q, t, \omega) = (4096, 63\text{-bits}, 2048, 8)$. According to computation procedure in [14], our parameters settings provide 364-bit security level to protect our protocols from some distinguishing attacks. Also, NIST [1] showed different security levels for many security algorithms and their corresponding validity periods. Furthermore, they declared that a minimum strength of 112-bit level security has a

Table 1. Performance of our protocols for 40-bit data

m (# of record)	k (# of conditions)	Timing (seconds)			Security level	
		Kim et al. [8]	Our protocol		Kim et al. [8]	Our protocol
			Conj	Disj		
100	10	8	2.948	3.058	93	364
1000	10	80	17.191	17.768	93	364

security lifetime up to 2030. They also disclosed that a security algorithm with a minimum strength of 128-bit level security has a security lifetime beyond 2030.

5.3 Implementation Details

Here Table 1 shows the performances of conjunctive and disjunctive query protocols compared to that of Kim et al. [8]. Here, we have implemented our protocols in C++ programming language with Pari C library (version 2.9.1) [13] and ran the programs on a single machine configured with 3.6 GHz Intel core-i5 processor and 8 GB RAM using Linux environment. For a database of 100 records (resp., 1000 records), our conjunctive query protocol took only 2.948 s (resp., 17.191 s). Also, our disjunctive query protocol took only and 3.058 s (resp., 17.768 s) for 100 records (resp., 1000 records). On the other hand, Kim et al. [8] needed 8 sec (resp., 80 s) for both conjunctive and disjunctive query processing over 100 records (resp., 1000 records). Furthermore, we achieve 364-bit security level for both our protocols whereas Kim achieved a security level of 93-bit. They also achieved a security level of maximum 125-bit which made computation time to twice of the timing with a 93-bit security level. Apart from the above advantages, we are also able to provide security to both values and attributes in the predicate of our queries whereas Kim et al. [8] provided only security to values appeared in the predicate of the query. Besides, Kim et al. [7] also tired to provide security to the attributes, but their performance was lower than that in [8] as shown in Table 4 of [7].

6   Conclusions

In this paper, we have shown two efficient protocols for processing private conjunctive and disjunctive queries over encrypted database using RLWE based somewhat homomorphic encryption in the semi-honest model. Our experiments proved that our protocols achieved a remarkable efficiency than Kim et al. [7,8] with a better security level. Furthermore, we have achieved the efficiency due to using low-cost equality circuits and batch technique with the packing methods. Moreover, our protocols can support a larger data size for both query and database by increasing the lattice dimension n.

Acknowledgment. This research is supported by KAKENHI Grant Numbers JP16H01705, JP17H01695, and JP24106008 for Scientific Research on Innovative Areas.

References

1. Barker, E.: Recommendation for key management. In: NIST Special Publication 800–57 Part 1 Rev. 4. NIST (2016)
2. Boneh, D., Gentry, C., Halevi, S., Wang, F., Wu, D.J.: Private database queries using somewhat homomorphic encryption. In: Jacobson, M., Locasto, M., Mohassel, P., Safavi-Naini, R. (eds.) ACNS 2013. LNCS, vol. 7954, pp. 102–118. Springer, Heidelberg (2013). doi:10.1007/978-3-642-38980-1_7
3. Brakerski, Z., Gentry, C., Vaikuntanathan, V.: (Leveled) fully homomorphic encryption without bootstrapping. ACM Trans. Comput. Theor. **6**(3), 13 (2014)
4. Cheon, J.H., Kim, M., Kim, M.: Search-and-compute on encrypted data. In: Brenner, M., Christin, N., Johnson, B., Rohloff, K. (eds.) FC 2015. LNCS, vol. 8976, pp. 142–159. Springer, Heidelberg (2015). doi:10.1007/978-3-662-48051-9_11
5. Cheon, J.H., Kim, M., Kim, M.: Optimized search-and-compute circuits and their application to query evaluation on encrypted data. IEEE Trans. Inf. Forensics Secur. **11**(1), 188–199 (2016). IEEE Press, New York
6. Gentry, C.: Fully homomorphic encryption using ideal lattices. In: Symposium on Theory of Computing - STOC 2009, pp. 169–178. ACM, New York (2009)
7. Kim, M., Lee, H.T., Ling, S., Ren, S.Q., Tan, B.H.M., Wang, H.: Better security for queries on encrypted databases. IACR Cryptology ePrint Archive, 2016/470 (2016)
8. Kim, M., Lee, H.T., Ling, S., Wang, H.: On the efficiency of FHE-based private queries. IEEE Trans. Dependable Secure Comput. 10.1109/TDSC.2016.2568182. (to appear)
9. Lauter, K., Naehrig, M., Vaikuntanathan, V.: Can homomorphic encryption be practical? In: ACM Workshop on Cloud Computing Security Workshop, CCSW 2011, pp. 113–124. ACM, New York (2011)
10. Rivest, R.L., Adleman, L., Dertouzos, M.L.: On data banks and privacy homomorphism. In: DeMillo, R.A., Dobkin, D.P., Jones, A.K., Lipton, R.J. (eds.) Foundations of Secure Computation, pp. 169–177. Academic Press, New York (1978)
11. Saha, T.K., Koshiba, T.: Private equality test using ring-LWE somewhat homomorphic encryption. In: 3rd Asia-Pacific World Congress on Computer Science and Engineering (APWConCSE), pp. 1–9. IEEE (2016)
12. Saha, T.K., Koshiba, T.: Private conjunctive query over encrypted data. In: Joye, M., Nitaj, A. (eds.) Progress in Cryptology - AFRICACRYPT 2017. Lecture Notes in Computer Science, vol. 10239, pp. 149–164. Springer, Cham (2017)
13. The PARI~Group, PARI/GP version 2.9.1, Bordeaux (2014). http://pari.math.u-bordeaux.fr/
14. Yasuda, M., Shimoyama, T., Kogure, J., Yokoyama, K., Koshiba, T.: Practical packing method in somewhat homomorphic encryption. In: Garcia-Alfaro, J., Lioudakis, G., Cuppens-Boulahia, N., Foley, S., Fitzgerald, W.M. (eds.) DPM/SETOP -2013. LNCS, vol. 8247, pp. 34–50. Springer, Heidelberg (2014). doi:10.1007/978-3-642-54568-9_3

Toward Group-Based User-Attribute Policies in Azure-Like Access Control Systems

Anna Lisa Ferrara[1], Anna Squicciarini[2(✉)], Cong Liao[2], and Truc L. Nguyen[1]

[1] Computer Science Department, University of Southampton, Southampton, UK
{al.ferrara,tnl2g10}@soton.ac.uk
[2] Information Sciences and Technology,
Pennsylvania State University, University Park, USA
{acs20,cl13}@psu.edu

Abstract. Cloud resources are increasingly pooled together for collaboration among users from different administrative units. In these settings, separation of duty between resource and identity management is strongly encouraged, as it streamlines organization of resource access in cloud. Yet, this separation may hinder availability and accessibility of resources, negating access to authorized and entitled subjects. In this paper, we present an in-depth analysis of group-reachability in user attribute-based access control. Starting from a concrete instance of an Access Control supported by the Azure platform, we adopt formal verification methods to demonstrate how it is possible to mitigate access availability issues, which may arise as per-attribute criteria groups are deployed.

1 Introduction

The increasing adoption of cloud computing has drawn attention to its security challenges [7,12]. This has led major cloud providers to incrementally add security features for increased costumers' confidence [1,12]. Access control, in particular, is acknowledged as one of the fundamental security features required to avoid unauthorized access to sensitive data and protect organizations assets.

Among prominent cloud providers, Microsoft Azure has stood out for its recent focus on security technologies for cloud resources [6]. In particular, Microsoft currently boasts features such as an extensive Security Center (backup features, dependability etc.), Automatic Monitoring, and support of Role-based Access Control (RBAC). The deployment of RBAC in Azure has been met by much praise, by both security researchers [23] and by practitioners [2].

The access control mechanism supported by Azure not only provides a suite of easy-to-use ways to implement RBAC, but also is integrated with the native identity management systems supported by Microsoft through Active Directory (referred to as AAD). One of the many peculiar features resulting from the integration is the ability to specify security groups, which can be used for role and privilege assignment. Security groups can be populated manually, on a per-user basis, or through dynamic triggers (aka dynamic groups).

© IFIP International Federation for Information Processing 2017
Published by Springer International Publishing AG 2017. All Rights Reserved
G. Livraga and S. Zhu (Eds.): DBSec 2017, LNCS 10359, pp. 349–361, 2017.
DOI: 10.1007/978-3-319-61176-1_20

Thus, Azure RBAC roles can be granted to Azure AD security groups with rule-based membership - effectively achieving *user attribute-based access control.* ABAC is deemed more desirable by researchers and practitioners since it provides flexibility and scalability for securely managing access to resources, particularly in collaborative environments.

Yet, implementing user attribute-based access control brings a set of new interesting challenges [1], particularly when dynamic security groups are used and managed by resource managers (and not identity managers).

Developers have no mean to check whether their attribute-based policies for resource (or resource group) access meet intended requirements and administrative units' policies. Identifying the right set of rules for security groups is a complicated and error prone process that should take into account resources' access control policies as well as user-attributes assignment policies.

Accordingly, automatic tools inspecting policy-compliant resource access in cloud - are desirable and needed to assist designers in policy development in order to avoid availability issues. Resource availability and accessibility are among the top security concerns of cloud adopters to date.

In this paper we propose to analyse user attribute-based policies developed under the Azure-like platform. Given dynamic groups implementing user-attribute policies to resources and administrative units policies, we first model the user attribute-based access control system as a state transition system that evolves via administrative actions to modify the user attributes and the user-groups membership; then we study the "group-reachability problem".

The Group-reachability problem. Given a subscription policy with unboundedly many users, a finite set of dynamic groups, a finite set of administrative units' policies, an initial configuration of the user attribute-based access control system, and a target security group *goal*, is there a reachable configuration of the access-control system where some user is assigned group *goal*?

We choose the GURA framework [9] to express user attributes assignments in Azure AD since it enables an elegant and succinct representation of user attributes administrative policies via RBAC. Our contributions include:

1. A proposal to develop user attribute-based access control policies for Azure management via a combination of AD dynamic security groups, GURA policies and the version of RBAC supported by Azure.
2. A proposal to analyse the group-reachability problem via security analysis of Administrative RBAC [3,16]. First, we show that the reachability problem in Azure-like AC systems is equivalent to the attribute-reachability problem in GURA systems [9]. We then reduce it to the role-reachability problem in ARBAC so that existing tools and techniques can be leveraged to address the group reachability problem.

We note that while we provide a concrete approach to verify user-attribute policies' for Azure-like platforms, our theoretical results on group reachability are agnostic to any implementation, as they could be applied in any user-attribute based access control system that supports dynamic groups.

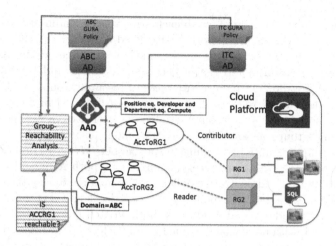

Fig. 1. Example scenario

2   Motivating Example

We now present our problem statement through a running example (inspired from discussions of existing cloud adopters). Our goal, through our proposed architecture is to ensure secure enforcement of organizations' access policies while ensuring *group-reachability*, that is to say, verify whether there is a reachable configuration where some group of users can eventually be granted access to a given resource or resource group, given (a) the available resources offered by a representative cloud provider such as Azure and (b) the administrative security requirements imposed by cloud collaborators and partners.

Consider a large-scale production project, InsightIT, involving multiple companies. Assume all such organizations share a production environment through an Azure Subscription (see Fig. 1). For simplicity, we focus on two of such collaborative units, ABC and ITC. ABC needs to set up a production environment for their client with PaaS Services like Azure App Service, Azure SQL Database (PaaS), Azure Storage. Assume that ABC employees are organized in 4 departments: storage, compute, networking, security. Each has corresponding responsibilities. ITC is doing development or testing work on some applications originally deployed by ABC. The applications are to be kept confidential to the public and the other partner collaborators, and ITC should only be given access to the environments needed for deployment and collaboration.

In order to enable the ITC Development team and the ABC on a single cloud platform (i.e. Azure Subscription), it is relatively easy (and recommended per Microsoft best practices [14]) to configure these PAAS services into separate containers, referred to as resource groups (see Fig. 1). ABC can configure its resources within two resource groups, RG1 and RG2, with RG2 including the company's Customer Databases and virtualized environment. RG1 can be used as a container for ITC and ABC collaborative effort. Access to RG2 and RG1 must be controlled carefully to comply with each organization's administrative policy (formalized in the GURA framework as discussed in Sect. 4). Access is granted to dynamic groups. These groups are specified by resource administrators (or resource owners) according to the project's access needs and the anticipated organizational workflow. In particular, access to resource groups can be granted by assigning designated roles to dynamic groups (discussed in Sect. 3.1). This enables automated access to resource groups per user-attributes. Let's refer to these dynamic groups as AccToRG1 and AccToRG2, respectively - and assume that membership is granted according to specific per-attribute conditions.

In addition to the resources access requirements, both ABC and ITC maintain distinct internal administrative policies, to be honored during the collaboration and used as a guidance for policy and access decisions. Typically, administrative policies include several conditions- both written (e.g. due to law compliance) and verbal, with no official documentation (reflecting the internal practices of an organization and its culture, e.g. interns are not allowed access to internal servers). For instance, part of ABC administrative policy mandates employees to have rotational periods in various departments. Testers are assigned to the Network department. Further, resources pooled for collaboration with partners may only be accessible to some ABC developers. Further, ABC domain is only granted to permanent workers. On the other hand, ITC, per its internal HR structure, maintains a policy stating that Tester is the official position of their Computing department.

Given a set of administrative policies like the ones above, checking consistency of group-based policies by simple inspection may be difficult. For instance, let's assume the RG1 owner, in an effort to accommodate both ITC and ABC administrative policies, decides that a user is assigned to a security group to access to RG1 only if he/she is both part of the Compute department and has a Tester position within the organization. However, this combination of attributes won't enable any ABC user to join the security group, as ABC Testers are mandated membership to Networking department as part of their rotational period. Similarly, ABC employees may be denied access to RG2 if the rule mandates only the ABC users to join the group, unaware of temporary domains assigned new employees or interns. These issues may be referred to as instances of the *group-reachability* problem.

3 Preliminaries

3.1 Access Control in Azure

Microsoft Azure is one of the dominant cloud IaaS platforms for enterprises. Azures core features include compute, storage, database and networking. An Azure account is referred to as *subscription*, and is overseen by one or more super admins, or owners. A subscription is essentially a container of the owner's Microsoft cloud environment. Azure access control include:

Azure Active Directory (AAD): provides a full suite of identity management capabilities, along with single sign-on (SSO) access to cloud SaaS Applications.

Users information within AAD is stored along with their attributes. Users can be organized in groups. Group owners or administrators can add users one-by-one, or in a criteria-based fashion. In the latter case, groups are referred to as dynamic groups. Users can be granted access to resources through individual role assignment or through group-role assignment.

Resources: are cloud assets which populate the subscriptions (e.g. virtual machines, databases, storage). Resource Groups (RG) are containers to segregate workloads and resources that require different access settings. Access to resources is assigned to RGs or to individual resources.

Roles (R): Users are assigned to a resource group with roles to get permissions to access to cloud resources. Azure supports a large set of built-in roles, and supports easy to add customized roles.

3.2 Generalized User Role Assignment

GURA model [9]: Let U be a finite set of users and $ATTR$ be a finite set of attributes. Each attribute is a function that takes users as input and returns a value from the attribute scope. Attributes can be atomic-valued or set-valued. An atomic-valued attribute will return a single value while a set-valued attribute will return a subset of values within its defined scope. Moreover, let AR be a finite set of administrative roles. Administrators are allowed to change attributes of a user according to some preconditions. A precondition is a logical formula expressed over user attributes that evaluates to **true** or **false**. Formally, administrative rules are tuples in the following relations where C is a set of preconditions:

- for each atomic-valued attribute $att \in ATTR$, permission to assign a particular value to att of a user is specified as: $assign \subseteq AR \times C \times ATTR \times SCOPE_{att}$. The meaning of an assign tuple $(admin, c, att, val) \in assign$ is that a member of the administrative role $admin \in AR$ can give value val to the atomic-valued attribute att of a user whose current user-attributes assignment satisfies precondition c, where $SCOPE_{att}$ is the scope of the attribute $att \in ATTR$;
- permission to add a particular value to the set-valued attribute att of a user is specified as: $add \subseteq AR \times C \times ATTR \times SCOPE_{att}$. The meaning of an add tuple $(admin, c, att, val) \in add$ is that a member of the administrative role $admin \in AR$ can add value val to the set of values for attribute att of a user whose current user-attributes assignment satisfies precondition c.
- permission to revoke the assignment of a value from any atomic-valued or set-valued attribute of a user is specified as: $delete \subseteq AR \times SCOPE_{att}$. The meaning of a tuple $(admin, val) \in delete$ is that value val can be deleted from attribute att of any user.

Management of membership in administrative roles is outside the scope of *GURA*. Thus, we assume that each role in AR always contains some administrator.

GURA Systems: A GURA system is a state transition system that evolves via administrative actions to modify user attributes. Formally, a GURA system is a tuple $S = \langle U, AR, ATTR, UATTR, assign, add, delete \rangle$ where $UATTR \subseteq U \times ATTR \times SCOPE_{att}$ is the user-attribute relation.

A *configuration* of S is any user-attribute assignment relation $UV \subseteq U \times ATTR \times SCOPE_{att}$. A configuration UV is *initial* if $UV = UATTR$.

Given two S configurations UV and UV', there is a *transition* from UV to UV' with rule $m \in (can\_assign \cup can\_add \cup can\_delete)$, denoted $UV \xrightarrow{m} UV'$, if one of the following holds:

[**assign move**] $m = (admin, c, att, val)$, the user-attribute assignment relation of a user u satisfy precondition c, att is an atomic-valued attribute, and $UV' = (UV \setminus \{(u, att, val') | (u, att, val') \in UV\}) \cup \{(u, att, val)\}$;

[**add move**] $m = (admin, c, att, val)$, the user-attribute assignment relation of a user u satisfy precondition c, att is a set-valued attribute, and $UV' = UV \cup \{(u, att, val)\}$;

[**delete move**] $m = (admin, att, val)$, and $UV' = UV \setminus \{(u, att, val)\}$;

A *run* of S is any finite sequence of S transitions $\pi = c_1 \xrightarrow{m_1} c_2 \xrightarrow{m_2} \ldots c_n \xrightarrow{m_n} c_{n+1}$ for some $n \geq 0$, where c_1 is an *initial* configuration of S. An S configuration c is reachable if c is the last configuration of an S run.

Definition 1 (ATTRIBUTE-REACHABILITY PROBLEM). *For any pair* (att, val) $\in ATTR \times SCOPE_{att}$, (att, val) *is reachable in* S *if there is an* S *reachable configuration* UV *such that* $(u, att, val) \in UATTR$, *for some* $u \in U$. *Given a GURA system* S *over the set of attributes* $ATTR$ *and a target pair* $(att, val) \in UV \times SCOPE_{att}$, *the attribute-reachability problem asks whether* (att, val) *is reachable in* S.

Restricted $GURA_1$ **system:** A restricted $GURA_1$ system ($rGURA_1$) is a $GURA$ system where a precondition can be expressed as a conjunction of literals, where each literal is either in positive form ℓ or in negative form $\neg \ell$, for some pair $\ell = (att, val) \in ATTR \times SCOPE_{att}$. A user u satisfies a positive pair (att, val) if $(att(u) = val)$ evaluates to **true**. On the contrary, she satisfies a negative pair if $\neg(att(u) = val)$ evaluates to **true**.

Since Azure AD policies always allow some administrator to revoke an attribute value, we assume that $rGURA_1$ policies contain a delete rule for each attribute-value.

3.3   Administrative Role Based Access Control

An RBAC policy is a tuple $\langle U, R, P, UA, PA \rangle$ where U, R and P are finite sets of *users*, *roles*, and *permissions*, respectively, $UA \subseteq U \times R$ is the *user-role assignment* relation, and $PA \subseteq P \times R$ is the *permission-role assignment* relation. A pair $(u, r) \in UA$ means that user u is a member of role r. Similarly, $(p, r) \in PA$ means that members of role r are granted the permission p.

The ARBAC-URA policy allows to change the user-role assignment UA by means of assignment/revocation rules carried out by administrators which are organized in a set AR of administrative roles.

Administrators are allowed to change roles of a user according to a precondition. A *precondition* is a conjunction of literals, where each literal is either in positive form r or in negative form $\neg r$, for some role r in R. A precondition can be partitioned in two sets denoted *Pos* and *Neg*, respectively corresponding to the set of roles that appear in positive and negative form in the precondition.

Permission to assign users to roles is specified as: $can\_assign \subseteq AR \times 2^R \times 2^R \times R$. The meaning of a can-assign tuple $(admin, Pos, Neg, r) \in can\_assign$ is that a member of the administrative role $admin \in AR$ can make a user whose current role memberships satisfies the precondition (Pos, Neg), a member of $r \in R$. In the rest of the paper we assume that $Pos \cap Neg = \emptyset$.

Permission to revoke users from roles is specified as: $can\_revoke \subseteq AR \times R$.

A tuple $(admin, r) \in can\_revoke$ means that a member of the administrative role $admin \in AR$, can revoke the membership of a user from a role $r \in R$.

ARBAC-URA Systems: An ARBAC-URA system is a state transition system that evolves via administrative actions to modify user role assignments. For a formal definition see [4].

Definition 2 (ROLE-REACHABILITY PROBLEM *[4]*). *For any role $r \in R$, r is reachable in an ARBAC-URA system S if there is an S reachable configuration UR such that $(u, r) \in UR$, for some $u \in U$. Given an URA system S over the set of roles R and a role* goal $\in R$, *the role-reachability problem asks whether* goal *is reachable in S.*

4 User Attribute-Based Access Control in Azure-Like Platforms and State Transition System

4.1 User Attribute-Based Access Control

We now briefly discuss how to enable User Attribute-Based Access Control in a RBAC deployment such as the one supported by Azure.

We define a practical Azure-like ABAC model. Let's assume a conventional formulation of users U, resources Res and Privilege sets P against resources in Res. Let a subscription denote a single Cloud domain where resources are maintained. Administrative units interface this subscription and corresponding resources by means of one Azure Active Directory. Users U are part of one or more administrative units connected to the AAD. The following elements define a User Attribute-based Access Control (UAA) model for an Azure-like platform.

- *Users and Attributes:* Each $u \in U$ is described by a unique identifier and a finite set of attributes $ATTR$. Each attribute is a function that takes users as input and returns a value from the attribute scope. Attributes can be atomic-valued or set-valued. An atomic-valued attribute will return a single value while a set-valued attribute will return a subset of values within its defined scope. An example of user is a User of id BobSmith, with attributes domain(BobSmith) = abc, Department(BobSmith) = $Network$ and Position(BobSmith) = $Tester$.

- *Administrative Units:* Users within a subscription are organized into one or more administrative units. Each administrative unit is responsible for managing attributes assignments for their users. In our framework, we express each administrative unit's policy by means of GURA policies. For instance, in our example of InsightIT, portions of the ABC and ITC administrative policies are reported in Tables 2 and 1, respectively (delete rules are excluded). We assume that each attribute can be revoked from a user regardless of any precondition. Administrative units are managed by designated administrators according to pre-defined administrative roles.
- *Dynamic Security Groups:* Users are organized in a finite set G of dynamic security groups- regardless of their original administrative unit. Moreover, *GLabel* is a function that maps a group to a logical formula expressed over user attributes that evaluates to **true** or **false**.
  Our example includes two dynamic groups AccToRG1 and AccToRG2. Examples of conditions of dynamic groups in our InsightIT example are reported in Table 3.
- *Privilege Assignment:* Resource administrators assign privileges to dynamic groups. Privileges are mediated by roles. That is to say, an access privilege is not assigned directly but obtained as part of a role-assignment.
  In our example, Bob Smith may be assigned *Contributor* role to resource group RG1 if he meets the requirements set for AccToRG1. Thus, users' access to resources is mediated by users' membership in dynamic groups.

We assume that resources are statically associated with dynamic groups, thus, in the rest of the paper we refer to an UAA policy as a tuple $\langle G, GLabel, GURAP \rangle$, where $GURAP = \langle U, AR, ATTR, UATTR, assign, add, delete \rangle$ is the *GURA* policy obtained by combining together all administrative units policies.

4.2 User Attribute-Based Systems

A User Attribute-based (UAA) system is a $GURA$ system where users can be organised in security groups according to their attributes. The system maintains the invariant that a user belongs to a group if and only if her attributes satisfies a given condition associated with the security group.

Formally, a *UAA* system is a state transition system defined as $S = \langle U, ARG, GLabel, GA, ATTR, UATTR, assign, add, delete \rangle$ where $\langle G, GLabel, GURAP \rangle$

Table 1. Examples of administrative rules for ITC

Rule	Admin	Pre-condition	Attr	Value
add	ITC Admin	Department(u) = Network $\land$ Position(u) = Developer	Department	Compute
add	ITC Admin	Position(u) = Tester	Department	Compute
assign	ITC Admin		Position	Tester

Table 2. Examples of administrative rules for ABC

Rule	Admin	Condition	Attr	Value
assign	ABC Admin	Department(u) = Network	Position	Tester
add	ABC Admin	Position(u) = Developer	Department	Compute
assign	ABC Admin		Position	Developer
assign	ABC Admin	Position(u) = guest	Domain	.abc
assign	ABC Admin	Position(u) = Intern	Domain	.abcBeta

Table 3. Two dynamic groups and corresponding conditions

Group	GLabel	Role and resource
AccToRG1	Position(u) = Tester and Department(u) = Compute	Contributor RG1
AccToRG2	Domain(u) = ABC	Reader RG2

is a UAA policy, $GURAP = \langle U, AR, ATTR, UATTR, assign, add, delete \rangle$ is the underlying $GURA$ system, and $GA \subseteq U \times G$ is the *user-group assignment* relation.

A *configuration* of $\mathcal{S}$ is any pair (UV, UG) where $UV \subseteq U \times ATTR \times SCOPE_{att}$ is a user-attribute assignment relation and $UG \subseteq U \times G$ is a user-group assignment relation. A configuration (UV, UG) is *initial* if $(UV, UG) = (UATTR, GA)$.

Given two $\mathcal{S}$ configurations (UV, UG) and (UV', UG'), there is a *transition* from (UV, UG) to (UV', UG') with rule $m \in (can\_assign \cup can\_add \cup can\_delete)$, denoted $(UV, UG) \xrightarrow{m} (UV', UG')$, if in the underlying $GURA$ system there is a transition $UV \xrightarrow{m} UV'$. Moreover, for each $(u, g) \in UG$ and $(u', g') \in UG'$, u satisfies $GLabel(g)$ and u' satisfies $GLabel(g')$.

A *run* of $\mathcal{S}$ is any finite sequence of $\mathcal{S}$ transitions $\pi = c_1 \xrightarrow{m_1} c_2 \xrightarrow{m_2} \dots c_n \xrightarrow{m_n} c_{n+1}$ for some $n \geq 0$, where c_1 is an *initial* configuration of $\mathcal{S}$. An $\mathcal{S}$ configuration c is reachable if c is the last configuration of an $\mathcal{S}$ run.

Definition 3 (GROUP-REACHABILITY PROBLEM). *For any group $g \in G$, g is reachable in $\mathcal{S}$ if there is an $\mathcal{S}$ reachable configuration (UV, UG) such that $(u, g) \in UG$, for some $u \in U$. Given an UAA system $\mathcal{S}$ and a target group $\mathbf{g} \in G$, the* group-reachability problem *asks whether $\mathbf{g}$ is reachable in $\mathcal{S}$.*

For instance, in our InsightIT example (Sect. 2) we can ask whether resource group RG2 can be reachable by ABC interns or whether group RG1 can be reachable by an ITC tester.

Restricted UAA (rUAA): A restricted UAA system is a UAA system where the underlying $GURA$ system is a $rGURA_1$ system.

5   Group, Attribute and Role Reachability

In this section, we first show that the group-reachability problem in Azure-like user attribute based access control is equivalent to the attribute-reachability problem in $GURA$. Then, we show that scalable techniques and tools that have been recently proposed to address the role-reachability problem in administrative RBAC can be employed to address the group-reachability problem for an interesting class of instances via a reduction from the attribute-reachability problem in $rGURA_1$ (see Sect. 3.2) to the ARBAC role-reachability problem. Proofs are omitted for lack of space.

Theorem 1. *The group-reachability problem in UAA is equivalent to the attribute-reachability problem in GURA.*

Corollary 1. *The group-reachability problem in rUAA is equivalent to the attribute-reachability problem in $rGURA_1$.*

We now show a reduction from the attribute-reachability problem in $rGURA_1$ to the role-reachability problem in $ARBAC\text{-}URA$.

Definition 4. (From $rGURA_1$ to ARBAC-URA). *Let $S = \langle U, AR,$ ATTR, UATTR, assign, add, delete$\rangle$ be a $rGURA_1$ system. We construct a corresponding ARBAC $-$ URA system $S = \langle U, R, AR, UA, can\_assign, can\_revoke\rangle$ as follows:*

1. *$U' = U$, and $AR' = AR$;*
2. *for each $(att, val) \in ATTR \times SCOPE_{att}$ add a role att_val to R;*
3. *for each tuple $(u, att, val) \in UATTR$ add a tuple (u, att_val) to UA;*
4. *for each $(admin, Pt, Nt, att, val) \in assign$ add $(admin, Pos, Neg, att\_val)$ to can_assign, where $Pos = \{att\_val \in R | (att, val) \in Pt\}$ and $Neg = \{att\_val \in R | (att, val) \in Nt\} \cup \{att\_val \in R | att$ is an atomic-valued attribute and $(att, val') \in Pt$ for some $val' \in SCOPE_{att}, val' \neq val\}$;*
5. *for each $(admin, Pt, Nt, att, val) \in add$ add $(admin, Pos, Neg, att\_val)$ to can_assign, where $Pos = \{att\_val \in R | (att, val) \in Pt\}$ and $Neg = \{att\_val \in R | (att, val) \in Nt\} \cup \{att\_val \in R | att$ is an atomic-valued attribute and $(att, val') \in Pt$ for some $val' \in SCOPE_{att}, val' \neq val\}$;*
6. *for each $(admin, att, val) \in delete$ add $(admin, att\_val)$ to can_revoke.*

Theorem 2. *Let $S = \langle U, AR, ATTR, UATTR, assign, add, delete\rangle$ be a $rGURA_1$ system. Let $S' = \langle U, R, AR, UA, can\_assign, can\_revoke\rangle$ be the corresponding ARBAC-URA system of Definition 4. The pair $(att^*, val^*) \in ATTR \times SCOPE_{att^*}$ is reachable in S iff the role $att^*\_val^*$ is reachable in S'.*

6   Related Work

Our work lies at the crossroad of two related research areas: access control in cloud and formal security analysis of administrative access control models.

Access Control in Cloud. Access control in cloud computing has gained great interest over the recent years [17,22], with emphasis on attribute based mechanisms [18,19,21]. Related work has also focused on secure information sharing in Cloud Computing environment [15]. Sandhu et al. have recently focused on models for sharing information and resources in various cloud systems, including Azure and Open-Stack [23,24]. In comparison to this body of work, we tackle the important issue of group-availability, which to our knowledge has not been addressed before. The main concept underlying such models is from Group-Centric Secure Information Sharing (g-SIS) [11], which introduces group-based information and resources sharing, allows sharing among a group of organizations.

Security Analysis. Li et al. [13] study role-based policies where role-membership rules may be added or removed by principals. Sasturkar et al. [20] prove reachability to be PSPACE-COMPLETE in ARBAC URA policies. Jayaraman et al. proposed Mohawk, a tool able to both finding shallow errors in complex ARBAC policies and proving correctness [8]. Ferrara et al. presented VAC, an automatic and scalable tool for the reachability problem of ARBAC policies with an unbounded number of users [3–5]. Ranise et al. proposed ASASPXL, which is able to analyse large ARBAC policies [16]. Sandhu et al. prove attribute-reachability to be PSPACE-COMPLETE in $rGURA_1$ policies [10].

7   Concluding Remarks

We presented an in-depth analysis of group-reachability in User-based access control systems. Starting from a concrete instance of an access control mechanism supported in the Azure platform, we demonstrated how it is possible to verify well-formedness in user-attribute access control. Well-formedness is analysed by addressing the group-reachability problem, which may arise as per-attribute criteria groups are used. We will provide a real-world architecture in the near future, and test it for scalability with respect to realistic scenarios.

Acknowledgements. Portions of Dr Squicciarini's work was funded from National Science Foundation Grant 1453080. Portions of Dr. Ferrara's work was supported by the EPSRC Grant no. EP/P022413/1. This research is also partly supported by the Microsoft Azure Internet of Things Research Award.

References

1. Beaver, K.: What admins should know about Microsoft Azure security and vulnerabilities.   http://searchwindowsserver.techtarget.com/tip/What-admins-should-know-about-Microsoft-Azure-security
2. Biz Tech: Why enterprises that value security trust Microsoft Azure. http://www.biztechmagazine.com/article/2016/10/why-microsoft-azure-essential-enterprises-value-security

3. Ferrara, A.L., Madhusudan, P., Nguyen, T.L., Parlato, G.: VAC - verifier of administrative role-based access control policies. In: Biere, A., Bloem, R. (eds.) CAV 2014. LNCS, vol. 8559, pp. 184–191. Springer, Cham (2014). doi:10.1007/978-3-319-08867-9_12

4. Ferrara, A.L., Madhusudan, P., Parlato, G.: Security analysis of role-based access control through program verification. In: Chong, S., (ed.) IEEE Computer Security Foundation, pp. 113–125. IEEE (2012)

5. Ferrara, A.L., Madhusudan, P., Parlato, G.: Policy analysis for self-administrated role-based access control. In: Piterman, N., Smolka, S.A. (eds.) TACAS 2013. LNCS, vol. 7795, pp. 432–447. Springer, Heidelberg (2013). doi:10.1007/978-3-642-36742-7_30

6. Freato, R.: Microsoft Azure Security. Packt Publishing Ltd, Birmingham (2015)

7. Inforworld: The dirty dozen: 12 cloud security threats (2016). http://www.infoworld.com/article/3041078/security/the-dirty-dozen-12-cloud-security-threats.html

8. Jayaraman, K., Tripunitara, M.V., Ganesh, V., Rinard, M.C., Chapin, S.J.: Mohawk: abstraction-refinement and bound-estimation for verifying access control policies. ACM Trans. Inf. Syst. Secur. 15(4), 18 (2013)

9. Jin, X., Krishnan, R., Sandhu, R.: A role-based administration model for attributes. In: First International Workshop on Secure and Resilient Architectures and Systems, pp. 7–12. ACM (2012)

10. Jin, X., Krishnan, R., Sandhu, R.: Reachability analysis for role-based administration of attributes. In: ACM Workshop on Digital Identity Management, pp. 73–84. ACM (2013)

11. Krishnan, R., Sandhu, R., Niu, J., Winsborough, W.: A conceptual framework for group-centric secure information sharing. In: 4th International Symposium on Information, Computer, and Communications Security, pp. 384–387. ACM (2009)

12. Krutz, R.L., Vines, R.D.: Cloud Security: A Comprehensive Guide to Secure Cloud Computing. Wiley Publishing, Hoboken (2010)

13. Li, N., Tripunitara, M.V.: Security analysis in role-based access control. In: 9th ACM SACMAT, pp. 126–135. ACM (2004)

14. Microsoft Azure Documentation. https://docs.microsoft.com/en-us/azure/active-directory/role-based-access-control-troubleshooting

15. Qiu, M., Gai, K., Thuraisingham, B., Tao, L., Zhao, H.: Proactive user-centric secure data scheme using attribute-based semantic access controls for mobile clouds in financial industry. Future Gener. Comput. Syst., 223–238 (2016)

16. Ranise, S., Truong, A.T., Armando, A.: Boosting model checking to analyse large ARBAC policies. In: Jøsang, A., Samarati, P., Petrocchi, M. (eds.) STM 2012. LNCS, vol. 7783, pp. 273–288. Springer, Heidelberg (2013). doi:10.1007/978-3-642-38004-4_18

17. Raykova, M., Zhao, H., Bellovin, S.M.: Privacy enhanced access control for outsourced data sharing. In: Keromytis, A.D. (ed.) FC 2012. LNCS, vol. 7397, pp. 223–238. Springer, Heidelberg (2012). doi:10.1007/978-3-642-32946-3_17

18. Riad, K., Yan, Z.: EAR-ABAC: an extended AR-ABAC access control model for SDN-integrated cloud computing. Environments 132(14), 9–17 (2015)

19. Riad, K., Yan, Z., Hu, H., Ahn, G-J.: AR-ABAC: a new attribute based access control model supporting attribute-rules for cloud computing. In: Collaboration and Internet Computing Conference (CIC), pp. 28–35. IEEE (2015)

20. Sasturkar, A., Yang, P., Stoller, S.D., Ramakrishnan, C.R.: Policy analysis for administrative role based access control. In: 19th IEEE Computer Security Foundations Workshop, (CSFW-19), pp. 124–138 (2006)

21. Wan, Z., Liu, J., Deng, R.: HASBE: a hierarchical attribute-based solution for flexible and scalable access control in cloud computing. IEEE Trans. Inf. Forensics Secur. **7**(2), 743–754 (2012)
22. Yu, S., Wang, C., Ren, K., Lou, W.: Achieving secure, scalable, and fine-grained data access control in cloud computing. In: IEEE Infocom, pp. 1–9 (2010)
23. Zhang, Y., Patwa, F., Sandhu, R.: Community-based secure information and resource sharing in Azure cloud IaaS. In: 4th International Workshop on Security in Cloud Computing, pp. 82–89. ACM (2016)
24. Zhang, Y., Patwa, F., Sandhu, R., Tang, B.: Hierarchical secure information and resource sharing in openstack community cloud. In: IEEE International Conference on Information Reuse and Integration, pp. 419–426. IEEE (2015)

Secure Storage in the Cloud

High-Speed High-Security Public Key Encryption with Keyword Search

Rouzbeh Behnia[✉], Attila Altay Yavuz, and Muslum Ozgur Ozmen

School of Electrical Engineering and Computer Science,
Oregon State University, Corvallis, USA
{behniar,attila.yavuz,ozmenmu}@oregonstate.edu

Abstract. Data privacy is one of the main concerns for clients who rely on cloud storage services. Standard encryption techniques can offer confidentiality; however, they prevent search capabilities over the encrypted data, thereby significantly degrading the utilization of cloud storage services. Public key Encryption with Keyword Search (PEKS) schemes offer encrypted search functionality to mitigate the impacts of privacy versus data utilization dilemma. PEKS schemes allow any client to encrypt their data under a public key such that the cloud, using the corresponding trapdoor, can later test whether the encrypted records contain certain keywords. Despite this great functionality, the existing PEKS schemes rely on extremely costly operations at the server-side, which often introduce unacceptable cryptographic delays in practical applications. Moreover, while data outsourcing applications usually demand long-term security, existing PEKS schemes do not offer post-quantum security.

In this paper, we propose (to the best of our knowledge) the first post-quantum secure PEKS scheme that is also significantly more computationally efficient than the existing (non-post-quantum) PEKS schemes. By harnessing the recently developed tools in lattice-based cryptography, the proposed scheme significantly outperforms the existing PEKS schemes in terms of computational overhead. For instance, the test (search) operation per item at the cloud side is approximately $36\times$ faster than that of the most prominent pairing-based scheme in the literature (for 192-bit security). The proposed PEKS scheme also offers faster encryptions at the client side, which is suitable for mobile devices.

Keywords: Public Key Encryption with Keyword Search · Cloud storage · Privacy · Lattice-based cryptography

1 Introduction

The emergence of cloud storage and computing services has revolutionized the IT industry. One of the most prominent cloud services is data storage outsourcing [3], which can drastically reduce the cost of data management via continuous service, expertise and maintenance for the resource-limited clients

© IFIP International Federation for Information Processing 2017
Published by Springer International Publishing AG 2017. All Rights Reserved
G. Livraga and S. Zhu (Eds.): DBSec 2017, LNCS 10359, pp. 365–385, 2017.
DOI: 10.1007/978-3-319-61176-1_21

(e.g., small/medium size businesses). Despite its merits, data outsourcing raises significant privacy concerns for users. Traditional data encryption techniques (e.g., symmetric ciphers) can be used to mitigate this concern. However, they abolish the data owner from performing any efficient search (and therefore retrieval) operation over the data that is remotely stored on the cloud. Various privacy enhancing technologies have been proposed towards addressing this limitation.

Searchable Encryption (SE) schemes allow a keyword-based search functionality over the encrypted data. SE schemes are generally applied to a client/server architecture, in which the client stores her encrypted data on a remote server. There are two main types of SE technologies: (i) Dynamic Symmetric SE (DSSE) (e.g., [30,41,43]) permits a client to perform encrypted search on her data via her own private key. DSSE is efficient as it relies on symmetric primitives, but it is rather suitable for a single client outsourcing/searching her own data. (ii) Public Key Encryption with Keyword Search (PEKS) [6] schemes allows *any* client to encrypt her data under a public key such that the server can later test whether the encrypted records contain certain keywords via search trapdoors produced by an entity holding the private key. PEKS is suitable for distributed applications (e.g., email, audit logging for Internet of Things), in which large number of entities generate encrypted data to be searched/analyzed by a particular auditor. *The focus of this paper is on PEKS schemes.* In Fig. 1, we provide some example applications of PEKS with their corresponding system model.

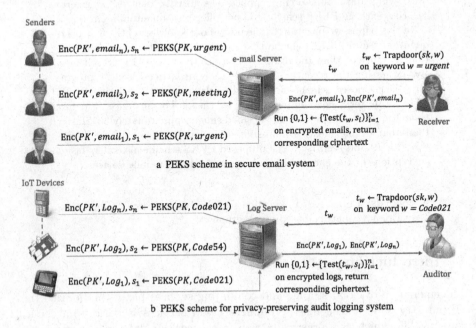

a PEKS scheme in secure email system

b PEKS scheme for privacy-preserving audit logging system

Fig. 1. Potential applications of PEKS schemes

One of the potential application scenarios for PEKS schemes is illustrated in Fig. 1-a. Alice has a number of devices (e.g., desktop, pager), and her email gateway is supposed to route her emails based on the keywords associated with each email. For instance, when one of the originators (i.e., Bob) sends her an email with keyword "urgent" the email should be routed to her pager. To achieve this, Bob encrypts his email using a standard public key encryption and uses PEKS algorithm to generate a searchable ciphertext of keyword $w = $ "*urgent*" to be associated with the email. Alice can then provide the server with trapdoor t_w computed for keyword w and enable the gateway to test whether any of the stored emails is associated with w via the Test algorithm of PEKS.

As depicted in Fig. 1-b, another possible scenario of employing PEKS schemes is for storing private log files on a remote server. In Internet of Things (IoT) applications, PEKS schemes can enable a set of heterogeneous devices to send their log files (encrypted under the auditor's public key) along with a searchable ciphertext of the keyword related to the logs to a storage server. To look for a specific event, the auditor can compute and send a trapdoor (of any keywords of his interest) to the server and receive all the files that contain the keyword.

1.1 Research Gap

There have been various PEKS schemes with additional features proposed in the literature [8,11,37,39]. We have identified two main research gaps that pose an obstacle towards the adoption of PEKS schemes in practice.

- Extreme Computational Overhead: Most of the proposed PEKS schemes are based on heavy pairing computations, and the schemes that are devised by pairing-free tools are even more costly than their pairing-based counterparts [15]. Despite their elegance, the existing constructions introduce a significant cryptographic delay as the server has to run a costly algorithm (i.e., Test(.) requiring at least one pairing operation per item) linear to the size of database.
- Lack of Long-term Security: When dealing with sensitive user data, most applications require long-term cryptographic security (e.g., sensitive medical data). However, due to algorithmic breakthroughs and the rise of powerful computers, the size of cryptographic keys are required to be steadily increased to ensure the same level of security. This causes the conventional crypto-graphic tools (e.g., RSA, ECC) to become increasingly inefficient. Therefore, there is a need for PEKS schemes that have a more efficient response against increasing key sizes. More importantly, with the predictions on the emergence of quantum computers, it is necessary to devise PEKS schemes that can achieve post-quantum security. However, current PEKS schemes are built on ECC or quadratic residuosity problems [15], which do not offer post-quantum security.

1.2 Our Contribution

Towards addressing aforementioned problems, we introduce the first (to the best of our knowledge) NTRU-Based PEKS scheme which we refer to as NTRU-PEKS hereafter. In the following, we outline our contributions.

- A New PEKS via NTRU: In the initial proposal of PEKS [6], Boneh et al. showed how one could derive a PEKS scheme from an IBE scheme. Abdalla et al. [1] supported this claim and provided requirements for the underlying IBE scheme to ensure the security and correctness of the derived PEKS. This led to the proposal of a large number of PEKS schemes (e.g., [15,31,44]) based on different IBE schemes. In this paper, we rigorously prove that Ducas et al.'s IBE scheme [17] meets these requirements and put forth the first NTRU-based PEKS scheme by leveraging this IBE. Furthermore, we prove the security and consistency of our PEKS scheme and suggest parameter sizes to avoid potential decryption errors.
- High Efficiency: We devise a highly efficient PEKS scheme that significantly reduces the cryptographic delay by harnessing the latest advancements in lattice-based cryptography, ring-LWE [38] and fast arithmetic operations over polynomial rings $\mathbb{Z}[x]/(x^N + 1)$. We implement our scheme[1] for 80-bit and 192-bit security and compare its efficiency with the most prominent PEKS schemes. As it is shown in Table 1, our scheme has significantly more efficient Test and PEKS algorithms than those in in [6,44]. The efficiency of Test algorithm is of vital importance, since it is executed by the server linearly with the total number encrypted keywords to be searched. The efficiency of the PEKS algorithm facilitates the implementation of PEKS schemes on battery-limited devices.
- Long-term Security: We develop the first practical NTRU-based PEKS that offers long-term security, while being (currently) secure against quantum computers, thanks to the security guarantees of lattice-based cryptography.

2 Preliminaries

In this section, we provide definitions and notations that are used by our scheme. For the sake of compliance, we use the same notation as in [17].

Notations. $a \xleftarrow{\$} \mathcal{X}$ denotes that a is randomly selected from distribution $\mathcal{X}$. H_i for $i \in \{1,\dots,n\}$ denotes a hash function which is perceived to behave as a random oracle in this paper. $\mathcal{A}^{\mathcal{O}_1\cdots\mathcal{O}_n}(.)$ denotes algorithm $\mathcal{A}$ is provided with access to oracles $\mathcal{O}_1\dots\mathcal{O}_n$. We denote scalars in plain (e.g., x) and vectors in bold (e.g., $\mathbf{x}$). The norm of a vector $\mathbf{v}$ is denoted by $\|\mathbf{v}\|$. $\lceil x \rfloor$ rounds x to the closest integer. $x \triangleq y$ means x is defined as y. The function $\gcd(x,y)$ returns the greatest common divisor of values x and y.

---

[1] The complete implementation can be found on https://github.com/Rbehnia/NTRUPEKS.git.

Table 1. Comparison of our NTRU-PEKS scheme with state-of-the-art

Schemes[a]		Test (ms)	PEKS (ms)	Trapdoor (ms)	QC[c] Resiliency
BCO[b]	$\kappa = 80$	3.38	2.53	0.36	χ
	$\kappa = 192$	43.39	46.02	2.69	χ
ZI[b]	$\kappa = 80$	8.12	16.37	1.05	χ
	$\kappa = 192$	118.65	194.42	5.61	χ
NTRU-PEKS	$\kappa = 80$	**0.34**	**0.69**	5.15	χ
	$\kappa = 192$	**1.23**	**2.50**	17.35	$\checkmark$

[a] Experimental setup and evaluation metrics are given is Sect. 5.
[b] BCO and ZI denote Boneh et al.'s scheme [6] and Zhang-Imai scheme [44],
respectively.
[c] QC stands for Quantum Computer.

2.1   NTRU-Based Cryptographic Tools

Ajtai [2] introduced the Short Integer Solution (SIS) problem and demonstrated
the connection between average-case SIS problem and worst-case problems over
lattices. Hoffstein et al. [26] proposed an efficient public key encryption scheme
called NTRU-based on polynomial rings. Regev [38] introduced the Learning
with Error (LWE) problem. The SIS and LWE problems have been used as the
building blocks of many lattice-based schemes.

NTRU encryption works over rings of polynomials $\mathcal{R} \triangleq \mathbb{Z}[x]/(x^N + 1)$ and
$\mathcal{R}' \triangleq \mathbb{Q}[x]/(x^N + 1)$ which are parametrized with N as a power-of-two integer.
$(x^N + 1)$ is irreducible, therefore, $\mathcal{R}'$ is a cyclotomic field. For $f = \sum_{i=0}^{N-1} f_i x^i$ and
$g = \sum_{i=0}^{N-1} g_i x^i$ as polynomials in $\mathbb{Q}[x]$, fg denotes polynomial multiplication in
$\mathbb{Q}[x]$ while $f * g \triangleq fg \mod (x^N + 1)$ is referred to as convolution product. For
an N-dimension anti-circulant matrix $\mathcal{A}_N$ we have $\mathcal{A}_N(f) + \mathcal{A}_N(g) = \mathcal{A}_N(f + g)$,
and $\mathcal{A}_N(f) \times \mathcal{A}_N(g) = (f * g)$.

Definition 1. *For prime integer q and $f, g \in \mathcal{R}$, $h = g * f^{-1} \mod q$, the NTRU
lattice with h and q is $\Lambda_{h,q} = \{(u, v) \in \mathcal{R}^2 | u + v * h = 0 \mod q\}$. $\Lambda_{h,q}$ is a full-
rank lattice generated by $\mathcal{A}_{h,q} = \begin{pmatrix} \mathcal{A}_N(h) & I_N \\ qI_N & 0_N \end{pmatrix}$, where I is an identity matrix.*

Note that one can generate this basis using a single polynomial $h \in \mathcal{R}_q$. However, the lattice generated from $\mathcal{A}_{h,q}$ has a large orthogonal defect which results
in inefficiency of standard lattice operations. As proposed by [25], another basis
(which is much more orthogonal) can be efficiently [17] generated by selecting
$F, G \in \mathcal{R}$ and computing $f * G - g * F = q$. The new base $\mathbf{B}_{f,g} = \begin{pmatrix} \mathcal{A}(g) & -\mathcal{A}(f) \\ \mathcal{A}(G) & -\mathcal{A}(F) \end{pmatrix}$
generates the same lattice $\Lambda_{h,q}$.

Definition 2 *(Gram-Schmidt norm [22]). Given $\mathbf{B} = (b_i)_{i \in I}$ as a finite basis
and $\tilde{\mathbf{B}} = (\tilde{b}_i)_{i \in I}$ as its Gram-Schmidt orthogonalization, the Gram-Schmidt norm
of $\mathbf{B}$ is $\left\| \tilde{\mathbf{B}} \right\| = \max_{i \in I} \|b_i\|$.*

Using Gaussian sampling, Gentry et al. [22] proposed a technique to use a short basis as trapdoor without disclosing any information about the short basis and prevent attacks similar as in [36].

Definition 3. *An n-dimensional Gaussian function $\rho_{\sigma,c} : (\mathbb{R} \to (0,1])$ is defined as $\rho_{\sigma,c}(x) \triangleq \exp(-\frac{\|x-c\|^2}{2\sigma^2})$. Given a lattice $\Lambda \subset \mathbb{R}^n$, the discrete Gaussian distribution over Λ is $D_{\Lambda,s,c}(\mathbf{x}) = \frac{\rho_{\sigma,c}(\mathbf{x})}{\rho_{\sigma,c}(\Lambda)}$ for all $\mathbf{x} \in \Lambda$.*

If we pick a noise vector over a Gaussian distribution with the radius not smaller than the *smoothing parameter* [34], and reduce the vector to the fundamental parallelepiped of our lattice, the resulting distribution is close to uniform. We formally define this parameter through the following definition.

Definition 4 *(Smoothing Parameter [34]). For any n-dimensional lattice Λ, its dual Λ^* and $\epsilon > 0$, the smoothing parameter $\eta_\epsilon(\Lambda)$ is the smallest $s > 0$ such that $\rho_{1/s\sqrt{2\pi},0}(\Lambda^* \backslash 0) \leqslant \epsilon$. A scaled version of the smoothing parameter is defined in [17] as $\eta'_\epsilon = \frac{1}{\sqrt{2\pi}}\eta_\epsilon(\Lambda)$.*

Gentry et al. [22] defined a requirement on the size of σ related to the smoothing parameter. In [17], Ducas et al. showed that using Kullback-Leibler divergence, the required width of σ can be reduced by factor of $\sqrt{2}$. Based on [17, 18, 22], for positive integers n, λ, $\epsilon \leqslant 2^{-\lambda/2}/(4\sqrt{2N})$, any basis $\mathbf{B} \in \mathbb{Z}^{N \times N}$ and any target vector $\mathbf{c} \in \mathbb{Z}^{1 \times n}$, the algorithm ($\mathbf{v_0} \leftarrow$ **Gaussian-Sampler**$(\mathbf{B}, \sigma, \mathbf{c})$) as defined in [17,22] is such that $\Delta(D_{\Lambda(\mathbf{B}),\sigma,\mathbf{c}}, \mathbf{v_0})$ is less than $2^{-\lambda}$.

In this paper, we will use the same algorithm in our **Trapdoor** algorithm.

Definition 5 *(Decision LWE Problem). Given $\mathcal{R} \triangleq \mathbb{Z}[x]/(x^N + 1)$ and an error distribution $\mathcal{X}$ over $\mathbb{R}$. For s as a random secret ring element, uniformly random a_i's $\in \mathcal{R}$ and small error elements $e_i \in \mathcal{X}$, the decision LWE problem asks to distinguish between samples of the form $(a_i, a_i s + e_i)$ and randomly selected $(a_i, b_i) \in \mathcal{R} \times \mathcal{R}$.*

Definition 6 *(A tool for computing Gram-Schmidt norm [17]). Let $f \in \mathcal{R}'$, we denote $\bar{f}$ as a unique polynomial in $f \in \mathcal{R}'$ such that $\mathcal{A}(f)^T = \mathcal{A}(\bar{f})$. If $f(x) = \sum_{i=0}^{N-1} f_i x^i$, then $\bar{f}(x) = f_0 - \sum_{i=1}^{N-1} f_{N-i} x^i$.*

2.2 Identity-Based Encryption

Definition 7. *An IBE scheme is a tuple of four algorithms IBE = (Setup, Extract, Enc, Dec) defined as follows.*

- *$(mpk, msk) \leftarrow$ Setup(1^k): On the input of the security parameter(s), this algorithm publishes system-wide public parameters params, outputs the master public key mpk and the master secret key msk.*
- *$sk \leftarrow$ Extract(id, msk, mpk): On the input of a user's identity $id \in \{0,1\}^*$, mpk, and msk, this algorithm outputs the user's private key sk.*

- $c \leftarrow$ Enc(m, id, mpk): On the input of a message $m \in \{0,1\}^*$, identity id, and mpk, this algorithm outputs a ciphertext c.
- $m \leftarrow$ Dec(c, sk): On the input of a ciphertext c, the receiver's private key sk and mpk, this algorithm recovers the message m from the ciphertext c.

Following the work of [1], the following definition defines anonymity in the sense of [24].

Definition 8. *Anonymity under chosen plaintext attack (IBE-ANO-RE-CPA) for an IBE scheme is defined as follows. Given an IBE scheme, we associate a bit $b \in \{0,1\}$ to the adversary $\mathcal{A}$ in the following experiment.*

$Experiment Exp_{IBE,\mathcal{A}}^{IBE\text{-}ANO\text{-}RE\text{-}CPA\text{-}b}(1^k)$

$idSet \leftarrow \emptyset, (mpk, msk) \xleftarrow{\$} $ Setup(1^k) KeyQuery(id)
for a random oracle H $idSet \leftarrow idSet \cup id$
$(id_0, id_1, m) \leftarrow \mathcal{F}^{\text{KeyQuery}(.), H}(find, mpk)$ $sk \leftarrow$ Extract(id, msk, mpk)
$c \leftarrow$ Enc$^H(m, id_b, mpk)$ *return sk*
$b' \leftarrow \mathcal{F}^{\text{KeyQuery}(.), H}(guess, c)$
if $\{id_0, id_1\} \cap idSet = \emptyset$ return b' else, return 0

$\mathcal{A}$'s advantage in the above experiment is defined as $Adv_{IBE,\mathcal{A}}^{IBE\text{-}ANO\text{-}RE\text{-}CPA}(1^k) = \Pr[Exp_{IBE,\mathcal{A}}^{IBE\text{-}ANO\text{-}RE\text{-}CPA\text{-}1}(1^k) = 1] - \Pr[Exp_{IBE,\mathcal{A}}^{IBE\text{-}ANO\text{-}RE\text{-}CPA\text{-}0}(1^k) = 0]$.

2.3 Public Key Encryption with Keyword Search

A PEKS scheme consists of the following algorithms.

Definition 9. *A PEKS scheme is a tuple of four algorithms PEKS = (KeyGen, PEKS, Trapdoor, Test) defined as follows.*

- $(pk, sk) \leftarrow$ KeyGen(1^k): On the input of the security parameter(s), this algorithm outputs the public and private key pair (pk, sk).
- $s_w \leftarrow$ PEKS(pk, w): On the input of user's public key pk and a keyword $w \in \{0,1\}^*$, this algorithm outputs a searchable ciphertext s_w.
- $t_w \leftarrow$ Trapdoor(sk, w): On the input of a user's private key sk and a keyword $w \in \{0,1\}^*$, this algorithm outputs a trapdoor t_w.
- $b \leftarrow$ Test(t_w, s_w): On the input of a trapdoor $t_w =$ Trapdoor(sk, w') and a searchable ciphertext $s_w =$ PEKS(pk, w), this algorithm outputs a bit $b = 1$ if $w = w'$, and $b = 0$ otherwise.

Definition 10. *Keyword indistinguishability against an adaptive chosen-keyword attack (IND-CKA) is defined as follows. Given a PEKS scheme, we associate a bit $b \in \{0,1\}$ to the adversary $\mathcal{A}$ in the following experiment.*

$Experiment Exp_{PEKS,A}^{PEKS-IND-CKA-b}(1^k)$

$wSet \leftarrow \emptyset$, $(pk, sk) \leftarrow \mathtt{KeyGen}(1^k)$	$\mathtt{TdQuery}(w)$
for a random oracle H	$\quad wSet \leftarrow wSet \cup w$
$(w_0, w_1) \leftarrow \mathcal{A}^{\mathtt{TdQuery}(.),H}(\mathit{find}, pk)$	$\quad sk \leftarrow \mathtt{Extract}(w, sk, pk)$
$s_w \leftarrow \mathtt{PEKS}^H(pk, w_b)$	$\quad \mathit{return}\ sk$
$b' \leftarrow \mathcal{A}^{\mathtt{TdQuery}(.),H}(\mathit{guess}, s_w)$	
if $\{w_0, w_1\} \cap wSet = \emptyset$ *return* b' *else, return* 0	

$\mathcal{A}$'s advantage in the above experiment is defined as $Adv_{PEKS,A}^{PEKS-IND-CKA}(1^k) = \Pr[Exp_{PEKS,A}^{PEKS-IND-CKA-1}(1^k) = 1] - \Pr[Exp_{PEKS,A}^{PEKS-IND-CKA-0}(1^k) = 0]$.

2.4   Consistency of PEKS

Due to the properties of NTRU-based encryption scheme, and following the work of [15], we investigate the consistency of our scheme from two aspects, namely, right-keyword consistency and adversary-based consistency [1]. Right-keyword consistency implies the success of a search query to retrieve records associated with keyword w for which the PEKS algorithm had computed a searchable ciphertext. On the other hand, adversary-based consistency [1] ensures the inability of an adversary to generate two distinct keywords that the Test algorithm returns 1 on the input of a trapdoor for one keyword, and the searchable ciphertext of the other. We define the adversary-based consistency [1] as follows.

Definition 11. *Adversary-based consistency of a PEKS scheme is defined in the following experiment.*
$Experiment Exp_{PEKS,A}^{PEKS-Consist}(1^k)$

$(pk, sk) \leftarrow \mathtt{KeyGen}(1^k)$
for a random oracle H
$(w_0, w_1) \leftarrow \mathcal{A}^H(pk)$, $s_{w_0} \leftarrow \mathtt{PEKS}^H(pk, w_0)$
$t_{w_1} \leftarrow \mathtt{Trapdoor}^H(pk, w_1)$
if $w_0 \neq w_1$ *and* $[\mathtt{Test}^H(pk, t_{w_1}, s_{w_0}) = 1]$ *return* 1 *else, return* 0
$\mathcal{A}$'s advantage in the above experiment is defined as $Adv_{PEKS,A}^{PEKS-Consist}(1^k) = \Pr[Exp_{PEKS,A}^{PEKS-Consist}(1^k) = 1]$.

3   Proposed Scheme

In this section, we present our scheme that consists of the following algorithms.

$(h, B) \leftarrow \mathtt{KeyGen}(q, N)$: Given a power-of-two integer N and a prime q, this algorithm works as follows.

1. Compute $\sigma_f \leftarrow 1.17\sqrt{\frac{q}{2N}}$ and select $f, g \leftarrow \mathcal{D}_{N,\sigma_f}$ to compute $\left\|\tilde{\mathbf{B}}_{f,g}\right\|$ and $Norm \leftarrow \max(\|(g, -f)\|, \left\|(\frac{g\bar{f}}{f*\bar{f}+g*\bar{g}}, \frac{g\bar{g}}{f*\bar{f}+g*\bar{g}})\right\|)$. If $Norm < 1.17\sqrt{q}$, proceed to the next step. Otherwise, if $Norm \geq 1.17\sqrt{q}$, this process is repeated by sampling new f and g.

2. Using extended euclidean algorithm, compute $\rho_f, \rho_g \in \mathcal{R}$ and $\mathcal{R}_f, \mathcal{R}_g \in \mathbb{Z}$ such that $\rho_f \cdot f = \mathcal{R}_f \mod (x^N + 1)$ and $\rho_g \cdot g = \mathcal{R}_g \mod (x^N + 1)$. Note that if $\gcd(\mathcal{R}_f, \mathcal{R}_g) \neq 1$ or $\gcd(\mathcal{R}_f, q) \neq 1$, start from the previous step by sampling new f and g.

3. Using extended euclidean algorithm, compute $u, v \in \mathbb{Z}$ such that $u \cdot \mathcal{R}_f + v \cdot \mathcal{R}_g = 1$. Compute $F \leftarrow q \cdot v \cdot \rho_g$, $G \leftarrow q \cdot u \cdot \rho_f$ and $k \leftarrow \lfloor \frac{F*\bar{f}+G*\bar{g}}{f*f+g*g} \rceil \in \mathcal{R}$ and reduce F and G by computing $F \leftarrow F - k*f$ and $G \leftarrow G - k*g$.

4. Finally, compute $h = g*f^{-1} \mod q$ and $\mathbf{B} = \begin{pmatrix} \mathcal{A}(g) & -\mathcal{A}(f) \\ \mathcal{A}(G) & -\mathcal{A}(F) \end{pmatrix}$ and output $(pk \leftarrow h, sk \leftarrow \mathbf{B})$.

$s_w \leftarrow \mathtt{PEKS}(pk, w)$: Given cryptographic hash functions $H_1 : \{0,1\}^* \rightarrow \mathbb{Z}_q^N$ and $H_2 : \{0,1\}^N \times \{0,1\}^N \rightarrow \mathbb{Z}_q^N$, the receiver's public key pk and a keyword $w \in \{0,1\}^*$ to be encrypted, the sender performs as follows.

1. Compute $t \leftarrow H_1(w)$ and pick $r, e_1, e_2 \overset{\$}{\leftarrow} \{-1, 0, 1\}^N$, $k \overset{\$}{\leftarrow} \{0,1\}^N$.
2. Compute $A \leftarrow r*h + e_1 \in \mathcal{R}_q$ and $B \leftarrow r*t + e_2 + \lfloor \frac{q}{2} \rfloor k \in \mathcal{R}_q$.
3. Finally, the algorithm outputs $s_w = \langle A, B, H_2(k, B) \rangle$.

$t_w \leftarrow \mathtt{Trapdoor}(sk, w)$: Given the receiver's private key sk, and a keyword $w \in \{0,1\}^*$, the receiver computes $t \leftarrow H_1(w)$ and using the sampling algorithm $\mathtt{Gaussian\text{-}Sampler}(B, \sigma, (t, 0))$, samples s and t_w such that $s + t_w * h = t$.

$b \leftarrow \mathtt{Test}(pk, t_w, s_w)$: On the input of a receiver's public key pk, a trapdoor t_w and a searchable ciphertext $s_w = \langle A, B, H_2(k, B) \rangle$, this algorithm computes $y \leftarrow \lfloor \frac{B - A*t_w}{q/2} \rceil$ and outputs $b = 1$ if $H_2(y, B) = H_2(k, B)$ and $b = 0$, otherwise.

3.1 Completeness and Consistency

In this section, we show the completeness and consistency of our scheme.

Lemma 1. *Given a public-private key pair $(h, B) \leftarrow \mathtt{KeyGen}(q, N)$, a searchable ciphertext $s_w \leftarrow \mathtt{PEKS}(pk, w)$, and a trapdoor generate by the receiver $t_w \leftarrow \mathtt{Trapdoor}(sk, w)$ our proposed scheme is complete.*

Proof. To show the completeness of our scheme for $s_w = \langle A, B, H_2(k, B) \rangle$, the Test algorithm should return 1 when $\lfloor \frac{B - A*t_w}{q/2} \rceil = k$. To affirm this, we work as follows.

$$B - A*t_w = (r*t + e_2 + \lfloor \frac{q}{2} \rfloor k) - (r*h + e_1)*t_w \in \mathcal{R}_q$$

$$= r*s + r*h*t_w + e_2 + \lfloor \frac{q}{2} \rfloor k - r*h*t_w - t_w*e_1 \in \mathcal{R}_q$$

$$= r*s + e_2 + \lfloor \frac{q}{2} \rfloor k - t_w*e_1 \in \mathcal{R}_q$$

Given r, e_1, e_2, t_w and s are all short vectors (due to the parameters of our sampling algorithm), all the coefficients of $r*s + e_2 - t_w*e_1$ will be in $(-\frac{q}{4}, \frac{q}{4})$, and therefore, $\lfloor \frac{B - A*t_w}{q/2} \rceil = k$. □

To address right-keyword consistency issues related to the decryption error of encryption over NTRU lattices, we need to make sure that all the coefficients of $z = r * s + e_2 - e_1 * tw$ are in the range $(-\frac{q}{4}, \frac{q}{4})$ and $q \approx 2^{24}$ for $\kappa = 80$ and $q \approx 2^{27}$ for $\kappa = 192$.

Theorem 1. *The NTRU-PEKS scheme is consistent in the sense of Definition 11.*

Proof. Upon inputting q and N, the challenger $\mathcal{C}$ initiates the experiment $(h, \mathbf{B}) \leftarrow \mathtt{KeyGen}(q, N)$. It passes h to the adversary $\mathcal{A}$ and keeps $\mathbf{B}$ secret.

$(w_0, w_1) \leftarrow \mathcal{A}^{H_1}(pk)$: $\mathcal{A}$ sends $\mathcal{C}$ two keywords (w_0, w_1).

$s_{w_0} \leftarrow \mathtt{PEKS}^H(pk, w_b)$: $\mathcal{C}$ computes $A = r * h + e_1$ and $B = r * H(w_0) + e_2 + \lfloor \frac{q}{2} \rfloor k$
for a random selection of $r, e_1, e_2 \stackrel{\$}{\leftarrow} \{-1, 0, 1\}^N$, $k \stackrel{\$}{\leftarrow} \{0, 1\}^N$, and sends $\langle A, B, H_2(k, B) \rangle$ to $\mathcal{A}$.

$t_{w_1} \leftarrow \mathtt{Trapdoor}^H(pk, w_b)$: $\mathcal{C}$ samples short vectors s, t_w such that $s + t_w * h = H(w_1)$ and returns t_w to $\mathcal{A}$.

Following Definition 11, $\mathcal{A}$ wins when $w_0 \neq w_1$, and the Test algorithm outputs 1 (i.e., $H_2(k, B) = H_2(y, B)$).

Note that in the above game, $\mathcal{A}$ wins when $w \neq w'$ and $H_2(z_1, z_1') = H_2(z_2, z_2')$. Let's assume $\mathcal{A}$ makes q_1 queries to H_1 and q_2 queries to H_2 oracles. Let E_1 be the event that there exist (x_1, x_2) such that $H_1(x_1) = H_1(x_2)$ and $x_1 \neq x_2$ and let E_2 be the event that there exist two pairs (z_1, z_1') and (z_2, z_2') such that $H_2(z_1, z_1') = H_2(z_2, z_2')$ for $z_1 \neq z_2$ and $z_1' \neq z_2'$. Then if $\Pr[\cdot]$ represents the probability of consistency definition,

$$Adv_{PEKS,\mathcal{A}}^{PEKS-Consist}(1^k) \leq \Pr[E_1] + \Pr[E_2] + \Pr[Exp_{PEKS,\mathcal{A}}^{PEKS-Consist} = 1 \wedge \bar{E}_1 \wedge \bar{E}_2]$$

Given the domain of our hash functions, the first and second terms are upper bounded by $(q_1+2)^2/N2^{\log_2 q}$ and $(q_2+2)^2/N2^{\log_2 q}$, respectively. For the last term, if $H_1(x_1) \neq H_1(x_2)$, then in our scheme, the probability that $B_1 = B_2$ is negligible due to the decryption error. Therefore, $H_2(y_1, B_1) \neq H_2(y_2, B_2)$, hence, the probability of the last term is also negligible.   □

3.2   Discussion on Alternative NTRU-Based Constructions

Bellare et al. [5] proposed a new variation of public key encryption with search capability called Efficiently Searchable Encryption (ESE). The idea behind ESE is to store a deterministically computed "tag" along with the ciphertext. To respond to search queries, the server only needs to lookup for a tag in a list of sorted tags. This significantly reduces the search time on the server. For ESE to provide privacy, the keywords need to be selected from a distribution with a high min-entropy. To compensate for privacy in absence of high min-entropy distribution for keywords, the authors suggested truncating the output of the hash function to increase the probability of collisions. However, this directly affects the consistency of the scheme and shifts the burden of decrypting unrelated responds to the receiver. As compared to PEKS schemes, in ESE schemes, the tag can be computed from both the plaintext and ciphertext. This highly differentiates the applications of these two searchable encryption schemes.

In this paper, we focused on PEKS scheme as it does not have consistency issues or min-entropy distribution requirement, and fits better for our target real-life applications (as discussed in Sect. 1). Nevertheless, for the sake of completeness, to extend the advantages of NTRU-based encryption [42] to ESE, we also instantiated an NTRU-based ESE scheme based on the encrypt-with-hash transformation proposed in [5]. We compared it with its counterpart which was instantiated based on El-Gamal encryption. Our implementations of NTRU-based ESE and El-Gamal ESE (developed on elliptic curves) were run on an Intel i7 6700HQ 2.6 GHz CPU with 12 GB of RAM. We observed that encryption for NTRU-based ESE takes 0.011 ms where encryption in El-Gamal ESE takes 2.595 ms. As for decryption, NTRU-based ESE takes 0.013 ms and El-Gamal ESE takes 0.782 ms. The differences are substantial, since the NTRU-base ESE is 236× and 60× faster in encryption and decryption, respectively.

4   Security Analysis

In this section, we focus on analyzing the security of our proposed schemes.

The security of lattice-based schemes is determined by hardness of the underlying lattice problem (in our case, ring-LWE). Based on [21], the hardness of lattice problems is measured using the root Hermite factor. For a vector $\mathbf{v}$ in an N-dimension lattice that is larger than the nth root of the determinant, the root Hermite factor is computed as $\gamma = \frac{\|\mathbf{v}\|}{\det(\Lambda_{h,q})^{1/n}}$. According to [16], for a short planted vector $\mathbf{v}$ in an NTRU lattice, the associated root Hermite factor is computed as $\gamma^n = \frac{\sqrt{N/(2\pi e)} \times \det(\Lambda)^{1/n} \|\mathbf{v}\|}{0.4 \times \|\mathbf{v}\|}$. Based on [13,21], $\gamma \approx 1.004$ guarantees intractability and provides at least 192-bit security.

Lemma 2. *If an IBE scheme is IBE-IND-CPA and IBE-ANO-RE-CPA-secure, then it is also IBE-ANO-CPA-secure.*

Proof. Please see appendix.

Following Lemma 2, to establish the security of our NTRU-PEKS scheme, we need to rely on the security of the underlying IBE scheme. Ducas et al. provided the proof of IBE-IND-CPA of their scheme in [17]. Therefore, we are left to prove the anonymity of their scheme via Theorem 2.

Theorem 2. *The IBE scheme of Ducas et al. is anonymous in the sense of Definition 8 under the decision ring-LWE problem.*

Proof. Since the output of the PEKS algorithm of our scheme corresponds to the encryption algorithm of [32,33], for $\mathcal{A}$ to determine s_w corresponds to which keyword with any probability $\Pr \geq \frac{1}{2} + \epsilon$ - for any non-negligible ϵ, it has to solve the decision ring-LWE. Our scheme works over the polynomial ring $\mathbb{Z}[x]/(x^N+1)$, for a power-of-two N and a prime $q \equiv 1 \mod 2N$. The ring-LWE based PEKS algorithm computes a pseudorandom ring-LWE vector $A = r * h + e_1$ (for a uniform $r, e_1 \overset{\$}{\leftarrow} \{-1, 0, 1\}^N$) and uses $H(w)$ to compute $B = r * H(w) + e_2 + \lfloor \frac{q}{2} \rfloor k$ that is also statistically close to uniform. Therefore, the adversary's view of $\langle A, B, H_2(A, k) \rangle$ is indistinguishable from uniform distribution under the hardness of decision ring-LWE. The pseudorandomness is preserved when t_w is chosen from the error distribution (by adopting the transformation to Hermite's normal form) similar to the one in standard LWE [35]. $\qquad\square$

Theorem 3. *If there exists an adversary $\mathcal{A}$ that can break IND-CKA of NTRU-PEKS scheme as in Definition 10, one can build an adversary $\mathcal{F}$ that uses $\mathcal{A}$ as subroutine and breaks the security of the IBE scheme as in Definition 8.*

Proof. The proof works by having adversaries $\mathcal{F}$ and $\mathcal{A}$ initiating the *find* phase as in Definitions 8 and 10 respectively.

Algorithm $\mathcal{F}^{\texttt{KeyQuery}(.),H}(find, mpk)$

- $(mpk, msk) \overset{\$}{\leftarrow} \texttt{Setup}(q, N)$: $\mathcal{F}$ receives mpk and passes it to $\mathcal{A}$.

Algorithm $\mathcal{A}^{TdQuery(.),H}(find, pk)$

- Queries on $\texttt{TdQuery}(.)$: Upon such queries, $\mathcal{F}$ queries $\texttt{KeyQuery}(.)$ which keeps a list *idSet* maintaining all the previously requested queries and responses. If the submitted query exists, the same response is returned, otherwise, to sample short vectors s, t_w, the oracle uses msk to run $(s, t_w) \overset{\$}{\leftarrow}$ $\texttt{Gaussian-Sampler}(msk, \sigma, (H(w), 0))$ and passes t_w to $\mathcal{F}$. $\mathcal{F}$ sends t_w to $\mathcal{A}$.

After the *find* phase, a hidden fair coin $b \in \{0, 1\}$ is flipped.

Execute$(w_0, w_1) \leftarrow \mathcal{A}^{\texttt{TdQuery}(.),H}(guess, pk)$

- Upon receiving (w_0, w_1), $\mathcal{F}$ selects a message $m \in \{0\}^N$ and calls $\texttt{Enc}(m, w_0, w_1)$ that runs encryption on (w_b, m) which works as in Definition 7 and outputs $s_w = \langle A, B, H_2(k, B) \rangle$. $\mathcal{F}$ relays s_w to $\mathcal{A}$.

Finally, $\mathcal{A}$ outputs its decision bit $b' \in \{0, 1\}$. $\mathcal{F}$ also outputs b' as its response. Omitting the terms that are negligible in terms of q and N, the upper bound on *IND-CKA* of NTRU-PEKS is as follows.

$$Adv_{\mathcal{A}}^{PEKS-IND-CKA}(q, N) \leq Adv_{\mathcal{F}}^{NTRU-IBE-ANO-CPA}(q, N)$$

$\qquad\square$

Secure Channel Requirement. Baek et al. [4] highlighted the requirement of a secure channel for trapdoor transmission between the receiver and the server and proposed the notion of Secure-Channel Free (SCF) PEKS schemes where the keywords are encrypted by both the server's and receiver's public key. Offline keyword-guessing attack, as introduced by Byun et al. [12], implies the ability of an adversary to find which keyword was used to generate the trapdoor. This inherent issue is due to low-entropy nature of the commonly selected keywords and public availability of the encryption key [10]. Since Byun et al.'s work [12], there have been many attempts in proposing schemes that are secure against keyword guessing attacks [20,27,28]. However, in all the proposals, once the trapdoor is received by the server, the keyword guessing attacks remain a perpetual problem [28]. Jeong et al. [28] showed the trade-off between the security of a PEKS scheme against keyword-guessing attacks and its consistency - by mapping a trapdoor to multiple keywords. For our scheme, we can assume a conventional or even post quantum secure [9] SSL/TLS connection between the receiver and the server. We believe such reliable protocols provide the best mean for communicating trapdoors to the servers. Establishing a secure line through SSL/TLS could be much more efficient than using any public key encryption as in SFC-PEKS. Since in such protocols, after the hand shake protocol, all communications are encrypted using symmetric encryption.

5 Performance Evaluation

We first describe our experimental setup and evaluation metrics. We then provide a detailed performance analysis of our scheme by also comparing its efficiency with the pairing-based schemes proposed in [6,44]. To the best of our knowledge, and based on [10], the selected pairing-based counterparts are the most efficient schemes proposed in random oracle and standard models.

5.1 Experimental Setup and Evaluation Metrics

We implemented our PEKS scheme in C++[2], using NTL [40] and GNU MP [23] libraries. The implementations of the pairing-based counterparts [6,44] were obtained from MIRACL library. We used the MIRACL suggested elliptic curves, MNT (with embedding degree $k = 6$) and KSS (with embedding degree $k = 18$) for 80-bit and 192-bit security, respectively. The implementations were done on an Intel Core i7-6700HQ laptop with a 2.6 GHz CPU and 12 GB RAM. Our evaluation metrics are computation, storage, and communication that are required by the sender, receiver and server.

5.2 Performance Evaluation and Comparisons

The Test algorithm of our scheme only requires one convolution product, which is much more efficient than the bilinear pairing operation required in all of the existing pairing-based PEKS schemes. Referring to Table 1, running the Test algorithm for one keyword and one record our scheme is 36× and 97× faster

---

[2] https://github.com/Rbehnia/NTRUPEKS.git.

Table 2. Analytical performance analysis and comparison.

Schemes	Computation			Storage			
	Test	PEKS	Trapdoor	PK size	SK size	SC size	TD size
NTRU-PEKS	$Conv$	$2Conv^{a}$	$GSamp$	$N\lvert q\rvert$	$2N$ $\log_2(2s\pi)^{b}$	$3N\lvert q\rvert$	$N\lvert q\rvert$
BCO [6]	bp	$1bp + sm$	sm	$2\lvert q'\rvert$	$\lvert q'\rvert$	$2\lvert q'\rvert + \lvert q'\rvert$	$2\lvert q'\rvert$
ZI [44]	$ex + bp$	$2sm + 2ex + 2bp$	$sm + 1pa$	$2\lvert q'\rvert$	$\lvert q'\rvert$	$2 \times 18\lvert q'\rvert$ $+2\lvert q'\rvert$	$2\lvert q'\rvert + \lvert q'\rvert$

For 192-bit security, we set $N = 1024$ and $q \approx 2^{27}$ which gives us a root Hermite factor $\gamma = 1.0042$ for our scheme and for BCO and ZI schemes, we set $q' \approx 2^{192}$.

PK and SK denote public key and private key, respectively. SC and TD refer to the searchable ciphertext and trapdoor, receptively. $Conv$ denotes convolution product as defined in Sect. 2. $GSamp$ denotes a Gaussian Sampling function as in [17]. bp denotes a bilinear pairing operation [7], pa and sm denote point addition and scalar multiplication in $\mathbb{G}$, respectively, and ex denotes exponentiation in $\mathbb{G}_T$.

Public key, private key and SC are stored on the sender, receiver and server's machines, respectively. PEKS, Trapdoor and Test algorithms are run by the senders, receiver and server machines, respectively.

[a] With a slight storage sacrifice, sender can pre-compute one of the convolution products.

[b] The value of s defines the norm of the Gram-Schmidt coefficient. In [17], the authors set the norm $s \approx \sqrt{\frac{q e}{2}}$, where e is the base of natural logarithm.

Table 3. Parameter sizes of our scheme and its pairing-based counterparts

Schemes	Public key size	Private key size	SC size	TD size
NTRU-PEKS	27.2 Kb	32 Kb	52 Kb	27 Kb
BCO [6]	0.38 Kb	0.19 Kb	0.57 Kb	0.38 Kb
ZI [44]	0.76 Kb	0.19 Kb	0.89 Kb	0.57 Kb

All schemes are implemented for 192 bits of security.

than BCO and ZI schemes, respectively. This gap significantly increases as the number of keywords/records increases, for instance, as depicted in Fig. 2, the search time for 10000 (with distinct keywords) records in database, is 10 s in our scheme, and 400 s and 1100 s for BCO and ZI schemes, respectively. For 10000 records (which is rather small comparing to the number of records in actual databases), our scheme is 40 times faster than Boneh et al.'s scheme. For real-world cases with a large database, our scheme seems to be the only practical solution at this moment. We believe that this is one of the main aspects of our scheme which makes it an attractive candidate to be implemented for real-world applications. As it is shown in Table 2, the dominant operations of the PEKS algorithm in our scheme are two convolution products of form $x_1 * x_2$. However, since one of the operands has very small coefficients (i.e., $r \xleftarrow{\$} \{-1, 0, 1\}^N$), the convolution products can be computed very rapidly. Specifically, in our case, since N has been selected as a power-of-two integer, the convolution product can be computed in $N \log N$ operations by Fast Fourier Transform. In Fig. 3, we compare the efficiency of the PEKS algorithm of our scheme with ones in [6,44]. Generating one searchable encryption in our scheme is 19× and 78× faster than that of BCO and ZI schemes, respectively. Therefore, in our scheme, the sender can generate 2000 searchable encryptions with distinct keywords in 4 s while this time is increased to 100 s and 400 s in BCO and ZI schemes, respectively. The

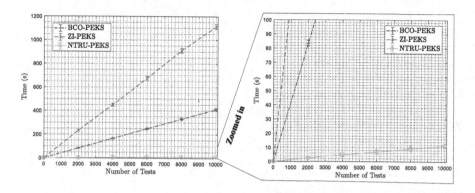

Fig. 2. Search time of server

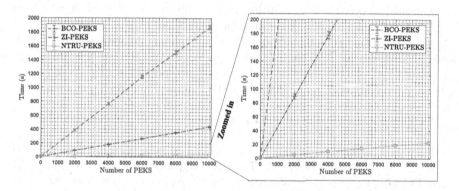

Fig. 3. Efficiency of PEKS algorithm

sender needs to store the receiver's public key of size $N|q|$, referring to Table 3, for 192-bit security, it can be up to 27.2 Kb. The resulting searchable encryption of our PEKS algorithm is to be sent to the server is of size 52 Kb, based on Table 3. This is larger than the searchable encryption size of BCO and ZI scheme. Due to the structure of our PEKS algorithm, the computation of A in searchable ciphertext can be done prior to having knowledge of the keyword. Therefore, with a slight storage sacrifice (i.e., storing $N|q|$ bits), the PEKS algorithm can become twice as fast.

The `Trapdoor` algorithm in our scheme requires a Gaussian Sampling similar as in [17,22]. This is the most costly operation in our scheme. As it is shown in Table 1, for 192-bit security, one trapdoor generation is 6.4× and 3× slower than those of BCO and ZI schemes, respectively. In Fig. 4 we compare the efficiency of the `Trapdoor` algorithm of our scheme with the ones in [6,44]. This algorithm in Boneh et al.s scheme only requires one scalar multiplication and consequently, it is the fastest. `Trapdoor` algorithm in

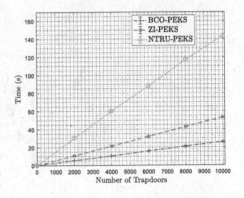

Fig. 4. Efficiency of `Trapdoor` algorithm

BCO scheme is capable of generating 2000 trapdoors for distinct keywords in 5 s, ZI scheme generates the same number of trapdoor in 10 s comparing to our scheme which takes 30 s. Each trapdoor in our scheme is 27 Kb which is much larger than those in BCO and ZI schemes. With the sacrifice of storage, the receiver can pre-compute and securely store the trapdoors locally. As discussed in Sect. 4, the trapdoors are to be transmitted to the server via a secure channel.

5.3   Discussion

To the best of our knowledge, our scheme is the first post-quantum secure PEKS scheme in the literature. Except the `Trapdoor` algorithm, which is only run $\mathcal{O}(1)$ times for each keyword, our algorithm enjoys from a very efficient `PEKS` and `Test` algorithms. While the `PEKS` algorithm is run $\mathcal{O}(1)$ times for each keyword and record, the efficiency of our `Test` algorithm, which is run $\mathcal{O}(L)$ times (for a database of size L), significantly decreases the search time on the server and minimizes the cryptographic end-to-end delay. Note that achieving a low end-to-end delay is of great importance, since even small delays (e.g., a few milliseconds) could incur significant financial costs for companies like Amazon [19].

One limitation of our scheme is that its searchable ciphertext sizes are larger than its pairing-based counterparts, as our scheme relies on NTRU. This incurs a larger storage overhead on the server. However, given its significantly efficiency advantage for `PEKS` and especially critical algorithm `Test`, and also high storage capability of the modern cloud servers with a relatively low storage cost, this can be considered as a highly favorable trade-off. Moreover, as discussed, a faster response time (i.e., lower end-to-end delay) seems a much critical ecumenical parameter for modern cloud services than having a relatively higher storage.

6   Related Work

Searchable encryption can be instantiated from both symmetric or asymmetric key settings. Song et al. [41] proposed the first SE scheme that relies on symmetric key cryptography. Kamara et al. [29] proposed the first DSSE scheme to address the limitation of its static ancestors. While being highly efficient, symmetric SE schemes are more suitable for applications that involve a single client who outsources her own data to the cloud relying on her private key.

In this paper, given the target applications that need multiple heterogeneous entities to create searchable encrypted data, our focus is on SE schemes instanced in asymmetric settings. In particular, we concentrate on PEKS, as it requires neither specific probability distributions on keywords nor performance/consistency trade-offs as dictated by some other asymmetric alternatives (e.g., ESE as discussed in Sect. 3.2). In PKES, decryption and trapdoor generation take place using the private key of the receiver, while any user can use the corresponding public key to generate searchable ciphertext. With a few exceptions, all of the proposed PEKS schemes are developed using costly bilinear pairing operations. The first instance of pairing-free PEKS schemes is constructed by Crescenzo and Saraswat [15] based on the IBE scheme in [14], which is constructed using quadratic residue for a composite modulus. Khader [31] proposed the first instance of such schemes in the standard model based on a k-resilient IBE, she also put forth a scheme which supports multiple-keyword search. Nonetheless, due to their costly operations, the proposed schemes are not practical to be implemented in real-world applications.

7   Conclusion

In this paper, we proposed (to the best of our knowledge) the first NTRU-based PEKS scheme, which harnesses some of the most recent cryptographic tools in lattice-based cryptography, IBE scheme based on ring-LWE and efficient polynomial arithmetics at the same time. We formally proved that our scheme is secure and consistent in IND-CKA model, and also showed that our base IBE scheme achieves anonymity property required by our PEKS construction. Our theoretical and experimental analysis confirmed that our NTRU-based PEKS scheme is significantly more computationally efficient than its most efficient pairing-based counterparts at the server and sender side, which offer the lowest end-to-end cryptographic delay among the existing PEKS schemes. In addition to its efficiency, our PEKS scheme demonstrated a much smoother performance for increasing key sizes and inherits the (current) post-quantum security properties of its underlying NTRU primitives. The high efficiency and long-term security of NTRU-based PEKS are expected to pave a path towards potential consideration of PEKS schemes for real-life applications.

Appendix

The following proof is obtained from [1].

Proof of Lemma 1.

Let $\mathcal{A}$ be an adversary on an IBE-ANO-CPA-secure scheme. We can build another adversaries $\mathcal{A}_1$ and $\mathcal{A}_3$ attacking the IBE-IND-CPA, and another adversary $\mathcal{A}_2$ attacking ANO-RE-CPA of the IBE scheme such that

$$\Pr[Exp_{IBE,\mathcal{A}}^{IBE\text{-}ANO\text{-}CPA\text{-}1}(k) = 1] - \Pr[Exp_{IBE,\mathcal{A}}^{IBE\text{-}ANO\text{-}RE\text{-}1}(k) = 1] \leqslant Adv_{IBE,\mathcal{A}_1}^{IBE\text{-}IND\text{-}CPA}(1^k)$$

$$\Pr[Exp_{IBE,\mathcal{A}}^{IBE\text{-}ANO\text{-}RE\text{-}1}(k) = 1] - \Pr[Exp_{IBE,\mathcal{A}}^{IBE\text{-}ANO\text{-}RE\text{-}0}(k) = 1] \leqslant Adv_{IBE,\mathcal{A}_2}^{IBE\text{-}ANO\text{-}RE}(1^k)$$

$$\Pr[Exp_{IBE,\mathcal{A}}^{IBE\text{-}ANO\text{-}RE\text{-}0}(k) = 1] - \Pr[Exp_{IBE,\mathcal{A}}^{IBE\text{-}ANO\text{-}CPA\text{-}0}(k) = 1] \leqslant Adv_{IBE,\mathcal{A}_3}^{IBE\text{-}IND\text{-}CPA}(1^k)$$

Adding the above equations will conclude this proof. $\square$

References

1. Abdalla, M., Bellare, M., Catalano, D., Kiltz, E., Kohno, T., Lange, T., Malone-Lee, J., Neven, G., Paillier, P., Shi, H.: Searchable encryption revisited: consistency properties, relation to anonymous IBE, and extensions. In: Shoup, V. (ed.) CRYPTO 2005. LNCS, vol. 3621, pp. 205–222. Springer, Heidelberg (2005). doi:10.1007/11535218_13

2. Ajtai, M.: Generating hard instances of lattice problems. In: Twenty-Eighth Annual ACM Symposium on Theory of Computing, Philadelphia, PA, USA. Proceedings, pp. 99–108. ACM (1996)

3. Armbrust, M., Fox, A., Griffith, R., Joseph, A.D., Katz, R.H., Konwinski, A., Lee, G., Patterson, D.A., Rabkin, A., Stoica, I., et al.: Above the clouds: a Berkeley view of cloud computing. Technical report. EECS Department, University of California, Berkeley (2009)

4. Baek, J., Safavi-Naini, R., Susilo, W.: Public key encryption with keyword search revisited. In: Gervasi, O., Murgante, B., Laganà, A., Taniar, D., Mun, Y., Gavrilova, M.L. (eds.) ICCSA 2008. LNCS, vol. 5072, pp. 1249–1259. Springer, Heidelberg (2008). doi:10.1007/978-3-540-69839-5_96

5. Bellare, M., Boldyreva, A., O'Neill, A.: Deterministic and efficiently searchable encryption. In: Menezes, A. (ed.) CRYPTO 2007. LNCS, vol. 4622, pp. 535–552. Springer, Heidelberg (2007). doi:10.1007/978-3-540-74143-5_30

6. Boneh, D., Crescenzo, G., Ostrovsky, R., Persiano, G.: Public key encryption with keyword search. In: Cachin, C., Camenisch, J.L. (eds.) EUROCRYPT 2004. LNCS, vol. 3027, pp. 506–522. Springer, Heidelberg (2004). doi:10.1007/978-3-540-24676-3_30

7. Boneh, D., Franklin, M.: Identity-based encryption from the weil pairing. In: Kilian, J. (ed.) CRYPTO 2001. LNCS, vol. 2139, pp. 213–229. Springer, Heidelberg (2001). doi:10.1007/3-540-44647-8_13

8. Boneh, D., Waters, B.: Conjunctive, subset, and range queries on encrypted data. In: Vadhan, S.P. (ed.) TCC 2007. LNCS, vol. 4392, pp. 535–554. Springer, Heidelberg (2007). doi:10.1007/978-3-540-70936-7_29

9. Bos, J.W., Costello, C., Naehrig, M., Stebila, D.: Post-quantum key exchange for the TLS protocol from the ring learning with errors problem. In: IEEE Symposium on Security and Privacy S&P, Â Fairmont, San Jose, California. Proceedings, pp. 553–570. IEEE Computer Society (2015)

10. Bösch, C., Hartel, P., Jonker, W., Peter, A.: A survey of provably secure searchable encryption. ACM Comput. Surv. **47**(2), 18:1–18:51 (2014)

11. Bringer, J., Chabanne, H., Kindarji, B.: Error-tolerant searchable encryption. In: IEEE International Conference on Communications, Dresden, Germany. Proceedings, pp. 1–6. IEEE (2009)

12. Byun, J.W., Rhee, H.S., Park, H.-A., Lee, D.H.: Off-line keyword guessing attacks on recent keyword search schemes over encrypted data. In: Jonker, W., Petković, M. (eds.) SDM 2006. LNCS, vol. 4165, pp. 75–83. Springer, Heidelberg (2006). doi:10.1007/11844662_6

13. Chen, Y., Nguyen, P.Q.: BKZ 2.0: better lattice security estimates. In: Lee, D.H., Wang, X. (eds.) ASIACRYPT 2011. LNCS, vol. 7073, pp. 1–20. Springer, Heidelberg (2011). doi:10.1007/978-3-642-25385-0_1

14. Cocks, C.: An identity based encryption scheme based on quadratic residues. In: Honary, B. (ed.) Cryptography and Coding 2001. LNCS, vol. 2260, pp. 360–363. Springer, Heidelberg (2001). doi:10.1007/3-540-45325-3_32

15. Crescenzo, G., Saraswat, V.: Public key encryption with searchable keywords based on Jacobi symbols. In: Srinathan, K., Rangan, C.P., Yung, M. (eds.) INDOCRYPT 2007. LNCS, vol. 4859, pp. 282–296. Springer, Heidelberg (2007). doi:10.1007/978-3-540-77026-8_21

16. Ducas, L., Durmus, A., Lepoint, T., Lyubashevsky, V.: Lattice signatures and bimodal Gaussians. In: Canetti, R., Garay, J.A. (eds.) CRYPTO 2013. LNCS, vol. 8042, pp. 40–56. Springer, Heidelberg (2013). doi:10.1007/978-3-642-40041-4_3

17. Ducas, L., Lyubashevsky, V., Prest, T.: Efficient identity-based encryption over NTRU lattices. In: Sarkar, P., Iwata, T. (eds.) ASIACRYPT 2014. LNCS, vol. 8874, pp. 22–41. Springer, Heidelberg (2014). doi:10.1007/978-3-662-45608-8_2

18. Ducas, L., Nguyen, P.Q.: Faster Gaussian lattice sampling using lazy floating-point arithmetic. In: Wang, X., Sako, K. (eds.) ASIACRYPT 2012. LNCS, vol. 7658, pp. 415–432. Springer, Heidelberg (2012). doi:10.1007/978-3-642-34961-4_26

19. Eaton, K.: How one second could cost Amazon $1.6 Billion in sales (2012). https://www.fastcompany.com/1825005/how-one-second-could-cost-amazon-16-billion-sales. Accessed 2017

20. Fang, L., Susilo, W., Ge, C., Wang, J.: Public key encryption with keyword search secure against keyword guessing attacks without random oracle. Inf. Sci. **238**, 221–241 (2013)

21. Gama, N., Nguyen, P.Q.: Predicting lattice reduction. In: Smart, N. (ed.) EUROCRYPT 2008. LNCS, vol. 4965, pp. 31–51. Springer, Heidelberg (2008). doi:10.1007/978-3-540-78967-3_3

22. Gentry, C., Peikert, C., Vaikuntanathan, V.: Trapdoors for hard lattices and new cryptographic constructions. In: Fortieth Annual ACM Symposium on Theory of Computing, STOC, Victoria, Canada. Proceedings, pp. 197–206. ACM (2008)

23. Granlund, T.: The GMP development team: GNU MP: the GNU multiple precision arithmetic library (2012). http://gmplib.org/

24. Halevi, S.: A sufficient condition for key-privacy. IACR Cryptology ePrint Archive 2005, p. 5 (2005)

25. Hoffstein, J., Howgrave-Graham, N., Pipher, J., Silverman, J.H., Whyte, W.: NTRUSign: digital signatures using the NTRU lattice. In: Joye, M. (ed.) CT-RSA

2003. LNCS, vol. 2612, pp. 122–140. Springer, Heidelberg (2003). doi:10.1007/3-540-36563-X_9

26. Hoffstein, J., Pipher, J., Silverman, J.H.: NTRU: a ring-based public key cryptosystem. In: Buhler, J.P. (ed.) ANTS 1998. LNCS, vol. 1423, pp. 267–288. Springer, Heidelberg (1998). doi:10.1007/BFb0054868

27. Hu, C., Liu, P.: A secure searchable public key encryption scheme with a designated tester against keyword guessing attacks and its extension. In: Lin, S., Huang, X. (eds.) CSEE 2011. CCIS, vol. 215, pp. 131–136. Springer, Heidelberg (2011). doi:10.1007/978-3-642-23324-1_23

28. Jeong, I.R., Kwon, J.O., Hong, D., Lee, D.H.: Constructing PEKS schemes secure against keyword guessing attacks is possible? Comput. Commun. 32(2), 394–396 (2009)

29. Kamara, S., Lauter, K.: Cryptographic cloud storage. In: Sion, R., Curtmola, R., Dietrich, S., Kiayias, A., Miret, J.M., Sako, K., Sebé, F. (eds.) FC 2010. LNCS, vol. 6054, pp. 136–149. Springer, Heidelberg (2010). doi:10.1007/978-3-642-14992-4_13

30. Kamara, S., Papamanthou, C., Roeder, T.: Dynamic searchable symmetric encryption. In: ACM Conference on Computer and Communications Security, CCS, Raleigh, NC, USA. Proceedings, pp. 965–976. ACM (2012)

31. Khader, D.: Public key encryption with keyword search based on K-resilient IBE. In: Gervasi, O., Gavrilova, M.L. (eds.) ICCSA 2007. LNCS, vol. 4707, pp. 1086–1095. Springer, Heidelberg (2007). doi:10.1007/978-3-540-74484-9_95

32. Lyubashevsky, V., Peikert, C., Regev, O.: On ideal lattices and learning with errors over rings. In: Gilbert, H. (ed.) EUROCRYPT 2010. LNCS, vol. 6110, pp. 1–23. Springer, Heidelberg (2010). doi:10.1007/978-3-642-13190-5_1

33. Lyubashevsky, V., Peikert, C., Regev, O.: A toolkit for ring-LWE cryptography. In: Johansson, T., Nguyen, P.Q. (eds.) EUROCRYPT 2013. LNCS, vol. 7881, pp. 35–54. Springer, Heidelberg (2013). doi:10.1007/978-3-642-38348-9_3

34. Micciancio, D., Regev, O.: Worst-case to average-case reductions based on Gaussian measures. SIAM J. Comput. 37(1), 267–302 (2007)

35. Micciancio, D., Regev, O.: Lattice-based cryptography. In: Bernstein, D.J., Buchmann, J., Dahmen, E. (eds.) Post-Quantum Cryptography, pp. 147–191. Springer, Heidelberg (2009)

36. Nguyen, P.Q., Regev, O.: Learning a parallelepiped: cryptanalysis of GGH and NTRU signatures. J. Cryptol. 22(2), 139–160 (2009)

37. Park, D.J., Kim, K., Lee, P.J.: Public key encryption with conjunctive field keyword search. In: Lim, C.H., Yung, M. (eds.) WISA 2004. LNCS, vol. 3325, pp. 73–86. Springer, Heidelberg (2005). doi:10.1007/978-3-540-31815-6_7

38. Regev, O.: On lattices, learning with errors, random linear codes, and cryptography. In: Thirty-Seventh Annual ACM Symposium on Theory of Computing, STOC, Hunt Valley, MD. Proceedings, pp. 84–93. ACM (2005)

39. Shi, E., Bethencourt, J., Chan, T.H.H., Song, D., Perrig, A.: Multi-dimensional range query over encrypted data. In: IEEE Symposium on Security and Privacy, S&P, Oakland, California, USA. Proceedings, pp. 350–364. IEEE Computer Society (2007)

40. Shoup, V.: NTL: a library for doing number theory (2003). http://www.shoup.net/ntl

41. Song, D.X., Wagner, D., Perrig, A.: Practical techniques for searches on encrypted data. In: IEEE Symposium on Security and Privacy S&P, Berkeley, California, USA. Proceedings, pp. 44–55. IEEE Computer Society (2000)

42. Whyte, W., Howgrave-Graham, N., Hoffstein, J., Pipher, J., Silverman, J.H., Hirschhorn, P.S.: IEEE P 1363.1 draft 10: draft standard for public key cryptographic techniques based on hard problems over lattices. IACR Cryptology EPrint Archive 2008, p. 361 (2008)
43. Yavuz, A.A., Guajardo, J.: Dynamic searchable symmetric encryption with minimal leakage and efficient updates on commodity hardware. In: Dunkelman, O., Keliher, L. (eds.) SAC 2015. LNCS, vol. 9566, pp. 241–259. Springer, Cham (2016). doi:10.1007/978-3-319-31301-6_15
44. Zhang, R., Imai, H.: Generic combination of public key encryption with keyword search and public key encryption. In: Bao, F., Ling, S., Okamoto, T., Wang, H., Xing, C. (eds.) CANS 2007. LNCS, vol. 4856, pp. 159–174. Springer, Heidelberg (2007). doi:10.1007/978-3-540-76969-9_11

HardIDX: Practical and Secure Index with SGX

Benny Fuhry[1]([⊠]), Raad Bahmani[2], Ferdinand Brasser[2], Florian Hahn[1],
Florian Kerschbaum[3], and Ahmad-Reza Sadeghi[2]

[1] SAP Research, Karlsruhe, Germany
{benny.fuhry,florian.hahn}@sap.com
[2] Technische Universität Darmstadt, Darmstadt, Germany
{r.bahmani,f.brasser,a.sadeghi}@trust.tu-darmstadt.de
[3] University of Waterloo, Waterloo, Canada
florian.kerschbaum@uwaterloo.ca

Abstract. Software-based approaches for search over encrypted data
are still either challenged by lack of proper, low-leakage encryption or
slow performance. Existing hardware-based approaches do not scale well
due to hardware limitations and software designs that are not specifically tailored to the hardware architecture, and are rarely well analyzed for their security (e.g., the impact of side channels). Additionally, existing hardware-based solutions often have a large code footprint in the trusted environment susceptible to software compromises.
In this paper we present HardIDX: a hardware-based approach, leveraging Intel's SGX, for search over encrypted data. It implements only the
security critical core, i.e., the search functionality, in the trusted environment and resorts to untrusted software for the remainder. HardIDX
is deployable as a highly performant encrypted database index: it is logarithmic in the size of the index and searches are performed within a few
milliseconds. We formally model and prove the security of our scheme
showing that its leakage is equivalent to the best known searchable
encryption schemes.

1 Introduction

Outsourcing the storage and processing of sensitive data to untrusted cloud
environment is still considered as too risky due to possible data leakage, government intrusion, and legal liability. The cryptographic solutions Secure Multiparty Computation (MPC) and in particular Fully Homomorphic Encryption
(FHE) [23] offer high degree of protection by allowing arbitrary computation
on encrypted data, but they are impractical for adoption in large distributed
systems [24].

Moreover, there are a number of useful applications that only require a small
set of operations. A prime example of such operations is the search and retrieval
in an encrypted databases without the need to download all data from the
cloud. For this task, different cryptographic schemes have been proposed such
as property-preserving encryption [6,8], or functional encryption [10] and its

© IFIP International Federation for Information Processing 2017
Published by Springer International Publishing AG 2017. All Rights Reserved
G. Livraga and S. Zhu (Eds.): DBSec 2017, LNCS 10359, pp. 386–408, 2017.
DOI: 10.1007/978-3-319-61176-1_22

special case searchable encryption [16,29,41]. In this context, performing efficient and secure *range* queries are commonly considered to be very challenging. CryptDB [36] resorts to order-preserving encryption for this purpose which is susceptible to simple ciphertext-only attacks as shown by Naveed et al. [34].

Many schemes for search over encrypted data supporting range queries require search time linear in the number of database records. Recently, schemes with polylogarithmic search time, based on an index structure, have been proposed [17,19,29]. In Sect. 7 and Table 2, we elaborate on the search time, query size, storage size and leakage problems of those approaches. Designing an *efficient* searchable encryption scheme with *minimal leakage on the queried ranges* remains an open challenge.

Another line of research [4,5] leverages the developments in hardware-assisted Trusted Execution Environments (TEEs) for search over encrypted data. Although Intel's recently introduced Software Guard Extension (SGX) [2,15,26, 31] has inspired new interest in TEEs, related technologies have been available before, e.g., in ARM processors known as ARM TrustZone [3] as well as in academic research [12,42]. Also, AMD has recently announced a TEE for their CPUs [27] rising the hope that TEEs will be widely available in x86 processors, and thus in many relevant environments such as clouds, in the near future. TEEs have to interact with untrustworthy components within the same computer system for various reasons. In order to achieve comprehensive security, information leakage through those channels has to be considered and taken care of. Previous SGX based solutions that allow search on encrypted data load and execute the entire unmodified database management system (DBMS) into an enclave [4,5]. They do not formally consider information leakage and do not scale well due to limited memory size of SGX's enclaves and the large footprint of the code they require in the TEE.

Our goal and contribution. We present an efficient scheme for search over encrypted data that can be deployed as a database index. SGX's protection characteristics are utilized to achieve an outstanding tradeoff between security, performance and functionality. The currently fastest software-based schemes that support range queries are [17] and [19]. Our solution significantly improves over these approaches in terms of performance and storage. Compared to the latest hardware-based schemes [4,5], we improve in terms of security and scalability. Our scheme organizes data in a B^+-tree structure that is frequently used for databases indexes in most database management systems (DMBSs) [37]. Our solution supports searches for single values and value ranges and it can easily be adapted to many other database (search) operations.

We implemented and extensively evaluated our two constructions on SGX-enabled hardware (see Sect. 6). Both have a very small code and memory footprint in the TEE compared to other hardware-based approaches [4,5]. Additionally, our solution scales to arbitrary index sizes as memory usage in the enclave is constant and untrusted resources are used to store the database itself. Our main contributions are as follows:

- Our scheme has logarithmic complexity in the size of index and searches are performed within a few milliseconds.
- We formally model and prove our scheme secure showing that its security (leakage) is comparable to the best known searchable encryption schemes.
- We provide an implementation and evaluate the performance and functional bottleneck of SGX on the basis of two different constructions that are designed specifically for SGX to reduce the Trusted Computing Base.

2   Background

2.1   Intel Software Guard Extensions (SGX)

SGX is an extension of the x86 instruction set architecture (ISA) introduced with the 6th Generation Intel Core processors (code name Skylake). We now present a high level overview of SGX's features utilized by HardIDX (see [2,15,26,31] for more details).

Memory Isolation. On SGX enabled platforms, programs can be divided into two parts, an *untrusted part* and an isolated, *trusted part*. The trusted part, called *enclave* in SGX terminology, is located in a dedicated portion of the physical RAM. The SGX hardware enforces additional protection on this part of the memory. In particular, all other software on the system, including privileged software like OS, hypervisor and firmware cannot access the enclave memory. The (untrusted) host process can invoke the enclave only through a well-defined interface. Furthermore, all isolated code and data is encrypted while residing outside of the CPU package. Decryption and integrity checks are performed when the data is loaded inside the CPU.

Memory Management. SGX dedicates a fixed amount of the system's main memory (RAM) for enclaves and related metadata. For current systems this memory is limited to 128 MB which is used for both, SGX metadata and the memory for the enclaves themselves. The enclaves can only be deployed in about 96 MB. The SGX memory is reserved in the early boot phase and is static throughout the runtime of the system. The OS can allocate (parts of) the memory to individual enclaves and change these allocation during the runtime of the enclaves. In particular, the OS can swap out enclave pages. SGX ensures integrity, confidentiality and freshness of swapped-out pages.

Attestation. SGX has a remote attestation feature which allows to verify the correct creation of an enclave on a remote system. During enclave creation the initial code and data loaded into the enclave are measured. This measurement can be provided to an external party to prove the correct creation of an enclave. The authenticity of the measurement as well as the fact that the measurement originates from a benign enclave is ensured by a signature, provided by SGX's attestation feature (refer to [2] for details). Furthermore, the remote attestation feature allows for establishing a secure channel between an external party and an enclave.

2.2   Side Channel Attacks

Side channel attacks allow an adversary to extract sensitive information without having direct access to the information source by observing effects of the processing of the sensitive information. They have been known for a long time and various variants have been studied in the past, e.g., hardware side channels, software timing side channels and cache timing side channels [13, 22, 45]. All these attacks are noisy and require repeated execution and measurements to extract the sensitive information.

In the context of SGX, there exist a new class of side channels, called deterministic side channel [44]. As the OS is untrusted, yet still manages the enclave's resources, it can observe the enclaves behavior. In particular, the OS can generate a precise trace of the enclave's code and data accesses at the granularity of pages. In [44] it is shown that this allows to extract sensitive information from an SGX enclave.

3   High Level Design

3.1   HardIDX Overview

The high level design of our solution is shown in Fig. 1. The design involves three entities: the client (who is the data owner and therefore trusted), the untrusted SGX enabled server and the trusted SGX enclave *within* the server.

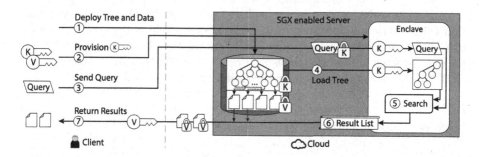

Fig. 1. High level design

Initially, a client prepares its data values by augmenting it with (index) search keys. We abbreviate data values as *values* and the search keys as *keys* throughout this paper. All other values and keys (e.g., cryptographic keys) are clearly differentiate if ambiguous. The values are stored at pseudo-random position. The keys are then inserted into a B^+-tree and the storage order of all nodes is also pseudo-random. The tree and values are linked by adding pointers to the leaves of the tree identifying the random position of the corresponding values. A value can be any data such as records in a relational database or files/documents in other database types. The client then encrypts all nodes of the tree with a secret key SK_k and all values with SK_v. The encrypted B^+-tree and encrypted values are deployed on the untrusted server in the cloud (see step ① in Fig. 1[1]).

The client uses the SGX attestation feature for authenticating the enclave and establishing a secure connection between the client and the enclave (see details in Sect. 2.1). Through this connection, the client provisions SK_k into the enclave (see step ②). This completes the setup of our scheme, which needs to be executed only once.

Now, the client can send (index) search queries to the server that are encrypted with a probabilistic encryption scheme under SK_k. Hence, the untrusted server cannot learn anything about the query, not even if the same query was send before. When a query arrives in the enclave, SK_k is used to decrypt the query (see step ③).

In step ④, the enclave loads the B^+-tree structure (tree nodes, but no values) from the untrusted storage into enclave memory and decrypts it. Given sufficient memory, the entire tree is loaded into the enclave and the search is performed afterwards (see step ⑤). As the tree size can exceed the memory available inside the enclave we provide a second design. In this case, only a subset of tree nodes is loaded into the enclave. The tree is traversed starting from the root node and nodes are fetched from the untrusted storage if necessary. In both cases the search algorithm eventually reaches a set of leaf nodes, which holds pointers to values matching the query. This list of pointers, representing the search result, is passed to the untrusted part (see step ⑥). The untrusted part learns nothing, except for the cardinality of the result set, from this interaction, because the values are stored in a randomized order.

The result of the index search could be processed further, e.g. in combination with additional SQL operators, in the SGX enclave at the server. In order to complete the end-to-end secure search, we assume that the server uses the pointers to fetch the encrypted values from untrusted storage and sends them to the client, where they are decrypted with SK_v (see step ⑦).

Notably, the plaintext values are never available on the server. They are encrypted with strong standard cryptography methods (AES-128 in GCM mode in our case) and never decrypted on the server, not even inside the SGX enclave. SK_v is only known to the client.

3.2   Assumptions and Attacker Model.

Due to SGX's protection, the attacker cannot directly access the enclave. However, side channels exist through which the attacker could potentially extract sensitive information. We assume the attacker has full control over all software on the system running HardIDX. (1) The attacker can observe all interaction of the enclave with resources outside the enclave. In particular, the attacker can observe the access pattern to B^+-tree nodes stored outside the enclave. (2) The attacker can use deterministic page-fault side channel to observe data

---

[1] For visualization purposes, the tree nodes and values are shown to be encrypted as a block. In reality each node and value is encrypted individually.

access inside the enclave at page granularity [44]. Through this side channel, the attacker can observe access patterns on the B^+-tree stored *inside* the enclave. (3) The attacker can use cache side channel to learn about code paths or data access patterns inside the enclave, as SGX does not protect against them [15].

Hardware attacks are out of scope in this paper. Furthermore, we consider denial of service (DoS) attacks on the cloud server and network out of scope. In this version, we only assume a passive attacker due to page constraints. We present mitigation strategies for an active attacker in the long version [20]. We furthermore assume as single user and the multi user case in Appendix C.

4   Notation and Definitions

4.1 B^+-tree

A B^+-tree is a balanced, n-ary search tree. So called search keys are utilized to index values. A B^+-tree can be used to search for single values, e.g., unique staff ids are used to find the corresponding database record (see Fig. 2) or for ranges, e.g., a salary index that allows to search for all employees falling in a specific salary range.

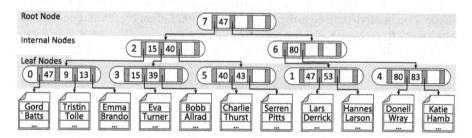

Fig. 2. B^+-tree example: the unique staff ids are used as keys and the values are the staff records (random storage position on the left).

Three node types are differentiated in a B^+-tree: the root node, internal nodes and leaf nodes. Every node x contains $x.\#\mathbf{k}$ keys that are stored in a nondecreasing order: $x.k_1 \leq ... \leq x.k_{(x.\#\mathbf{k})}$. At every inner node x (including the root if not the only node), the keys separate the key domain into $(x.\#\mathbf{k}+1)$ subtrees that are reachable by $(x.\#\mathbf{k}+1)$ child pointers: $\{x.p_0, ..., x.p_{(x.\#\mathbf{k})}\} = x.\mathbf{p}$. Every key $x.k_i$ has a corresponding pointer $x.p_i$ that points to a node containing elements greater than or equal to $x.k_i$ and smaller than any other tag $x.k_j \ \forall j \in [i+1, x.\#\mathbf{k}]$. $x.p_0$ points to a node containing only keys, which are smaller than $x.k_1$. No internal node is linked to a value. Instead, every leaf node x stores $x.\#\mathbf{k}$ keys and a pointer to its corresponding value ($x.p_0$ is not used at the leaves). Every node x in the tree has a unique id $x.id$ and a flag $x.isLeaf$

that stores if the x is a leaf. We denote the B^+-tree without the values as B^+-tree *structure* and p_{x_i} as the storage position of x_i, i.e., the physical memory address.

We use unchained B^+-trees, i.e., the leafs are not connected. Linked leaves would increase the search performance, but it would severely deteriorate the security. The reason is that a range query would directly leak the relationship among leaves if links are followed during a query.

With HardIDX, it is not necessary to define the key domain D in advance as in many other approaches. D can be an arbitrary domain with a defined order relation and a defined minimal and a maximal element recognizable by the algorithms. These two elements, denoted as $-\infty$ and ∞, fulfill the following: $-\infty < x.k < \infty \; \forall x.k \in D$.

The branching factor b specifies a B^+-tree by defining the maximal number of pointers. b also defines the minimal number of pointer for the different node types, but we do not further elaborate on details. For ease of exposition, we assume that every key and pointer fits in an 32 bit block, but this is no prerequisite for our constructions.

4.2   Probabilistic Symmetric Encryption

A probabilistic authenticated symmetric encryption consists of three probabilistic polynomial-time algorithms $\texttt{PASE} = \big(\texttt{PASE\_Gen}(1^\lambda), \texttt{PASE\_Enc}(SK, v),$ $\texttt{PASE\_Dec}(SK, C)\big)$ with the usual definitions of functionality. $\texttt{PASE}$ has to be an authenticated IND-CCA secure encryption, e.g., AES-128 in GCM mode.

4.3   Hardware Secured B^+-tree (HSBT)

Based on the presented definition of a B^+-tree, we define the notion of a Hardware Secured B^+-tree (HSBT) as follows. We assume that the B^+-tree should store a set $\mathbf{s}$ of n key-value pairs: $\mathbf{s} = ((k_1, v_1), ..., (k_n, v_n))$. This set consists of n values $\mathbf{v} = (v_1, ..., v_n)$ and their corresponding keys $\mathbf{k} = (k_1, ...k_n)$.

Definition 1 (HSBT). *A secure hardware B^+-tree scheme is a tuple of six polynomial-time algorithms* $\big(\texttt{HSBT\_Setup}, \texttt{HSBT\_Enc}, \texttt{HSBT\_Tok}, \texttt{HSBT\_Dec}, \texttt{HSBT\_}$ $\texttt{SearchRange}, \texttt{HSBT\_SearchRange\_Trusted}\big).$

Algorithms executed at the client:

$SK \leftarrow \texttt{HSBT\_Setup}(1^\lambda)$: *Take the security parameter λ as input and output a secret key SK.*

$\gamma \leftarrow \texttt{HSBT\_Enc}(SK, \mathbf{s})$: *Take the secret key SK and a set $\mathbf{s}$ of key-value pairs as input. Output an encrypted B^+-tree γ.*

$\tau \leftarrow \texttt{HSBT\_Tok}(SK, R)$: *Take the secret key SK and a range $R = [R_s, R_e]$ as input. Output a search token τ.*

$\mathbf{v}' \leftarrow \texttt{HSBT\_Dec}(SK, C')$: *Take the secret key SK and a set of ciphertext C' as input. Decrypt the ciphertexts and output plaintext values $\mathbf{v}'$.*

Executed at the server on untrusted hardware:

$C' \leftarrow$ HSBT_SearchRange(τ, γ): *Take a search token τ and an encrypted tree γ as input and call the secure hardware function HSBT_SearchRange_Trusted. Output a set of encrypted values C'.*

Executed at the server on secure hardware:

$\mathbf{P} \leftarrow$ HSBT_SearchRange_Trusted$(\tau, \mathbf{X})$: *Take a search token τ as input. Output a set of pointers $\mathbf{P}$.*

Definition 2 (Correctness). *Let $\mathcal{D}$ denote a HSBT-scheme consisting of the six algorithms described in Def. 1. We say that $\mathcal{D}$ is correct if for all $\lambda \in \mathbb{N}$, for all SK output by HSBT_Setup(1^λ), for all key-value pairs $\mathbf{s}$ used by HSBT_Enc$(SK, \mathbf{s})$ to output γ, for all R used by HSBT_Tok(SK, R) to output τ, for all C' output by HSBT_SearchRange(τ, γ), the values $\mathbf{v'}$ output by HSBT_Dec(SK, C') are all values in $\mathbf{s}$ for which the corresponding keys $\mathbf{k'}$ fall in R, i.e., $\mathbf{v'} = \{v_i | (k_i, v_i) \in \mathbf{s} \wedge k_i \in [R_s, R_e] = R\}$.*

Our security model, which we define next, is based on a proof framework introduced by Curtmola et al. in [16]. A description about the proof technique can be found in Appendix A. At security models of searchable encryption schemes so far, the leakage only covers the transaction between the client and server. In our scenario, there is an additional transaction between the server and the secure hardware that can be viewed by the adversary. Therefore, we extend the CKA2-security to CKA2-HW-security by introducing a new type of leakage denoted as $\mathcal{L}_{hw}$. It consists of the inherent leakage of the used secure hardware and the inputs/outputs to/from the secure hardware.

Definition 3 (CKA2-HW-security). *Let $\mathcal{D}$ denote a HSBT-scheme consisting of the six algorithms described in Definition 1. Consider the probabilistic experiments $\mathbf{Real}_{\mathcal{A}}(\lambda)$ and $\mathbf{Ideal}_{\mathcal{A},\mathcal{S}}(\lambda)$, whereas $\mathcal{A}$ is a stateful adversary and $\mathcal{S}$ is a stateful simulator that gets the leakage functions $\mathcal{L}_{enc}$ and $\mathcal{L}_{hw}$.*

$\mathbf{Real}_{\mathcal{A}}(\lambda)$: *the challenger runs HSBT_Setup(1^λ) to generate a secret key SK. $\mathcal{A}$ outputs a set of key-value pairs $\mathbf{s}$. The challenger calculates $\gamma \leftarrow$ HSBT_Enc$(SK, \mathbf{s})$ and passes γ to $\mathcal{A}$. Afterwards, $\mathcal{A}$ makes a polynomial number of adaptive queries for arbitrary ranges R. The challenger returns search tokens τ to $\mathcal{A}$ after calculating $\tau \leftarrow$ HSBT_Tok(SK, R). $\mathcal{A}$ can use γ and the returned tokens at any time to make queries to the secure hardware. The secure hardware returns a set of pointers $\mathbf{P}$. Finally, $\mathcal{A}$ returns a bit b that is the output of the experiment.*

$\mathbf{Ideal}_{\mathcal{A},\mathcal{S}}(\lambda)$: *the adversary $\mathcal{A}$ outputs a set of key-value pairs $\mathbf{s}$. Using $\mathcal{L}_{enc}$, $\mathcal{S}$ creates γ and passes it to $\mathcal{A}$. Afterwards, $\mathcal{A}$ makes a polynomial number of adaptive queries for arbitrary ranges R. The simulator $\mathcal{S}$ creates tokens τ and passes them to $\mathcal{A}$. The adversary $\mathcal{A}$ can use γ and the returned tokens at any time to make queries to $\mathcal{S}$ (that simulates the secure hardware). $\mathcal{S}$ is given $\mathcal{L}_{hw}$ and returns a set of pointers $\mathbf{P}$. Finally, $\mathcal{A}$ returns a bit b that is the output of the experiment.*

We say $\mathcal{D}$ is $(\mathcal{L}_{enc}, \mathcal{L}_{hw})$-secure against adaptive chosen-keyword attacks if for all probabilistic polynomial-time algorithms $\mathcal{A}$, there exists a probabilistic polynomial-time simulator $\mathcal{S}$ such that

$$|\Pr[\mathbf{Real}_{\mathcal{A}}(\lambda) = 1] - \Pr[\mathbf{Ideal}_{\mathcal{A},\mathcal{S}}(\lambda) = 1]| \leq \mathrm{negl}(\lambda)$$

5   Search Algorithms

In this section, we will present two different constructions that enable a client to the search for a single value or a range of values based on keys. We use B^+-trees in both constructions to achieve logarithmic search and SGX to protect the confidentiality and integrity of the data.

5.1   Construction 1

We sketch our first correct (according to Definition 2) and secure (according to Definition 3) construction in this section. A more detailed construction can be found in the long version of the paper [20]. The guiding idea of the construction is that the entire data should be stored and processed inside the enclave. The client constructs the B^+-tree locally, encrypts the B^+-tree structure and the values with SK_k and SK_v, respectively (see 5.2 for details). Both are sent to the cloud provider and the SGX application consisting of a trusted and an untrusted part get deployed there.

Software measurement as described in Sect. 2.1 is used for remote attestation, i.e., to prove to the client that the correct software is deployed on an SGX enabled CPU. During the deployment, the application reserves an SGX protected memory region (see Sect. 2.1). A secure transfer protocol between client and server (see Sect. 2.1) is used to deploy SK_k inside the enclave. Thus SK_k is only known by the enclave's process and the client. The cloud provider and all other processes cannot access this key at any point in time. The next step is to load the B^+-tree structure (tree nodes, but no values) from an untrusted to the isolated memory region and use SK_k to decrypt the tree nodes. It is important to note that the tree is still protected, because all data inside the enclave is secured by SGX. The values are stored outside of the enclave to save enclave memory. This has no security implications as the values are encrypted with an authenticated IND-CCA secure encryption scheme and they are stored in a randomized order. The enclave is then ready to receive search query tokens τ from the client.

The untrusted part relays τ to the trusted part. There, the token gets decrypted and a B^+-tree traversal is performed. All pointers that lead to values falling in the queried range are returned in random order. The untrusted part receives pointers to the values, loads the values from memory or disk and passes them to the client.

This construction suffers from the substantial problem mentioned before: the memory reserved for SGX is limited to 128 MB, and only about 96 MB can be used for data and code. SGX supports a larger enclave size, but enclave pages have to be swapped in and out in this case. Our evaluation (see Sect. 6) shows that even $50,000,000$ values are possible. A B^+-tree, however, is only a small part of a full encrypted DBMS based on our constructions. Other components would occupy large regions of the restricted enclave memory and thus further limit the available space.

Construction 1 provides a very low leakage (see [20] for details), but it is not usable in a big data cloud scenario, because of the described limitation. Therefore, we present a second construction in the next section.

5.2 Construction 2

In this section, we describe our second correct (according to Definition 2) and secure (according to Definition 3) construction. Instead of loading all nodes into the enclave, the main idea is to only load the nodes required to traverse the tree. The challenge is to optimize the communication bottleneck between the untrusted part and the enclave. We performed extensive benchmarking and algorithm engineering in order to identify and minimize the most important run-time consuming tasks, such as switching between the untrusted part and the enclave. The decisive advantage of our second construction is that the required memory space inside the enclave is $O(1)$ for a tree of arbitrary size. The trade-off is that all nodes are stored encrypted inside untrusted main memory or on the untrusted disk and thus have to be decrypted by the enclave before processing. This also leads to a slightly larger leakage than in the first construction, namely a finer-granular access pattern on node instead of page level (details are described by a formal model and proof later).

The setup of the HSBT-scheme is slightly different than in the first construction in order to implement the described features. As before, the B^+-tree is constructed and encrypted at the client and is then transferred to the cloud provider. However, the application does not reserves memory for the whole B^+-tree structure inside the enclave. Instead, it only reserves a fixed space, denoted as reservedSpace, for on the fly processing. The remote attestation and secure key deployment are performed as in the previous construction.

We now describe an HSBT-scheme consisting of six algorithms (HSBT2_Setup, HSBT2_Enc, HSBT2_Tok, HSBT2_Dec, HSBT2_SearchRange, HSBT2_SearchRange_-Trusted) that utilizes a pseudorandom permutation $\Pi : \{0,1\}^\lambda \times \{0,1\}^{log_2 \#x} \rightarrow \{0,1\}^{log_2 \#x}$ in more detail. Note that all but HSBT2_SearchRange and HSBT2_SearchRange_Trusted exactly match the algorithms utilized for Construction 5.1.

$SK \leftarrow$ HSBT2_Setup(1^λ): Use input λ to execute PASE_Gen two times and output $SK = (SK_k, SK_v)$. SK_v and SK_k are kept secret at the client. SK_k is additionally shared with the server enclave using a secure transport and deployment protocol. The enclave stores SK_k inside the isolated enclave.

$\gamma \leftarrow$ HSBT2_Enc($SK, \mathbf{s}$): Take SK and $\mathbf{s} = ((k_1, v_1), ..., (k_n, v_n))$ as input. Start by storing all values $\mathbf{v} = (v_1, ..., v_n)$ in a random order. An almost standard B^+-tree insertion is used for all keys. One difference is that every newly created node x gets an id according to the creation order, i.e., the first node gets id 0 ($x.id = 0$), the second id 1 ($x.id = 1$) et cetera. After each pair is inserted, the empty position for keys and pointers in the tree get filled up. More specifically, a node x that contains $x.\#\mathbf{k}$ keys from the domain gets filled with $(b - 1 - x.\#\mathbf{k})$ keys ∞ and $(b - x.\#\mathbf{k})$ dummy pointers. Then, all keys and pointers are padded to a length of 32 bit (this is no prerequisite of our solution). The ids are used by the algorithm to store the nodes at pseudorandom positions: $p_x = \Pi(SK_k, x.id)$. Now, we have a B^+-tree in which every node occupies the same storage space and the order of the nodes and values is random. Finally, $\text{PASE\_Enc}(SK_v, \cdot)$ is used to encrypt every value and $\text{PASE\_Enc}(SK_k, \cdot)$ is used to encrypt every node. The encrypted nodes and values form the encrypted tree γ, which is protected by an authenticated IND-CPA secure encryption.

$\tau \leftarrow \text{HSBT2\_Tok}(SK_k, R)$: Use input SK_k and $R = [R_s, R_e]$ to calculate $\tau \leftarrow \text{PASE\_Enc}(SK_k, R_s \| R_e)$ and output τ. Queries for all elements below R_e or all elements above R_s can be created by using $R_s = -\infty$ or $R_e = \infty$, respectively.

$\mathbf{v}' \leftarrow \text{HSBT2\_Dec}(SK_v, \mathbf{C}')$: Use input SK_v to decrypt the encrypted values $\mathbf{C}' = (C_0, ..., C_j)$: $\mathbf{v}' = (\text{PASE\_Dec}(SK_v, C_0), ..., \text{PASE\_Dec}(SK_v, C_j))$. Output $\mathbf{v}'$.

$\mathbf{C}' \leftarrow \text{HSBT2\_SearchRange}(\tau, \gamma)$: Take the search token τ and the encrypted tree γ as input. At the beginning, pass only the root node to the trusted part and receive pointers to nodes that should be traversed next. The trivial solution is to pass one node after another. A problem with this design is that every context switch from the untrusted to the trusted part or back causes an overhead. We therefore optimized the number of context switches: transfer as many nodes as currently in the queue, but not more than fit into reservedSpace. We denote the maximal number of nodes as maxAmount, which is directly influenced by reservedSpace: $\text{maxAmount} = \text{reservedSpace}/(o \cdot 128\text{bit})$ where o is the number of AES-blocks used by each node. Nodes are passed until no further are requested. Then output $\mathbf{C}'$ by dereferencing pointers to the values. See Algorithm 1 for details.

$\mathbf{P} \leftarrow \text{HSBT2\_SearchRange\_Trusted}(\tau, \mathbf{X})$: Take a search token τ and nodes $\mathbf{X}$ as input. During the setup phase, SK_k was deployed inside the secure hardware. Therefore, the algorithm is able to decrypt all nodes and the token that are encrypted with SK_k. Then, search all keys falling in the query range, whereby *all* keys are accessed. Finally, return the corresponding pointers in a random order together with the plaintext *isLeaf* tag for each pointer. The tag is necessary, because the untrusted part has no direct access to this encrypted node information. See Algorithm 2 for details.

The construction is correct according to Definition 2, because it is based on a textbook B^+-tree traversal. The difference to the textbook algorithm is that the nodes are loaded inside the enclave after another and that each node is encrypted. These changes do not influence the correctness, because each node remains accessible to the enclave and the encryption (at the client) and the decryption (inside the enclave) are based on a correct PASE-scheme.

Algorithm 1. HSBT2_SearchRange(τ, γ)

1: $\mathbf{X} = \emptyset$ ▷ FIFO queue	9: **for** (*isLeaf, p*) in $results_{tmp}$ **do**
2: $\mathbf{X}$.ENQUEUE(root)	10: **if** *isLeaf* **then**
3: $results = \emptyset$	11: $results$.ADD(*p)
4: **while** $\mathbf{X} \: != \emptyset$ **do**	12: **else**
5: **for** i=0; i < $\mathbf{X}$.*size* && i < maxAmount; $i++$ **do**	13: $\mathbf{X}$.ENQUEUE(*p)
	14: **end if**
6: $\mathbf{X}_{tmp} = \mathbf{X}$.DEQUEUE( )	15: **end for**
7: **end for**	16: **end while**
8: $results_{tmp} = $ HSBT2_SEARCHRANGE_TRUSTED($\tau, \mathbf{X}_{tmp}$)	17: **return** $results$

Algorithm 2. HSBT2_SearchRange_Trusted($\tau, \mathbf{X}$)

1: $\tau_{Plain} = $ PASE_Dec(SK_k, τ)	10: **if** ($x.k_i \leq R_s < x.k_{i+1}$) ||
2: parse τ_{Plain} as (R_s, R_e)	($x.k_i \leq R_e < x.k_{i+1}$) ||
3: $\mathbf{X}_{tmp} = \{$PASE_Dec($SK_k, $*$X_0$), PASE_Dec($SK_k, $*$X_1$), ...$\}$	($R_s \leq x.k_i$ && $x.k_{i+1} \leq R_e$) **then**
	11: $\mathbf{P}$.ADD($x.p_i$)
4: $\mathbf{P} = \emptyset$	12: **end if**
5: **for** x in $\mathbf{X}_{tmp}$ **do**	13: **end for**
6: **if** not $x.isLeaf$ and $R_s < x.k_1$ **then**	14: **if** $R_e \geq x.k_{b-1}$ **then**
7: $\mathbf{P}$.ADD($x.p_0$)	15: $\mathbf{P}$.ADD($x.p_{b-1}$)
8: **end if**	16: **end if**
9: **for** $i = 1, i < b - 1, i++$ **do**	17: **end for**
	18: $\mathbf{P} = $ random permutation of $\mathbf{P}$
	19: **return** (*$\mathbf{P}_0.isLeaf$, $\mathbf{P}_0$), (*$\mathbf{P}_1.isLeaf$, $\mathbf{P}_1$), ...

Next, we will prove the security of Construction 5.2. The first step is to define the leakage functions that are based on the attack model described in Sect. 3.

$\mathcal{L}_{enc}(\mathbf{s})$: Given the key-value pairs $\mathbf{s} = ((k_1, v_1), ...(k_n, v_n))$, this function outputs the amount n of values, the size of each value and the amount of B^+-tree nodes $\#\mathbf{x}$.

$\mathcal{L}_{hw}(\mathbf{s}, T, R, t)$: Given the key-value pairs $\mathbf{s}$, the plaintext B^+-tree T, the search range R and given point in time t, this function outputs the nodes access pattern $\mathcal{X}(\mathbf{s}, T, R, t)$ and the value pointers access pattern $\Delta(\mathbf{s}, T, R, t)$.

The nodes access pattern $\mathcal{X}(\mathbf{s}, T, R, t)$ is a tree that contains the storage positions of all nodes in T that get accessed when searching for the range R. For a more formal definition, we denote the set of leaf nodes that contain keys from the range as M, i.e., $M = \{x \mid x \in T \wedge x.isLeaf \wedge x.k_j \in R, j \in [1, b-1]\}$. Additionally, we define $x \to parent_1$ as the parent node of x and $x \to parent_j$ denotes the node that is reached by moving j layers up in the tree starting from x. We denote a node that only contains the storage position of a node x_i as y_i. Now, we can specify the node set $\mathbf{Y}$ of $\mathcal{X}$ as $\mathbf{Y} = \{y_i \mid x_i \in M\} \cup \{y_i \mid x_i \in T \wedge x \in M \wedge x_i == x \to parent_j, j \in [1, h-1]\}$. The set of directed edges in $\mathcal{X}$ is $\{(y_i, y_j) \mid y_i, y_j \in \mathbf{Y} \wedge \exists x_i, x_j \in T : x_i == x_j \to parent_1\}$. The time parameter t defines a snapshot of the random (but fixed) order of sibling nodes at a given point in time. See Fig. 3 for an illustrative example.

The value pointers access pattern $\Delta(\mathbf{s}, T, R, t)$ is defined as the pointers to the result values together with the leaf nodes in which these pointers are stored. More formally, $\Delta(\mathbf{s}, T, R, t) = \{(x, \mathbf{P}_x) \mid x \in T \wedge x.isLeaf \wedge \exists x.k_j \in R, j \in [1, b-1] \wedge \mathbf{P}_x = \{x.p_l \mid x.k_l \in R, l \in [1, b-1]\}\}$. The time parameter t defines a random but fixed order of the pointers.

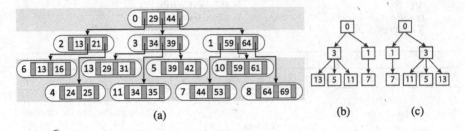

Fig. 3. Illustration of nodes access pattern leakage: (a) example B^+-tree T (storage positions on the left), (b) leakage $\mathcal{X}(\mathbf{s}, T, R, t)$ for $R = [33, 55]$ and B^+-tree T at t_1, (c) leakage $\mathcal{X}(\mathbf{s}, T, R, t)$ for $R = [33, 55]$ and B^+-tree T at t_2

Theorem 1 (Security). *The secure hardware B^+-tree construction presented above is $(\mathcal{L}_{enc}, \mathcal{L}_{hw})$-secure according to Definition 3.*

The idea of the security proof is to describe how a polynomial-time simulator $\mathcal{S}$ simulates the encrypted B^+-tree γ, the token τ and the secure hardware so that a PPT adversary $\mathcal{A}$ can distinguish between $\mathbf{Real}_{\mathcal{A}}(\lambda)$ and $\mathbf{Ideal}_{\mathcal{A},\mathcal{S}}(\lambda)$ with at most negligible probability. The detailed proof can be found in Appendix B.

The leakage definition and security proofs for Construction 5.1 are similar. The main difference is the granularity of the tree and value pointers access pattern. In Construction 5.2, the attacker is able to reveal accesses on a node level. In contrast, the attacker in Construction 5.1 is able to reveal accesses on page level, because SGX inherently leaks the page access pattern.

Note that the page access leakage is an upper bound in the first construction. Each page allocates 4 KB and every encrypted node consists of one or multiple AES-blocks. Up to $k = 4\,KB/(o \cdot 128bit)$ nodes are contained in one page if o AES-blocks are used by each node. Experiments showed that 102 AES-blocks are used for each node if $b = 100$ and 32 bit keys and pointers are used. Therefore, even multiple of those huge nodes fit within a single page.

5.3 Side Channels

Our implementation is concerned with three types of (side) channels: external resource access, page-fault side channel and cache timing side channel (see Sect. 3). By means of all three channels an adversary can observe access patterns to memory with the goal of inferring sensitive information from the observed access patterns.

By monitoring access to external resources, the attacker tries to gain information about the tree structure and ultimately the order of values stored in the database. The only external resources accessed by Construction 5.1 are the encrypted values stored at random positions, which does not leak information. Construction 5.2 also accesses B^+-tree nodes but this is explicitly covered in its leakage.

The page-fault side channel allows the attacker to reliably observe memory access patterns at a granularity of 4 KB. All accesses within the same page are indistinguishable for the attacker and, thus, are not exploitable. The implementation of Construction 5.1 explicitly considers the leakage of the tree structure through this side channel. In Construction 5.2, this side channel does not leak additional information, as nodes are smaller than memory pages and the nodes access pattern is leaked anyway by storing the B^+-tree outside of the enclave.

Cache timing side channel allow finer grained memory access observations while being less reliable. Nevertheless, assuming an adversary who is able to observe accesses within a node, the attacker needs to determine which links to child nodes are followed. Our algorithm, however, accesses every key and pointer, whether the pointer is followed or not. By this and other fine grained implementation details, we achieve data independent accesses and thwart the cache timing side channel.

Leakage of cryptographic keys are thwarted for page-fault and cache timing side channel by using leakage resilient implementations and hardware features [7]. For instance, the AES implementation used in HardIDX holds the S-Boxes in CPU registers instead of RAM to hamper cache side channel attacks [32].

6 Performance Evaluation

In this section, we present our evaluation results collected in a number of experiments with real SGX hardware. Our evaluation system was equipped with an Intel Core i7-6700 processor at 3.40 Ghz and 32 GB DDR4 RAM. 64-bit Ubuntu 14.04.1 extended with SGX support was used as operating system.

6.1 Construction 1 vs. Construction 2

First, we compare the performance of our two constructions. The parameters of the B^+-tree are held constant for this comparison: the branching factor is 10 and the tree contains 1,000,000 key-value pairs. Queries with five different sizes of the result set are used: $2^0, 2^4, 2^8, 2^{12}, 2^{16}$. The search ranges were selected uniformly at random and every result size is tested with 1,000 different ranges. Figure 4a depicts the results of this evaluation, whereby the x-axis shows the size of the result set and the y-axis shows the median of the run-times in ms.

The performance difference can be explained by the following effects:

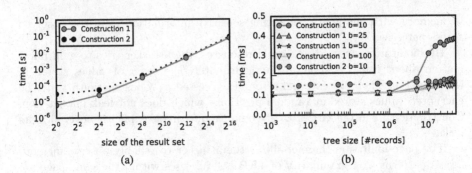

Fig. 4. (a) Comparison of constructions and (b) effect of different branching factors

- **Processor mode switch.** Before executing inside an enclave, the processor has to switch into "enclave mode". This includes, e.g., storing the current CPU context on the host process' stack and loading the CPU context of the enclave. In Construction 5.1 only one switch is required, whereas in the Construction 5.2 $O(\log_b n)$ switches are performed, as at least each level of the B^+-tree is loaded into the enclave.
- **Data transfer.** In Construction 5.1, the data transfer between trusted and the untrusted code is limited to the result set and the query whereas in Construction 5.2 also part of the B^+-tree is transferred between the two components.
- **Access to plain data.** In Construction 5.1, decryption is a one-time effort after loading the entire B^+-tree into the enclave. During query processing, it has access to plaintext nodes of the B^+-tree. Construction 5.2 incrementally loads the B^+-tree nodes from untrusted storage. All processed nodes need to be decrypted during query processing.

Construction 5.2, therefore, is slower than Construction 5.1 by a small factor at any result size. For an increasing size of the result set, both algorithms search a linearly increasing part of the tree. Figure 4a shows that the run-times of our two constructions converge (on a logarithmic scale). This shows that the effects described above diminish compared to the search time of the algorithm.

6.2 Memory Management

In order to identify the limiting parameters in the memory management of our two constructions, we evaluate B^+-trees with different tree sizes (amounts of key-value pairs) and branching factors. On each tree we ran 1,000 randomly chosen queries with result set size of 100 and tested with the branching factors 10, 25, 50 and 100. The results of these evaluation are depicted in Fig. 4b. The x-axis shows the size of the B^+-tree and the y-axis shows the median run-time of the queries.

We see a sharp increase of the run-time above a tree size of 10^6 records. This is due to the exhausted memory in SGX and the virtual memory mechanism of the operating system that swaps pages in and out. This is not security critical, since pages remain encrypted and integrity protected by the SGX system, even when they are swapped out of the SGX protected memory. The figure also reveals a significant difference in the impact of paging between different branching factors. The reason becomes clear by considering the number of required page swaps. The lower the branching factor, the higher the number of nodes in a B^+-tree. The higher the number of nodes, the higher the number of accesses to different memory pages. The higher the number of different page accesses, the higher the probability of a swapped out page.

We also see that Construction 5.2 is not affected by paging, albeit supporting an unlimited tree size. Our data also shows that, as expected, higher branching factors result in better performance and the runtime approaches the runtime of Construction 5.1.

6.3 Comparison with Related Work

In this section, we compare our Construction 5.2 against the currently fastest approach with comparable security features and a security proof presented by Demertzis et al. in [17]. The authors present seven different constructions that support range queries. The constructions have different tradeoffs regarding security, query size, search time, storage and false positives. We do not compare against the highly secure scheme with prohibitive storage cost and also exclude the approaches with false positives as our construction does not lead to false positives. Instead, we compare against the most secure approach without these problems: Logarithmic-URC.

We assume that the OXT construction from [14] is used as underlying symmetric searchable encryption scheme (SEE) by Logarithmic-URC. Fundamentally, the SSE scheme is changeable, but the authors of [17] also utilize OXT for the security and performance evaluation. One has to note that a quite equal construction as Logarithmic-URC was presented independently by Faber et al. in [19]. We implemented the algorithm of [17], but a security and performance comparison to [19] would lead to comparable results.

Table 1 compares our Construction 5.2 and Logarithmic-URC. In this evaluation, we use a branching factor of 100 for Construction 5.2 and search for a randomly chosen range that contains 100 results. Every test for the four different tree sizes (100, 1,000, 10,000, 100,000) was performed 1,000 times and the table shows the mean.

Table 1. Time comparison of random range queries with Logarithmic-URC [17] and our Construction 5.2

Tree size	100	1,000	10,000	100,000
Logarithmic-URC	0.015 s	0.020 s	0.051 s	1.052 s
Construction 5.2 ($b = 100$)	0.119 ms	0.121 ms	0.124 ms	0.125 ms

Our construction runs in about a tenth of a millisecond and with very moderate increase for all tree sizes. In contrast, Logarithmic-URC requires at least multiple milliseconds up to a seconds for bigger trees. A reason for the performance difference might be that OXT construction itself is less efficient then our construction. Furthermore, the search time of OXT depends on the number of entries. Logarithmic-URC fills the OXT construction with elements from a binary tree over the domain for every stored key. An increasing domain severely increases the tree height of a binary tree and thus the number of entries for OXT. In contrast, the height of the B^+-tree in our construction increases much slower with the tree size.

It is not trivial to compare Logarithmic-URC and Construction 5.2 regarding security. The access pattern leakage and the leakage of the internal data structure of Logarithmic-URC is comparable to our access pattern leakages. However, Logarithmic-URC additionally leaks the domain size, the search range size and the search pattern. The search pattern reveals whether the same search was performed before, which might be sensitive information.

7   Related Work

7.1   Searchable Encryption

Searchable encryption scheme supporting range queries are rare. Table 2 shows a comparison of different searchable encryption schemes and other schemes that support range queries. Note that all existing range-searchable encryption schemes leak the access pattern – including ours. The first range-searchable scheme by Boneh and Waters in [11] encrypt every entry linear in the size of the plaintext domain. The first scheme with logarithmic storage size per entry in the domain was proposed by Shi et al. in [40]. Their security model is somewhat weaker than standard searchable encryption. The construction is based on inner-product predicate encryption which has been made fully secure by Shen et al. in [39]. All of these schemes have linear search time.

Lu built the range-searchable encryption from [39] into an index in [29]. He enabled polylogarithmic search time, but his encrypted inverted index tree reveals the order of the plaintexts and is hence only as secure as order-preserving encryption. Wang et al. [43] proposed a multi-dimensional extension of Lu [29], but it suffers from the same problem of order leakage. There is no known searchable encryption schemes for ranges – until ours – that has polylogarithmic search time and leaks only the access pattern.

ORAM can in principle be used to hide the access pattern of searchable encryption. However, Naveed shows that the combination of the two is not straightforward [33]. Special ORAM techniques, like TWORAM [21], are needed.

7.2   Encrypted Databases

Encrypted databases, such as CryptDB [36], use property-preserving encryption for efficient search. Property-preserving encryption has very low deployment

Table 2. Comparison of range-searchable encryption schemes. n is the number of keys, D is the size of the plaintext domain and R is the query range size.

Scheme	Search time	Query size	Storage size	Search pattern leakage	Order leakage
Boneh, Waters [11]	$O(nD)$	$O(D)$	$O(nD)$	yes	no
Shi et al. [40]	$O(n \log D)$	$O(\log D)$	$O(n \log D)$	yes	no
Shen et al. [39]	$O(n \log D)$	$O(\log D)$	$O(n \log D)$	no	no
Lu [29]	$O(\log n \log D)$	$O(\log D)$	$O(n \log D)$	no	yes
Demertzis et al. [17] Faber et al. [19]	$O(\log R)$	$O(\log R)$	$O(n \log D)$	yes	no
This papers	$O(\log n)$	$O(1)$	$O(n)$	no	no

and runtime overhead due to the ability to use internal index structures of the database engine in the same way as on plain data. Order-preserving encryption [1,8,9,28] allows range queries on the ciphertexts as on the plaintexts. However, Naveed et al. [34] initiated the research direction of practical ciphertext-only attacks on property-preserving encryption, in particular order-preserving encryption, which recover the plaintext in many cases with very high probability (close to 100%) and further attacks followed [18,25].

Cash et al. [14] introduce a new protocol called OXT that allows evaluation of boolean queries on encrypted data. Faber et al. [19] extend this data structure to support range queries but either leak additional information on the queried range or the result set contains false positives. In [17], Demertzis et al. present several approaches for range queries. We provide an experimental and detailed comparison in Sect. 6.3.

7.3   TEE-Based Applications

Trusted Database System (TDB) uses a trusted execution environment (TEE) to operate the entire database in a hostile environment [30]. While TDB encrypts the entire database storage and metadata, it is not concerned with information leakage from the TEE. Neither does TDB aim at hiding access patterns nor does it consider side channels attacks against the TEE. Furthermore, since the entire DB operates in the TEE the trusted computing base is very large exposing a very large attack surface.

Haven is an approach to shield application on an untrusted system using SGX [5]. The goal of Haven is to enable the execution of unmodified applications inside an SGX enclave. This technique could be used to isolated off-the-shelf databases with SGX, however, Haven does not consider information leakages through memory access patterns or interactions with the untrustworthy outside world. Furthermore, this approach limits the size of the database due to limited enclave size.

VC3 adapts the MapReduce computing paradigm to SGX [38]. There the data flows between Mapper and Reducer can leak sensitive information and it is excluded from their adversary model. In contrast, we focus on information leakage in the interaction of an enclave with other entities.

In [35], SGX protected machine learning algorithms have been adapted to prevent the exploitation of side channels by the usage of data-oblivious primitives. Access to external data is addressed by randomizing the data and always accessing all data, i.e., their solution has an complexity of $O(n)$, even for tree searches.

8 Conclusion

In this paper, we introduce HardIDX – an approach to search for ranges and values over encrypted data using hardware support making it deployable as a secure index in an encrypted database. We provide a formal security proof explicitly including side channels and an implementation on Intel SGX. Our solution compares favorably with existing software- and hardware-based approaches. We require few milliseconds even for complex searches on large data and scale to almost arbitrarily large indices. We only leak the access pattern and our trusted code protected by SGX hardware is very small exposing a small attack surface.

Acknowledgments. This research was co-funded by the German Science Foundation, as part of project P3 within CRC 1119 CROSSING, the European Union's Horizon 2020 Research and Innovation Programme under grant agreement No. 644412 (TREDISEC) and No. 643964 (SUPERCLOUD), and the Intel Collaborative Research Institute for Secure Computing (ICRI-SC).

A   Proof Framework

In [16], Curtmola et al. introduced a three step framework to proof the security of searchable encryption. The first step is to formulate a leakage, i.e., an upper bound of the information that an adversary can gather from the protocol. Secondly, one defines the $\mathbf{Real}_{\mathcal{A}}(\lambda)$ and a $\mathbf{Ideal}_{\mathcal{A},\mathcal{S}}(\lambda)$ game for an adaptive adversary $\mathcal{A}$ and a polynomial time simulator $\mathcal{S}$. $\mathbf{Real}_{\mathcal{A}}(\lambda)$ is the execution of the actual protocol and $\mathbf{Ideal}_{\mathcal{A},\mathcal{S}}(\lambda)$ utilizes $\mathcal{S}$ to simulate the real game by using only the formulated leakage. An adaptive adversary can use information learned in previous protocol iterations for its queries. Third, a scheme is CKA2-secure if one can show that $\mathcal{A}$ can distinguish the output of the games with probability negligibly close to 0. This in turn means that $\mathcal{A}$ does not learn anything besides the leakage stated in the first step, because otherwise he could use this additional information to distinguish the games.

B $(\mathcal{L}_{enc}, \mathcal{L}_{hw})$-Security Proof

In Sect. 5.2, we provide the description of Construction 5.2 and the leakage of the construction. We now prove Theorem 1 by describing a polynomial-time simulator S for which a PPT adversary $\mathcal{A}$ can distinguish between $\mathbf{Real}_{\mathcal{A}}(\lambda)$ and $\mathbf{Ideal}_{\mathcal{A},S}(\lambda)$ with negligible probability.

Proof. The simulator S works as follows:

- *Setup:* S creates a new random key $\widetilde{SK} = \texttt{PASE\_Gen}(1^{\lambda})$ and stores it.
- *Simulating γ:* S gets $\mathcal{L}_{enc}$ and creates $\#\mathbf{x}$ nodes $\mathbf{X} = (x_1, ..., x_{\#\mathbf{x}})$ filled with random keys, random pointers and increasing node ids. These nodes are stored in the pages $(\omega_1, ... \omega_{\#\omega})$. Additionally, S generates n encryptions of random values $\boldsymbol{C} = (C_1, ..., C_n)$ using $\texttt{PASE\_Enc}$, the number of values and the size of the values. Every encrypted value is given a distinct index. S outputs $\gamma = (\mathbf{X}, C)$

  All described operations can be executed by S, because the information required for the encryption of values is included in the leakage. The simulated γ has the same size as the output of $\mathbf{Real}_{\mathcal{A}}(\lambda)$ and the IND-CCA security of $\texttt{PASE}$ makes the nodes and values indistinguishable from the output of $\mathbf{Real}_{\mathcal{A}}(\lambda)$.

- *Simulating τ:* The simulator S creates two random values r_1 and r_2 and encrypts them: $\tau \leftarrow \texttt{PASE\_Enc}(SK_k, R_s \| R_e)$. S outputs τ.

  The simulated τ is indistinguishable from the output of $\mathbf{Real}_{\mathcal{A}}(\lambda)$ as a result of the IND-CCA security of $\texttt{PASE}$.

- *Simulating secure hardware:* At time t, the simulator S receives encrypted nodes (denoted as $\mathbf{X}$), a token τ and $\mathcal{L}_{hw}$. It has to simulate the output of the secure hardware enclave. The simulator decrypts every node $x_i \in \mathbf{X}$ with $\texttt{PASE\_Dec}(\widetilde{SK}, x_i)$.

  We differentiate between two cases for every x_i:
  1. x_i is not leaf: S reads the id of x_i and searches the corresponding y_i in $\mathcal{X}(\mathbf{s}, T, R, t)$. It returns a pointer to all children in the order defined by t. $\mathcal{A}$ cannot distinguish between the output of $\mathbf{Real}_{\mathcal{A}}(\lambda)$ and the simulated output, because the pointers point to indistinguishable nodes according to the IND-CCA security of $\texttt{PASE}$. Furthermore, the results are consistent for different requests of the same range as the nodes access pattern delivers deterministic results and the pseudorandom permutation creates unambiguous positions for the simulated nodes. The same argument applies for queries of distinct or overlapping ranges.
  2. x is leaf: S uses the leakage $\Delta(\mathbf{s}, T, R, t)$ to output all result pointers $\mathbf{P} = \bigcup_x \mathbf{P}_x, \forall (x, \mathbf{P}_x) \in \Delta$ in the order defined by t.

     This output is indistinguishable from the output of $\mathbf{Real}_{\mathcal{A}}(\lambda)$ as the number of result pointers matches and the pointers are consistent because $\Delta(\mathbf{s}, T, R, t)$ is unambiguous. The values pointed on are indistinguishable, because they are protected by IND-CCA secure encryption. $\qquad\square$

C Multiple Users

So far, we considered a setup comprising one user, but multiple user are directly supported by HardIDX. Multiple users are able to concurrently query data without limitations, as concurrent tree traversals do not influence each other. The only requirement is that each user has access to the key SK_k to create query tokens and SK_v to decrypt the result. It is also possible that each user shares a different key SK_k with the enclave. This would hide the search pattern of one user from all other users, but it requires a small modification in the protocol: the token has to be accompanied by client information, because the enclave has to identify the key to use for the token decryption. The nodes can be encrypted by any key that is known to the enclave. Particularly, it is not required to be a key shared with any user.

References

1. Agrawal, R., Kiernan, J., Srikant, R., Xu, Y.: Order preserving encryption for numeric data. In: ACM International Conference on Management of Data, SIG-MOD (2004)
2. Anati, I., Gueron, S., Johnson, S.P., Scarlata, V.R.: Innovative technology for CPU based attestation and sealing. In: Workshop on Hardware and Architectural Support for Security and Privacy, HASP (2013)
3. Limited, A.R.M.: ARM Security Technology - Building a Secure System using TrustZone Technology (2009)
4. Bajaj, S., Sion, R.: TrustedDB: A trusted hardware-based database with privacy and data confidentiality. IEEE Trans. Inf. Forensics Secur. **26**, 752–765 (2014)
5. Baumann, A., Peinado, M., Hunt, G.: Shielding applications from an untrusted cloud with Haven. In: 11th USENIX Symposium on Operating Systems Design and Implementation, OSDI (2014)
6. Bellare, M., Boldyreva, A., O'Neill, A.: Deterministic and efficiently searchable encryption. In: Menezes, A. (ed.) CRYPTO 2007. LNCS, vol. 4622, pp. 535–552. Springer, Heidelberg (2007). doi:10.1007/978-3-540-74143-5_30
7. Bernstein, D.J., Lange, T., Schwabe, P.: The security impact of a new cryptographic library. In: Hevia, A., Neven, G. (eds.) LATINCRYPT 2012. LNCS, vol. 7533, pp. 159–176. Springer, Heidelberg (2012). doi:10.1007/978-3-642-33481-8_9
8. Boldyreva, A., Chenette, N., Lee, Y., O'Neill, A.: Order-preserving symmetric encryption. In: Joux, A. (ed.) EUROCRYPT 2009. LNCS, vol. 5479, pp. 224–241. Springer, Heidelberg (2009). doi:10.1007/978-3-642-01001-9_13
9. Boldyreva, A., Chenette, N., O'Neill, A.: Order-preserving encryption revisited: improved security analysis and alternative solutions. In: Rogaway, P. (ed.) CRYPTO 2011. LNCS, vol. 6841, pp. 578–595. Springer, Heidelberg (2011). doi:10.1007/978-3-642-22792-9_33
10. Boneh, D., Sahai, A., Waters, B.: Functional encryption: definitions and challenges. In: Ishai, Y. (ed.) TCC 2011. LNCS, vol. 6597, pp. 253–273. Springer, Heidelberg (2011). doi:10.1007/978-3-642-19571-6_16
11. Boneh, D., Waters, B.: Conjunctive, subset, and range queries on encrypted data. In: Vadhan, S.P. (ed.) TCC 2007. LNCS, vol. 4392, pp. 535–554. Springer, Heidelberg (2007). doi:10.1007/978-3-540-70936-7_29
12. Brasser, F., El Mahjoub, B., Koeberl, P., Sadeghi, A.R., Wachsmann, C.: TyTAN: Tiny Trust Anchor for Tiny Devices. In: Design Automation Conference. DAC (2015)
13. Brumley, B.B., Tuveri, N.: Remote timing attacks are still practical. In: Atluri, V., Diaz, C. (eds.) ESORICS 2011. LNCS, vol. 6879, pp. 355–371. Springer, Heidelberg (2011). doi:10.1007/978-3-642-23822-2_20

14. Cash, D., Jarecki, S., Jutla, C., Krawczyk, H., Roşu, M.-C., Steiner, M.: Highly-scalable searchable symmetric encryption with support for boolean queries. In: Canetti, R., Garay, J.A. (eds.) CRYPTO 2013. LNCS, vol. 8042, pp. 353–373. Springer, Heidelberg (2013). doi:10.1007/978-3-642-40041-4_20

15. Costan, V., Devadas, S.: Intel SGX Explained. Technical report, IACR Cryptology ePrint Archive, Report 2016/086

16. Curtmola, R., Garay, J., Kamara, S., Ostrovsky, R.: Searchable symmetric encryption: improved definitions and efficient constructions. In: 13th ACM Conference on Computer and Communications Security, CCS (2006)

17. Demertzis, I., Papadopoulos, S., Papapetrou, O., Deligiannakis, A., Garofalakis, M.: Practical Private Range Search Revisited. In: International Conference on Management of Data, SIGMOD (2016)

18. Durak, F.B., DuBuisson, T.M., Cash, D.: What else is revealed by order-revealing encryption?. In: Conference on Computer and Communications Security, CCS (2016)

19. Faber, S., Jarecki, S., Krawczyk, H., Nguyen, Q., Rosu, M., Steiner, M.: Rich queries on encrypted data: beyond exact matches. In: Pernul, G., Ryan, P.Y.A., Weippl, E. (eds.) ESORICS 2015. LNCS, vol. 9327, pp. 123–145. Springer, Cham (2015). doi:10.1007/978-3-319-24177-7_7

20. Fuhry, B., Bahmani, R., Brasser, F., Hahn, F., Kerschbaum, F., Sadeghi, A.R.: HardIDX: Practical and Secure Index with SGX (2017)

21. Garg, S., Mohassel, P., Papamanthou, C.: $textbsansTWORAM$: Efficient Oblivious RAM in Two Rounds with Applications to Searchable Encryption. In: Robshaw, M., Katz, J. (eds.) CRYPTO 2016. LNCS, vol. 9816, pp. 563–592. Springer, Heidelberg (2016). doi:10.1007/978-3-662-53015-3_20

22. Genkin, D., Pipman, I., Tromer, E.: Get your hands off my laptop: physical side-channel key-extraction attacks on PCs. J. Cryptographic Eng. **5**, 95–112 (2015)

23. Gentry, C.: Fully homomorphic encryption using ideal lattices. In: Symposium on Theory of Computing, STOC (2009)

24. Gentry, C., Halevi, S., Smart, N.P.: Homomorphic evaluation of the AES circuit. In: Safavi-Naini, R., Canetti, R. (eds.) CRYPTO 2012. LNCS, vol. 7417, pp. 850–867. Springer, Heidelberg (2012). doi:10.1007/978-3-642-32009-5_49

25. Grubbs, P., Sekniqi, K., Bindschaedler, V., Naveed, M., Ristenpart, T.: Leakage-Abuse Attacks against Order-Revealing Encryption. Technical report, IACR Cryptology ePrint Archive, Report 2016/895

26. Hoekstra, M., Lal, R., Pappachan, P., Phegade, V., Del Cuvillo, J.: Using innovative instructions to create trustworthy software solutions. In: Workshop on Hardware and Architectural Support for Security and Privacy, HASP (2013)

27. Kaplan, D., Powell, J., Woller, T.: AMD Memory Encryption (2016). http://amd-dev.wpengine.netdna-cdn.com/wordpress/media/2013/12/AMD_Memory_Encryption_Whitepaper_v7-Public.pdf

28. Kerschbaum, F., Schröpfer, A.: Optimal average-complexity ideal-security order-preserving encryption. In: 21st ACM Conference on Computer and Communications Security, CCS (2014)

29. Lu, Y.: Privacy-preserving logarithmic-time search on encrypted data in cloud. In: 19th Network and Distributed System Security Symposium, NDSS (2012)

30. Maheshwari, U., Vingralek, R., Shapiro, W.: How to build a trusted database system on untrusted storage. In: 4th Conference on Symposium on Operating System Design and Implementation, OSDI (2000)

31. McKeen, F., Alexandrovich, I., Berenzon, A., Rozas, C.V., Shafi, H., Shanbhogue, V., Savagaonkar, U.R.: Innovative instructions and software model for isolated execution. In: Workshop on Hardware and Architectural Support for Security and Privacy, HASP (2013)

32. Mowery, K., Keelveedhi, S., Shacham, H.: Are AES x86 cache timing attacks still feasible?. In: ACM Workshop on Cloud Computing Security Workshop, CCSW (2012)

33. Naveed, M.: The fallacy of composition of oblivious RAM and searchable encryption. Technical report, IACR Cryptology ePrint Archive, Report 2015/668

34. Naveed, M., Kamara, S., Wright, C.V.: Inference attacks on property-preserving encrypted databases. In: 22nd ACM Conference on Computer and Communications Security, CCS (2015)

35. Ohrimenko, O., Schuster, F., Fournet, C., Meht, A., Nowozin, S., Vaswani, K., Costa, M.: Oblivious multi-party machine learning on trusted processors. In: 25th USENIX Security Symposium. USENIX Security (2016)

36. Popa, R.A., Redfield, C.M.S., Zeldovich, N., Balakrishnan, H.: CryptDB: protecting confidentiality with encrypted query processing. In: Proceedings of the 23rd ACM Symposium on Operating Systems Principles, SOSP (2011)

37. Ramakrishnan, R., Gehrke, J.: Database Management Systems, 3rd edn. McGraw-Hill, (2002)

38. Schuster, F., Costa, M., Fournet, C., Gkantsidis, C., Peinado, M., Mainar-Ruiz, G., Russinovich, M.: VC3: Trustworthy data analytics in the cloud using SGX. In: IEEE Symposium on Security and Privacy, S&P (2015)

39. Shen, E., Shi, E., Waters, B.: Predicate privacy in encryption systems. In: Reingold, O. (ed.) TCC 2009. LNCS, vol. 5444, pp. 457–473. Springer, Heidelberg (2009). doi:10.1007/978-3-642-00457-5_27

40. Shi, E., Bethencourt, J., Chan, H.T.H., Song, D.X., Perrig, A.: Multi-dimensional range query over encrypted data. In: IEEE Symposium on Security and Privacy, S&P (2007)

41. Song, D.X., Wagner, D., Perrig, A.: practical techniques for searches on encrypted data. In: IEEE Symposium on Security and Privacy, S&P (2000)

42. Strackx, R., Piessens, F., Preneel, B.: Efficient isolation of trusted subsystems in embedded systems. In: SecureComm (2010)

43. Wang, B., Hou, Y., Li, M., Wang, H., Li, H.: Maple: scalable multi-dimensional range search over encrypted cloud data with tree-based index. In: 9th ACM Symposium on Information, Computer and Communications Security, ASIACCS (2014)

44. Xu, Y., Cui, W., Peinado, M.: Controlled-channel attacks: deterministic side channels for untrusted operating systems. In: IEEE Symposium on Security and Privacy, S & P (2015)

45. Yarom, Y., Falkner, K.: FLUSH+RELOAD: a high resolution, low noise, L3 cache side-channel attack. In: 23rd USENIX Security Symposium. USENIX Security (2014)

A Novel Cryptographic Framework for Cloud File Systems and CryFS, a Provably-Secure Construction

Sebastian Messmer[2], Jochen Rill[1(✉)], Dirk Achenbach[1],
and Jörn Müller-Quade[2]

[1] FZI Forschungszentrum Informatik, Karlsruhe, Germany
mail@smessmer.de, {rill,achenbach}@fzi.de
[2] Karlsruhe Institute of Technology (KIT), Karlsruhe, Germany
joern.mueller-quade@kit.edu

Abstract. Using the cloud to store data offers many advantages for businesses and individuals alike. The cloud storage provider, however, has to be trusted not to inspect or even modify the data they are entrusted with. Encrypting the data offers a remedy, but current solutions have various drawbacks. Providers which offer encrypted storage themselves cannot necessarily be trusted, since they have no open implementation. Existing encrypted file systems are not designed for usage in the cloud and do not hide metadata like file sizes or directory structure, do not provide integrity, or are prohibitively inefficient. Most have no formal proof of security. Our contribution is twofold. We first introduce a comprehensive formal model for the security and integrity of cloud file systems. Second, we present CryFS, a novel encrypted file system specifically designed for usage in the cloud. Our file system protects confidentiality and integrity (including metadata), even in presence of an actively malicious cloud provider. We give a proof of security for these properties. Our implementation is easy and transparent to use and offers performance comparable to other state-of-the-art file systems.

1 Introduction

In recent years, cloud computing has transformed from a trend to a serious competition for traditional on-premise solutions. Elastic cost models and the availability of virtually infinite resources present an alternative to offers of a preset volume. The more bandwidth is available to consumers, the more economically reasonable it is to replace an on-premise solution with a cloud solution. In the wake of the PRISM disclosures, it seems naïve to trust in the security of one's data in the cloud, however. The scientific challenge for security researchers is to solve this dilemma by finding solutions without sacrificing the economic benefits of cloud technology.

Cryptographic research offers methods that guarantee the confidentiality and integrity of data in the presence of an adversary. The principle of cryptographic *proof* eliminates trust requirements by highlighting precisely which

Published by Springer International Publishing AG 2017. All Rights Reserved
G. Livraga and S. Zhu (Eds.): DBSec 2017, LNCS 10359, pp. 409–429, 2017.
DOI: 10.1007/978-3-319-61176-1_23

guarantees hold under which assumptions. A proof of security makes use of a formal model in formulating security properties. Cryptographic schemes can then also be expressed in the terms of the formal model. Formal proofs of security constructively establish how a scheme achieves a security property (under given assumptions). This is a significant difference to the "ad-hoc security" method of eliminating vulnerabilities from a scheme until one can no longer conceive of any more attacks.

Provably-secure schemes are rarely adopted in practice. The abstract computational models that form the basis of cryptographic frameworks don't usually facilitate a straightforward implementation. Also, the concept of efficiency in these models differs from practical efficiency notions, so that many asymptotically efficient schemes are rather inefficient in practice. In contrast, there are many practical solutions to security challenges. They are deployed widely, but seldomly lend themselves to a formal security analysis and are thus analysed in an "ad-hoc" fashion.

Returning to the cloud scenario from before, a particular case in this area of conflict is outsourced file system data. Encrypting snapshots of file systems (backups) as one single block is certainly a mastered task. To update a single file in a huge file system, one were to re-encrypt and re-upload the whole snapshot. It is a different challenge altogether to efficiently deduplicate and compress encrypted remote backups to conserve bandwidth and storage space. In a similar fashion, it is not immediately obvious how to allow fine-grained access to single files in a file system hierarchy while *provably* protecting metadata and at the same time conserving efficiency. Indeed, we are not aware of any efficient cryptographic cloud file system in literature.

1.1   Our Contribution

Our contribution is twofold. We first give a formal security model for encrypted file systems and cloud file systems in particular. Our model covers both integrity and confidentiality for chosen ciphertext attacks, as well as chosen plaintext attack scenarios. Our model is designed to be as generic as possible to be useful for analysing the security of other cloud file systems beyond the scope of this paper.

Second, we design and implement CryFS[1], a provably secure encrypted file system for the cloud which is easy to use and acts completely transparent to the user. In addition to hiding file contents, we also hide file metadata, like sizes and permission bits, and the directory structure. Our file system is designed to be used by multiple users. When used only by a single user, CryFS also protects the integrity of the file system in the sense that no malicious storage provider can change the file system (for example delete, undelete or roll back files) without being noticed. We achieve good network performance by keeping ciphertext data in small same-sized blocks, which are organised in a special tree data structure and are synchronised individually. Local changes only cause few blocks to be

---

[1] https://www.cryfs.org.

synchronised. We prove that our file system is secure in our security model. The performance of our reference implementation is already comparable to other state-of-the-art encrypted file systems. It is open source and available on github[2].

1.2 Related Work

There are various commercial and free solutions for secure cloud storage. Providers like SpiderOak[3], tresorit[4] and boxcryptor[5] offer cloud storage space in combination with a proprietary client application to synchronise data. They claim that all data is encrypted on the client and stored securely on the servers. However, these services do not disclose the specification of their protocols. Thus, they presume a certain level of trust in their service that is not much different from trusting a popular cloud provider in the first place. Traditional encrypted file systems like EncFS[6], eCryptFs[7] and NCryptFS [12] are open and theoretically usable in a cloud setting, however, they lack important security features: By encrypting files individually, they protect the content but leave metadata like the directory structure unencrypted. Using this, an attacker can easily distinguish a music CD collection (which has about 20 files per directory, 3MB each) from a folder containing only documents. Other solutions like the now-discontinued TrueCrypt[8], VeraCrypt[9], and dm-crypt[10], hide the directory structure by encrypting the whole file system into one big container. However, these solutions cannot be used in a cloud setting efficiently, as changing one small file in the file system causes the whole container to be re-encrypted and thus to be re-uploaded.

What is more, none of the presented solutions have a formal proof of security. There has been research into how to model the security of file systems, however, most of this research is directed at disk encryption schemes. Damgård et al. [5] for example introduce a formalisation of encryption schemes for file systems that is based on the Universal Composability framework. However, there are many artefacts in their model which are not relevant in the cloud setting (e.g. they explicitly model physical and logical sectors). Their model also misses components on which our security is based (i.e. different states for client and server) and thus is not well suited for our setting. Kristian Gjøsteen [8] and more recently Khati et al. [11] both introduce a game-based security model, which, however, is also only suited for modeling full disk encryption.

Modeling the security of outsourced data in general has been mainly investigated in the context of *searchable encryption* and *proofs of data possession*

---

[2] https://github.com/cryfs/cryfs.
[3] https://spideroak.com.
[4] http://tresorit.com.
[5] http://www.boxcryptor.com.
[6] http://www.arg0.net/#!encfs/c1awt.
[7] http://www.ecryptfs.org.
[8] http://truecrypt.sourceforge.net.
[9] https://veracrypt.codeplex.com.
[10] https://gitlab.com/cryptsetup/cryptsetup/wikis/DMCrypt.

(PDP), as well as *proofs of retrievability* (POR). For searchable encryption, there are many different security models (e.g. by Chase et al. [4], Goh [9] and others) which are specifically designed for the corresponding scheme and cannot easily be applied to other settings and schemes. In addition, keeping the queries private is an important goal in the context of searchable encryption and is thus almost always included in the security model. For cloud based file systems, this is not as important. Achenbach et al. [1] introduce a more general security framework for modeling the security of outsourcing schemes but their model does not consider integrity. However, our framework is in part inspired by their ideas. There is a rich body of work regarding outsourcing schemes and corresponding security models which provide proofs of data possession and retrievability (e.g. Zhang et al. [13], Erway et al. [7] and Cash et al. [3]). Similar to our goals, all these schemes provide integrity for outsourced data. However, their requirements are fundamentally different. The goal of a PDP scheme is for a cloud provider to be able to prove that he has all of the outsourced data and that he did not modify it maliciously without requiring the user to hold a copy of the data himself and without having to download it. This is very useful if the server performs computations on the outsourced data without interaction of the user and the user wants to verify if all the data is still correct. In our case however, the server is only used for storage and users interact with the data only locally. Thus, all integrity checks can be performed by the user on the data itself. In order to achieve these particular integrity guarantees, PDP schemes require design and performance trade offs, which are also reflected in their security models. This makes the schemes incomparable to our scheme and the security models hard to adapt to our case.

2   A Security Model for Cryptographic File Systems

In this chapter, we introduce a novel formal security model for cloud file systems which covers both security and integrity in a non-adaptive as well as in an adaptive setting. We first give security definitions in the chosen plaintext attack scenario and then show how to extend them to the chosen ciphertext attack scenario. Further, we show that chosen ciphertext security for file systems can be achieved by combining plaintext security and integrity. Note that throughout this work we use $\cdot$ to denote a free parameter, which can be chosen by the adversary.

2.1   Basic Definitions

In general, encrypted file systems use a symmetric encryption scheme as underlying primitive. We give a formal definition of such an encryption scheme.

Definition 1 (Symmetric Encryption Scheme). *A symmetric encryption scheme $\mathcal{E}$ is a tuple $\mathcal{E} := (\mathsf{Gen}, \mathsf{Enc}, \mathsf{Dec})$ with*

- $\mathsf{Gen} : 1^k \to \{0,1\}^k$ *is a PPT algorithm which given a security parameter k, outputs a key K.*

- Enc : $\{0,1\}^k \times \{0,1\}^n \to \{0,1\}^m$ is a probabilistic polynomial time (PPT) algorithm which given a key K and a plaintext outputs the corresponding ciphertext.
- Dec : $\{0,1\}^k \times \{0,1\}^m \to (\{\bot\} \cup \{0,1\}^n)$ is a PPT algorithm which given a key K and a ciphertext outputs the corresponding plaintext. It outputs $\bot$ if K is wrong or the ciphertext was not valid.

Security and integrity of these basic encryption schemes are modeled by the standard security notions *indistinguishability under chosen plaintext* (IND-CPA) and *integrity of ciphertexts* [2] (INT-CTXT) respectively.

Security Game 1 (IND-CPA$^{\mathcal{A}}(k)$)
- The experiment chooses a key $K \leftarrow \text{Gen}(1^k)$ and a random bit $b \leftarrow \{0,1\}$.
- The adversary is given oracle access to $\text{LR}(K, m_0, m_1)$, which outputs an encryption of m_b under K, if $|m_0| = |m_1|$.
- $\mathcal{A}$ submits a guess b' for b.

The result of the experiment is 1, if $b' = b$, and 0 else.

Security Game 2 (INT-CTXT$^{\mathcal{A}}(k)$)
- The experiment chooses a key $K \leftarrow \text{Gen}(1^k)$.
- The adversary is given oracle access to $\text{Enc}(K, \cdot)$.
- The adversary is given oracle access to $\text{Dec}(K, \cdot)$.

The result of the experiment is 1, if for any Dec oracle query: $\text{Dec}(K, c) \neq \bot$ and c was never *output* by the Enc oracle.

Note that there are several equivalent formalisations for IND-CPA security [10]. We use the formalisation with a *left-or-right oracle* to reduce the complexity of our proofs. If a classic encryption oracle is needed, we can simulate it easily by setting both inputs to LR to be equal.

We now give a formal definition of an encrypted file system. In general, a file system needs four algorithms: one for initialising the file system (like setting up data structures), one for updating the file system (like adding and removing files), one for decrypting the file system and one for generating the cryptographic keys. The file system, and all algorithms which interact with it, are stateful.

Definition 2 (Encrypted File System). *Let $\mathbb{F}$ be the set of plaintext file systems, $\mathbb{C}$ the set of ciphertext file systems, and $\mathbb{S}$ the set of client states. Let $\mathbb{K} = \{0,1\}^k$ be the set of keys and $\mathcal{E} = (\text{Gen}', \text{Enc}', \text{Dec}')$ be a symmetric encryption scheme. An encrypted file system $\mathcal{C}$ is a tuple $\mathcal{C} := (\text{Gen}, \text{Init}, \text{Update}, \text{Dec}, \mathcal{E})$ with*

- Gen : $\{1\}^k \to \mathbb{K}$ is a PPT algorithm which generates a key K.
- Init : $\mathbb{K} \to \mathbb{C} \times \mathbb{S}$ is a PPT algorithm which takes the key K and initialises an empty ciphertext file system C, and the client state s.
- Update : $\mathbb{K} \times \mathbb{C} \times \mathbb{F} \times \mathbb{S} \to (\{\bot\} \cup \mathbb{C}) \times \mathbb{S}$ is a PPT algorithm used to update the file system. It is given the key K, an old ciphertext file system C, a new plaintext file system F and a client state s. It outputs $\bot$ if the decryption of C fails, else a new ciphertext file system C', and a new client state s'.
- Dec : $\mathbb{K} \times \mathbb{C} \times \mathbb{S} \to (\{\bot\} \cup \mathbb{F}) \times \mathbb{S}$ is a PPT algorithm which is given a key K, a ciphertext file system C, and the client state s and outputs $\bot$ if the decryption fails, else the decrypted file system F, and a new client state s.

2.2   Modelling Non-adaptive Security

Traditionally, security against non-adaptive adversaries requires that an adversary cannot gain any information from a scheme which they did not observe or interact with before. In the case of file systems however, we additionally require that the adversary could have interacted with other encrypted file systems using the same key. We allow the adversary to create an arbitrary but constant number of file systems, which are available before and after he chooses the challenge. Also, we do not require the client state to be kept secret. We allow the challenges to be restricted by a relation R_d (e.g. both file systems must store the same amount of data). This means that from looking at a freshly encrypted file system, an attacker cannot deduce any information even if he observed modifications on different file systems using the same key. In particular, this requires the file system to introduce measures to be secure under key reuse (e.g. a user encrypting two different file systems with the same password). We call this security notion *indistinguishability under non-adaptive chosen file system attacks* (IND-naCFA).

Security Game 3 (IND-naCFA$^{\mathcal{A},R_d}(k)$)

- The experiment chooses a key $K \leftarrow \mathsf{Gen}(1^k)$ and a random bit $b \leftarrow \{0,1\}$.
- The adversary is given oracle access to $\mathsf{Init}(K)$. The j-th query returns a new ciphertext file system (C_j, s_j) using the same key, and the following oracle to interact with it:
    - $(C'_j, s'_j) \leftarrow \mathsf{Update}_j(K, C_j, \cdot, s_j)$. The game sets $(C_j, s_j) := (C'_j, s'_j)$.
  The number of Init queries is bounded by an adversary-chosen constant q_{Init}.
- The adversary outputs two file systems F^0 and F^1 with $(F^0, F^1) \in R_d$.
- The experiment generates $(C, s) \leftarrow \mathsf{Init}(K)$.
- The experiment computes $(C', s') \leftarrow \mathsf{Update}(K, C, F^b, s)$.
- $\mathcal{A}$ is given (C, s) and (C', s').
- $\mathcal{A}$ submits a guess b' for b.

The result of the experiment is 1 if $b' = b$, and 0 else.

Definition 3 (Nonadaptive Security). *A file system is* IND-naCFA *secure, if*

$$\forall \mathcal{A}, c \in \mathbb{N} \exists k_0 \in \mathbb{N} \forall k > k_0 : |\mathrm{Pr}[\mathsf{IND\text{-}naCFA}^{\mathcal{A},R_d}(k) = 1]| \leq \frac{1}{2} + k^{-c}$$

2.3   Modelling Adaptive Security

Intuitively, while IND-naCFA models security of a file system directly after creation, adaptive security models the security of a file system later in its life. To achieve this, we allow the adversary to choose a file system as challenge with which he already interacted. We then require that he cannot distinguish which of two modifications he chose is performed. Again, we allow to restrict the adversary's choice of challenge by a relation R_d. We call this security notion *indistinguishability under adaptive chosen file system attacks* and it is a direct extension of IND-naCFA.

Security Game 4 (IND-aCFA$^{\mathcal{A},R_d}(k)$)

- The experiment chooses a key $K \leftarrow \mathsf{Gen}(1^k)$ and a random bit $b \leftarrow \{0,1\}$.
- The adversary is given oracle access to $\mathsf{Init}(K)$, which on the j-th query initialises $F_j = \perp$ (empty file system), returns a new ciphertext file system (C_j, s_j) using the same key and an oracle to interact with it.
  - $(C'_j, s'_j) \leftarrow \mathsf{Update}_j(K, C_j, \cdot, s_j)$. The game remembers the most recent input F_j and sets $(C_j, s_j) := (C'_j, s'_j)$.

  The number of Init queries is bounded by a constant q_{Init} chosen by the adversary.
- The adversary outputs j and two file systems F^0, F^1 with $(F_j, F^0, F^1) \in R_d$.
- The experiment computes $(C'_j, s'_j) \leftarrow \mathsf{Update}_j(K, C_j, F^b, s_j)$ and passes (C'_j, s'_j) to the adversary.
- $\mathcal{A}$ submits a guess b' for b.

The result of the experiment is 1 if $b' = b$ and 0 else.

Definition 4 (Adaptive Security). *A file system is* IND-aCFA *secure, if*

$$\forall \mathcal{A}, c \in \mathbb{N} \exists k_0 \in \mathbb{N} \forall k > k_0 : |\Pr[\mathsf{IND\text{-}aCFA}^{\mathcal{A},R_d}(k) = 1]| \leq \frac{1}{2} + k^{-c}$$

2.4 Modelling Integrity

To provide integrity, a cloud file system must ensure that a malicious server cannot alter the file system in any way, even though the server can observe every modification made to this file system and to other file systems using the same key. In particular, a server must not be able to provide the client with old states of the file system. This results in the following security model, which we call *integrity of file systems*.

Security Game 5 (INT-FS$^{\mathcal{A}}(k)$)

- The experiment chooses a key $K \leftarrow \mathsf{Gen}(1^k)$.
- The adversary is given oracle access to $\mathsf{Init}(K)$. The j-th query returns a new ciphertext file system (C_j, s_j) using the same key, and the following oracles to interact with it:
  - $(C'_j, s'_j) \leftarrow \mathsf{Update}_j(K, C_j, \cdot, s_j)$. The game sets $(C_j, s_j) := (C'_j, s'_j)$.
  - $(F, s'_j) \leftarrow \mathsf{Dec}_j(K, \cdot, s_j)$. The game sets $s_j := s'_j$ for the next query.

  The number of Init queries is bounded by an adversary-chosen constant q_{Init}.

The result of the experiment is 1 if for any of the decryption oracle queries $\mathsf{Dec}_j(K, C', s_j) \neq \perp$, $C_j \neq C'$.

Definition 5 (Integrity). *A file system is* INT-FS *secure, if*

$$\forall \mathcal{A}, c \in \mathbb{N} \exists k_0 \in \mathbb{N} \forall k > k_0 : |\Pr[\mathsf{INT\text{-}FS}^{\mathcal{A}}(k) = 1]| \leq k^{-c}$$

2.5   Security Against Chosen Ciphertext Attacks

Like IND-CCA security is an extension of IND-CPA security, we extend IND-naCFA to IND-naCCFA and IND-aCFA to IND-aCCFA. The security games are identical to their chosen plaintext counterparts, except that Init returns an additional decryption oracle $\mathsf{Dec}_j(K, \cdot, s_j)$, which is modeled like in the INT-FS game.

For basic encryption schemes, ciphertext security (IND-CCA) can be achieved by combining plaintext security (IND-CPA) with integrity (INT-CTXT) [2]. We show that this is also true for file systems within our security framework.

Lemma 1. *A file system* $\mathcal{F} = (\mathsf{Gen}, \mathsf{Init}, \mathsf{Update}, \mathsf{Dec})$ *is* IND-(n)aCCFA *secure, if it is* IND-(n)aCFA *and* INT-FS *secure.*

Proof. Assume a modified version of IND-(n)aCCFA, where the Dec_j oracle only works for the most recent output of the corresponding Update_j oracle, or (if Update has not been called yet) for the output of Init. For all other queries, it returns $\bot$. We call this modified game IND-(n)aCCFA'. It is straightforward to reduce an adversary against IND-(n)aCCFA' to an adversary against IND-(n)aCFA by remembering the most recent Update_j queries and answer the decryption query accordingly. We now show that any adversary with non-negligible success probability against IND-(n)aCCFA also has a non-negligible success probability against IND-(n)aCCFA'. Assume towards a contradiction an adversary A with a non-negligible different success probability in playing IND-(n)aCCFA and IND-(n)aCCFA'. We transform this adversary into an adversary A' against INT-FS. When A requests access to the Init oracle, A' forwards the calls to the Init oracle provided by INT-FS, returning (C_j, s_j) and the Update_j oracle. When A requests access to the Dec_j oracle, A' calls the Dec_j oracle provided by INT-FS, but ignores the response and implements the behaviour described for the IND-(n)aCCFA' game by remembering the most recent Update_j query. For IND-aCCFA', the challenge (C_j', s_j') is generated by another call to the Update_j oracle. For IND-naCCFA', the challenge (C, s, C', s') is generated by calling Init and then using the freshly returned Update_j oracle. Since the success probability of A is non-negligibly different for IND-(n)aCCFA and IND-(n)aCCFA', and the only difference in the games is the behaviour of Dec_j oracle queries that are not the most recent output of the Update_j oracle but decrypts successfully, we know such a query must happen with non-negligible probability. This query can be used directly to win the INT-FS game.   □

3   CryFS: An Encrypted File System for the Cloud

CryFS is an overlay file system that can be mounted to a virtual folder. Everything the user stores in this virtual folder is encrypted in the background. The ciphertexts are stored on the hard disk (through the underlying file system) and can be picked up by third party synchronisation clients like Dropbox and uploaded to a cloud storage. This allows for a flexible use on top of any file system or cloud storage provider. In contrast to many other encrypted file systems, we do hide file contents as well as metadata like file sizes, file permissions and directory structure. We achieve this by splitting all file system data into same-size blocks. These blocks are then individually encrypted using an authenticated cipher. Using a specifically tailored data structure, we ensure that all file operations are still fast and we induce little space overhead, even though all files are segmented into small blocks (see Sect. 3.1). To prevent malicious storage providers from violating the integrity of the file system, we introduce additional measures to prevent rollback, deletion and re-introduction of deleted blocks (see Sect. 3.3). We point out that we decided against using hash trees to protect integrity: The primary reason behind this decision is our goal to support concurrent access to the file system. Hash trees induce changes from the affected block up to the root node, thus increasing the chance of edit conflicts. The second reason for avoiding hash trees are performance considerations. Although hash trees have only logarithmic overhead in the size of the file system, any non-constant overhead is prohibitive for file systems with many frequent changes in many small files. Even though these integrity protections are only fully effective when the file system is used by a single user, CryFS is designed to work well with multiple users. See Appendix C for details. As most other encrypted file systems, CryFS uses two keys: a file system key for encrypting the file system blocks and a master key for encrypting the filesystem key. This makes it easy to change passwords for example.

3.1   Data Structures, Blocks and Files

As already mentioned, CryFS does not encrypt files individually. Rather, it splits every file into same-sized blocks, which are then encrypted. A tree data structure then associates blocks to files and files to directories. We base our construction on Dielissen et al.'s work on left-perfect binary trees [6] and generalise their definition to *left-max-data trees*.

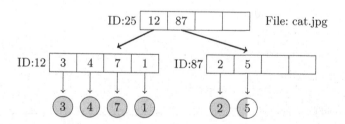

Fig. 1. The tree for an exemplary file "cat.jpg". Each tree node is one same-sized block in CryFS. The actual file data is stored in the leaves, whereas inner nodes store only pointers. For determining the file size, one only has to descend into the right-most branch of the tree and examine how much data is stored in the right-most leaf. Since all leaves are at the same depth and only the right-most elements are allowed to contain a less-than-maximum amount of data, this descend suffices to know how many blocks the file contains and thus the total file size.

The main idea for this data structure is that all nodes in the tree are as far left as possible. The actual binary file data is always stored in the left-most leaves of the file system tree and in-order. All leaves in the tree are at the same depth, and with exception of the right-most one, store exactly the same amount of data. This allows to represent arbitrary file sizes. Internal nodes contain only pointers to other blocks. If the block size is chosen appropriately (and thus the number of available pointers in each block), even large files can be represented by a tree with little depth. Every block is identified by a unique id, which is randomly chosen each time a block is created. See Fig. 1 for an example file represented as a left-max-data tree. This structure leads to very efficient algorithms for file system access. When trying to read a certain position in a file, one only needs to compute the respective block number from the total number of blocks in this file and the fixed block data size. Also, small changes to a file are particularly efficient: only a small block has to be changed (and synchronised to the cloud) not the whole file. Increasing the file size is described in Algorithm 1, decreasing is similar. Since only the right-most leaf can contain a less-than-maximum amount of data, determining the file size can also be achieved without reading all blocks by determining the amount of data in the right-most leaf. In our reference implementation with 32 KB blocks and 16 byte block ids, this data structure induces a space overhead of roughly 0.05% for inner nodes plus an additive overhead of at most one leaf node's size if the right-most leaf is not full.

Algorithm 1. Grow an existing tree by one leaf

```
function GROWTREE(treeRoot, newBlock)
    ℓ ← LOWESTNONFULLINNERNODE(treeRoot)
    if ℓ = ⊥ then /* All nodes are full. We need to add a level. */
        ℓ ← NEWINNERNODE() /* Create a new root block */
        ℓ.ADDCHILD(treeRoot)
        treeRoot ← ℓ
    end if
    while depth(ℓ) < depth(leaves) − 1 do
        n ← NEWINNERNODE()
        ℓ.ADDCHILD(n)
        ℓ ← n
    end while
    ℓ.ADDCHILD(NEWLEAF(newBlock))
    return treeRoot
end function
```

3.2   Directory Structure

Directories in CryFS are basically files themselves. Directories, however, do not store binary data but store a list of the directory's entries—i.e. pointers to the root block of files and directories. To allow for an efficient listing of all directory entries without having to descend into all individual file trees, we store the

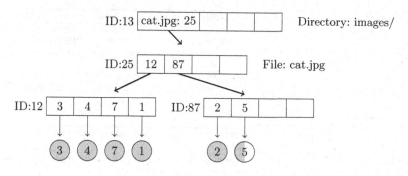

Fig. 2. The file "cat.jpg" is contained in a directory "images". To list all files of a directory efficiently, the name of each file is included with the respective pointer. As it is the case with files, once the number of entries in a directory exceeds the size of one block, the directory itself is represented as a tree.

name of each entry, as well as all file metadata (like permission bits) along with the corresponding pointer in the directory structure. This layout allows for fast modifications of the directory structure. Moving a large directory only requires re-encrypting both the old and the new parent directory. See Fig. 2 for an example of a file system tree with one directory and one file.

3.3 Encryption and Integrity

Encryption is on the block level—i.e. each block is encrypted individually. This allows for good performance because blocks can be encrypted in parallel. We use a cipher with an authenticated operation mode (e.g. AES-GCM) to prevent an adversary from altering the content of the blocks themselves. However, this is not yet sufficient to protect the integrity of the file system as a whole, since the connections between different blocks are not protected. An adversary can still try to reorder blocks, replace newer blocks with older versions, delete or re-add already deleted blocks.

We use a number of different mechanisms to prevent these attacks. First, we store the block ID in the header of the block, where it is integrity-protected by the authenticated encryption scheme. This ensures that an attacker cannot assign a different ID to a block (by changing the name of the file storing the block) and therefore prevents reordering. To prevent an attacker from replacing a block with a previous version of the same block, a block also stores a version counter in its header. Clients store a local list of all known blocks with a flag whether the block still exists, and their corresponding version numbers and check that it does not decrease. This list is also used to prevent an attacker from deleting or re-adding already deleted blocks without the client noticing. Additionally, the clients remember the master-key-encrypted file system key to prevent an adversary from replacing the whole file system including the key. In Sect. 4, we formally prove that this approach achieves the desired security goals. See Algorithms 1–5 for a description of relevant file system algorithms in pseudo-code.

Algorithm 2. Returns a new block with a unique id and the version number set to 0	**Algorithm 3.** Add a file or a folder tree to a directory

Algorithm 2. Returns a new block with a unique id and the version number set to 0

> **function** CREATEBLOCK
> $i \leftarrow$ GENERATEUNIQUEID()
>   **return** $(i, i || 0)$
> **end function**

Algorithm 3. Add a file or a folder tree to a directory

> **function** ADDTODIRECTORY($directory, newEntry$)
>   **if** RIGHTMOSTLEAF($directory$).ISFULL() **then**
>   GROWTREE($directory$, CREATEBLOCK())
>   **end if**
>   RIGHTMOSTLEAF($directory$).ADDDATA($newEntry$)
> **end function**

Algorithm 4. Creates a tree data structure from a file and returns the root node

> **function** CREATEFILE($file$)
> $D := (d_0, \ldots, d_n) \leftarrow$ SPLITDATA($file$)
> $t \leftarrow$ CREATEBLOCK()
> t.ADDDATA(d_0)
>   **for all** other $d_i \in D$ **do**
> $b_i \leftarrow$ CREATEBLOCK()
> b_i.ADDDATA(d_i)
> $t \leftarrow$ GROWTREE(t, b_i)
>   **end for**
>   **return** t
> **end function**

Algorithm 5. Creates the data structure for a complete file system

> **function** CREATEFILESYSTEM($sourceFileSystemRoot$)
> $rootBlock \leftarrow$ CREATEBLOCK()
>   **for all** Directories dir in $sourceFileSystemRoot$ **do**
> $rootBlock$.ADDTODIRECTORY(CREATEFILESYSTEM(dir))
>   **end for**
>   **for all** Files $file$ in $sourcFileSystemRoot$ **do**
> $rootblock$.ADDTODIRECTORY(CREATEFILE($file$))
>   **end for**
>   **return** $rootBlock$
> **end function**

4 Proving the Security of CryFS

In this section, we prove the adaptive and non-adaptive security of CryFS and show that it also provides integrity. Further, we show that CryFS also achieves ciphertext indistinguishability. We first give a formal description of CryFS. To simplify notation, we represent the tree structure of CryFS as a set of node blocks.

Definition 6 (CryFS). *Let $\mathbb{I}$ be the space of block IDs, $\mathbb{I} \times \{0,1\}^n$ the set of plaintext blocks, and $\mathbb{I} \times \{0,1\}^m$ the set of ciphertext blocks. $\mathsf{CryFS}^{\mathcal{E}_1, \mathcal{E}_2}$ is an encrypted file system $(\mathsf{Gen}, \mathsf{Init}, \mathsf{Update}, \mathsf{Dec})$ with $\mathcal{E}_1 = (\mathsf{Gen}_1, \mathsf{Enc}_1, \mathsf{Dec}_1)$ and $\mathcal{E}_2 = (\mathsf{Gen}_2, \mathsf{Enc}_2, \mathsf{Dec}_2)$. The client state $\mathbb{S} \subseteq 2^{\mathbb{I} \times \mathbb{N} \times \{0,1\}} \times \{0,1\}^{k'}$ stores a set of all known blocks with their id $i \in \mathbb{I}$, current version $v \in \mathbb{N}$ and a flag whether the block still exists (1) or was deleted in the past (0). The state also stores $c_{\mathsf{fs}} \in \{0,1\}^{k'}$, an encrypted version of the file system key. For the sake of clarity of the exposition, we first define intermediate functions:*

- Repr : $\mathbb{F} \rightarrow 2^{\mathbb{I} \times \{0,1\}^n}$: *Takes a plaintext file system and generates its representation as a set of plaintext blocks.*
- EncBlock : $\mathbb{K} \times (\mathbb{I} \times \{0,1\}^n) \times \mathbb{N} \rightarrow (\mathbb{I} \times \{0,1\}^m)$: *Takes a key K_{fs}, a plaintext block (i, b) and a version number $v \in \mathbb{N}$. Prepends block ID and version number to the data and encrypts it. Outputs (i, c) with $c := \mathsf{Enc}_2(K_{\mathsf{fs}}, i || v || b)$.*
- DecBlock : $\mathbb{K} \times (\mathbb{I} \times \{0,1\}^m) \rightarrow \{\bot\} \cup [(\mathbb{I} \times \{0,1\}^m) \times \mathbb{N}]$: *Takes a key K_{fs} and a ciphertext block (i, c). Decrypts it to $i' || v || b := \mathsf{Dec}_2(K_{\mathsf{fs}}, c)$. If decryption fails or $i \neq i'$, returns $\bot$. Otherwise, returns the plaintext block (i, b) and the version number v.*

Now we define the functions forming an encrypted file system.

- $\mathsf{Gen}(1^k) \mapsto (K_{\mathrm{master}})$: *Uses Gen_1 to generate a master key K_{master}.*

- $\mathsf{Init}(K_{\mathrm{master}}) \mapsto (C, s)$: *Takes K_{master} and generates $K_{\mathrm{fs}} \leftarrow \mathsf{Gen}_2(1^k)$. Encrypts it with the master key to $c_{\mathrm{fs}} = \mathsf{Enc}_1(K_{\mathrm{master}}, K_{\mathrm{fs}})$. Computes $B := \mathsf{Repr}(F) = \{(i_0, b_0), \ldots, (i_n, b_n)\}$, a set of blocks representing an empty file system F.*
  Sets $C := (c_{\mathrm{fs}}, \mathsf{EncBlock}(K_{\mathrm{fs}}, (i_0, b_0), 0), \ldots, \mathsf{EncBlock}(K_{\mathrm{fs}}, (i_n, b_n), 0))$ and $s := (\{(i_0, 0, 1), \ldots, (i_n, 0, 1)\}, c_{\mathrm{fs}})$ and outputs (C, s).
- $\mathsf{Dec}(K_{\mathrm{master}}, C, s) \mapsto (F, s)$: *Reads c_{fs} from C and compares it with the c_{fs} stored in s. If they differ, returns $\perp$. Otherwise, decrypts it to $K_{\mathrm{fs}} := \mathsf{Dec}_1(K_{\mathrm{master}}, c_{\mathrm{fs}})$. Then, computes $D := \{((i', b), v) \mid ((i', b), v) = \mathsf{DecBlock}(K_{\mathrm{fs}}, (i, c)), (i, c) \in C\}$. Outputs $\perp$ in the following cases:*
  - *Dec_1 fails to decrypt c_{fs} (wrong key or an integrity violation).*
  - *$\mathsf{DecBlock}$ fails to decrypt c (wrong key, an integrity violation, or $i \neq i'$).*
  - *There is an $((i, b), v) \in D$ for which there is no $(i, v', 1) \in s$*
  - *There is an $((i, b), v) \in D$ for which there is an $(i, v', 1) \in s$ with $v < v'$*
  - *There is an $(i, v, 1) \in s$ for which there is no $((i, b), v') \in D$*

  Otherwise, computes the plaintext file system $F := \mathsf{Repr}^{-1}(\{(i_0, b_0), \ldots, (i_n, b_n)\})$ and outputs (F, s). The client state is not changed.
- $\mathsf{Update}(K_{\mathrm{master}}, C, F', s) \mapsto (C', s')$: *Decrypts the old file system state to $F := \mathsf{Dec}(K_{\mathrm{master}}, C, s)$. Then, reads c_{fs} from C and decrypts it to K_{fs}. If either decryption fails, returns $\perp$. Initializes $s' := s$. Compares $\mathsf{Repr}(F)$ and $\mathsf{Repr}(F')$ and does the following:*
  - *For each block $(i, b) \notin \mathsf{Repr}(F)$, $(i, b') \in \mathsf{Repr}(F')$:*
    - *If $(i, v, 0) \in s$, replace it in s' with $(i, v + 1, 1)$. Else, add $(i, 0, 1)$ to s'*
    - *Note: if $(i, v, 1) \in s$, Dec would have failed above.*
  - *For each block $(i, b) \in \mathsf{Repr}(F)$, $(i, b') \in \mathsf{Repr}(F'), b \neq b'$*
    - *Replace $(i, v, 1)$ in s' with $(i, v' + 1, 1)$, where v' is the version number returned from $\mathsf{DecBlock}$ on decryption.*
    - *Note: $(i, v, 1) \in s \wedge v' \geq v$, otherwise Dec would have failed above.*
  - *For each block $(i, b) \in \mathsf{Repr}(F)$, $(i, b') \notin \mathsf{Repr}(F')$*
    - *Replace $(i, v, 1)$ with $(i, v, 0)$ in s'.*
    - *Note: $(i, v, 1) \in s$ otherwise Dec would have failed above.*

  Then, encrypts F' using $\mathsf{EncBlock}$ with updated version numbers and outputs the new ciphertext file system C' (including c_{fs}), and the modified state s'.

We now show that CryFS exhibits non-adaptive security according to Definition 3. We set R_d to restrict the challenge file systems to be representable using the same number of blocks. Formally, this means

$$R_d = \{(F^0, F^1) \in \mathbb{F} \times \mathbb{F} : |\mathsf{Repr}(F^0)| = |\mathsf{Repr}(F^1)|\}$$

Theorem 1 (Nonadaptive Security of CryFS). CryFS$^{\mathcal{E}_1, \mathcal{E}_2}$ = (Gen, Init, Update, Dec) *is IND-naCFA secure, if $\mathcal{E}_1$ = (Gen$_1$, Enc$_1$, Dec$_1$) and $\mathcal{E}_2$ = (Gen$_2$, Enc$_2$, Dec$_2$) are IND-CPA secure encryption schemes.*

Proof. We prove the claim by reduction using two steps. First, we modify IND-naCFA to IND-naCFA' such that when the adversary gets the challenge $(C, s), (C', s')$, it does not contain an encryption of K_{fs}, but an encryption of 0s instead. We prove that an adversary which has a different advantage in IND-naCFA and IND-naCFA' can be used to break the IND-CPA security of $\mathcal{E}_1$. Second, we give a reduction from IND-naCFA' to the IND-CPA security of $\mathcal{E}_2$.

Consider the following modification to IND-naCFA: When the adversary expects the challenge (C', s'), replace the encrypted file system key $\mathsf{Enc}_1(K_{master}, K_{fs})$ in state and ciphertext with $\mathsf{Enc}_1(K_{master}, 0)$. We call this modified game IND-naCFA'. Now, assume towards a contradiction an adversary A with a probability of success p against IND-naCFA and p' against IND-naCFA', where $p = p' + d$ for a positive non-negligible d. This adversary can be used to construct an adversary B with a non-negligible advantage of $\frac{d}{2}$ against the IND-CPA security of $\mathcal{E}_1$. The reduction works as follows: The IND-CPA game draws $K_{master} \leftarrow \mathsf{Gen}_1(1^k)$ and a random bit b. When A uses the Init oracle, B generates $K'_{fs} \leftarrow \mathsf{Gen}_2(1^k)$ and (C_j, s_j) using the algorithms described in Definition 6 and uses the encryption oracle of IND-CPA to generate c'_{fs} as an encryption of K'_{fs}. Since B knows K'_{fs} it can also build the Update$_j$ oracle. When the adversary outputs F^0, F^1, B generates another independent $K_{fs} \leftarrow \mathsf{Gen}_2(1^k)$, and passes 0 and K_{fs} as challenge to the IND-CPA game. The game returns c_{fs}. When $b = 0$, this is an encryption of 0. When $b = 1$, this is an encryption of K_{fs}. B then draws a random bit a, and knowing K_{fs}, can build the challenge (C, s) and (C', s') as an encryption of F^a. It replaces the encrypted file system key in C, s, C' and s' with the c_{fs} and returns the result to A. If A outputs a, A wins and B outputs 1 to the IND-CPA game. If A loses, B outputs 0. For $b = 0$, this was a perfect simulation of the IND-naCFA' game. B has success probability $\Pr[a \nleftarrow A \mid b = 0] = 1 - p'$. For $b = 1$, this was a perfect simulation of the IND-naCFA game. B has success probability $\Pr[a \leftarrow A \mid b = 1] = p = p' + d$. Together, B has success probability $\Pr[b \leftarrow B] = \frac{1}{2}(1 - p') + \frac{1}{2}(p' + d) = \frac{1}{2} + \frac{d}{2}$. Since d is non-negligible, B has a non-negligible advantage in the IND-CPA game which is a contradiction.

Now, assume towards another contradiction that A' is a successful attacker on IND-naCFA'. We transform A' into a successful attacker B' on IND-CPA security of $\mathcal{E}_2$: The game draws K_{fs} and a random bit b. B' draws $K_{master} \leftarrow \mathsf{Gen}_1(1^k)$. When A' uses Init, B' generates a new K'_{fs}, encrypts it with K_{master}, and creates an empty ciphertext file system. Knowing K_{master}, the Update$_j$ oracle can be implemented easily.

Upon receiving challenges F^0 and F^1 from A', B' first generates an empty file system, and encrypts it to (C, s) using the encryption oracle and prepending $c'_{fs} = \mathsf{Enc}_1(K_{master}, 0)$. Then, B' updates it with F^0 and F^1 respectively, and uses the LR-oracle provided by IND-CPA successively for each pair of blocks in $\mathsf{Repr}(F^0)$ and $\mathsf{Repr}(F^1)$. This is possible, since we require $(F^0, F^1) \in R_d$ (i.e. both have the same number of blocks), Repr can be implemented to choose the same block ids for F^0 and F^1, and all blocks are of the same size. B' remembers all encrypted blocks returned by the oracle, prepends c'_{fs} to get C', and passes it to A' together with a generated file system state s' in which all block ids in have version number 1.

This is a correct simulation of the IND-naCFA' game. When A' submits a guess for b, B' forwards it and thus inherits its success probability. This is a contradiction to the assumption that $\mathcal{E}_2$ is IND-CPA-secure. $\qquad\square$

Theorem 2 shows that CryFS is also adaptively secure according to Definition 4. Since block IDs are public and CryFS only re-encrypts blocks for which the plaintext changed (for performance reasons), we set R_d to restrict both challenge file systems add, delete or modify blocks with the same block IDs. Theorem 3 shows that CryFS exhibits integrity according to Definition 5.

Theorem 2 (Adaptive Security of CryFS). $\mathsf{CryFS}^{\mathcal{E}_1,\mathcal{E}_2} = (\mathsf{Gen}, \mathsf{Init}, \mathsf{Update},$ $\mathsf{Dec})$ *is* IND-aCFA *secure, if* $\mathcal{E}_1 = (\mathsf{Gen}_1, \mathsf{Enc}_1, \mathsf{Dec}_1)$ *and* $\mathcal{E}_2 = (\mathsf{Gen}_2, \mathsf{Enc}_2, \mathsf{Dec}_2)$ *are* IND-CPA *secure encryption schemes.*

Theorem 3 (Integrity of CryFS). $\mathsf{CryFS}^{\mathcal{E}_1,\mathcal{E}_2} = (\mathsf{Gen}, \mathsf{Init}, \mathsf{Update}, \mathsf{Dec})$ *is* INT-FS *secure, if* $\mathcal{E}_1$ *is* IND-CPA *and* $\mathcal{E}_2$ *is* INT-CTXT *secure.*

The proofs for Theorems 2 and 3 can be found in Appendices A and B.

Lastly, we show that CryFS can also be secure against chosen ciphertext attacks.

Theorem 4 (Chosen Ciphertext Attacks). $\mathsf{CryFS}^{\mathcal{E}_1,\mathcal{E}_2} = (\mathsf{Gen}, \mathsf{Init}, \mathsf{Update},$ $\mathsf{Dec})$ *is* IND-naCCFA *and* IND-aCCFA *secure, if* $\mathcal{E}_1 = (\mathsf{Gen}_1, \mathsf{Enc}_1, \mathsf{Dec}_1)$ *is* *an* IND-CPA *and* $\mathcal{E}_2 = (\mathsf{Gen}_2, \mathsf{Enc}_2, \mathsf{Dec}_2)$ *an* IND-CPA *and* INT-CTXT *secure encryption scheme.*

Proof. This follows directly from Theorems 1 and 3 and Lemma 1. $\qquad\square$

5 Performance

In this section, we present results of our performance evaluation for our reference implementation of CryFS. We tested various performance factors in comparison to other popular file systems. Even though our implementation is preliminary and still has potential for optimisation, our experiments show that our file system has performance comparable to existing encrypted file systems and is practical.

CryFS is implemented using C++ and can be compiled with either GCC or Clang. For cryptography, the Crypto++[11] library is used, but the code is written in a way that allows for easy switching to another library. We tested CryFS 0.10-m2, EncFS 1.8.1, TrueCrypt 7.1a, and VeraCrypt 1.19. CryFS was built with GCC 5.3.1 using optimization level Ofast. In all cases, the underlying file system was Ext4. For comparision we also tested the performance of Ext4 itself without using a cryptographic file system on top. CryFS was configured to use aes-256-gcm and run with a block size of 32 KB. EncFS was also set to

---

[11] https://www.cryptopp.com/.

Table 1. Experimental results for file system operations using the bonnie++ 1.03e benchmark. Bonnie++ tests sequential read and write speed, both bytewise and block-wise, and of a Rewrite run, which iteratively loads a block from the file, modifies it, and writes it back. It tests the performance of random seeks, creations and deletions. In parentheses next to each value, the average CPU utilization is reported.

		CryFS	EncFS	TrueCrypt	VeraCrypt	Plain Ext4
Sequential output	Bytewise (MB/s)	40.5 (35%)	39.3 (38%)	29.2 (26%)	28.1 (25%)	70.9 (64%)
	Blockwise (MB/s)	53.8 (3%)	63.2 (8%)	34.7 (3%)	35.1 (3%)	71.0 (5%)
Sequential input	Bytewise (MB/s)	20.9 (23%)	65.7 (52%)	66.1 (59%)	67.4 (61%)	64.8 (69%)
	Blockwise (MB/s)	23.7 (1%)	67.8 (3%)	68.5 (3%)	69.0 (3%)	66.4 (4%)
Rewrite	Blockwise (MB/s)	19.3 (3%)	28.9 (4%)	31.4 (4%)	31.4 (4%)	31.6 (3%)
Random seeks (/s)		79.4 (0%)	53.5 (0%)	111.5 (0%)	108.3 (0%)	155.9 (0%)
Random create (/s)		2701 (6%)	4208 (12%)	4071 (99%)	4036 (99%)	–
Random delete (/s)		4070 (4%)	24250 (19%)	9424 (99%)	9457 (99%)	–

aes-256. For TrueCrypt and VeraCrypt, a container with 50 GB size was created, also using aes-256. We used a machine with Intel(R) Core(TM) i5-2500K CPU @ 3.30 GHz QuadCore, 16 GB DDR3-RAM on Ubuntu 16.10, Linux 4.8.0-49 x86_64. As hard-drive, a Samsung HD 204UI was used. The experiments were run using the benchmarking tool bonnie++ 1.03e[12]. To minimize the influence of cache effects, bonnie++ runs the read/write tests with a test file size that is twice the size of main memory (32 GB in our case). For create/stat and delete tests, we used $16 * 1024$ files with 10 KB each. Each experiment was run three times to ensure a low standard deviation, and we report the average value. The benchmark script is available online.[13]

We found that writes by CryFS on HDDs are 15% slower than EncFS, while random seeks are faster by 45%. Read performance is slower by about a factor of three. All operations are still fast enough to be used in practice, however. CryFS uses less CPU time for all operations. Table 1 includes measurements for all tested file systems and shows the measured performance in detail.

6   Conclusion and Future Work

In this work, we introduced a novel formal model for the security and integrity of cloud file systems. Our model is generic and designed to be applicable for a wide range of file systems. We also introduced CryFS, a novel encrypted file system specifically designed for the cloud. It has low communication and storage overhead. It ensures the confidentiality of the file system by hiding file contents as well as metadata like file sizes and directory structure. It ensures the integrity of the file system even against a malicious storage provider when used by a single user, but can also be used efficiently by multiple users when integrity is not important. We proved the security of CryFS in our new framework. Our

---

[12] http://www.coker.com.au/bonnie++/.
[13] https://github.com/cryfs/benchmark/tree/0.10-m2.

benchmarks show that CryFS offers comparable performance to other state-of-the-art file systems even though our implementation is preliminary and has room for improvements. Our implementation is available on github.

Regarding our framework, there are a few open questions to be addressed in the future. First, even though we establish basic relations between our security notions, it remains open to show other relations or separations to get a better understanding of the requirements for secure cloud file systems. Second, we show that if a basic encryption primitive is IND-CPA and INT-CTXT secure, it can be used to construct a IND-CCFA secure file system. It remains an open question, if IND-CCA security (which is a weaker notion) would also be sufficient. Last, extending our formal model to a multi-user setting as well as extending CryFS itself to provide integrity for multiple users is left for future work.

A   Adaptive Security of CryFS

Theorem 2 (Adaptive Security of CryFS). CryFS$^{\mathcal{E}_1,\mathcal{E}_2}$ = (Gen, Init, Update, Dec) *is* IND-aCFA *secure, if* $\mathcal{E}_1$ = (Gen$_1$, Enc$_1$, Dec$_1$) *and* $\mathcal{E}_2$ = (Gen$_2$, Enc$_2$, Dec$_2$) *are* IND-CPA *secure encryption schemes.*

Proof. Consider the following modification to IND-naCFA: When the adversary queries Init or the Update$_j$ oracles or expects output (C, s), instead of getting Enc$_1(K_{\mathrm{master}}, K_{\mathrm{fs}})$ they instead get Enc$_1(K_{\mathrm{master}}, 0)$. Now, assume towards a contradiction an adversary A with a success probability of p against IND-aCFA and a success probability of p' against IND-aCFA', where $p = p' + d$ for a positive non-negligible d. This adversary can be used to construct an adversary B with a non-negligible advantage of $\frac{d}{2}$ which breaks the IND-CPA security of $\mathcal{E}_1$. The game draws $K_{\mathrm{master}} \leftarrow \mathrm{Gen}_1(1^k)$ and a random bit b. When A uses the Init oracle, B generates a new file system key $K_{\mathrm{fs}} \leftarrow \mathrm{Gen}_2(1^k)$ and uses the LR oracle of the IND-CPA game to get c_{fs} as either an encryption of 0 or of K_{fs}, depending on the value of b. Then it generates a new empty file system (C_j, s_j) but replaces the encryption of K_{fs} with c_{fs}. A expects access to an Update$_j$ oracle which can be built by using K_{fs} to decrypt and encrypt blocks. Again, B replaces all encryptions of K_{fs} with c_{fs}. When the adversary outputs j, F^0, F^1, B draws a random bit a. It uses Update$_j$ to build the challenge (C', s') as an encryption of F^a. If A outputs a (A wins), B outputs 1. If A loses, B outputs 0. For $b = 0$, this was a perfect simulation of the IND-aCFA' game. B has success probability $\Pr[a \nleftarrow A \mid b = 0] = 1 - p'$. For $b = 1$, this was a perfect simulation of the IND-aCFA game. B has success probability $\Pr[a \leftarrow A \mid b = 1] = p = p' + d$. Together, B has success probability $\Pr[b \leftarrow B] = \frac{1}{2}(1 - p') + \frac{1}{2}(p' + d) = \frac{1}{2} + \frac{d}{2}$. Since d is non-negligible, B has a non-negligible advantage against IND-CPA.

Now, assume towards another contradiction that A' is a successful attacker on IND-aCFA'. We transform A' into a successful attacker B' on the IND-CPA security of $\mathcal{E}_2$. Intuitively, B' selects a random file system created by A' and uses A' to break its security. Since the number of file systems is a fixed constant, this only reduces the success probability by a constant amount. The reduction works as follows. The game draws K_{fs} and a random bit b. B' draws $K_{master} \leftarrow Gen_1(1^k)$ and draws a random $j^* \leftarrow \{1, \ldots, q_{Init}\}$. When A' uses Init for the j-th time and $j \neq j^*$, B' generates a new K'_{fs}, encrypts it with K_{master}, and creates an empty ciphertext file system. Knowing K_{master}, the $Update_j$ oracle can easily be implemented. In every output, $Enc_1(K_{master}, K_{fs})$ is replaced with an encryption of 0. When A' uses Init for the j^*-th time, B' generates a new empty file system by using the encryption oracle of the IND-CPA experiment to encrypt all blocks. Again, B' prepends $Enc_1(K_{master}, 0)$. B' also saves the current plaintext file system F_j (which is empty). If A' uses their access to the $Update_j$-oracle, B' updates the saved plaintext according to the input to the oracle. It uses the encryption oracle to encrypt added or modified blocks and exchanges them in the saved ciphertext. B' updates the saved file system F_j and the state s_j. Upon receiving challenge j, F^0 and F^1 from A', B' updates the corresponding plaintext F_j for both F^0 and F^1 respectively and passes the added and modified blocks of $Repr(F^0)$ and $Repr(F^1)$ (when compared to $Repr(F_j)$) to the LR oracle of the IND-CPA experiment. It now has an encryption of either the modified blocks in F^0 or in F^1. Since it is required that $(F_j, F^0, F^1) \in R_d$ (i.e. they add, remove, and modify blocks with the same ID), B' knows which ciphertext blocks it has to add, remove and replace with their new versions in order to generate the correct ciphertext file system, even though it does not know which change was selected by the experiment. B' prepends $Enc_1(K_{master}, 0)$ to the generated ciphertext and passes it to A' along with the updated state. This is a correct simulation of the IND-aCFA' game. When A' submits a guess for b, B' forwards it to the game and thus inherits its success probability. This is a contradiction to the assumption that $\mathcal{E}_2$ is IND-CPA secure. $\qquad\square$

B   Integrity of CryFS

Theorem 3 (Integrity of CryFS). $CryFS^{\mathcal{E}_1, \mathcal{E}_2} = (Gen, Init, Update, Dec)$ *is* INT-FS *secure, if* $\mathcal{E}_1$ *is* IND-CPA *and* $\mathcal{E}_2$ *is* INT-CTXT *secure.*

Proof. Again, we first change INT-FS to INT-FS' by replacing $Enc_1(K_{master}, K_{fs})$ with $Enc_1(K_{master}, 0)$ in the output of all oracles. Assume towards a contradiction that an adversary A with success probability of p against INT-FS and success probability of p' against INT-FS' exists (where $p = p' + d$ for a positive non-negligible d). This adversary can be used to construct an adversary B with an advantage of $\frac{d}{2}$ against the IND-CPA security of $\mathcal{E}_1$ by using the following reduction: When A uses Init, B generates $K_{fs} \leftarrow Gen_2(1^k)$ and uses the LR oracle of the IND-CPA game to get c_{fs} as either an encryption of 0 or of K_{fs}. It generates (C_j, s_j) using K_{fs} but replaces the encrypted file system key with c_{fs}. B builds the $Update_j$ and Dec_j oracles using K_{fs} to decrypt and encrypt blocks. Each output contains c_{fs} instead of the encrypted file system key. When Dec_j is used, B checks whether decryption was successful for $C \neq C'$, i.e. whether A was successful. If A was successful, B outputs 1, otherwise it outputs 0. If $b = 0$, this was a perfect simulation of the INT-FS' game. B has success probability $Pr[0 \leftarrow B \mid b = 0] = 1 - p'$. If $b = 1$, this was a perfect simulation of the INT-FS game. B has success probability $\Pr[1 \leftarrow B \mid b = 1] = p = p' + d$ Together, B has success probability $\Pr[b \leftarrow B] = \frac{1}{2}(1 - p') + \frac{1}{2}(p' + d) = \frac{1}{2} + \frac{d}{2}$. Since d is non-negligible, this is a non-negligible advantage for B against IND-CPA.

Now, assume towards another contradiction that A' is a successful attacker on INT-FS'. We give a reduction which transforms A' into a successful attacker B' on INT-CTXT. The game draws $K_{fs} \leftarrow \mathsf{Gen}_2(1^k)$ and B' draws $K_{master} \leftarrow \mathsf{Gen}_1(1^k)$. B' draws a random $j^* \leftarrow \{1, \ldots, q_{Init}\}$. When A' uses Init for the j-th time with $j \neq j^*$, B' generates a new independent K'_{fs} and creates a new ciphertext file system with this key. Knowing K'_{fs}, implementing Update$_j$ and Dec$_j$ oracles is straightforward. In every output, $\mathsf{Enc}_1(K_{master}, K_{fs})$ gets replaced by $\mathsf{Enc}_1(K_{master}, 0)$. When A' uses Init for the j^*-th time, B' creates a new empty file system but uses the encryption oracle provided by INT-CTXT to encrypt all blocks. It also builds Update$_j$ and Dec$_j$ but uses the decryption and encryption oracles of the INT-CTXT game to decrypt and encrypt. Instead of prepending $\mathsf{Enc}_1(K_{master}, K_{fs})$, which B' does not know, it prepends $\mathsf{Enc}_1(K_{master}, 0)$.

Since A' is successful, there is an oracle query $\mathsf{Dec}_j(K, C', s_j)$ which decrypts successfully with $C_j \neq C'$. With non-negligible probability $\frac{1}{q_{Init}}$, this happens for $j = j^*$, where B' implemented Init using the INT-CTXT experiment. C_j and C' have the same set of block IDs, otherwise $\mathsf{Dec}_j(K_{master}, C', s_j) = \perp$. So there has to be a block in C' which is different from the corresponding block in C_j, i.e. $\exists i, c_i, c'_i : (i, c_i) \in C_j, (i, c'_i) \in C', c_i \neq c'_i$. This block c'_i was input to the decryption oracle of the IND-CTXT game when decrypting C'. We argue that c'_i wins the INT-CTXT game. First note that INT-FS' decrypts with $c_{fs} = \mathsf{Enc}_1(K_{master}, K_{fs})$ from the state, not with the $c'_{fs} = \mathsf{Enc}_1(K_{master}, 0)$ passed to the adversary. Therefore c'_i decrypts successfully with the key from the INT-CTXT experiment. We now have to argue that c'_i was never output by the INT-CTXT encryption oracle. Recall that this oracle is only used for encrypting the output of the j-th query of the Init oracle and for the outputs of the Update$_j$ oracle. Since C' decrypts successfully, we know that the plaintext $((i', b'_i), v'_i) := \mathsf{DecBlock}(K, (i, c'_i))$ has ID $i = i'$ and a version number $v'_i \geq v^s_i$ where v^s_i is the version number in the state. All previous Update'$_j$ oracle queries for this block ID encrypted a block with version number $v_i \leq v^s_i$, and $v_i = v^s_i$ only for c_i where we know $c'_i \neq c_i$. So we know c'_i was not output of the Update$_j$ oracle. If (i, c'_i) was in the j-th output of the Init oracle, then $v'_i = 0$. In this case, either block i was never modified, which is a contradiction to $c_i \neq c'_i$, or block i was modified, which means $v^s_i > 0$ and is a contradiction to successful decryption. Taking everything into account, we know that c'_i was never output by the INT-CTXT encryption oracle and thus wins the game. This is a contradiction to the assumed security of $\mathcal{E}_2$. $\square$

C   Achieving Multi-user-Compatibility

CryFS provides confidentiality, integrity and fast file system operations in a single-user context. However, the design presented so far does not work well when used by multiple users for multiple reasons. For example, we cannot distinguish whether an integrity violation was caused by an attacker rolling back a block, or by a second client synchronising modifications on top of an outdated version. We resolve these problems by introducing a number of measures, which ensure that CryFS can be used with multiple users without integrity guarantees while maintaining integrity in the single-user setting.

First, in addition to having a pointer from the directory block to the root of a file, we also add a pointer from each file root node back to the directory it belongs to. That is, the whole directory structure is stored twice, once bottom-up in these pointers and once top-down through the file system tree. Using this, we can recover from a race condition where two users both add a different file to the same directory by periodically scanning for "dangling" pointers and reintegrate the corresponding files into the directory block.

Second, we extend the header of each block to also contain a unique client ID of the client who last modified the block along with the version counter. Further, each client saves the newest version for every block ID and client combination, and remembers the last updating client. Now, when a client reads a block that still has the same client ID as in his local state, the version number is checked to be non-decreasing otherwise it has to be increasing.

Third, instead of explicitly flagging deleted blocks in the local state, we set their last updating client ID to $\perp$. This allows clients to reintroduce deleted blocks as long as they increase the version number. Last, we allow to disable the check for missing blocks since there is no mechanism for a client to communicate, that he deleted a block, which will cause other clients to think that an attacker has deleted it.

References

1. Achenbach, D., Huber, M., Müller-Quade, J., Rill, J.: Closing the gap: a universal privacy framework for outsourced data. In: Pasalic, E., Knudsen, L.R. (eds.) BalkanCryptSec 2015. LNCS, vol. 9540, pp. 134–151. Springer, Cham (2016). doi:10.1007/978-3-319-29172-7_9
2. Bellare, M., Namprempre, C.: Authenticated encryption: relations among notions and analysis of the generic composition paradigm. J. Crypt. 21(4), 469–491 (2008)
3. Cash, D., Küpçü, A., Wichs, D.: Dynamic proofs of retrievability via oblivious ram. J. Cryptol. 30(1), 22–57 (2017)
4. Chase, M., Shen, E.: Substring-searchable symmetric encryption. Cryptology ePrint Archive, Report 2014/638 (2014). http://eprint.iacr.org/2014/638
5. Damgård, I., Dupont, K.: Universally composable disk encryption schemes. Cryptology ePrint Archive, Report 2005/333 (2005). http://eprint.iacr.org/
6. Dielissen, V.J., Kaldewaij, A.: A simple, efficient, and flexible implementation of flexible arrays. In: Möller, B. (ed.) MPC 1995. LNCS, vol. 947, pp. 232–241. Springer, Heidelberg (1995). doi:10.1007/3-540-60117-1_13
7. Erway, C., Küpçü, A., Papamanthou, C., Tamassia, R.: Dynamic provable data possession. In: Proceedings of the 16th ACM Conference on Computer and Communications Security, CCS 2009, pp. 213–222. ACM, New York (2009)
8. Gjøsteen, K.: Security notions for disk encryption. In: Vimercati, S.C., Syverson, P., Gollmann, D. (eds.) ESORICS 2005. LNCS, vol. 3679, pp. 455–474. Springer, Heidelberg (2005). doi:10.1007/11555827_26

9. Goh, E.J.: Secure indexes. Cryptology ePrint Archive, Report 2003/216 (2003). http://eprint.iacr.org/2003/216

10. Katz, J., Lindell, Y.: Introduction to Modern Cryptography. Chapman and Hall/CRC Cryptography and Network Security. Chapman and Hall/CRC, Boca Raton (2008)

11. Khati, L., Mouha, N., Vergnaud, D.: Full disk encryption: bridging theory and practice. In: Handschuh, H. (ed.) CT-RSA 2017. LNCS, vol. 10159, pp. 241–257. Springer, Cham (2017). doi:10.1007/978-3-319-52153-4_14

12. Wright, C.P., Martino, M.C., Zadok, E.: NCryptfs: a secure and convenient cryptographic file system. In: Proceedings of the 2003 USENIX Annual Technical Conference, San Antonio, TX, pp. 197–210, June 2003

13. Zhang, Y., Blanton, M.: Efficient dynamic provable possession of remote data via update trees. Trans. Storage **12**(2), 9:1–9:45 (2016)

Secure Systems

Keylogger Detection Using a Decoy Keyboard

Seth Simms, Margot Maxwell, Sara Johnson, and Julian Rrushi$^{(\boxtimes)}$

Department of Computer Science, Western Washington University,
Bellingham, WA 98226, USA
{simmss2,maxwelm,johns782}@students.wwu.edu
julian.rrushi@wwu.edu

Abstract. Commercial anti-malware systems currently rely on signatures or patterns learned from samples of known malware, and are unable to detect zero-day malware, rendering computers unprotected. In this paper we present a novel kernel-level technique of detecting keyloggers. Our approach operates through the use of a decoy keyboard. It uses a low-level driver to emulate and expose keystrokes modeled after actual users. We developed a statistical model of the typing profiles of real users, which regulates the times of delivery of emulated keystrokes. A kernel filter driver enables the decoy keyboard to shadow the physical keyboard, such as one single keyboard appears on the device tree at all times. That keyboard is the physical keyboard when the actual user types on it, and the decoy keyboard during time windows of user inactivity. Malware are detected in a second order fashion when data leaked by the decoy keyboard are used to access resources on the compromised machine. We tested our approach against live malware samples that we obtained from public repositories, and report the findings in the paper. The decoy keyboard is able to detect 0-day malware, and can co-exist with a real keyboard on a computer in production without causing any disruptions to the user's work.

Keywords: Decoy I/O · 0-knowledge anti-malware · Kernel drivers

1 Introduction

Malware development, dispersion, and infection are an ever-present threat to system security and user privacy. Among the most insidious and difficult to detect are 0-day malware, as well as those using code and data structure mutation. With nearly limitless access to system services, drivers, and modules, these types of malware have some distinct advantages over current detection methods – not only must they have been previously encountered and analyzed in order to be caught, but they can interfere with and elude the software trying to track them. Keyloggers are a common component of malware, able to spy on a user's activity and gather information like passwords, account numbers, and credit card numbers.

© IFIP International Federation for Information Processing 2017
Published by Springer International Publishing AG 2017. All Rights Reserved
G. Livraga and S. Zhu (Eds.): DBSec 2017, LNCS 10359, pp. 433–452, 2017.
DOI: 10.1007/978-3-319-61176-1_24

There are several methods used by keyloggers to capture information from the keyboard in the Windows operating system. User level keyloggers can implement a software hook that can intercept keystroke events, perform cyclical querying of the keyboard device to determine state changes, or inject special code into running processes that has access to the messages being passed. Kernel level keyloggers work by using filter drivers that can view the I/O traffic bound for a keyboard. There are varying methods of implementing keyloggers, and while many are documented, it is likely that there are some that are not. New keystroke interception techniques are discovered continuously.

Contribution. We present a novel defensive deception approach, which uses a decoy keyboard to receive contact from keyloggers in a way that leads to their detection unequivocally. The decoy keyboard emulates the presence and operation of its real counterpart with the help of drivers in the kernel. The decoy keyboard can be installed on a full production system in active use, and is able to not interfere with the activities of a normal user. The decoy keyboard appears as a standard USB keyboard in the Windows device tree. It is effectively invisible to the user, and indistinguishable from a real keyboard to malware. The defender's objective is to proactively misdirect malware into intercepting keystrokes emulated by the decoy keyboard. An attack surface is created by sending emulated keystrokes to a decoy process (dprocess), which runs in user space. Malware are detected when they use the data that they had intercepted from the decoy keyboard on the compromised computer. Some malware are detected upon contact with the decoy keyboard. The decoy keyboard can be combined with a decoy mouse and a decoy monitor for consistency reasons, using techniques similar to those discussed in this paper.

Novelty. The decoy keyboard is a 0-knowledge detector, meaning that it can detect malware without any prior knowledge of their code and data. 0-day malware, polymorphic malware, metamorphic malware, data-structure mutating malware, are all detectable by this work. A shadowing mechanism enables the decoy keyboard to coexist with its real counterpart on a computer in production. A timing model helps the decoy keyboard expose emulated keystrokes on the attack surface in a realistic and consistent fashion. Data intercepted from a decoy keyboard have the same timing characteristics as data intercepted from a real keyboard. The decoy keyboard is a usable security tool. It runs automatically with very little computational overhead, and requires no user input or any other involvement. Overall, our approach creates an active redirection capability that is effective in trapping malware.

Threat Model. Our approach currently works against keyloggers that run in user space. Kernel-level malware are planned as future work. Malware could use any exploits to land on a computer, and could have any form of internal design to intercept keystrokes. Our approach works independently of the exploitation techniques and the inner workings of malware. We did this work with reference to the Windows operating system, as we were seeking a proof of concept to show the potential of our approach. Similar principles can be applied to other modern time-shared operating systems as well.

Paper organization. The remainder of this paper is organized as follows. Section 2 describes the proposed approach in four parts, namely modeling human keystroke dynamics, low-level deceptive driver, keyboard shadowing, and a discussion of how our approach attains malware detection. In Sect. 3 we discuss the evaluation of the proposed approach against live malware. In Sect. 4 we discuss related work in malware detection, deception tactics, and keystroke dynamics. Section 5 summarizes our findings and concludes the paper.

2   Approach

2.1   Modeling Human Keystroke Dynamics

Just as we humans have unique patterns in our handwriting (to the level that it can be used for identification), we have similar patterns in the way that we type on computer keyboards. This has been extensively studied [35] as a biometric means of user authentication and identification, with generally good results. However, it is important to note that we are not using this model for authentication, but rather for *generation* of keystrokes. This presents a different set of unique challenges, and eliminates some of the specific considerations needed for user verification. Notably, we are proceeding with the assumption that the malware is not specifically targeting a user or users that they already have acquired a large amount of recorded keystroke data and built complex models for. We used the large body of previous research to help define a model that we could use to send emulated keystrokes that would appear to come from a unique human user, while simultaneously meeting our goals.

One primary goal was having low overhead, since we are working at the driver and operating system level, and additional processing time could add unwanted and unintentional delays that could cause our deception to become visible. Low overhead implies a measure of simplicity in the model, with the additional consideration that a more complex and specific model is difficult to implement without error and likely easier to detect, as well as being unsuitable for the initial stage of research. Of course, model accuracy was also important, keeping in mind that the accuracy of keystroke dynamic authentication may not map directly to our generation of keystrokes. Based on these considerations and requirements, we chose to implement a model consisting of the time between each key in a *digraph*, or two consecutive keys. For example, in the word 'the', there are two digraphs: 'th' and 'he'.

This is similar to the models used by Gaines [33] and Umphress [32] (among others), and has shown to be quite effective in user authentication for text such as passwords. Importantly, Gaines found there was little difference in the digraph times between english text, random words, or random phrases, and the data was found to be normally distributed. A histogram of a single example digraph is shown in Fig. 1.

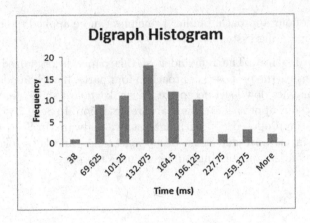

Fig. 1. Distribution of timing data for a single digraph.

The model itself consists of the mean time in milliseconds and the sample standard deviation for each possible key combination. Each unique digraph can be represented as a normal curve with those two values. With a relatively small finite number of possible digraphs, this can be quickly implemented with a lookup table containing the two variables for each digraph, and a statistical function used to generate a random number (for milliseconds of key delay) that falls within the normal curve. Thus we are very likely to get a value that is close to the mean, but occasionally we will get a delay that is less than or more than usual, as is common with human behavior. We developed the model using data from some publicly available datasets [36–38], but for the operation of the software we are using data recorded from individual members of the research group, in order to present a unique user.

While our software is still a prototype, future production and distribution across numerous machines could enable the possibility of malware analyzing data from different sources that are all operating the software, and thereby the ability to expose the deception if they all had nearly identical timing values. Thus we must provide a unique user, and we evaluate and test the software using models created from real people. Working with these datasets exposed us to the issue of outliers when using recorded data. After implementing and testing our model, we discovered that there were several instances of unnaturally long delays between generated keys, due to unrealistically high standard deviation values. This was due to the fact that any sort of pause in the middle of typing, even to the extent of leaving the computer for several minutes, was reflected in the timing data; these outliers were affecting the model and were not representative of what we were trying to model.

We explored two methods of outlier removal, the simple but naive method (removing values that fall outside of a certain number of standard deviations from the mean), and a more robust method using the median absolute deviation (MAD). Leys et al. [39] describes this method in detail, but the results are evident and can be seen in Table 1.

Table 1. Comparison of outlier removal methods for a single digraph

	Baseline	Naive	MAD
Mean	261.7162	158.2817	131.8971
Median	127	124	120.5
Std. dev.	562.0665	142.0196	57.17455

To generate the models used in testing the deceptive driver against real malware, we recorded our own typing data for a full page of text and generated a
model for each user, associating every digraph with a mean time in milliseconds
and standard deviation. This is also how the decoy keyboard learns the typing
profile of a user on a machine in operation. Significant outliers were removed, and
digraphs for which there was insufficient data (less than three occurrences) were
discarded. When encountering digraphs for which there is no entry in the model,
the deceptive driver will use a predetermined mean value with high variance to
represent the unpredictability of such rare occurrences; regardless, a user's typing patterns can be identified using a limited number of common digraphs [33],
and rare occurrences are not of much use in generating or validating a model. To
illustrate the effectiveness of the model in distinguishing users, the timing values
for the ten most common digraphs of three of the authors are shown in Fig. 2.
Each data point consists of the mean timing value for one of the ten digraphs,
where the distance from the center represents the time in milliseconds. We can
see that the two users in the middle have roughly the same typing speed, but
they have unique characteristics for each digraph. The user on the outside is
significantly different just by nature of a slower typing speed.

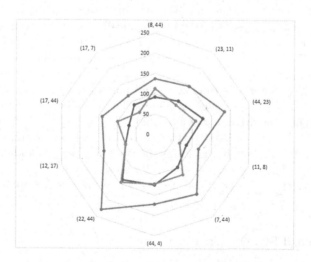

Fig. 2. Comparing the digraph model of three users.

2.2   Low-Level Deceptive Driver

This is the driver that emulates keystrokes. We refer to this driver as the decoy keyboard (DK) driver. It is a customized Windows USB Human Interface Device (HID) function driver, based on an open-source project called VMulti [43], i.e., a virtual multiple HID driver using the Kernel Mode Driver Framework (KMDF). We modified and extended VMulti to turn it into a decoy keyboard driver. KMDF is the same framework used by many vendor drivers for real hardware. It allows us to follow the same requirements and standards as a real keyboard. Thus, it appears in the device tree and Device Manager just as any other device. The location of the decoy function driver within the Windows device tree can be seen in Fig. 3. Normally, a driver does not initiate messages or data traffic itself, but responds to signals either from the hardware device (like keys being pressed) or from the operating system (turning on lights for Caps Lock, etc.).

In our case, with no physical hardware to generate signals, the keystrokes must originate from within the driver itself. That is, keystrokes are processed as normal key events through the driver so as to be indistinguishable from a real keyboard. These events have no explicit time stamps. They are retrieved using a first in, first out queue. However, any software that intercepts or records the events can easily associate a time value to each event for the purposes of analysis. Keystrokes sent without regard to timing would simply go at machine speed, that is, far faster than any human could possibly type. This is why we use the statistical model that we described previously, to regulate the time each event is sent, emulating human typing behavior. The model of a user's typing profile is represented internally as an array that functions as a lookup table, providing the core statistical values for each pair of keys.

On loading, the DK driver processes the statistical model and thus initializes a normal distribution function for each digraph, using the mean and standard deviation in combination with a cryptographically secure random number generator. Keystrokes are sent by the driver one at a time. Each key event is an individual action. Text/keycodes are taken as input and then are processed by the DK driver. For each pair of keys encountered, the appropriate distribution is used to randomly generate a time value that falls within the range as defined by the statistical model. That value is used as the delay between sending the two key events. When the driver decides to send a specific keystroke, it will immediately process it per the USB HID protocol, queuing a keyboard report as a USB interrupt transfer request that contains the key being pressed and any modifiers (shift, alt, etc.). The operating system will periodically poll the interrupt transfer pipe and retrieve the data, delivering it to any requesting applications, i.e., dprocess in our case.

2.3   Keyboard Shadowing

The operation of a decoy keyboard in parallel with a real keyboard creates an outlier configuration. Malware could simply browse all I/O devices on the

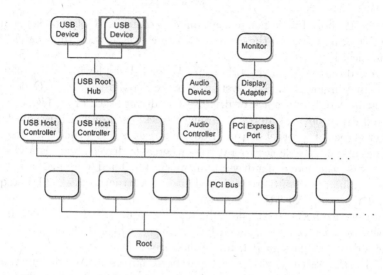

Fig. 3. Location of the decoy keyboard in the device tree. (Source: Microsoft Hardware Dev Center [42])

computer, and then check if multiple keyboards are attached to the compromised computer. Computers with two physical keyboards are not common, consequently the decoy keyboard needs to operate when the user is not typing any keys on the real keyboard. We have devised a technique, which we refer to as the keyboard shadowing mechanism (Kshadow), to detect windows of time when the user is not using the keyboard. The duration of those time windows of inactivity varies from several seconds to several minutes, and at times may last one or more hours depending on the work situation a user is in. For example, the user may be reading a document, and occasionally use the real keyboard to perform keyword searches on it. The user may be attending an hour-long presentation, in which the user mostly listens and sometimes writes down notes. The user may also be away from the computer for extended periods of time, such as when going to lunch or when participating in a meeting. Note that, since Kshadow runs in the kernel, there is now running process for it in user space.

Kshadow signals the DK driver when a time window of inactivity begins, and again when it ends. A go signal gives the DK driver a green light to start sending keystrokes to a dprocess. A stop signal notifies the DK driver that the physical keyboard is now producing keystrokes. It is time that the DK driver quickly wraps up the communication with a dprocess, and goes to sleep until the next signal from Kshadow arrives. The DK driver may choose to not use a time window of inactivity entirely, especially if the time window is long. In those cases, the DK driver selects specific portions of the time window in which to send keystrokes to a dprocess. The remaining fractions of the time window are left to be inactive. Keyboard shadowing is transparent to the user, who does not see or interact with the decoy keyboard. The user types on the physical keyboard as if the decoy keyboard were not present on the computer. Only one keyboard is discoverable at any time on the computer, with characteristics that match those of the physical keyboard.

We now discuss the inner workings of keyboard shadowing, with reference to Fig. 4. In Windows, the I/O system is packet driven [40]. An I/O device in general is operated by a stack of drivers. The driver stack of an I/O device is an ordered list of device objects, i.e., device stack, each of which is linked with the driver object of a kernel driver. A device object is a C struct that describes and represents an I/O device to a driver, whereas a driver object is a C struct that represents the image of a driver in memory [40]. The I/O requests that read keystrokes from a keyboard are packaged into data structures called I/O request packets (IRPs) [40]. An IRP is self contained, in the sense that it contains all the data that are necessary to describe an I/O request. IRPs originate from a component of the I/O system called I/O manager, which is also responsible for enabling a driver to pass an IRP to another driver.

An IRP traverses the device stack top to bottom. It is processed along the way by the drivers in the driver stack using the I/O manager as an intermediary. Once an IRP is fully processed by those drivers and thus reaches the bottom of the driver stack, the lowest driver populates its payload with scancodes. A scancode is a byte that corresponds to a specific key on the keyboard being pressed or released. The IRP may climb back up the driver stack. At the end, the I/O manager responds to the caller thread in user space by passing the keystrokes to it. The keyboard class driver referenced in Fig. 4 does IRP processing that applies generally to all keyboards, regardless of their hardware, low-level design, and type of hardware connection to the computer. When a process in user space requests to read keystrokes, the I/O manager converts the request into an IRP and sends it to the driver located at the top of the stack. Here is how.

The I/O operation performed by an IRP is indicated by a field called major function code, which may be accompanied by a minor function code. The I/O manager accesses the device object located at the top of the stack, and from there follows a pointer to the corresponding driver object. At that point, the I/O manager uses the major function code as an index into a dispatch table of the driver object, and obtains the address of a driver routine to call. The routine belongs to the keyboard class driver, which can now process the IRP and subsequently pass it down the driver stack. All the other drivers in the stack are given an opportunity to process IRPs this way as well. Without any keyboard shadowing in place, the keyboard class driver passes the IRP to a function driver. Generally speaking, the function driver of an I/O device has the most knowledge of how the device operates. The function driver presents the interface of the device to the I/O system in the kernel.

The function driver of the physical keyboard is an HID driver, which is referred to as keyboard HID client mapper driver. It is written in an independent way from the actual transport. Possible transports can be USB, Inter-Integrated Circuit (I2C), Bluetooth, and Bluetooth Low Energy. In the case of the computers that we used for code development and research in this work, the HID class driver serves as a bridge between the keyboard HID client mapper driver and a USB bus. The reader is referred to [20] for a thorough description of HID concepts and architecture. Keyboard shadowing is implemented as a filter driver, which is positioned between the keyboard class driver and the keyboard HID client mapper driver, as depicted in Fig. 4. Kshadow creates a filter device object (FiDO). This FiDO is similar to the functional device object (FDO) and physical device object (PDO) created by the other drivers in the stack. These device objects are only different in the type of drivers to which they represent an I/O device.

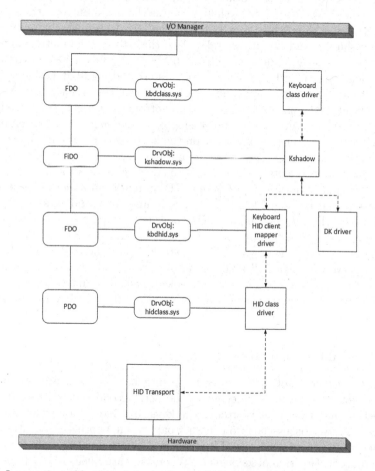

Fig. 4. Integration of keyboard shadowing with the driver stack of an HID keyboard.

By registering with the I/O manager as a filter driver, Kshadow gets to see all IRPs bound for the keyboard, physical or decoy. Kshadow has access to the process ID of dprocess. Furthermore, for each IRP, Kshadow retrieves the process ID of the thread that requested the I/O operation. This is how Kshadow knows whether an IRP originated in dprocess or in another process. Regardless of the source, Kshadow receives IRPs from the keyboard class driver. If an IRP originated in a process other than dprocess, Kshadow passes it down to the keyboard HID client mapper driver. If an IRP originated in dprocess, Kshadow simply passes it to DK driver, which populates it with the scancodes of emulated keystrokes. Kshadow acquires high resolution time stamps to measure the interval between the current time and the time an IRP was last seen going down the driver stack. The Windows kernel provides APIs such as KeQueryPerformanceCounter() that are highly accurate.

If no IRPs from processes other than dprocess arrive for an interval of time that exceeds a given threshold, Kshadow marks the beginning of a time window of inactivity and thus sends a go signal to the DK driver. All communications between Kshadow and DK driver take place through direct function calls, and do not involve the I/O manager. This is because DK driver is not part of the driver stack, consequently no device object is created for it. It is slightly more challenging for Kshadow to spot the end of a time window of inactivity. We deem that inactivity ends when the driver at the bottom, which is also referred to as a miniport driver, completes its processing of a pending IRP. The completion event is due to the user pressing a key on the keyboard, and the miniport driver subsequently reading the scancode byte and placing it on the IRP. We leverage the fact that the completed IRP could be made to climb up the driver stack.

When Kshadow notices a completed IRP coming from underneath, it knows that the physical keyboard has become active and thus sends a stop signal to the DK driver. However, a completed IRP does not just climb up the driver stack by itself. Kshadow registers one of its functions as an IoCompletion routine for an IRP, before passing it down the driver stack. When the IRP is complete, the IoCompletion routines of all higher-level drivers are called in order. When the IoCompletion routine of Kshadow is invoked, Kshadow marks the end of the time windows of inactivity and DK driver stops sending emulated keystrokes to dprocess.

2.4   First and Second Order Detection

First order detection of a keylogger happens when the malware attempts to intercept keystrokes and is detected in the act. Second order detection refers to detection at a time that postdates the keystroke interception mounted by a keylogger. As we discuss later on in this paper, our approach can attain first order detection of various forms of keyloggers. Nevertheless, its main strength is in second order detection of keyloggers. When we started this research, our goal was to deliver effective second order detection. First order detection is primarily consequential, and was noticed mostly during the practical tests against live malware. We base this work on a simple observation: keyloggers intercept data for attackers to use. Of course, not all intercepted data will be used, but some of those data will. In fact, malicious use of intercepted data is the reason behind the very existence of keyloggers.

The idea that underpins this work is to use a decoy keyboard to generate emulated and hence decoy keystrokes for keyloggers to intercept. Keyloggers commonly encrypt the data they intercept on a compromised computer, and then send those data to the attacker over the network. Furthermore, keyloggers commonly come as one of the modules of larger malware. Other modules open up backdoors into the compromised computer, and enable the attacker to access any resources. Yet other modules intercept filesystem traffic, spy on a user over the webcam, or authenticate to other computers using stolen credentials. The malware are detected when they use decoy data leaked by the decoy keyboard. In our prior work we have created decoy network services, which require authentication. In this case, we need to leak an account by using the decoy keyboard to make it appear as if a user is typing his or her username and password to authenticate to those services.

More specifically, we have explored a decoy network interface card (DNIC), which makes a computer appear to have access to an internal network [41]. Neither the DNIC nor the internal network exist for real. Furthermore, they are both masked to the user, consequently the user does not see and hence does not access them. We have developed an Object Linking and Embedding (OLE) for Process Control (OPC) server, which provides power grid data after authenticating the client. All of these decoys are implemented in the kernel, therefore no network packet ever leaves the computer for real. The decoy keyboard leaks an OPC server account, which includes the IP address of the server computer, the name of the OPC server, and of course a username and password. The IP address in question is reachable over the imaginary network. Once malware intercepts the OPC server account from the decoy keyboard and thus uses it to access the OPC server over the DNIC, we attain second order detection.

Another example involves the leakage of credentials that provide access to a virtual private network (VPN), which is reachable only over the DNIC. VPNs are of particular interest to attackers, since they commonly lead to protected resources. For instance, the BlackEnergy malware that compromised a regional power distribution company in Ukraine was able to access electrical substation networks through a VPN using stolen VPN credentials. We can display an imaginary VPN over the DNIC. The imaginary VPN is accessible via credentials leaked by the decoy keyboard, leading to an unequivocal second order detection. It is of paramount importance to this work that the attacker does not discard the data that the keylogger had intercepted from the decoy keyboard. This is why the model of human keystroke dynamics, which we discussed earlier in this paper, is so critical to this work. If we can leak data in a way that resembles those produced by a real keyboard, we give the attacker no reason not to use the decoy data. For consistency reasons, the decoy keyboard can provide decoy non-sensitive data for other decoy processes. For example, the decoy keyboard can fill a webform through a decoy web browser process.

3   Evaluation

The decoy keyboard approach was installed and run on a 64-bit Windows 10 virtual machine in our lab. The lab was isolated both logically and physically from any physical computer networks. Firstly, we tested the ability of the decoy keyboard to co-exist with a real keyboard. We simply used a computer equipped with a decoy keyboard for several days to do our usual work. We paid attention to all details related to the keyboard use. We observed no visible delays when typing on the physical keyboard. There were numerous periods of keyboard inactivity, in which we were away from the computer. The logs show that the decoy keyboard had been in operation multiple times, however we saw no anomalies with the use of the real keyboard when we returned to work on the computer. The only observable was that the screen saver and the power saving mode appeared to be affected by the operation of the decoy keyboard. As a dprocess requests keystrokes, and the DK driver generates them, normal computer activity is created, consequently Windows assumes that there is a user working on the computer.

The delay in IRP processing caused by Kshadow, as the user types on the physical keyboard, is characterized by the data of Fig. 5. The lower chart indicates the delays that occur when Kshadow passes the IRPs down the driver stack as soon as they arrive. This graph is given solely for comparison reasons. The higher graph shows the delays when Kshadow works to determine if a time window of inactivity has begun. The greater delays are due to Kshadow registering one of its functions as an IoCompletion routine, acquiring high-resolution time stamps to measure time intervals, and retrieving the process ID for the thread that originally requested to read a keystroke. Other overhead is due to starting and maintaining a kernel thread.

The delays caused by the overall approach when a user returns to his or her workplace and presses a key on the keyboard are given in Fig. 6. This is the situation in which the decoy keyboard has been operating for some time, and now the user needs the physical keyboard back in operation. The data pertain to 10 separate occurrences of the aforementioned situation. The lower chart shows the delays that occur when the DK driver is able to respond to the stop signal immediately. This only happens when the DK driver is not emulating keystrokes, and dprocess does not exist or is suspended and hence is not reading keystrokes. Otherwise the DK driver needs some time to wrap up its keystroke emulation, halt dprocess, and altogether get out of the way. The duration of these finalizations depends on the data that the decoy keyboard is leaking through the attack surface and hence may vary. There is also some delay occurring when the keyboard transitions in the opposite direction, namely from physical to decoy. Nevertheless, the user is not affected, since he or she is not typing on the physical keyboard at that time.

We examined the arrival times of keystrokes in user space for the purpose of testing the statistical model. Without incorporating the model in the DK driver, keystrokes arrive at an almost constant rate. The delay between any two keys is the same, and quite minimal. Clearly this is an obvious indication that the

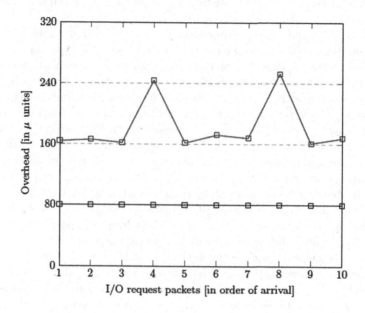

Fig. 5. I/O filtering overhead while the physical keyboard is in use.

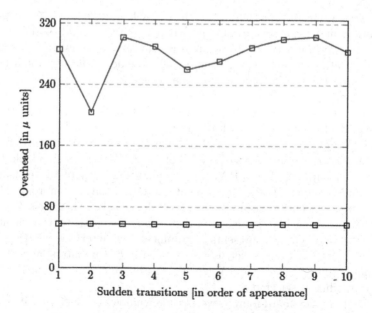

Fig. 6. Decoy-to-physical transition overhead.

keyboard is a decoy. With the statistical model in use within the DK driver, the keystrokes arrive at times that match the typing profile of an actual user. We have not included the corresponding charts due to room limitations.

We now discuss the evaluation of the effectiveness of our approach. We tested the decoy keyboard against live malware. The malware corpus was comprised of samples that were obtained from public malware repositories, namely Open Malware, AVCaesar, Kafeine, Contagio, StopMalvertising, and unixfreaxjp. Those repositories provide malware samples for academic research. The malware samples were known, and included remote access tools (RATs), worms and trojans, and also viruses. We used the IDA Pro tool to analyze the malware samples to the greatest degree that we could in order to identify and remove goodware. We also did analysis work to identify malware samples that are of the same malware, but appear different because of polymorphic or metamorphic techniques. At the end, the malware corpus consisted of 50 malware samples. Most of those malware samples have a history of involvement in malware campaigns in the recent years, therefore are valid and pertinent for testing purposes.

Some of the malware samples use keylogger modules that intercept keystrokes by probing their target keyboard. Those probes resulted in IRPs bound for the target keyboard. When the decoy keyboard was in operation, the malicious IRPs reached Kshadow and DK driver, causing a first order detection of the malware. The other malware samples that escaped first order detection needed to be tested against second order detection. In that regard, we did not operate the decoy keyboard against real human attackers on the Internet. We are a University and thus are not in the position to run real-world cyber operations at this time. Nevertheless, we tested whether or not the decoy keystrokes were intercepted by the malware, and the malware subsequently attempted to send those keystrokes to attackers over the network. If we succeeded in causing malware to attempt to send decoy data to attackers, then the use of those decoy data by the attackers, and hence a second order detection of the malware by our approach, is only a matter of time.

Our testing differed depending on the type of malware. Let us first discuss the case of RATs. We started with testing the decoy keyboard approach against QuasarRAT and jRAT, and later realized that they resemble the software architecture and interception tactics of several other RATs such as Bandook, Ozone, Poisonivy, and njRAT. These RATs are comprised mainly of two executables, a client and a server. The server executable is run on a compromised computer, whereas the client executable is run on the attacker's machine. The server executable performs keylogging, and then sends the data to the client. The client executable typically presents a graphical user interface (GUI) panel to the attacker, which the attacker uses to select target computers, as well as the operations to perform on them. A typical control panel is shown in Fig. 7. The keylogger module is selected.

As the decoy keyboard found several time windows of inactivity, it sent keystrokes to dprocess. The keystrokes were intercepted by the server executable, which sent them to the client executable. The client executable in turn displayed

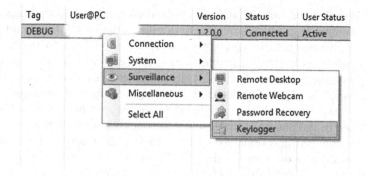

Fig. 7. The control panel of QuasarRAT.

the intercepted keystrokes on the GUI panel. The decoy data that were leaked by the decoy keyboard were now listed on the GUI panel. In the case of the other malware samples, the testing was challenging as there were no GUI panels to display the intercepted data. In those cases, we used a kernel debugger to set breakpoints on instructions of the malware that accessed intercepted keystrokes. We dumped the intercepted data on the debugger's console, and verified that some of those data were indeed data generated by the DK driver.

In conclusion, all of the samples in the malware corpus intercepted keystrokes generated by the decoy keyboard, and either sent them to the client executable running on a computer on the local network, or attempted to send them to external IP addresses. In the latter case, the network packets were eventually dropped due to the testbed isolation. There were no false alarms raised, because no goodware ever requested to read keystrokes after the user went away from the computer and a decoy keyboard was made active.

4 Related Work

Decoy I/O devices are intrinsically a form of deception, and there has been a great deal of prior work in the field. Cohen's Deception Toolkit [9] in 1997 was one of the first publicly available programs designed to deceive and identify attackers by presenting fake information and vulnerabilities [10]. Honeypots and honeyfiles, decoy machines and files designed to entice and trap attackers, have also been explored quite heavily. In 1989, Stoll published a book that detailed some of the earliest uses of honeypots and honeyfiles as he worked to catch a hacker that infiltrated the computer systems at Lawrence Berkley National Laboratory [11]. By Presenting Attractive yet fake files to the attacker, he was able to catch them "in the act," eventually leading to their arrest. These concepts have since been applied towards attracting automated software instead of real people, but the underlying idea remains the same.

A decoy I/O device as presented in this research has two main differences from existing deception tactics and decoys such as honeypots and honeyfiles.

The first is that it can (and is intended to) be deployed on an active, production machine, running in the background transparent to the user. The second is that it is specifically designed to emulate a real I/O device down to the hardware level, making it as indistinguishable from a real device as the complexity of the decoy communications allow. A typical honeypot, such as Sebek [12], is a system designed solely for logging and reporting attacks that presents itself as a potential target through the activity of a typical system in production use. However, due to the nature of the design, the honeypot itself cannot be in active use; any incoming connections are either malicious in nature or from someone who has stumbled upon it by accident.

Sebek and other high-interaction honeypots will imitate a complete machine, and thus require the full resources of one, and are expensive to maintain. Low-interaction honeypots like PhoneyC [13] are designed to only present certain vulnerable services, and require less resources to operate. In order to operate multiple honeypots on one physical computer, virtual machines can be utilized, but malware such as Conficker [14] can detect that they are being run virtually and change their behavior, or simply avoid the honeypot altogether. Along that vein, Rowe proposed using fake honeypots as a defensive tactic by making a standard production machine appear to be a honeypot [15], scaring away any would-be attackers by emulating the signatures and anomalies that are typically associated with honeypots.

Anagnostakis et al. proposed a hybrid technique called a "shadow honeypot" [16] that utilizes actual production applications with embedded honeypot code. Incoming connections are first filtered through an anomaly detection process, which will redirect possible malicious traffic to the shadow honeypot, and normal traffic to the standard server. However, the shadow honeypot is running the same server application with the same internal state, with the added honeypot code serving to analyze the behavior of the request. The shadow honeypot can then send valid requests to the real server, and any change in state from malicious activity is thrown out. Spitzner discussed the issue of insider threats [17], and how they target actual information that they can use, thus the honeypot could provide fake data that appears attractive. This data could be business documents, passwords, and so on, and would be a "honeytoken" designed to redirect the attacker to the honeypot. This is similar to the research presented here in that fake data attracts an attacker to the decoy I/O device, but is on an entirely different scale and still requires a high-interaction honeypot to implement.

The honeytokens proposed by Spitzner are a type of honeyfile, a concept explored by Yuill et al. [18] which are commonly used on honeypots but have also been placed on systems in actual use. Software can then analyze traffic to and from the files with the assumption that any activity is malicious in nature, as those files are not in use by legitimate applications. This concept of a Canary File, or a honeyfile placed amongst real files on a system, was proposed by Whitman [19], and he also discusses the automatic generation and management of the Canary Files. However, the content of automatically generated files is difficult to present as authentic upon inspection, and malware that resides at the operating system level is able to examine file access patterns to ignore files that are not being used.

5   Conclusion

The decoy keyboard is a novel anti-malware approach that is transparent to both normal users and attackers. The approach requires no prior knowledge of malware code and data structures to be able to detect them, and can operate on computers in production. The filter driver is able to shadow a physical keyboard. It can reliably guide a low-level deceptive driver in the generation of decoy keystrokes for malware to intercept. The decoy keystrokes are delivered to malware according to a timing model that makes both the decoy data and their delivery quite consistent with the behavior of a physical keyboard. The decoy keyboard is safe to operate, and does not interfere with the user's work. The decoy keyboard showed to be effective against a large malware corpus, attaining approximated second order detections and some first order detections as well.

Acknowledgments. This material is based on research sponsored by U.S. Air Force Academy under agreement number FA7000-16-2-0002. The U.S. Government is authorized to reproduce and distribute reprints for Government purposes notwithstanding any copyright notation thereon.

    The views and conclusions contained herein are those of the authors and should not be interpreted as necessarily representing the official policies or endorsements, either expressed or implied, of U.S. Air Force Academy or the U.S. Government.

Appendix A: Malware Detection

Current malware detection (including keyloggers) is based on both static and dynamic analysis. Malware detection through static analysis consists of scanning an executable file for specific strings or instruction sequences that are unique to a specific malware sample [1,2,21]. This is a primary type of detection employed by commercial anti-virus software packages; in addition to simply scanning files present on a computer, the keyboard and other device stacks are inspected and interceptions by known malware are flagged and reported, but lists of known signatures must be kept and frequently updated. The main limitation of the static analysis techniques is that malware can change its appearance by means of polymorphism, metamorphism, and code obfuscation. Other static analysis research focused on higher-order properties of executable files, such as the distribution of character n-grams [22,23], control flow graphs [3,24], semantic characteristics [4], and function recognition [25–27]. Dynamic analysis is another field of ongoing research that studies the execution flow of a malware binary using methods such as function call monitoring and information flow tracking [5]. Tools like Detours [6] allow an analyst to hook into the function calls of a piece of malware and execute their own code for investigative purposes before returning control to the original program. This type of hook can be performed on binary files located on disk as well as by modifying the memory space of a currently running process. Using these techniques, researchers are able to learn various indicators, which are then used to detect that malware. Detection indicators include disk access patterns [28], malspecs [29], sequences of system calls and system call parameters [30], and behavior graphs [31].

All of these methods are rooted in and constrained by the concept of analyzing existing malware. The main advantage and differentiator of the decoy keyboard approach over such a large body of malware detection research is that a decoy keyboard does not require any prior analysis of a malware sample in order to detect it. A decoy keyboard can detect malware encountered for the very first time, whereas the techniques from the mentioned large body of malware detection research require that an instance of malware be given to them as input for analysis. A specific instance of the malware needs to be detected by other means – someone has to provide the malicious code, along with an explicit and validated statement that it is malware. Those techniques will then be used to analyze the program, which at that time is known to be malware, and thus will learn indicators such as those discussed in the previous paragraph. Those indicators are subsequently used to detect other instances of the malware. This introduces a period of time between the release and discovery of new malware and the update of any software to protect against it during which targeted systems are vulnerable. Additionally, kernel-based malware has such a degree of access to the system and underlying processes that it may very well be able to find and circumvent any installed security software.

Some work has been done to counteract the deficiencies of static and dynamic analysis by using machine learning techniques such as behavioral clustering [7] and classification [8], but the accuracy of these techniques is dependent on the quality of the training set; malware that is significantly different in operation from the majority may still manage to evade detection.

References

1. Christodeorescu, M., Jha, S.: Static Analysis of Executables to Detect Malicious Patterns. Department of Computer Sciences, Wisconsin Univ-Madison (2006)
2. Griffin, K., Schneider, S., Hu, X., Chiueh, T.C.: Automatic generation of string signatures for malware detection. In: Kirda, E., Jha, S., Balzarotti, D. (eds.) RAID 2009. LNCS, vol. 5758, pp. 101–120. Springer, Heidelberg (2009). doi:10.1007/978-3-642-04342-0_6
3. Bruschi, D., Martignoni, L., Monga, M.: Detecting self-mutating malware using control-flow graph matching. In: Büschkes, R., Laskov, P. (eds.) Detection of Intrusions and Malware, and Vulnerability Assessment. LNCS, vol. 4064, pp. 129–143. Springer, Heidelberg (2006)
4. Feng, Y., Anand, S., Dillig, I., Aiken, A.: Apposcopy: semantics-based detection of android malware through static analysis. In: Proceedings of the 22nd ACM SIGSOFT International Symposium on Foundations of Software Engineering, pp. 576–587 (2014)

5. Egele, M., Scholte, T., Kirda, E., Kruegel, C.: A survey on automated dynamic malware-analysis techniques and tools. ACM Comput. Surv. **44**(2), 6 (2012)
6. Hunt, G., Brubacher, D.: Detours: binary interception of Win32 functions. In: 3rd Usenix Windows NT Symposium (1999)
7. Bailey, M., Oberheide, J., Andersen, J., Mao, Z.M., Jahanian, F., Nazario, J.: Automated classification and analysis of internet malware. In: Kruegel, C., Lippmann, R., Clark, A. (eds.) RAID 2007. LNCS, vol. 4637, pp. 178–197. Springer, Heidelberg (2007). doi:10.1007/978-3-540-74320-0_10
8. Rieck, K., Trinius, P., Willems, C., Holz, T.: Automatic analysis of malware behavior using machine learning. J. Comput. Secur. **19**(4), 639–668 (2011)
9. Cohen, F.: The deception toolkit. Risks Digest, vol. 19 (1998)
10. Cohen, F.: A note on the role of deception in information protection. Comput. Secur. **17**(6), 483–506 (1998)
11. Stoll, C.: The Cuckoo's Egg: Tracing a Spy through the Maze of Computer Espionage. Doubleday, New York (1989)
12. Balas, E.: Know your enemy: Sebek. The Honeynet Project (2003)
13. Nazario, J.: PhoneyC: A Virtual Client Honeypot. LEET, vol. 9, pp. 911–919 (2009)
14. Leder, F., Werner, T.: Know your enemy: Containing conficker. The Honeynet Project (2009)
15. Rowe, N.C., Duong, B.T., Custy, E.J.: Defending cyberspace with fake honeypots. J. Comput. 2(2) (2007)
16. Anagnostakis, K.G., Sidiroglou, S., Akritidis, P., Xinidis, K., Markatos, E.P., Keromytis, A.D.: Detecting targeted attacks using shadow honeypots. In: Usenix Security (2005)
17. Spitzner, L.: Honeypots: catching the insider threat. In: 19th Annual Computer Security Applications Conference, pp. 170–179 (2003)
18. Yuill, J., Zappe, M., Denning, D., Feer, F.: Honeyfiles: deceptive files for intrusion detection. In: Proceedings from the Fifth Annual IEEE SMC Information Assurance Workshop, pp. 116–122 (2004)
19. Whitham, B.: Canary files: generating fake files to detect critical data loss from complex computer networks. In: The Second International Conference on Cyber Security, Cyber Peacefare and Digital Forensic, pp. 170–179 (2013)
20. Microsoft Device and Driver Technologies: HID drivers (2016). https://msdn.microsoft.com/en-us/windows/hardware/drivers/hid/index
21. Szor, P.: The Art of Computer Virus Research and Defense. Pearson Education, Indianapolis (2005)
22. Li, W.-J., Stolfo, S., Stavrou, A., Androulaki, E., Keromytis, A.D.: A study of malcode-bearing documents. In: Hämmerli, B., Sommer, R. (eds.) DIMVA 2007. LNCS, vol. 4579, pp. 231–250. Springer, Heidelberg (2007). doi:10.1007/978-3-540-73614-1_14
23. Li, W.J., Wang, K., Stolfo, S.J., Herzog, B.: Fileprints: identifying file types by n-gram analysis. In: Information Assurance Workshop, Proceedings from the Sixth Annual IEEE SMC 2005, pp. 64–71 (2005)
24. Kruegel, C., Kirda, E., Mutz, D., Robertson, W., Vigna, G.: Polymorphic worm detection using structural information of executables. In: International Workshop on Recent Advances in Intrusion Detection, pp. 207–226 (2005)
25. Kruegel, C., Robertson, W., Vigna, G.: Detecting kernel-level rootkits through binary analysis. In: 20th Annual Computer Security Applications Conference, pp. 91–100 (2004)

26. Kinder, J., Katzenbeisser, S., Schallhart, C., Veith, H.: Detecting malicious code by model checking. In: International Conference on Detection of Intrusions and Malware, and Vulnerability Assessment, pp. 174–187 (2005)
27. Christodorescu, M., Jha, S., Seshia, S.A., Song, D., Bryant, R.E.: Semantics-aware malware detection. In: 2005 IEEE Symposium on Security and Privacy, pp. 32–46 (2005)
28. Felt, A., Paul, N., Evans, D., Gurumurthi, S.: Disk level malware detection. In: Poster: 15th Usenix Security Symposium (2006)
29. Christodorescu, M., Jha, S., Kruegel, C.: Mining specifications of malicious behavior. In: Proceedings of the 1st India Software Engineering Conference, pp. 5–14 (2008)
30. Canali, D., Lanzi, A., Balzarotti, D., Kruegel, C., Christodorescu, M., Kirda, E.: A quantitative study of accuracy in system call-based malware detection. In: Proceedings of the 2012 International Symposium on Software Testing and Analysis, pp. 122–132 (2012)
31. Kolbitsch, C., Comparetti, P.M., Kruegel, C., Kirda, E., Zhou, X.Y., Wang, X.: Effective and efficient malware detection at the end host. In: USENIX Security Symposium, pp. 351–366 (2009)
32. Umphress, D., Williams, G.: Identity verification through keyboard characteristics. Int. J. Man Mach. Stud. **23**(3), 263–273 (1985)
33. Gaines, R.S., Lisowski, W., Press, S.J., Shapiro, N.: Authentication by keystroke timing: Some preliminary results. No. RAND-R-2526-NSF. RAND CORP (1980)
34. KeyTrac Keyboard Biometrics. http://www.keytrac.net
35. Banerjee, S.P., Woodard, D.L.: Biometric authentication and identification using keystroke dynamics: a survey. J. Pattern Recogn. Res. **7**(1), 116–139 (2012)
36. Roth, J., Liu, X., Metaxas, D.: On continuous user authentication via typing behavior. IEEE Trans. Image Process. **23**(10), 4611–4624 (2014)
37. Feit, A.M., Weir, D., Oulasvirta, A.: How we type: movement strategies and performance in everyday typing. In: Proceedings of the 2016 Chi Conference on Human Factors in Computing Systems, pp. 4262–4273 (2016)
38. Choi, Y.: Keystroke patterns as prosody in digital writings: a case study with deceptive reviews and essays. In: Empirical Methods on Natural Language Processing, p. 6 (2014)
39. Leys, C., Ley, C., Klein, O., Bernard, P., Licata, L.: Detecting outliers: do not use standard deviation around the mean, use absolute deviation around the median. J. Exp. Soc. Psychol. **49**(4), 764–766 (2013)
40. Russinovich, M.E., Solomon, D.A., Ionescu, A.: Windows Internals, Part 1 and 2, 6th edn. Microsoft Press, Redmond (2012)
41. Rrushi, J.: NIC displays to thwart malware attacks mounted from within the OS. Comput. Secur. **61**(C), 59–71 (2016)
42. Microsoft Hardware Dev Center: Device nodes and device stacks. https://msdn.microsoft.com/en-us/windows/hardware/drivers/gettingstarted/device-nodes-and-device-stacks
43. Newton, D.: Virtual Multiple HID Driver (multitouch, mouse, digitizer, keyboard, joystick). https://github.com/djpnewton/vmulti

The Fallout of Key Compromise
in a Proxy-Mediated Key Agreement Protocol

David Nuñez[✉], Isaac Agudo, and Javier Lopez

Network, Information and Computer Security (NICS) Laboratory,
Computer Science Department, University of Málaga, Málaga, Spain
{dnunez,isaac,jlm}@lcc.uma.es

Abstract. In this paper, we analyze how key compromise affects the protocol by Nguyen et al. presented at ESORICS 2016, an authenticated key agreement protocol mediated by a proxy entity, restricted to only symmetric encryption primitives and intended for IoT environments. This protocol uses long-term encryption tokens as intermediate values during encryption and decryption procedures, which implies that these can be used to encrypt and decrypt messages without knowing the corresponding secret keys. In our work, we show how key compromise (or even compromise of encryption tokens) allows to break forward security and leads to key compromise impersonation attacks. Moreover, we demonstrate that these problems cannot be solved even if the affected user revokes his compromised secret key and updates it to a new one. The conclusion is that this protocol cannot be used in IoT environments, where key compromise is a realistic risk.

1 Introduction

Nguyen, Oualha, and Laurent presented at ESORICS 2016 an authenticated key agreement protocol based on an ad-hoc variant of symmetric proxy re-encryption [16], called AKAPR. This protocol is intended for highly-constrained IoT devices, and therefore, deliberately excludes the use of asymmetric primitives such as Diffie-Hellman key exchange due to its reliance on modular exponentiations. The proposed protocol is implemented on top of block ciphers, MACs, and simple modular arithmetic. The authors provide a formal verification of the mutual authentication property using ProVerif.

In this paper we identify some deficiencies in the proposed protocol that were not considered by the formal analysis and describe several attacks that exploit them. In particular, our attacks exploit the fact that encryption and decryption operations do not use secret keys directly, but intermediate secret values (which we call "encryption tokens"). These secrets are relatively exposed throughout the protocol and can be used to compromise the security of their corresponding owners (and in certain cases, other users). We show how compromise of these encryption tokens (or directly, of secret keys) makes it possible to break forward security and lead to key compromise impersonation attacks. An interesting

G. Livraga and S. Zhu (Eds.): DBSec 2017, LNCS 10359, pp. 453–472, 2017.
DOI: 10.1007/978-3-319-61176-1_25

(and devastating) consequence of the nature of these attacks is that these cannot be prevented even if the affected user renews his secret key, given that the compromised secret can be linked to the encryption token of other users.

This paper constitutes an example of how providing a formal analysis of a protocol or system is often not enough, since this may be incomplete, incorrectly designed or directly flawed. In the case of the analyzed protocol, the authors did not consider the possibility of key compromise. However, in IoT environments (which is the target application of the protocol), key compromise is a realistic risk due to the broad attack surface of most devices, which includes physical attacks (e.g., use of JTAG and UART interfaces, power analysis, etc.), software vulnerabilities (e.g., buffer over-reads like Heartbleed in OpenSSL), privilege escalation through backdoor accounts, etc.

Related Work. The protocol we analyze in this paper is an example of authenticated key exchange (AKE), one of the most recurring topics in the literature when it comes to security protocols. It is also the basis for most Internet applications that relay on secure channels, as it is embedded in TLS and IPSEC. We can clearly distinguish two research trends in this area, one based on public-key cryptography, where Diffie-Hellman [7] is the current "standard", and the other on secret key cryptography. With respect to the latter, most proposals rely on a Key Distribution Center (KDC) that shares a secret key with all users in the system and supports them on agreeing on a session key. Since 1978, when the Needham-Schroeder symmetric protocol [14] was proposed, several authors have attempted to propose better symmetric AKE protocols, usually showing how previous ones can be attacked and fixing their weaknesses. For example, one of the weaknesses of the original Needham-Schroeder protocol was the inability to prove the freshness of the session key. Denning and Sacco [6] proposed the use of Timestamps as a way fix this problem. Later, the Kerberos protocol [15], also built on the idea of using timestamps, was proposed as the current "standard" AKE solution in the symmetric set-up. In the last years, new assumptions on the capabilities of the attackers have been defined as well as new application scenarios, security requirements and constrains that make this problem still an interesting research topic, particularly in those environment where public-key cryptography is not viable. Some recent works show that Elliptic Curve Cryptography (ECC) can be run in most wireless sensor platforms [12], although there are still some highly-constrained devices, such as passive RFID, that are not yet capable of using PKC and would then require the use of symmetric AKE protocols.

The analyzed protocol uses techniques from proxy re-encryption (PRE). This is a research topic with increasing popularity, with new use cases arising in different contexts (e.g., data sharing in the cloud, key management, etc.). Although the vast majority of PRE schemes are based on public-key cryptography [18], there have been some proposals based on symmetric cryptography. Syalim et al. [21] propose a symmetric PRE scheme based on the All-Or-Nothing Transform, although it assumes that both sender and receiver share a common secret. Cook and Keromytis [5] present a solution based on a double-encryption approach, but as in the previous case, a priori shared keys are needed. The key-homomorphic PRF primitive by Boneh et al. [2] can be used to construct symmetric PRE schemes without the shared key requirement, although as noted by Garrison et al. [10], its computational cost is comparable or greater than traditional public-key cryptography. Sakazaki et al. [19] propose a symmetric PRE scheme that is essentially equivalent to the one of this paper; this is discussed further in Sect. 2.1.

Organization. The rest of this paper is organized as follows: In Sect. 2, we describe the AKAPR protocol in detail, including actors, protocol flow and the underlying symmetric proxy re-encryption primitive. In Sect. 3, we present several attacks and weaknesses of the AKAPR protocol. In Sect. 4, we discuss some possible alternatives to the protocol. Finally, Sect. 5 concludes the paper.

2 Description of the Authenticated Key Agreement Protocol

In this section we briefly describe the protocol AKAPR (Authenticated Key Agreement mediated by a Proxy Re-Encryptor). The goal of this protocol is to provide an authenticated key exchange between two entities, using a proxy entity as a mediator. As in the general case of proxy re-encryption, although the proxy entity assists in the process, it should not learn any information. A critical restriction that guides the design of this protocol is that it assumes that the authenticating entities may not be capable of using common public-key cryptography primitives, such as Diffie-Hellman key exchange and digital signatures, due to the its reliance on modular exponentiations. Instead, AKAPR is designed solely on top of block ciphers, MACs, and simple modular arithmetic.

In the remaining of this section we first briefly explain the underlying symmetric proxy re-encryption primitive. Next, we describe the general setting of the protocol, which includes the description of actors, network architecture and trust assumptions. Finally, we detail the protocol flow.

2.1 Symmetric Proxy Re-encryption Primitive

The authors propose a symmetric proxy re-encryption algorithm, which is composed of the following five functions:

- KeyGen(λ): On input the security parameter λ, the key generation algorithm KeyGen outputs the secret key sk and identity id. As an example, for user A, these are sk_A and id_A.
- ReKeyGen(sk_A, id_A, sk_B, id_B): On input the secret keys sk_A and sk_B, and identities id_A and id_B, the re-encryption key generation algorithm ReKeyGen computes the re-encryption key between users A and B as $rk_{A \to B} = h(sk_A \| id_B)^{-1} \cdot h(sk_B \| id_A)^{-1}$, where $h : \{0,1\}^* \to \mathbb{Z}_p$ is a hash function.

- Enc(sk_A, id_B, M): On input the secret key sk_A, the identifier of B, and a message M, the encryption algorithm Enc samples t from $\mathbb{Z}_p$, derives a fresh key K from t using a key derivation function (KDF), and outputs ciphertext $C_A = (\mathsf{SymEnc}_K(M), t \cdot h(sk_A \| id_B))$, where SymEnc is a symmetric encryption algorithm.
- ReEnc$(rk_{A \to B}, C_A)$: On input a re-encryption key $rk_{A \to B}$ and a ciphertext $C_A = (C_{A,1}, C_{A,2})$, the re-encryption algorithm ReEnc outputs ciphertext $C_B = (C_{A,1}, C_{A,2} \cdot rk_{A \to B})$.
- Dec(sk_B, id_A, C_B): On input the secret key sk_B, the identity of the original recipient A, and a ciphertext $C_B = (C_{B,1}, C_{B,2})$, the decryption algorithm Dec computes $t \leftarrow h(sk_B \| id_A) \cdot C_{B,2}$, derives key K from t using the KDF and decrypts the message M from $C_{B,1}$ using symmetric decryption algorithm SymDec.

We note that there is a previous proposal by Sakazaki et al. [19], further discussed in [23], which is essentially equivalent to this symmetric proxy re-encryption scheme. The only difference is that the original proposal by Sakazaki et al. uses the XOR operator instead of the modular multiplication. If instantiated correctly, both of them are equivalent, security-wise; however, Sakazaki et al.'s choice of the XOR operator can be implemented more efficiently than modular multiplication.

Additionally, we also remark that this symmetric proxy re-encryption scheme is not consistent with the prevalent idea of proxy re-encryption, where the sender does not need to know a priori the identity of the intended recipient. In this scheme, the sender fixes the recipient identity during encryption, and therefore, the resulting ciphertext can only be re-encrypted to this identity. In contrast, in the traditional idea of proxy re-encryption (regardless if it is public-key or symmetric), ciphertexts can be re-encrypted for any possible recipient, as long the corresponding re-encryption key exists. Although, for simplicity, we will continue to refer to this scheme as symmetric proxy re-encryption, we believe that it is actually some kind of multiparty encryption, where the sender and the proxy jointly create a ciphertext decryptable by an a priori fixed recipient.

2.2 Protocol Setting

There are three main types of actors in the AKAPR protocol, which are the following:

- The Initiator (I) and the Responder (R), which are two entities located in separate subnetworks and that want to establish an authenticated session between them. It is assumed that these entities may lack the capacity to use asymmetric primitives such as Diffie-Hellman, which requires modular exponentiation, and therefore will only use symmetric techniques. Note, however, that a regular device can also participate in this protocol.

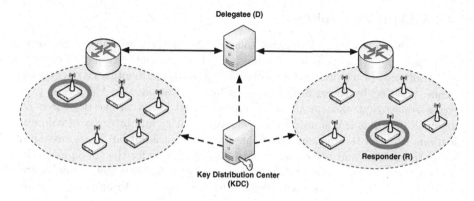

Fig. 1. Network architecture (adapted from [16])

- The Delegatee[1] (D), which mediates in the key agreement protocol between the Initiator and the Responder, without being able to learn the negotiated session keys.
- The Key Distribution Center (KDC), which initially generates all the secret keys and necessary re-encryption keys, and distributes them (secret keys to entities, and re-encryption keys to the delegatee).

Figure 1 depicts the network setting assumed by this protocol, as well as its main actors.

2.3   Trust Assumptions

The protocol has the following trust assumptions:

- The Delegatee is assumed to be honest-but-curious, which means it behaves correctly with respect to the protocol, but at the same time, it has an interest in reading the underlying information. We will latter show that if this assumption is relaxed (e.g., an attacker gains control of the delegatee, or simply, the delegatee cooperates with him), it makes several attacks possible.
- The Key Distribution Center is fully trusted. It is clear that, given that the KDC knows all the keys involved, it can gain full control of the network. The existence of an omniscient KDC can be seen as a single point of failure, precisely for this reason [13, Remark 13.3].

---

[1] Note that, in the proxy re-encryption literature, the term "Delegatee" is usually referred to the recipient of a re-encrypted ciphertext. Hence, a re-encryption from user A to B can be seen as the delegation of decryption rights from a "delegator" A to a "delegatee" B. However, for consistency with the analyzed protocol, we will also refer to it as "delegatee".

2.4   AKAPR Protocol Flow

We begin this subsection by defining some of the notation used in the protocol. Recall that one of the principal requirements is to use only symmetric encryption primitives. In particular, the protocol requires an authenticated encryption scheme, denoted by AEnc, and a message authentication code MAC. The protocol assumes that each principal X has a shared key with the delegatee D, namely K_{XD}. This key is used to provide authenticity and integrity of communications between the principal and the delegatee. In addition, it also assumes that both initiator and responder maintain two counters for the communication between them, in order to protect against replay attacks. These counters are labeled CT_{IR} and CT_{RI}, respectively. Additionally, nonces N_I and N_R are used to ensure freshness of messages; a session identifier SID also contributes towards this goal. Table 1 summarizes the notation used in this paper.

Table 1. Notation

Symbol	Description
I	Initiator
R	Responder
D	Delegatee
KDC	Key Distribution Center
CT_{IR}, CT_{RI}	Counters
SID	Session identifier
N_I, N_R	Nonces
H	Hash function
KDF	Key derivation function
AEnc, ADec	Authenticated encryption/decryption primitives
MAC	Message authentication code
K_{ID}, K_{RD}	Pre-shared keys with the delegatee
$\mathbb{Z}_p$	Multiplicative group of integers modulo p (i.e., $\{1, ..., p-1\}$)

The AKAPR protocol is composed of 4 messages. In the the first two messages, the delegatee acts as intermediary between the initiator and responder, while the third and fourth messages are directly between them. Figure 2 shows the flow of messages of the protocol. Next, we describe these messages in detail.

Message 1 ($I \rightarrow D$). The first message of the protocol, M_1, is created by an initiator I and sent to the responder R, via the delegatee D. The initiator performs the following steps:

1. $CT_{IR} \leftarrow CT_{IR} + 1$
2. $SID \leftarrow H(id_I \| id_R \| w)$, for a random w

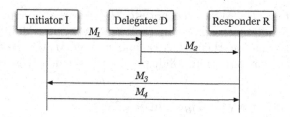

Fig. 2. Protocol flow

3. $N_I \xleftarrow{R} \mathbb{Z}_p, \quad t \xleftarrow{R} \mathbb{Z}_p$
4. $AK \leftarrow \mathsf{KDF}(id_I, id_R, t)$
5. $AE_1 \leftarrow \mathsf{AEnc}_{AK}(SID\|id_I\|id_R\|N_I\|CT_{IR})$
6. $C_1 \leftarrow t \cdot h(sk_I\|id_R)$
7. $\bar{M}_1 \leftarrow SID\|id_I\|id_R\|AE_1\|C_1$
8. $M_1 \leftarrow \bar{M}_1\|\mathsf{MAC}_{K_{ID}}(\bar{M}_1)$

Message 2 ($D \rightarrow R$). Once the delegatee D receives a message from an initiatior I, he performs the following procedure in order to verify the validity of the message and to generate the message for responder R, namely M_2.

1. Verify that SID is not repeated and that MAC in M_1 with key K_{ID}; if failed, abort the protocol.
2. $C_2 \leftarrow rk_{I \rightarrow R} \cdot C_1$
3. $\bar{M}_2 \leftarrow SID\|id_I\|id_R\|AE_1\|C_2$
4. $M_2 \leftarrow \bar{M}_2\|\mathsf{MAC}_{K_{RD}}(\bar{M}_2)$

Note that after the re-encryption performed by the delegatee, the value of C_2 is $t \cdot h(sk_R\|id_I)^{-1}$.

Message 3 ($R \rightarrow I$). When the responder R receives the previous message from the delegatee D, he also verifies its validity, proceeds to decrypt it and extract the necessary information, as described below. Finally, he produces a new message M_3 that is sent directly to the initiator I, without intermediaries. Note that at the end of this process, the responder already knows the agreed session key K_S.

1. Verify MAC in M_2 with key K_{RD}; if failed, abort the protocol.
2. $t \leftarrow h(sk_R\|id_I) \cdot C_2$
3. $AK \leftarrow \mathsf{KDF}(id_I, id_R, t)$
4. $SID\|id_I\|id_R\|N_I\|CT_{IR} \leftarrow \mathsf{ADec}_{AK}(AE_1)$
5. Check that $CT_{IR} \geq CT_{RI}$; if failed, abort the protocol.
6. $CT_{RI} \leftarrow CT_{IR} + 1$
7. $N_R \xleftarrow{R} \mathbb{Z}_p$
8. $K_S \leftarrow \mathsf{KDF}(CT_{RI}, id_I, id_R, N_I, N_R)$
9. $M_3 \leftarrow \mathsf{AEnc}_{AK}(SID\|id_R\|id_I\|N_I\|t\|N_R\|CT_{RI})$

Message 4 ($I \rightarrow R$). When the initiator I receives the response from the responder R, he follows the procedure below to verify its validity and extract the necessary information. He produces a final message M_4 that is sent directly to the responder R, who verifies its validity using the previously generated session key K_S.

1. $SID\|id_R\|id_I\|N_I\|t\|N_R\|CT_{RI} \leftarrow \mathsf{ADec}_{AK}(M_3)$
2. Check that SID, N_I and t correspond to the original ones sent in M_1, and that $CT_{RI} = CT_{IR} + 1$; if failed, abort the protocol.
3. $K_S \leftarrow \mathsf{KDF}(CT_{RI}, id_I, id_R, N_I, N_R)$
4. $M_4 \leftarrow SID\|\mathsf{MAC}_{K_S}(SID\|id_I\|id_R\|N_I\|N_R)$

As a final remark, we note that the authors do not describe any procedure for the initial key distribution.

3   Attacks to the AKAPR Protocol

In this section we describe several attacks and weaknesses of the AKAPR protocol. Most of them stem from the use of encryption tokens as intermediate values for encryption and decryption, as well as the possibility of obtaining an encryption token associated to another user by means of the delegatee or analysis of past protocol traffic. This is particularly problematic in the event of compromise of long-term secret keys, to the point that even key revocation and update does not allow the affected user to recover the guarantees of secrecy and authentication with respect to other users. In the Appendix, we also show how the choice of an insecure keyed-hash construction can lead to length-extension attacks, which combined with our previous attacks, can be used to mount complex attack scenarios.

3.1   Breaking Forward Secrecy

A first and simple attack to this protocol is to recover previous session keys from past traffic, once the long-term secret of a user is leaked. The security goal we are breaking in this case is *forward secrecy*, formally defined as follows:

Definition 1 (Forward secrecy [3]**).** *A protocol provides forward secrecy if compromise of the long-term keys of a set of principals does not compromise the session keys established in previous protocol runs involving those principals.*

For this type of attack, we assume that an attacker has collected the messages of the protocol for one or several runs, and that, at a latter stage, he is able to get access to the secret key of the responder. The goal is to recover previous session keys. In the case of the AKAPR, the attacker proceeds as follows:

1. The attacker stores the protocol messages for one or several rounds. For simplicity, let us assume he stores the traffic for only the run he wants to extract its session key.

2. At some point, the attacker compromises the responder R and obtains his long-term secret key sk_R.
3. From message M_2 of the stored protocol run, he parses $t \cdot h(sk_R \| id_I)^{-1}$ and extracts t, since he can compute $h(sk_R \| id_I)$. The value t is used to compute the encryption key $AK \leftarrow \mathsf{KDF}(id_I, id_R, t)$.
4. From message M_3, the attacker can decrypt the values N_R, N_I and CT_{IR} using the key AK computed in the previous step.
5. The attacker computes the session key K_S used during the stores run as described in the protocol:

$$K_S = \mathsf{KDF}(CT_{RI}, id_I, id_R, N_I, N_R)$$

Therefore, it can be seen this protocol does not fulfill forward secrecy. We note that if the attacker compromises the initiator I instead, then the attack is the same except for step 3, where he uses sk_I to extract t from $t \cdot h(sk_I \| id_R)$, contained in message M_1.

Countermeasure. The messages M_1 and M_2 are transmitted over an authenticated channel using MACs, with pre-shared MAC keys K_{ID} and K_{RD}, respectively. A possible countermeasure to the previous attack is to assume additional encryption keys for securing the confidentiality of the channel, given that the assumption of pre-shared keys already exists. However, if attackers were able to compromise the secret key of one user, it is reasonable to assume that the corresponding MAC key is potentially leaked too.

3.2   Key Compromise Impersonation Attacks

The previous subsection was devoted to attacking forward secrecy, a common concern of protocol designers. There are, however, other kinds of attacks that are not that well known, but can be potentially more hazardous. Key compromise impersonation (KCI) attacks occur when an adversary gains access to the secret key of a principal, and uses it to establish a session with him impersonating a different user. The attacker may use this session to actively gain new knowledge about the victim or causing him harm (e.g., by sending him malware), as the victim believes he is communicating with a legitimate user [4]. Therefore, KCI attacks can be more dangerous than breaking forward secrecy, which is limited to passive eavesdropping of past and future traffic. The following is a more formal definition of KCI, adapted from [20].

Definition 2 (Key compromise impersonation). *A key agreement protocol is vulnerable to key compromise impersonation (KCI) if compromise of the long-term key of a specific user allows the adversary to establish a session key with that user by masquerading as a different user.*

Here we describe three types of KCI attacks, although it may be possible that other variations exist. For the first two KCI attacks, we assume that the attacker compromises the secret key of one user and tries to impersonate an initiator I

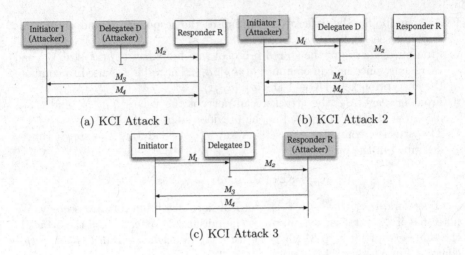

(a) KCI Attack 1   (b) KCI Attack 2

(c) KCI Attack 3

Fig. 3. Flow of KCI attacks

(which may be any user in the system chosen by the attacker); therefore, the victim acts as the responder R. Since the attacker needs to direct the attack via the delegatee, he needs at least one of the pre-shared MAC keys: either the key K_{RD} between the delegatee D and the responder R, or the key K_{ID} between the delegatee and the impersonated user I. The two first KCI attacks differ on which MAC key is compromised. Finally, we also identify a third KCI attack that occurs in the opposite situation, when the attacker tries to impersonate a responder. Although in this case the attacker has to wait for a key agreement request from the victim (which can be forced by an out-of-band action), this attack is much easier to achieve, since it does not require any MAC key.

KCI Attack 1. If the attacker knows sk_R and K_{RD} then the strategy to impersonate an initiator I is to produce a message M_2 (i.e., the second message of the AKAPR protocol, which is between the delegatee and the responder), since this message does not depend on any secret from user I; the only requirement is to chose a counter value high enough. The responder R will accept this message as coming from user I via the delegatee, as depicted by Fig. 3b, since its authenticity can be checked with K_{RD}. The attacker now only has to continue the protocol to establish a session key between him and R, although R thinks the session is between him and I. This attack works even in the attacker does not know sk_R, but only $h(sk_R\|id_I)$. The initial assumption of the attacker knowing K_{RD} is reasonable given he has access to sk_R.

KCI Attack 2. If the attacker knows sk_R and K_{ID} then the attacker needs to know also $h(sk_I\|id_R)$ in order to impersonate an initiator I. There are different ways to achieve this:

– A first option is to use past protocol traffic between R and I in order to extract this value, following a strategy similar to the forward secrecy attack:

Knowing sk_R enables the attacker to compute the t value from the messages of a protocol run, which in turn, can be used to extract $h(sk_I\|id_R)$ from message M_1.

- A second option is to trick the delegatee into delivering this value by means of the re-encryption function, basically using it as a re-encryption oracle[2]. In order to do this, the attacker initiates a key agreement protocol between R and I, where C_1 is the component of message M_1 with value $t \cdot h(sk_R\|id_I)$. Next, he captures the response M_2 of the delegatee and parses the re-encrypted component C_2. Finally, he computes $h(sk_I\|id_R) = t \cdot (C_2)^{-1}$.
- A third option is to assume that the attacker colludes with the delegatee (or even that the attacker *is* the delegatee). In this case, $h(sk_I\|id_R)$ can be computed from the re-encryption keys since $rk_{I \to R} = h(sk_I\|id_R)^{-1}h(sk_R\|id_I)^{-1}$. In addition, the assumption of knowing K_{ID} is natural.

Once the attacker knows $h(sk_I\|id_R)$, he initiates a normal key agreement protocol with R via the delegatee, as shown in Fig. 3b, using K_{ID} for computing the MAC for the first message. Note that the attacker can participate successfully in this key agreement since he does not need to know the secret sk_I, but the encryption token $h(sk_I\|id_R)$. As a final comment on this attack, note that it works even if the attacker initially does not know sk_R, but only the encryption token $h(sk_R\|id_I)$.

KCI Attack 3. Previous KCI attacks supposed that the leaked secret was of a user who later acted as the responder. Suppose now that the secret key of an initiator I is leaked. Figure 3c shows the protocol flow of this attack, where the attacker acts as the responder R. An attacker that knows the secret key sk_I of the initiator can impersonate any user R if he is able to obtain $h(sk_R\|id_I)$. As in the KCI attack 2, there are several ways to do this:

- Previous traffic between I and R. Knowing sk_I enables the attacker to compute the t value from the messages of a protocol run, which in turn, can be used to compute $h(sk_R\|id_I)$ from message M_2.
- Using the delegatee as a re-encryption oracle. This requires knowledge of K_{ID}.
- Assuming that the attacker controls the delegatee, colludes with it, or the corresponding re-encryption key is leaked somehow.

Once the attacker knows $h(sk_R\|id_I)$, he can respond to normal key agreement requests from I. As in previous KCI attacks, the attacker succeeds without actually knowing the secret sk_R, but the encryption token $h(sk_R\|id_I)$; similarly, this attack works even if the attacker only knows the encryption token $h(sk_I\|id_R)$ at the beginning.

As a final remark, KCI attacks 1 and 2 required the additional knowledge of one of the pre-shared MAC keys, since in both of them the attacker impersonated an initiator, whose messages are required to be validated, and therefore,

---

[2] Note that in the traditional proxy re-encryption literature (i.e., in the public-key setting), the re-encryption oracle is considered as a delicate point. See for example [17], where several generic attacks using the re-encryption oracle are discussed.

is performing a proactive impersonation. On the contrary, in this KCI attack the attacker impersonates a responder, which implies that the impersonation is reactive in this case (i.e., it requires that the initiator I starts the key agreement). This can be achieved either by waiting for a key agreement request to occur or by forcing it using some out-of-band mechanism (e.g., hard-resetting the initiator's device, social engineering, etc.).

3.3   Limited Scope of Key Revocation and Update

An interesting, yet not immediate, takeaway of the previous attacks is that they exploit the following undesired properties of the protocol: (1) encryption tokens can replace long-term secret keys, and (2) associated encryption tokens (e.g., $h(sk_I\|id_R)$ and $h(sk_R\|id_I)$) can be linked with each other through valid protocol messages. These issues are problematic when long-term secret keys are compromised, as illustrated by the attacks we identified.

A natural action that the affected principal performs once he becomes aware of the compromise of his secret key, is to initiate some kind of key revocation/update procedure. In principle, one can believe that key revocation is an effective countermeasure against a long-term key compromise event. However, we show next that this is not the case: once the long-term key of a user is compromised, and even after key revocation and update is realized, the protocol still remains vulnerable with respect to forward secrecy and key compromise impersonation, for all users whom the affected user communicated before. This is a consequence of the two undesired properties described above.

Breaking Forward Secrecy. As an illustration, suppose that the responder R updates his secret key to sk'_R after his previous long-term secret sk_R is exposed, revokes previous re-encryption keys and asks for their update. In order for the protocol to be correct, the new re-encryption keys should be of the form $rk'_{I \to R} = h(sk_I\|id_R)^{-1} \cdot h(sk'_R\|id_I)^{-1}$, for all possible initiators I. Note that the term $h(sk_I\|id_R)$ does not change with respect to the previous re-encryption key, and therefore, the initiator I still uses this same encryption token when participating in a key agreement with the responder R. The consequence of this is that an attacker that was able to break forward secrecy before can extract $h(sk_I\|id_R)$ from previous rounds of the protocol, since $h(sk_I\|id_R)$ has not changed. Using this encryption token, the attacker can compute the corresponding t value from the first message of the protocol (i.e., M_1) of any future round, and break forward secrecy again.

Key Compromise Impersonation. Suppose now that it is an initiator I who updates his secret key to sk'_I. Analogously to the previous case, it is still possible for an attacker to recover $h(sk_R\|id_I)$ from previous traffic between I and any other user R. However, in this case, the attacker does not limit himself to passively eavesdrop protocol messages as above (i.e., breaking forward secrecy), but can actively impersonate the responder R, given that he knows the value $h(sk_R\|id_I)$ necessary to decrypt message M_2. Notice how updating the secret key of I does not have any effect on the encryption token that protects the second message of the protocol.

Countermeasures. The only possible countermeasures against this problem are either to issue a new identity for the affected user or to revoke also the users that established communications with him previously. Both options do not seem adequate nor practical.

4 Discussion

The previous attacks demonstrated that the main weaknesses of the AKAPR protocol is that the encryption tokens are, in fact, long-term secrets, just as the secret keys, and that the encryption tokens between two users can be linked to each other. These problems make it possible for an attacker to perpetually threaten the security goals of the protocol once he gains control of an encryption token. In this section we informally discuss some possible amendments to this protocol.

First, we note that if we assume that the initiator I is also capable of reading Message 2 (from the delegatee D to the responder R), then this means he can extract the encryption token $h(sk_R \| id_I)$. This is a plausible assumption in several settings compatible with this protocol (e.g., wireless environments), and it is also consistent with the philosophy of the Dolev-Yao model [8], by which one can consider the network to be open and all its messages public and subject to scrutiny by other entities. Therefore, it can be seen that this is functionally equivalent to the initiator I knowing this encryption token. Given that this encryption token is implicitly known by the responder (since he can generate it), then it acts as a sort of shared key between them. Therefore, a possible variation of the protocol is to simply distribute this encryption token to the initiator when he wants to commence a key agreement with responder R. This can be realized by transforming the delegatee D into a mere key server that distributes these encryption tokens to initiators. Therefore, instead of the delegatee re-encrypting Message 1 into Message 2, which requires it to know a re-encryption key linking two encryption tokens, he stores the "shared encryption tokens".

Note that this variation is still compatible with achieving the goal of secrecy with respect to the delegatee, simply by encrypting these encryption tokens with the secret key of the corresponding initiator. Therefore, re-encryption keys stored by the delegatee D are replaced by encryptions of encryption tokens. Specifically, each $rk_{I \to R}$ is substituted by an encrypted key $ek_{I \to R} = \mathsf{Enc}_{sk_I}(h(sk_R \| id_I))$. Note that this implies that $ek_{I \to R} \neq ek_{R \to I}$, and therefore, this doubles the number of keys managed by the delegatee D; nevertheless, this number is still of quadratic order (since it is necessary to store a key between each pair of users of the system), dominated by the number of users. This new type of keys eliminates the link between encryption tokens, reducing this way the applicability of some of the attacks we propose (although not completely). An interesting insight about this variation is that it is extremely similar to the traditional Needham-Schroeder symmetric key agreement protocol [14], with the main difference that in this case the key server (i.e., the delegatee) does not know the session keys because these are encrypted by a separate key server (i.e., the KDC).

However, it is important to note that this variant still suffers from the long-term nature of encryption tokens. An improvement to this respect could be to include a timestamp in the encryption token generation, in a similar way than Kerberos [15] improves over Needham-Schroeder protocol. Thus, encryption tokens would be of the form $h(sk_A\|id_B\|T_i)$, where T_i represents a time period. Although this could mitigate the attacks to a certain extent (since compromised encryption tokens would only be useful for a given time period), it also implies that the re-encryption keys should be recomputed by the KDC and distributed to the delegatee on each time period. This can represent a great inconvenience in some settings.

5   Conclusions

Proxy re-encryption (PRE) is a hot research topic nowadays, with new use cases arising in different contexts. Most PRE schemes are based on public-key cryptography, mostly because public key cryptography has become omnipresent in most environments. PRE schemes using symmetric cryptography would make an excellent contribution for IoT scenarios where devices are highly constrained, but it is true that properly capturing the essence of PRE and providing secure solutions in the same sense as in the public-key world is challenging.

In this paper we show how difficult is to properly define and implement PRE in the symmetric world by exploring the weaknesses of a particular scheme proposed at ESORICS 2016. This protocol tries to adapt previous proposals for symmetric key PRE to the IoT scenario, focusing not only on providing a lighter solution (not involving public-key cryptography) but also trying to avoid redundant messages that are needed in well-know and deeply-studied authentication and key exchange protocols (e.g., Needham-Schroeder).

Unfortunately the paper fails to accomplish its goals, mainly because of the use of two immutable encryption tokens for every pair of communicating parties that depend only on the secret keys of the initiator and the identity of the responder (and vice versa). The way the protocol is designed implies that compromising the secrets of one party allows to obtain the encryption token of the other party, which can be used to mount key compromise impersonation attacks and break forward security. Moreover, revocation of compromised keys turns out to be ineffective, which makes the protocol unusable in it present form.

Acknowledgments. This work was partly supported by the Junta de Andalucía through the project FISICCO (P11-TIC-07223) and by the Spanish Ministry of Economy and Competitiveness through the PERSIST project (TIN2013-41739-R). The first author is supported by a contract from the Regional Ministry of Economy and Knowledge of Andalucía.

Appendix: Length-Extension Attacks

We now describe a completely different attack strategy, based on a potentially dangerous design choice in the AKAPR protocol. The symmetric proxy re-encryption primitive that underlies the key agreement protocol (see Sect. 2.1), uses a keyed hash function to derive the encryption tokens. Specifically, these values are of the form $h(sk_A \| id_B)$, where the identity of the recipient is used as the message of a keyed hash, with $h : \{0,1\}^* \to \mathbb{Z}_p$ as the hash function[3]. The authors state that the hash function is required to behave as a random oracle, which can be a rather strong assumption. However, it is known that real instantiations of hash functions do not necessarily behave as random oracles. In fact, widely used hash functions such as SHA-1 and SHA-256 (as well as others based on the Merkle-Damgård construction) suffer from length-extension vulnerabilities [22] that make possible to create new encryption tokens without knowing sk_A, under the assumption that identifiers can have variable length.

First, let us recall the idea of length-extension attacks. In this type of attacks, knowledge of the hash value $h(m_1)$ and the length of m_1 can be used to compute $h(m_1 \| pad \| m_2)$, for any message m_2 chosen by the attacker and some required internal padding pad. It is known that hash functions based on the Merkle-Damgård construction (e.g., MD5, SHA-1, SHA-256) are vulnerable to this type of attacks. This problem has been shown to permeate to real systems, such as the Flickr's API signature forgery vulnerability [9].

In the AKAPR protocol, if we assume that identifiers can have different lengths (i.e., $id \in \{0,1\}^*$), then it may be possible that, for some identity id_B, there exists an identity id_{BX} where id_B is a prefix. For example, let us assume that $id_B =$ "Bob" and that the attacker knows $h(sk_A \| id_B)$, but that sk_A remains secret. Then, by the length-extension vulnerability he can produce $h(sk_A \| id_{BE})$, where $id_{BE} =$ "Bob Esponja"[4]. For the sake of illustration, let us ignore the internal padding pad for the moment.

As discussed in Sect. 3.2, an attacker that knows $h(sk_A \| id_{BE})$ can impersonate A in a key agreement with BE without knowing sk_A, regardless if he acts as an initiator or a responder. The length-extension attack is particularly worrisome when the attacker has access to previous traffic, or the delegatee is compromised or deceived by the attacker; in this case, the delegatee can be used in combination with the attack strategies presented in previous sections in order to compromise the security of non-involved users. For example, let us suppose that an attacker gains access to the private key of user B. Figure 4 shows a combination of some of the attacks described in this paper that eventually compromise the security of key agreements between users A and BE. An attacker that knows sk_B can

---

[3] Note that when we assume that h is implemented with a traditional hash function, it is also necessary an additional encoding from its output domain to $\mathbb{Z}_p$, as required in Sect. 2.1; usual encodings of this type (e.g., taking modulo p) are easily invertible, so we will omit this for simplicity.

[4] "Bob Esponja" is the Spanish name for "SpongeBob SquarePants" [24]. Not associated with the sponge-based hash function SHA3 [1], which, interestingly, is not susceptible to length-extension attacks.

trivially generate encryption tokens $h(sk_A\|id_B)$ associated to any other user A; knowing this, he can compute the opposite encryption tokens using strategies discussed in previous attacks (e.g., using the delegatee as an encryption oracle, compromising the delegatee, analyzing previous traffic between A and B). The resulting encryption tokens can be used as input to length-extension attacks, undermining this way the security of other users. The output of the length-extension attacks are also encryption tokens, which in turn, can be used again to obtain more encryption tokens. Therefore, it can be seen how the initial leakage of B's secret can potentially affect key agreements between other users, since once the attacker obtains $h(sk_A\|id_{BE})$ and $h(sk_{BE}\|id_A)$, he can potentially control all their past and future key agreements. Note that, although BE is some user whose identity has id_B as prefix (i.e., the choice of BE is not completely free), A is an arbitrary user, which is a serious threat to the application of this protocol in a real setting.

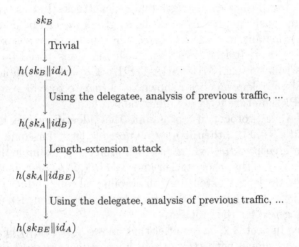

Fig. 4. Abstract flow of a length-extension attack

We next describe some possible attack scenarios that apply this strategy in combination with others of the previously identified attacks. In particular, we illustrate two scenarios: the first is reminiscent of sybil attacks in distributed networks, where a malicious entity gains control of multiple identities in order to increase its control over a system or network; the second is similar to a spear phishing attack, where the attacker targets specific victims aided by relevant, yet fake, contextual information that gains the trust of the victim.

Sybil-Like Attack Scenario. The goal of the attacker in this scenario is to gain as much encryption tokens as possible, each of them associated to a different identity, and to start multiple key agreements with a victim A, who will believe is communicating with different entities. In particular, the idea is to obtain

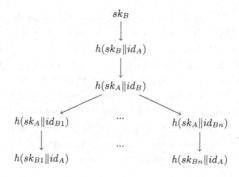

Fig. 5. Abstract flow of a sybil attack

encryption tokens between A and fresh "pseudonyms" (all of them with the same prefix in order to exploit the length-extension vulnerability, as described before).

We initially assume either that the attacker is a legitimate user (e.g., user B), or that he gained access to the secret key sk_B of some user B; this does not matter from the point of view of the attack. Now, he can set up a sybil attack against any user A of his choosing as follows, also represented in Fig. 5. First, knowledge of sk_B allows to compute $h(sk_B\|id_A)$ trivially; next, using the techniques described for previous attack strategies, the attack obtains the opposite encryption token $h(sk_A\|id_B)$. Now he proceeds to launch the length-extension attack against this hash value, by appending random i values to the end of id_B, obtaining as a result hashes of the form $h(sk_A\|id_B\|pad\|i)$. Therefore, fresh pseudonyms will be identities of the form $id_{Bi} = id_B\|pad\|i$. Note that in this case the internal padding pad that is necessary for the length-extension attack is not necessarily a problem (e.g., if identities have a numerical format); the only important requirement from the point of view of the sybil attack is that identities id_{Bi} are different from each other.

Finally, using once again the techniques for obtaining the opposite encryption token, the attacker eventually obtains a set of encryption tokens of the form $h(sk_{Bi}\|id_A)$, for different i values, which enable him to mount a sybil-like attack against entity A. The attacker can now start multiple key agreements with the victim A, making him believe that there are several independent entities B_i, which in fact are controlled uniquely by the attacker.

Spear Phishing Scenario. Suppose that an attacker gains access to sk_B, where id_B is a common but dangerous prefix (e.g., "Bank"). Once again, following the previously discussed strategies, the attacker can obtain the encryption tokens between a victim A and user B, namely $h(sk_A\|id_B)$, and later use this hash value to derive new encryption tokens involving a tailored bait whose name starts with "Bank" (e.g., "Bank of America"). Recall that the attacker does not need to know the secret of "Bank of America", but uses instead our attack strategies to obtain an encryption token that compromises the security with respect to the victim A. This type of attack also works if, instead of compromising user B, we suppose B does not initially exist and the attacker is capable of registering it with the KDC as a new user.

There is, however, a small problem with the internal padding pad in this case. Following our example, the length-extension attack would allow the attacker to compute the encryption token $h(sk_A\|\text{"Bank"}\|pad\|\text{"of America"})$. Therefore, the forged identity is actually "Bank"$\|pad\|$"of America", which can be problematic if the padding allows the victim to detect the attack. We will, however, illustrate with a real example how this problem can be overcome, and experimentally demonstrate the viability of the attack.

Table 2. Spear phishing based on a length-extension attack

Original encryption token $h(sk_A\|\text{"Bank"})$	
Hash input (string)	"????????????????????????????????Bank"
Hash input (hex)	3F 3F 3F 3F 3F 3F 3F 3F 3F 3F 3F 3F 3F 3F 3F 3F 3F 3F 3F 3F 3F 3F 3F 3F 3F 3F 3F 3F 3F 3F42 61 6E 6B
Hash output (hex)	D2 7A D9 E5 FD FC 84 3E 8F 73 74 05 72 04 1D 80 72 48 F2 58 09 06 04 BF 5A 38 AA B7 B5 74 C8 CB
"Extended" encryption token $h(sk_A\|\text{"Bank"}\|pad\|\text{"of America"})$	
"Bank"$\|pad\|$"of America" (hex)	42 61 6E 6B 80 00 00 00 00 00 00 00 00 00 00 00 00 00 00 00 00 00 00 00 00 00 00 00 00 00 00 00 01 20 6F 66 20 41 6D 65 72 69 63 61
"Bank"$\|pad\|$"of America" ('latin1')	"Bank of America"
Hash output (hex)	4F FA C6 1B A9 9F F2 AD 28 93 92 21 BD 7D 12 CF F4 01 BA C5 5F 41 C3 FB 64 41 F7 45 EE 17 5C A8

Suppose that the protocol uses SHA-256 as hash function and that secret keys are of 256 bits. In our experimental attack, the secret key is the hexadecimal value 3F (which corresponds to the character '?') repeated 32 times; note that the specific value of this key is irrelevant for the attack and that it always remains unknown to the attacker. Table 2 shows the concatenation of the secret key and the identity "Bank", represented both as text and as hexadecimal strings, as well as the output of the hash function for this input. We suppose now that the attacker knows this output value (i.e., the encryption token $h(sk_A\|\text{"Bank"})$); knowing this, the attacker can use the length-extension vulnerability of SHA-256 to compute the encryption token $h(sk_A\|\text{"Bank"}\|pad\|\text{"of America"})$.

The concrete result for the length-extension attack applied to the previous values is also presented in Table 2. Note that the first four bytes of the extended identity correspond to the string "Bank" (42 61 6E 6B). Next, there is the internal padding *pad* (80 00 ... 00 01 20), followed by the characters that correspond to the extension string "of America" (6F 66 ... 63 61). It should be noted that the intermediate padding may be rendered differently depending on the implementation of the user agent (e.g., browsers, mobile apps, stand-alone GUI, etc.), and some options may be exploitable. For example, when the extended identity "Bank"$\|pad\|$"of America" is rendered with the ISO 8859-1 encoding (also called 'latin1'), the intermediate padding may be ignored, depending on the implementation, since it is composed of NUL characters (00) and invalid codes (80). This is the case of Python 2.7.10 implementation, which we used for reproducing the length-extension attack, and that displays "Bank of America" as the extended identity when 'latin1' encoding is selected. This can be used to deceive users into believing they are interacting with the real Bank of America, since the only difference between the legitimate identity and the displayed extended identity are non-printable characters (i.e., the user cannot visually distinguish one string from another). This also can be reproduced with UTF-8 encoding if errors are ignored when printing UTF-8 encoded strings (in Python this is achieved by specifying the option `ignore`).

We note that this example, although somewhat artificial, only makes relatively common assumptions, such as the use of SHA256, secret keys of 256 bits, 'latin1' encoding of strings, variable-size identities, etc.

Countermeasures. As a simple and effective countermeasure to the length-extension attack, the protocol can require the use of a hash function resistant to length-extension attacks (e.g., SHA3), or even better, use an HMAC or a key derivation function (e.g., HKDF [11]). HMACs and KDFs are specifically designed to take a secret as input, and to be secure against length-extension attacks.

References

1. Bertoni, G., Daemen, J., Peeters, M., Van Assche, G.: The Keccak SHA-3 submission. NIST (Round 3) **6**(7), 16 (2011, to be submitted)
2. Boneh, D., Lewi, K., Montgomery, H., Raghunathan, A.: Key homomorphic PRFs and their applications. In: Canetti, R., Garay, J.A. (eds.) CRYPTO 2013. LNCS, vol. 8042, pp. 410–428. Springer, Heidelberg (2013). doi:10.1007/978-3-642-40041-4_23
3. Boyd, C., Mathuria, A.: Protocols for Authentication and Key Establishment. Springer, Heidelberg (2013)
4. Chalkias, K., Baldimtsi, F., Hristu-Varsakelis, D., Stephanides, G.: Two types of key-compromise impersonation attacks against one-pass key establishment protocols. In: Filipe, J., Obaidat, M.S. (eds.) ICETE 2007. CCIS, vol. 23, pp. 227–238. Springer, Heidelberg (2008). doi:10.1007/978-3-540-88653-2_17
5. Cook, D.L., Keromytis, A.D.: Conversion functions for symmetric key ciphers. J. Inf. Assur. Secur. **1**(2), 119–128 (2006)
6. Denning, D.E., Sacco, G.M.: Timestamps in key distribution protocols. Commun. ACM **24**(8), 533–536 (1981)

7. Diffie, W., Hellman, M.E.: New directions in cryptography. IEEE Trans. Inf. Theor. **22**(6), 644–654 (1976)
8. Dolev, D., Yao, A.: On the security of public key protocols. IEEE Trans. Inf. Theor. **29**(2), 198–208 (1983)
9. Duong, T., Rizzo, J.: Flickr's API signature forgery vulnerability (2009)
10. Garrison, W.C., Shull, A., Myers, S., Lee, A.J.: On the practicality of cryptographically enforcing dynamic access control policies in the cloud. In: 2016 IEEE Symposium on Security and Privacy (SP), pp. 819–838, May 2016
11. Krawczyk, D.H., Eronen, P.: HMAC-based extract-and-expand key derivation function (HKDF). RFC. 5869, October 2015
12. Liu, Z., Huang, X., Hu, Z., Khan, M.K., Seo, H., Zhou, L.: On emerging family of elliptic curves to secure internet of things: ECC comes of age. IEEE Trans. Dependable Secur. Comput. **14**(3), 237–248 (2016). doi:10.1109/TDSC. 2016.2577022. ISSN: 1545-5971
13. Menezes, A.J., Van Oorschot, P.C., Vanstone, S.A.: Handbook of Applied Cryptography. CRC Press, Boca Raton (1996)
14. Needham, R.M., Schroeder, M.D.: Using encryption for authentication in large networks of computers. Commun. ACM **21**(12), 993–999 (1978)
15. Neuman, B.C., Ts'o, T.: Kerberos: an authentication service for computer networks. IEEE Commun. Mag. **32**(9), 33–38 (1994)
16. Nguyen, K.T., Oualha, N., Laurent, M.: Authenticated key agreement mediated by a proxy re-encryptor for the internet of things. In: Askoxylakis, I., Ioannidis, S., Katsikas, S., Meadows, C. (eds.) ESORICS 2016. LNCS, vol. 9879, pp. 339–358. Springer, Cham (2016). doi:10.1007/978-3-319-45741-3_18
17. Nuñez, D., Agudo, I., Lopez, J.: A parametric family of attack models for proxy re-encryption. In: Proceedings of the 28th IEEE Computer Security Foundations Symposium, CSF 2015, pp. 290–301. IEEE Computer Society (2015)
18. Nuñez, D., Agudo, I., Lopez, J.: Proxy re-encryption: analysis of constructions and its application to secure access delegation. J. Netw. Comput. Appl. **87**, 193–209 (2017)
19. Sakazaki, H., Anzai, K., Hosoya, J.: Study of re-encryption scheme based on symmetric-key cryptography. In: 31st Symposium on Cryptography and Information Security (SCIS 2014) (2014)
20. Strangio, M.A.: On the resilience of key agreement protocols to key compromise impersonation. In: Atzeni, A.S., Lioy, A. (eds.) EuroPKI 2006. LNCS, vol. 4043, pp. 233–247. Springer, Heidelberg (2006). doi:10.1007/11774716_19
21. Syalim, A., Nishide, T., Sakurai, K.: Realizing proxy re-encryption in the symmetric world. In: Abd Manaf, A., Zeki, A., Zamani, M., Chuprat, S., El-Qawasmeh, E. (eds.) ICIEIS 2011. CCIS, vol. 251, pp. 259–274. Springer, Heidelberg (2011). doi:10.1007/978-3-642-25327-0_23
22. Tsudik, G.: Message authentication with one-way hash functions. ACM SIGCOMM Comput. Commun. Rev. **22**(5), 29–38 (1992)
23. Watanabe, D., Sakazaki, H., Miyazaki, K.: Representative system and security message transmission using re-encryption scheme based on symmetric-key cryptography. J. Inf. Process. **25**, 67–74 (2017)
24. Wikipedia: SpongeBob SquarePants – Wikipedia, the free encyclopedia (2016). https://en.wikipedia.org/wiki/SpongeBob_SquarePants. Accessed 18 Oct 2016

Improving Resilience of Behaviometric Based Continuous Authentication with Multiple Accelerometers

Tim Van hamme$^{(\boxtimes)}$, Davy Preuveneers, and Wouter Joosen

imec-DistriNet-KU Leuven, Leuven, Belgium
{tim.vanhamme,davy.preuveneers,wouter.joosen}@cs.kuleuven.be

Abstract. Behaviometrics in multi-factor authentication schemes continuously assess behavior patterns of a subject to recognize and verify his identity. In this work we challenge the practical feasibility and the resilience of accelerometer-based gait analysis as a behaviometric under sensor displacement conditions. To improve misauthentication resistance, we present and evaluate a solution using multiple accelerometers on 7 positions on the body during different activities and compare the effectiveness with Gradient-Boosted Trees classification. From a security point of view, we investigate the feasibility of zero and non-zero effort attacks on gait analysis as a behaviometric. Our experimental results with data from 12 individuals show an improvement in terms of EER with about 2% (from 5% down to 3%), with an increased resilience against observation attacks. When trained to defend against such attacks, we observe no decrease in classification performance.

1 Introduction

Multi-factor authentication schemes are adopting behavioral biometrics (or behaviometrics) [3] to continuously verify in the background the identity of users by leveraging information about the user's device [21,22], context or the user's behavior [5,11] within that context. These trends are often referred to as *Active Authentication*, also known as *Context-aware* [9], *Continuous* [19] or *Implicit* [20] *Authentication*. The key challenges that these authentication schemes aim to address are (1) the ability to conveniently and reliably authenticate the identity of a user, and (2) to continuously assess the confidence in the user's identity.

One well-known behaviometric is gait recognition [10,17] using accelerometer data to analyze motion patterns. While this technique is hardly new, several challenges from a practical feasibility and security point of view remain: (1) there has been little research that investigates the practical resilience of such schemes against sensor displacement, (2) reported high recognition rates were only achieved in a controlled setup where the test subjects are known to walk, making it difficult to ascertain the accuracy − or even the misauthentication resistance − under other conditions and motion activities, and (3) the feasibility and effectiveness of zero and non-zero effort attacks against gait analysis.

G. Livraga and S. Zhu (Eds.): DBSec 2017, LNCS 10359, pp. 473–485, 2017.
DOI: 10.1007/978-3-319-61176-1_26

To the best of our knowledge, we are the first to evaluate the effectiveness of using multiple accelerometers to collectively further improve this type of authentication scheme. Additionally, in this work, we enhance the resilience against the above threats with the following contributions:

- We investigate the effectiveness of accelerometers on 9 different places on the body, and analyze the impact of different human activities on the EER.
- We research whether multiple accelerometers can enhance misauthentication resistance, report on the use of different machine learning algorithms, and discuss which combination of on-body positions is the most effective.
- We evaluate a solution that relies on a common set of features, rather than a unique set for each type of activity, to improve classification robustness under diverse circumstances and motion activities.
- We evaluate our authentication scheme against zero-effort and non-zero effort attacks, and compare the results against single accelerometer schemes.

We evaluate our research on the public REALDISP benchmark dataset[1] that was previously collected to evaluate sensor displacement in activity recognition [2,4]. Based on a study with data from 12 individuals, our results show a recognition improvement reducing the equal error rate (EER) from 5% down to 3%, with an increased resilience against observation and spoofing attacks.

The remainder of this paper is structured as follows. In Sect. 2, we discuss related work on accelerometer based gait authentication. Section 3 presents our multi-sensor approach. In Sect. 4, we describe the experiments and results. We conclude in Sect. 5 summarizing our main insights and discussing future work.

2   Related Work

This section reviews relevant research on gait authentication schemes, summarizing the EER and recognition accuracy results in Table 1.

Table 1. Comparisng the EER and recognition rate of gait authentication schemes

Related work	# Sensors	Body position	EER	RR
Mantyjarvi et al., 2005 [13]	1	Hip	7%	88%
Gafurov et al., 2007 [7]	1	Hip (back pocket)	7.3%	86.3%
Annadhorai et al., 2008 [1]	1	Ankle	?	84%
Derawi et al., 2010 [6]	1	Hip	20%	?
Nickel et al., 2011 [16]	1	Hip (phone in pouch)	10.3%	?
Ngo et al., 2014 [15]	5 (1 at a time)	Hip, back	14.3%	?
Lu et al., 2014 [12]	1	Different positions	14%	?

---

[1] https://archive.ics.uci.edu/ml/datasets/REALDISP+Activity+Recognition+Dataset.

Mantyjarvi et al. [13] investigated the feasibility of using gait signals for identification using correlation, frequency domain and distribution statistics. For 36 subjects wearing the accelerometer on 2 different days, correlation proved to be the best method, obtaining a 7% EER and a 88% recognition rate (RR). Similar work by Gafurov et al. [7] compared absolute distance, correlation, histogram, and higher order moments to evaluate performance of the system both in authentication and identification modes. Their analysis on 50 subjects showed that the distance metric had the best performance with an EER of 7.3%, and a recognition rate of 86.3%. Annadhorai et al. [1] identified subjects from gait cycles using k-Nearest Neighbor classification. Features were extracted for each gait cycle from accelerometer (3D), pitch and roll data. A subject was identified with an accuracy of 84%. However, these results were obtained on a relatively small data set, with only 2 walks from 4 different subjects. Derawi et al. [6] tested the feasibility of gait as a behaviometric by using the accelerometer in a smartphone. During an enrollment phase the average gait cycle is determined. Two gait cycles are compared using Dynamic Time Wrapping (DTW). An EER of 20% was achieved on a dataset containing 51 subjects, with two walks per subject. Contrary to previous works, Nickel et al. [16] did not rely on extracting gait cycles to calculate feature vectors, but used Hidden Markov Models to classify gait patterns of 48 subjects. They reported a False Reject Rate (FRR) of 10.42% at a False Acceptance Rate (FAR) of 10.29% (or an EER of ≈10.3%). A large scale experiment was conducted by Ngo et al. [15] with 744 subjects between 2 to 78 years old, walking under different ground slope conditions. They verified four different gait based authentication methods. The authors conclude that the maturity of the subject's walking ability and the slope greatly influence the performance of gait based user authentication. Lu et al. [12] describe a gait verification system based on Gaussian Mixture Model - Universal Background Model (GMM-UBM) framework. The design objective was to adapt the gait model for mobile phones such that it can account for different body placements and over time variance in the user's gait pattern. The UBM was trained using data from 47 different subjects, the user gait model was tested for 12 subjects. The reported EER was 14%.

3   Challenges with Gait Authentication Schemes

This section identifies challenges and the gap that we aim to bridge when using accelerometer based gait recognition as a behaviometric in real life scenarios.

3.1   Different Body Positions and Sensor Displacement

Most people own at least one mobile device, with different types of sensors. In the future even more sensors will be attached to our body, in the form of smart watches, activity trackers, smart shoes or even smart clothes. Therefore, there is an opportunity to research what positions on the body are the most characterizing and effective for authentication purposes.

However, most of these devices are not fixed at all times to a certain place on our body. They do have an area where they are normally located, but their exact placement varies from time to time, e.g. changing your smartphone from your left to your right pocket, or wearing pants with entirely different pockets. These subtle sensor displacements in the real world, will have an impact on the classification accuracy, and hence the effectiveness of accelerometer based gait authentication schemes.

3.2   Misauthentication Resistance Under Different Motion Activities

Walking is not the only predominant activity in human life. We are sitting, running, cycling, climbing stairs, etc. as well. A behaviometric should be able to deal with different types of activities. The related work showed that most techniques (1) assume that people are walking and do not consider other activities; (2) explicitly exploit gait cycles to extract features: their first step is always to discover the gait cycle and extract it from the data sample. While the first assumption is reasonable for completely different activities (Wilson et al. [24] achieved an activity classification accuracy of 95%), this is not valid for the latter. It is useless to extract gait cycles for sitting, and not straightforward to find patterns similar to gait cycles for activities like rowing, going to the gym or cycling. Moreover, the related work seemed to struggle when the walking conditions changed slightly (i.e. changing the walking speed, the type of shoes used, the amount of weight being carried, the type of surface and the inclination of the ground). While it might be possible to classify whether the wearer of the accelerometer is running or walking, and maintaining different models for both cases, it certainly is not practical to repeat this for a range of different speeds.

We therefore investigate the feasibility of a common feature set − rather than special features fitted to every particular activity − and the added value of using multiple accelerometers for behaviometric-based authentication. We will use data where the activity is known beforehand. This is reasonable because of high accuracies achieved for activity recognition in other work [24]. Based on our previous research in the field of activity recognition [18], our hypothesis is that we can obtain even higher accuracies for our use case and setting, because our solution does not rely on a fine-grained distinction between activities, as discussed in Sect. 4.4.

3.3   Security Threats and Attacker Model

To evaluate the effectiveness of the proposed scheme, we consider the impact of two different types of attacks:

- **Zero-effort attack**: the adversary is simply another subject in the database that acts as a casual impostor
- **Non-zero effort attack**: the adversary actively masquerades as someone else by mimicking and spoofing the gait pattern of the claimed identity

In the zero-effort attack, we use the data of the other subjects as negative examples for a given user to get insights into the probability of misauthentication.

A non-zero effort attack would occur when the attacker tries to obtain activity patterns of the subject (i.e. observation and spoofing). The attacker attempts to act like the subject by walking at the same pace or mimicking the characteristic activity, as investigated in [8,14], or he can try to sneak an accelerometer into the coat of the subject. To make these attacks harder to perform, we combine multiple sensors on different places on the body. This way, we collect more data to learn a subject's movement patterns, with an opportunity to further decrease the EER.

4   Evaluation

This section reports on the experiments conducted to test the concerns expressed in Sect. 3. We use the public REALDISP benchmark dataset [2,4] to enable the reproducibility of our research results. It contains 17 subjects, all performing 33 different actions, among which: walking, jogging, running, cycling, rowing, etc. All subjects wore 9 sensors on different positions. They performed the set of exercises twice: once with the sensors adjusted carefully by the makers of the dataset; and once adjusting the sensors themselves. The data collected consists of 3D accelerometer, 3D gyroscope, 3D magnetic field measurements and an estimation of the orientation using quaternions. The sampling rate is 50 Hz.

4.1   Activity-Agnostic Behaviometrics

We do not make any assumptions on a particular motion pattern (e.g. presence of gait cycles) so that our behaviometrics can be used for different types of activities.

The REALDISP dataset contains 33 different activities. For each of them we extracted features using the same approach: the data was split in intervals of 128 samples (which is ≈2.5 s). For each interval we calculated some straightforward features, in both the time and frequency domain. Among them: mean, standard deviation, kurtosis, mean average derivation, energy in the signal, average resultant vector. This led to a feature vector of length 224. Only the activities walking, jogging, running and cycling had a meaningful amount of samples (≈23) per subject. Only 12 subjects appeared to have walking, running and jogging data, of them only 9 cycled. Each subject had performed all actions twice, once with self sensor placement and once with ideal sensor placement.

With authentication in mind, we trained a model for each subject and each activity. We constructed a set which consists for 50% of samples belonging to the subject and for 50% of samples from other subjects. The samples belonging to the other subjects were sampled equally among the total distribution w.r.t. subjects. For each subject adjacent samples w.r.t. time were taken. This set was split in a training and test set using n-fold cross-validation. This process splits the set in n temporal adjacent chunks, in a stratified manner, thus taking

Table 2. Comparison of EER with different machine learning classifiers

SVM with sigmoid kernel	SVM with rbf kernel	kNN	GBT	Bagging	AdaBoost	Random forest
0.511	0.488	0.198	0.068	0.094	0.047	0.036

Table 3. EERs of ideally placed sensors

	Walking	Jogging	Running	Cycling
BACK	0.036	0.010	0.040	0.020
LC	0.034	0.033	0.022	0.010
LLA	0.076	0.060	0.029	0.032
LT	0.033	0.021	0.011	0.012
LUA	0.062	0.045	0.019	0.031
RC	0.026	0.025	0.024	0.009
RLA	0.057	0.082	0.061	0.056
RT	0.016	0.018	0.015	0.011
RUA	0.071	0.064	0.031	0.022

into account to what user the samples belong. A model is trained n times, each time leaving a different chunk out for testing. The others are used for training. The number of false positives (fp), false negatives (fn), true positives (tp) and true negatives (tn) are accumulated over the different iterations. This process is repeated for every subject in the dataset. We compared different classification algorithms by calculating the average EER of all body positions for the walking activity (see Table 2). Support Vector Machines produced bad results due to the small amount of training samples compared to the dimension of the feature space. Ensemble methods like AdaBoosting, Random Forests, Bagging and Gradient-Boosted Trees performed a lot better. Because of the robustness w.r.t. outliers in other machine learning experiments and its ability to handle heterogeneous features, we decided upon Gradient-Boosted Trees. We will use this model throughout the following experiments.

4.2 Optimal Sensor Positions on the Body

The REALDISP dataset contains data from sensors placed on different positions. This allowed us to evaluate which are the most relevant ones for authentication. We used the approach described above to train a model for each subject. The FAR and FRR can be tuned by demanding a minimal certainty before accepting a sample as genuine. In a first experiment, we used the data collected during walking under an ideal sensor placement. The results were evaluated using 8-fold cross validation. 9 body positions were considered: the back (BACK), the left (LUA) and right upper arm (RUA), the left (LLA) and right lower arm (RLA),

Table 4. EERs of slightly displaced sensors (training on both self-placement and ideal placement data) for each body position and different activities

Position	Walking	Jogging	Running	Cycling
BACK	0.054	0.035	0.059	0.051
LC	0.050	0.036	0.039	0.023
LLA	0.100	0.077	0.052	0.046
LT	0.050	0.043	0.042	0.021
LUA	0.090	0.049	0.035	0.033
RC	0.042	0.039	0.040	0.030
RLA	0.086	0.067	0.054	0.072
RT	0.051	0.027	0.033	0.028
RUA	0.074	0.059	0.050	0.028

Table 5. EERs when training on data from both sides of the body

Position	Walking	Jogging	Running	Cycling
RUA & LUA	0.079	0.052	0.055	0.042
RLA & LLA	0.081	0.066	0.064	0.056
RC & LC	0.044	0.052	0.053	0.030
RT & LT	0.055	0.039	0.048	0.025

the left (LC) and right calf (RC), the left (LT) and right thigh (RT). First, we note that the results are very promising, with really low EERs: ≈8% in the worst case scenario, and reducing further down to ≈2%. Second, the lower body seems to be more relevant than the upper body.

We repeated the same experiment for the other activities: jogging, running and cycling. The EERs are shown in Table 3. The conclusion that the lower body is more informative than the upper body remains valid for the activities considered. This observation holds in all subsequent experiments. Due to the limited amount of data, we cannot make more fine-grained conclusions.

4.3   Impact of Sensor Displacements

In real life scenarios, a sensor will never be worn on the exact same position. Therefore we investigated the effect of small sensor displacements.

In a first experiment the model was trained with walking samples where the sensors were administered by a professional, while testing with walking data when the sensors were self placed, and vice versa. The results plummeted, with best case EERs of ≈45%, worst case up to ≈50%. This can be explained by the lack of walks under sensor displacement in the training set.

Table 6. EERs training on different activities

Position	Without cycling	With cycling
BACK	0.025	0.025
LC	0.020	0.015
LLA	0.055	0.031
LT	0.028	0.016
LUA	0.045	0.029
RC	0.035	0.017
RLA	0.055	0.037
RT	0.024	0.014
RUA	0.043	0.029

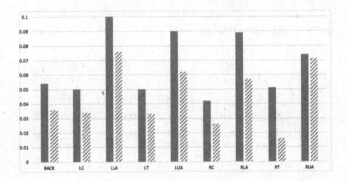

Fig. 1. EERs for walking. The left bar corresponds to Table 4 and the right to Table 3.

A second experiment uses data from both (ideal and self placement) walks as training data. The best EER, for the RC sensor, is ≈5%. The worst EER is ≈10%.

The same experiment was repeated for the other activities as well. The results are shown in Table 4. Our earlier conclusion that the upper body seems to be less suited for authentication than the lower body still holds. On top of that, the lower arm consistently has worse EERs than the upper arm. The increase in amount of training data makes our results more consistent.

Figure 1 illustrates our conclusions. It shows the results w.r.t. EER for the walking activity of the previous experiment (left bar) and the experiment described in Sect. 4.2 corresponding with Table 3 (right bar). It can clearly be seen that in both cases the upper body is less suited for authentication then the lower body and the back. Furthermore, when data of both ideal and self sensor placement (left bar) is used, the EERs suffer a bit, when compared to using data of only ideal sensor placement.

In a third experiment the training set contained the data from the right part of the body and tests were executed with the corresponding left side of the body.

Table 7. EERs when combining multiple accelerometers

	BACK	LC	LLA	LT	LUA	RC	RLA	RT	RUA
BACK	0.054	0.033	0.048	0.032	0.056	0.04	0.066	0.049	0.073
LC	0.037	0.052	0.049	0.061	0.067	0.036	0.043	0.049	0.062
LLA	0.051	0.05	0.1	0.05	0.081	0.054	0.083	0.059	0.055
LT	0.033	0.061	0.058	0.053	0.047	0.04	0.048	0.035	0.061
LUA	0.056	0.067	0.08	0.043	0.089	0.065	0.079	0.047	0.068
RC	0.04	0.037	0.057	0.043	0.074	0.044	0.046	0.047	0.061
RLA	0.066	0.042	0.082	0.047	0.08	0.045	0.086	0.051	0.066
RT	0.049	0.052	0.056	0.033	0.047	0.049	0.054	0.053	0.06
RUA	0.071	0.062	0.055	0.064	0.066	0.063	0.063	0.056	0.072

4 fold cross validation yielded EERs of approximately 45%. The clarification is probably similar to the one in the first experiment: there is not enough training data for this type of brutal sensor displacements.

In a fourth experiment the model was trained using data from both the right and left sensor. The tests were conducted using 4 fold cross validation. The results are shown in Table 5. We conclude that this type of sensor displacement has no additional measurable impact.

4.4   Impact of Other Motion Activities

Earlier we argued that having a model for every activity is infeasible. Even more, engineering optimal features for each activity under different circumstances is impossible. We conduct an experiment where we train our model using different activities. In a first experiment we use walking, jogging and running data, for self and ideal sensor placements. The results, using 4 fold cross validation, are shown in Table 6. Compared to training the model using only one activity, as shown in Table 4, the results have improved. The best EER is ≈2%, while the worst is ≈5.5%. We assume that using training data at different speeds improves the EER. This needs to be verified using more fine grained data w.r.t. speed.

In the second experiment we added cycling to the dataset, which does not seem to affect the results significantly (see Table 6). The EERs are lower than in the first experiment, but cycling gave better results in previous experiments as well. Furthermore, only 9 subjects were available for cycling.

4.5   Resilience Against Observation Attacks

To improve the EER and the resilience against observation and spoofing attacks, we combined the data of two accelerometers. A feature vector of length 448 (2*224) is obtained. The experiment is similar as before, using walking data for both self and ideal sensor placement. The results are shown in Table 7.

As expected, combining the same sensor yields no new information. The values on the diagonal of Table 7 are similar to the results shown in Table 4. On top of that, the order in which sensors are combined does not matter, since $EER_{i,j} \approx EER_{j,i}$. The best result is achieved by combining the sensors adjusted on the back and the left thigh, which gives an EER of $\approx 3\%$. This is an improvement of $\approx 2\%$ compared to using both sensors separately (see Table 4). However, if we consider for each body position, the results for left and right sensor together, the combination of a sensor placed on a calf with a sensor on the back yields the best results. Furthermore, for each sensor placement, a combination with the back sensor is among the best scoring. Combining two sensors does not always lead to an improvement in performance, i.e. combining the right upper arm and back sensor leads to an EER of $\approx 7\%$, this is higher then the $\approx 5\%$ EER obtained using the back sensor by itself.

For completeness we investigated what would happen if three sensors were used together. This lead to a feature vector of length 672 (3*224). We conclude that adding a third sensor leads to a minor improvement, but definitely not in all circumstances. This is illustrated in Fig. 2. The left bar shows the EER corresponding to each body position (as shown in Table 3). For each body position we add the sensor that leads to the best combined EER. The EERs for two sensors are shown by the middle bar. Then the third sensor, leading to the best EER is added, which illustrated by the right bar.

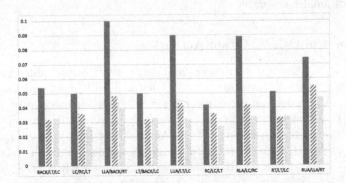

Fig. 2. Each bar represents the EER of at least one sensor. The left bar shows the EER when using only the first sensor from the description. The middle bar represents EER of the first two sensors. The right bar, is the EER of all three sensors.

An attacker can execute an observation attack by collecting accelerometer data through the HTML5 APIs of a mobile browser. An authentication system relying on only one sensor would now be compromised. When two sensors are used, we need to test the feasibility of misauthentication when the attacker constructs a trace, using the obtained data and his own data for the second sensor. We assume that the attacker knows the location of the second sensor.

To test the above use case we trained the system as we did before; using data from the walking activity, where the sensors are placed on the back and the left calf. Positive training samples are combinations of traces from the subject itself. Negative training samples consist of back and left calf data from other subjects. We test the system with genuine combinations of traces and with constructed combinations of traces by an attacker. The attacker combines an obtained a back trace and his own left calf accelerometer data. This leads to bad results, an EER of ≈17%. At a FRR (false rejection rate) of ≈10% a FAR (false acceptance rate) of ≈37% is achieved. At the threshold used to achieve the result in Table 4, the FAR is ≈43%.

In a last experiment we added some samples of the attack to the training data. This lead to an EER of ≈4%, which is only slightly more then before (0.037). The FAR is ≈1% when the FRR is ≈10%. When the old threshold is used, the FAR is ≈4%. We conclude that the model has to be trained with the observation attack in mind in order to be resilient against it.

The above results were obtained with a public dataset to guarantee the reproducibility of our research results. However, this means there might be some validity threats to generalize our research results. To address this concern, we are collecting bigger datasets to confirm our findings.

5   Conclusion

In this work, we evaluated the resilience of the accelerometer as a behaviometric for gait authentication, and more specifically the effect on the equal error rate (EER) when using multiple sensors at different places on the body.

Our experiments on gait authentication with a single accelerometer using Gradient-Boosted Trees and a fairly elaborate feature vector showed low EERs and recognition accuracies that go beyond the state-of-the-art. For data collected from 12 subjects and on 9 different places on the body, we obtained EER values between 2 to 8% for ideally positioned sensors. However, further experiments demonstrated that the accuracy dropped significantly after subtle sensor displacements, with the EER worsening from single digits to about 45%. We obtained similar results when sensors were displaced from one side to the other side of the body. When we incorporate data from displaced sensors in our training data set, the accuracy improves again with EERs between 5 to 10%, i.e. slightly worse compared to our first experiment. We also measured the impact of other motion activities, including jogging, running and cycling, and their effect on misauthentication

We evaluated the effectiveness of multiple accelerometers for gait authentication and the impact on the classification accuracy, demonstrating further improvements in the EER of ≈2%. Additionally, as a hacker can carry out an observation attack and obtain accelerometer data with the HTML5 APIs, we investigated the resilience of our multi-sensor scheme against spoofing attacks. Our experimental results show our scheme is more robust against such attacks under the condition that not all sensors are compromised.

As future work, we will investigate to what extent motion patterns are independent. We will execute additional attack experiments based on an adversary leveraging his own sensor data to reproduce traces of the victim in order to ascertain whether such attacks are feasible and practical. Furthermore, we will investigate whether it is feasible to modularize our multi-sensor behaviometric-based authentication scheme, not by fusing the different data sets before training but rather fusing individual decisions based on each data set [23], allowing for more flexibility to combine different behaviometrics at runtime.

Acknowledgments. This research is partially funded by the Research Fund KU Leuven and DiskMan. DiskMan is a project realized in collaboration with imec. Project partners are Sony, IS4U and Televic Conference, with project support from VLAIO (Flanders Innovation and Entrepreneurship).

References

1. Annadhorai, A., Guenterberg, E., Barnes, J., Haraga, K., Jafari, R.: Human identification by gait analysis. In: Proceedings of the 2nd International Workshop on Systems and Networking Support for Health Care and Assisted Living Environments. p. 11: 1–11: 3. HealthNet '08, NY, USA. ACM, New York (2008)
2. Baños, O., Damas, M., Pomares, H., Rojas, I., Tóth, M.A., Amft, O.: A benchmark dataset to evaluate sensor displacement in activity recognition. In: Proceedings of the 2012 ACM Conference on Ubiquitous Computing. pp. 1026–1035 (2012)
3. Bailey, K.O., Okolica, J.S., Peterson, G.L.: User identification and authentication using multi-modal behavioral biometrics. Computers & Security **43**, 77–89 (2014)
4. Baños, O., Tóth, M.A., Damas, M., Pomares, H., Rojas, I.: Dealing with the effects of sensor displacement in wearable activity recognition. Sensors **14**(6), 9995–10023 (2014)
5. Crossler, R., Johnston, A., Lowry, P., Hu, Q., Warkentin, M., Baskerville, R.: Future directions for behavioral information security research. Computers and Security **32**, 90–101 (2013)
6. Derawi, M.O., Nickel, C., Bours, P., Busch, C.: Unobtrusive user-authentication on mobile phones using biometric gait recognition. In: 2010 Sixth International Conference on Intelligent Information Hiding and Multimedia Signal Processing. pp. 306–311 (2010)
7. Gafurov, D., Snekkenes, E., Bours, P.: Gait authentication and identification using wearable accelerometer sensor. In: 2007 IEEE Workshop on Automatic Identification Advanced Technologies. pp. 220–225 (2007)
8. Gafurov, D., Snekkenes, E., Bours, P.: Spoof attacks on gait authentication system. IEEE Transactions on Information Forensics and Security **2**(3), 491–502 (2007)
9. Hayashi, E., Das, S., Amini, S., Hong, J., Oakley, I.: Casa: Context-aware scalable authentication. In: Proceedings of the Ninth Symposium on Usable Privacy and Security. p. 3: 1–3: 10. SOUPS '13, NY, USA. ACM, New York (2013)
10. Kale, A., Cuntoor, N., Yegnanarayana, B., Rajagopalan, A.N., Chellappa, R.: Gait analysis for human identification. In: Kittler, J., Nixon, M.S. (eds.) AVBPA 2003. LNCS, vol. 2688, pp. 706–714. Springer, Heidelberg (2003). doi:10.1007/3-540-44887-X_82
11. Kayacik, H.G., Just, M., Baillie, L., Aspinall, D., Micallef, N.: Data driven authentication: On the effectiveness of user behaviour modelling with mobile device sensors. CoRR abs/1410.7743 (2014)

12. Lu, H., Huang, J., Saha, T., Nachman, L.: Unobtrusive gait verification for mobile phones. In: Proceedings of the 2014 ACM International Symposium on Wearable Computers. pp. 91–98. ISWC '14, NY, USA. ACM, New York (2014)
13. Mantyjarvi, J., Lindholm, M., Vildjiounaite, E., Makela, S.M., Ailisto, H.A.: Identifying users of portable devices from gait pattern with accelerometers. In: Proceedings. (ICASSP '05). IEEE International Conference on Acoustics, Speech, and Signal Processing, 2005. vol. 2, pp. ii/973-ii/976 Vol. 2 (2005)
14. Mjaaland, B.B.: The plateau: imitation attack resistance of gait biometrics. In: Leeuw, E., Fischer-Hübner, S., Fritsch, L. (eds.) IDMAN 2010. IAICT, vol. 343, pp. 100–112. Springer, Heidelberg (2010). doi:10.1007/978-3-642-17303-5_8
15. Ngo, T.T., Makihara, Y., Nagahara, H., Mukaigawa, Y., Yagi, Y.: The largest inertial sensor-based gait database and performance evaluation of gait-based personal authentication. Pattern Recognition 47(1), 228–237 (2014)
16. Nickel, C., Busch, C., Rangarajan, S., Möbius, M.: Using hidden markov models for accelerometer-based biometric gait recognition. In: 2011 IEEE 7th International Colloquium on Signal Processing and its Applications. pp. 58–63 (2011)
17. Ntantogian, C., Malliaros, S., Xenakis, C.: Gaithashing: A two-factor authentication scheme based on gait features. Computers & Security 52, 17–32 (2015)
18. Ramakrishnan, A.K., Preuveneers, D., Berbers, Y.: A modular and distributed Bayesian framework for activity recognition in dynamic smart environments. In: Augusto, J.C., Wichert, R., Collier, R., Keyson, D., Salah, A.A., Tan, A.-H. (eds.) AmI 2013. LNCS, vol. 8309, pp. 293–298. Springer, Cham (2013). doi:10.1007/978-3-319-03647-2_27
19. Shepherd, S.: Continuous authentication by analysis of keyboard typing characteristics. In: European Convention on Security and Detection. pp. 111–114 (1995)
20. Shi, E., Niu, Y., Jakobsson, M., Chow, R.: Implicit authentication through learning user behavior. In: Burmester, M., Tsudik, G., Magliveras, S., Ilić, I. (eds.) ISC 2010. LNCS, vol. 6531, pp. 99–113. Springer, Heidelberg (2011). doi:10.1007/978-3-642-18178-8_9
21. Spooren, J., Preuveneers, D., Joosen, W.: Mobile device fingerprinting considered harmful for risk-based authentication. In: 8th European Workshop on System Security, EuroSec 2015, France, April 21, 2015. p. 6: 1–6: 6 (2015)
22. Spooren, J., Preuveneers, D., Joosen, W.: Leveraging battery usage from mobile devices for active authentication. Mobile Information Systems 2017, 14 (2017)
23. Tao, Q., Veldhuis, R.: Threshold-optimized decision-level fusion and its application to biometrics. Pattern Recogn. 42(5), 823–836 (2009)
24. Wilson, J., Najjar, N., Hare, J., Gupta, S.: Human activity recognition using lzw-coded probabilistic finite state automata. In: 2015 IEEE International Conference on Robotics and Automation (ICRA). pp. 3018–3023 (2015)

Security in Networks and Web

A Content-Aware Trust Index for Online Review Spam Detection

Hao Xue[⊠] and Fengjun Li

The University of Kansas, Lawrence, KS, USA
{haoxue,fli}@ku.edu

Abstract. Online review helps reducing uncertainty in the pre-purchasing decision phase and thus becomes an important information source for consumers. With the increasing popularity of online review systems, a large volume of reviews of varying quality is generated. Meanwhile, individual and professional spamming activities have been observed in almost all online review platforms. Deceptive reviews with fake ratings or fake content are inserted into the system to influence people's perception from reading these reviews. The deceptive reviews and reviews of poor quality significantly affect the effectiveness of online review systems. In this work, we define novel aspect-specific indicators that measure the deviations of aspect-specific opinions of a review from the aggregated opinions. Then, we propose a three-layer trust framework that relies on aspect-specific indicators to ascertain veracity of reviews and compute trust scores of their reviewers. An iterative algorithm is developed for propagation of trust scores in the three-layer trust framework. The converged trust score of a reviewer is a credibility indicators that reflects the trustworthiness of the reviewer and the quality of his reviews, which becomes an effective trust index for online review spam detection.

Keywords: Trust · Online review · Opinion mining

1 Introduction

With the increasing popularity of online social-collaborative platforms, people get more connected and share various types of information to facilitate others' decision-making processes. A vast amount of user-generated content (UGC) has been made available online. For example, TripAdvisor.com, which specializes in travel-related services, has reached 315 million unique monthly visitors and over 200 million reviews. Yelp.com, which is known for restaurant reviews, has a total of 71 million reviews of businesses and a monthly average of 135 million unique visitors to the site. This plethora of data provides a unique opportunity for the formation of "aggregated opinions", from which people make reasonable judgments about the quality of a service or a product from an unknown provider.

Published by Springer International Publishing AG 2017. All Rights Reserved
G. Livraga and S. Zhu (Eds.): DBSec 2017, LNCS 10359, pp. 489–508, 2017.
DOI: 10.1007/978-3-319-61176-1_27

However, the quality of UGC is problematic. For example, it has been observed that a non-negligible portion of online reviews is unfairly biased or misleading. To make things worse, deceptive reviews have been purposely planted into online review systems by individual or professional spammers [6,12,18,39]. For example, recent studies on Yelp show that about 16% restaurant reviews are considered suspicious and rejected by Yelp internally [20]. Opinions expressed in deceptive reviews deviate largely from the fact to mislead the consumers to make unwise decisions. This is known as *online review spamming*, which was first identified by Jindal et al. in [12].

Many deterrence-based and reputation-based approaches have been adopted to address the spamming problem in online review systems. For example, the "Verified Purchase" mechanism of Amazon.com labels reviews that are posted by consumers who actually have purchased the reviewed items. This label is often perceived by consumers as a positive indicator of the trustworthiness of the review. A more generally adopted approach is the "review of the review", which allows users to rate a review or vote for its "helpfulness". Readers then use the ratings and vote counts as a measure to assess the quality and the trustworthiness of the review. While these mechanisms provide additional information about how helpful or trustworthy a review is, their limitations are obvious. Similar to reviews, the "review of review" is a subjective judgment that can be easily gamed by purposeful spammers. Moreover, it suffers from inadequate user participation. Surveys show that only a small portion of users provides reviews online. Furthermore, among thousands of users who read a review, only a few provides feedback. In most online review systems, a large amount of reviews does not have any helpfulness or usefulness rating at all.

Detection-based mechanisms are considered more effective approaches to address the review spamming problem. Many learning-based schemes have been proposed to identify deceptive reviews and spamming reviewers from textual features [12,23,25], temporal features [6,41], individual or group behavior patterns of spammers [39], and sentiment inconsistency [18,24]. The rationale behind these approaches is two-fold. Along the first direction, the detection models rely on the deviation in rating behaviors. Since the objective of opinion spammers is to alter users' perception of the quality of a target, spammers often generate a large amount of reviews, seemingly from different users, with extreme ratings. In this way, spammers can significantly distort the mean rating. Such detection focuses on rating-based features that reflect the deviation from the aggregated rating (or the majority vote) [1,18] and other rating behaviors (e.g., change of average rating over time, change of average ratings across groups of users, etc.). While these approaches have been used with success, they can be easily gamed by avoiding extreme rating behaviors. Along the second direction, detection schemes rely on text-based features, by identifying duplicated messages in multiple reviews [12,23], or psycholinguistic deceptive characteristics [25]. These approaches involve training classifiers with manually labeled reviews, which is expensive and time-consuming. Combining review rating and textural features at the same time, some approaches detect spam reviews whose ratings are inconsistent with the opinions expressed in the review text [18,24].

Inspired by these approaches, we propose a content-based trust index for review spam detection, which is based on a set of aspect-specific opinions extracted from review content and iteratively computed in a three-layer trust propagation framework. First, we have observed two types of spams – (1) "Low-quality spams" that usually have short, poorly-written, sometimes irrelevant content. The low quality spams are generated at low cost so they often come in large quantity. These spams can be easily identified by existing detection mechanisms. (2) "High-quality spams" that are long, carefully composed, and well-written deceptive opinions. It is costly to generate such spams. However, it is also very difficult to detect them, especially using text features.

In this work, we target the second type of spamming reviews, whose content is carefully composed with bogus opinions. To engage the reader, these spams may include fake information cues or social motivational descriptions, which make them difficult to be detected by schemes using simple text-based features. To tackle this problem, we propose aspect-specific indicators that measure the deviation of an aspect-specific opinion of a review from the opinion aggregated across all reviews on that aspect of the target, based on majority vote. Although majority vote has some limitations in extreme cases, such as inertia against sudden change of quality, we assume in most cases the aspect-specific majority vote effectively reflects the fact. This is because in our approach, we first attempt to remove the low-quality reviews, regardless of benign or deceptive, as they do not contribute any meaningful *opinion*. With the remaining "meaningful" reviews, we can reasonably assume the benign reviews always outnumber the high-quality spams that are costly to generate. In extreme cases, where the number of high-quality spams is larger than or comparable to the number of truthful reviews, the review system is considered broken and no detection scheme could work. The rationale behind our approach is that if every review carries aspect-specific opinions, the majority vote on a common aspect should reflect the factual quality of the target, so that the agreement between the opinion of a review and the aggregated opinion reflects the quality of that review. Then, our scheme considers multiple aspect-specific indicators and integrates the deviations across all aspects.

To effectively integrate the aspect-specific indicators, we adopt a three-layer trust propagation framework, which was first described in [38]. It calculates trust scores for reviews, reviewers, and the aspect-specific opinions of the target (defined as "statement"). To do this, we first apply opinion mining techniques to extract aspect-specific opinions from the reviews, and then input them into the three-layer trust propagation model that iteratively computes the trust scores by propagating the scores between reviews, reviewers, and statements. As a result, the converged trust score of a reviewer reflects his overall deviation from the aggregated opinion across all aspects and all targets that he has reviewed. This is a strong indicator of trust to distinguish benign reviewers and high-quality spammers.

We summarize our contributions as: (i) We propose a novel aspect-specific opinion indicator as a content-based measure to quantify the quality and trustworthiness of review content. And (ii) We develop an iterative three-layer trust propagation framework to compute trust scores for users, reviews, and statements as a measure of users' trustworthiness and stores' reputation.

2   Related Work

Opinion Mining. Opinion mining has been used to analyze the opinions, sentiments, and attitudes expressed in a textual content towards a target entity. It typically includes work from two related areas, opinion aspect extraction and sentiment analysis. Aspect extraction aims to extract product features from opinionated text. Many work considered it as a labeling task, thus rule based methods are used extensively [10,19,31–33]. To group the extracted aspect into categories, lexical tools like WordNet [22] is often used. Topic modeling-based approaches are also very popular [3,4,13,17,36], as they are able to extract and group aspects simultaneously. On the other hand, the goal of is to analyze the polarity orientation of the sentiment words towards a feature or a topic of the product. One of most common way is to use some sentiment lexicons directly, such as MPQA Subjectivity Lexicon [40] and SentiWordNet [2]. However, just like WordNet, these tools have their own limitations. Another common practice is to infer the polarity of target words using a small group of seed terms with known polarity [5,11,37]. In addition, supervised learning algorithms are often applied in previous work [15,27,28,34]. In this work, we adopt the supervised learning method as our opinion mining technique. Note that opinion mining is not our focus here. The difference between our goal and typical opinion mining work is that we are not trying to improve the performance of extracting opinions. Instead, our purpose of applying opinion technique is to use the extracted aspects as deviation indicators for trustworthiness analysis.

Trust Propagation. Trust and trust propagation have been extensively studied in literature. The general idea of reinforcement based on graph link information has been proved effective. HITS [16] and PageRank [26] are successful examples in link-based ranking computation. [39] applied graph-based reinforcement model to compute trustworthiness scores for users, reviews, and stores. However, these approaches did not consider content information. [38] proposed a content-driven framework for computing trust of sources, evidence, and claims. The difference between this model and ours is that we extract more fine-grained information from content, while the model in [38] mainly used the similarities between content in general. In [38], the inter-evidence similarity plays an important role to make sure that similar evidences get similar scores. However, the consensuses of opinions used in our model already represent such similarity, so we did not add the inter-evidence similarity. Besides, we also redefined the computational rules in the context of our problem. In many work, trust is often generated and transmitted in a graph of trust, such in [8,14,21,42]. Trust can also be inferred from rating deviations [35]. Different from previous approaches, our model derives trust from the consistence between an individual's opinion and the majority opinion.

3   Aspect-Specific Opinion Indicator

Existing content-based detection approaches take textual content of a review as input, which often use word-level features (e.g., n-grams) and known lexicons (e.g., WordNet [22] or psycholinguistic lexicon [18]) to learn classifiers that identify a review as spam or non-spam. To train the classifier, costly and time-consuming manually labeling of reviews is required. Due to subjectiveness of human judgment and personal preferences, there is no readily available ground truth of opinions. Therefore, a high-quality labeled dataset is difficult to obtain. Some existing work adopt crowdsourcing platforms such as Amazon Mechanical Turk to recruit human labeler, however, it is pointed out the quality of the labeled data is very poor. Different from these approaches, our opinion spam detection scheme focuses on deviation from the majority opinion. Although biased opinions always exist in UGC, we argue that a majority of users may be *biased but honest*, instead of maliciously deceptive. This is based on an overarching assumption regarding reviewer behaviors – that is the majority of reviews are posted by honest reviewers, as recognized by many existing work on opinion spam detection [1,18,24]. If this assumption does not hold, online peer review systems will be completely broken and useless. As a result, we propose to use the majority opinions as the "ground truth".

3.1   Aspect Extraction

Existing work on opinion mining studies opinions and sentiments expressed in review text at document, sentence or word/phrase levels. Typically, the overall sentiment or subjectiveness of a review (document-level) or a sentence of a review is classified and used as a text-based feature in spam detection. However, we consider these opinions are either too coarse or too fine-grained. For example, it is common that opposite opinions are expressed in an individual review – it may be positive about one aspect of the target entity but negative about another. This is difficult to capture using the document-level sentiment analysis. Therefore, the derived review-level majority opinion is inaccurate and problematic. Another direction of approaches proposes to use opinion features that associate opinions expressed in a review with specific aspects of the target entity [9,24]. Intuitively, opinion features are nouns or noun phrases that typically are the subjects or objects of a review sentence. For example, in the below review, the underlined words/phrases can be extracted as opinion features.

"This place is the <u>bomb</u> for <u>milkshakes</u>, <u>ice cream sundaes</u>, etc. <u>Onion rings</u>, <u>fries</u>, and all other "basics" are also fantastic. <u>Tuna melt</u> is great, so are the <u>burgers</u>. Classic old <u>school</u> <u>diner</u> <u>ambiance</u>. <u>Service</u> is friendly and fast. Definitely come here if you are in the <u>area</u> ..."

Obviously, users may comment on a large number of very specific aspects about the target entity. The derived opinion features are thus too specific and too fine-grained to form a majority opinion on each feature, since other reviews about the same target may not comment on these specific features. However, from the above example, we can see that opinion features such as "milkshakes", "fries", and "burgers" are all related to an abstract aspect "food". If we define a set of aspect categories, opinion features about a same or a similar high-level concept can be grouped together.

Consider a set of reviews ($\mathbf{R}$), which are written by a group of users ($\mathbf{U}$) about a set of entities ($\mathbf{E}$). Each review $r \in \mathbf{R}$ consists of a sequence of words $\{w_1, w_2, ..., w_{n_r}\}$. Then, we can define a set of m abstract aspects $a = \{a_1, a_2, ...a_m\}$, and sentiment polarity label $l = \{l_1, l_2, ...l_k\}$. As a result, for each review r, we can extract a set of aspect-sentiment tuple, denoted as $ao_i = <a_i, l_i>$, to represent the aspect-specific opinions of a user u towards a target entity e.

Typical sentiment polarity labels include "positive", "negative", "neutral", and "conflict" [7,30]. Since "conflict" captures inconsistencies within a review but does not contribute to inter-review consistence, we do not include this label in our model. Abstract aspect categories are more difficult to define since they are domain-specific and thus need to be carefully tuned for a given domain. In this work, we use Yelp reviews as our dataset to study the credibility of users and their reviews. Therefore, we define a small set of aspect categories including four meaningful aspects *food, price, service,* and *ambience* for restaurant reviews. We consider the opinion extractions as a classification problem and adopt the support vector machine supervised learning model for opinion extraction. In this way, we classify each sentence to a specific aspect, and group sentences in a review about a same aspect as an aspect-specific "statement" (denoted as s_i). We use the SemEval dataset [30], which is a decent-sized set of labeled data for restaurant reviews, to train our classifier (see more details in Sect. 5). Our goal is to identify an adequate number of aspects that are commonly addressed by all reviews so that we can construct a credibility indicator from the aggregated opinions. In fact, too many over-specific aspects complicate the credibility computing model instead of improving it. Therefore, we combine all other aspect category labels in the SemEval dataset as a fifth category "miscellaneous". This is different from previous work that considered all aspects in the classification [7].

Next, we conduct the aspect-specific sentiment classification upon the classified aspect-specific statements to obtain aspect-specific sentiment polarities. For each category, the classification is conducted independently. For example, to determine the sentiment polarities of the "food" category, we conduct a sentiment classification upon all statements that have been classified into the category "food", and determine the aspect-sentiment tuples: "food-positive", "food-negative", and "food-neutral".

3.2   Opinion Vector and Quality Vector

To use the extracted opinions for further analysis, we define an opinion vector $\mathbf{o} = [o_1, ..., o_5]$ to capture aspect-specific opinions and their sentiment polarities. Each element of the opinion vector corresponds to an aspect of food, price, service, ambience, and miscellaneous, respectively. Sentiment polarities are represented by element values, where a positive sentiment is denoted by "+1", a negative sentiment is denoted by "-1", and neutral is denoted by "0". Since a statement may not necessary to express an opinion about an aspect, we distinguish no opinion expressed from a neutral opinion by defining a corresponding opinion status vector **os**. For example, if a statement expresses three opinions, positive about food, neutral about price, and negative about service, its opinion vectors are $\mathbf{o} = [1, 0, -1, 0, 0]$ and $\mathbf{os} = [1, 1, 1, 0, 0]$.

With the opinion vectors, we can aggregate the opinions on multiple aspects from all reviewers of an entity to form four aspect-specific aggregated opinions. While aspect-specific opinions are subject judgements and thus can be biased, the aggregated aspect-specific sentiments are highly likely to reflect the true quality of the entity from a specific aspect. This is because individual biases are typically smaller aspect level than at document level, which is more affected by the weights subjectively assigned by individuals to multiple aspects. In this sense, aspect-level bias can be corrected by the majority view if the review amount is adequate. Furthermore, comparing with rating, aspect-specific opinions are more difficult to be tampered by opinion spammers, whose review text are likely to be pointless, wrong focused, or brief. Finally, the aggregated sentiments are robust to correct the inaccuracy introduced by opinion mining models. Opinion mining often suffers from precision problems, but our goal is to decide if the overall aspect-specific opinion is positive, neutral, or negative. Although each individual input incurs a small uncertainty, the chance to affect overall value is very small. Based on these considerations, we derive the aggregated aspect-specific opinion vectors as $\mathbf{o_{agg}} = [o_{agg_1}, ..., o_{agg_5}]$ and $\mathbf{os_{agg}} = [os_{agg_1}, ..., os_{agg_5}]$, where

$$o_{agg_i} = \begin{cases} 1, & avg_{i \in A_i}(o_i) \geq \theta_p \\ -1, & avg_{i \in A_i}(o_i) \leq \theta_n \\ 0, & \text{otherwise.} \end{cases} \qquad (1)$$

In this way, the aggregated sentiment polarity of each aspect is mapped to the positive, neutral, and negative labels based on the averages. The aggregated aspect-specific opinion vector is considered a *quality vector*, which can be used to determine the credibility of a user statement. Intuitively, a statement is more credible and of higher quality, if it expresses a consistent opinion with the aggregated opinion about one aspect of the target entity, and thus the reviewer is considered more honest and trustworthy.

4   Content-Based Trust Computation

We compute the aggregated aspect-specific opinion vector as a quality measure and use individual aspect-specific opinion vector as a credibility (or trust) measure. To integrate trust measures across multiple users and multiple entities, trust propagation models are commonly used [38,43]. Therefore, in this work, we adopt a three-layer trust propagation model to compute iteratively the trust-related scores for users, reviews, and aspect-specific statements.

The three-layer propagation model was first introduced in [38] to compute trustworthiness of data sources of free-text claims online. Most of the previous work is based on a bi-partite graph structure, which ignores the content and the context in which the content is expressed. The intermediate layer in the three-layer model can include the content context (i.e., the reviews) and capture the intertwined relationships between users, reviews, and opinions expressed in specific statement. Therefore, we define three types of nodes, *users*, *reviews*, and *statements*, and compute the trustworthiness scores from the obtained opinion vectors. Each user is connected to the reviews she posts. Each review is connected to the statements expressed in the review itself. The statement is defined as an opinion expressed on a target in the review system, e.g. restaurant1-food-positive. The structure is shown in Fig. 1. In the figure, u_i, r_i, and s_i represent a user node, a review node, and a statement node respectively. $h(u_i)$, $f(r_i)$, and $t(s_i)$ are defined as the score of a user, a review, and a statement respectively. The value $p(r_i, s_i)$ is a weight on the link from a review to a statement. These values will be described in rest of this subsection shortly.

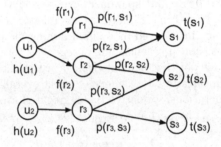

Fig. 1. Structure of the model

For each type of node, a score is defined, namely *honesty* for users, *faithfulness* for reviews, and *truthfulness* for statements. Different with the original model in [38], we defined a support weight on the links between reviews and statements, which measures the how supportive a review is for a statement. The support weight is defined as the sentiment consistency between the review and the statement it expressed. As mentioned above, there are three predefined sentiment polarities, positive, negative, and neutral. Here, if the sentiment polarity expressed in the review on a specific aspect category is the same as the sentiment polarity of the statement, then we say the review fully supports the statement. On the other hand, if the sentiment polarities between a review and a statement are totally opposite, i.e. positive and negative, or negative and positive, we say the review rejects the statement. For all other cases, we say the reviews partially support the statement. With these definitions, the values of support between review r_i and statement s_i are defined as:

$$p(r_i, s_i) = \begin{cases} 1, & r_i \text{ fully supports } s_i \\ 0, & r_i \text{ rejects } s_i \\ 0.5, & r_i \text{ partially supports } s_i \end{cases} \tag{2}$$

As mentioned before, each type of node has a type of score that measures the extent of trustworthiness. The honesty score for a user is defined in the following equation:

$$h^{n+1}(u_i) = \alpha h^n(u_i)$$
$$+ (1 - \alpha) \frac{\sum\limits_{r_j \in \mathcal{R}(u_i)} \sum\limits_{s_k \in \mathcal{S}(r_j)} [p(r_j, s_k) \times t^n(s_k)]}{|\mathcal{R}(u_i)| \times |\mathcal{S}(r_j)|} \tag{3}$$

Here, different with the original model in [38], under our definition, the honesty score of a user consists of two parts. The first part is the honesty score from the last round. We added this part because a user's honesty does not entirely depend on the feedback from his/her statements. The second part is the feedback from the reviews and statements related to the user. $\mathcal{R}(u_i)$ is the collection of reviews that user u_i posts. $\mathcal{S}(r_j)$ is collection of statements review r_j expresses. $t^n(s_k)$ is the truthfulness score of statement s_k. $p(r_j, s_k)$ is a support value of review r_j to statement s_k. The second part is essentially a weighted average of truthfulness scores of all statements that reviews of user u_i expresses. The support value serves as a factor that controls the feedback. If a statement supports a statement with high truthfulness score, the contribution from this statement will be high. Otherwise, a user will be penalized for supporting a statement with low truthfulness score or rejecting a statement with high truthfulness score. The parameter α controls the ratio between the two parts. The faithfulness score for a review is defined as:

$$f^{n+1}(r_i) = \mu f^n(r_i) + (1 - \mu)h^n(u(r_i)) \tag{4}$$

The faithfulness score of a review also comes from two parts, the faithfulness score from the previous round and the honesty score of the author. The parameter μ is also used to control the ratio between the two parts. Here $u(r_i)$ represents the user who writes review r_i. The truthfulness score for a statement is defined as:

$$t^{n+1}(s_i) = \frac{\sum\limits_{r_j \in \mathcal{R}(s_i)} [f^n(r_j) \times h^n(u(r_j)) \times p(r_j, s_i)]}{|\mathcal{R}(s_i)|} \tag{5}$$

$\mathcal{R}(s_i)$ is the collection of reviews that express statement s_i. The truthfulness score of statement s_i is essentially a weighted average of honesty scores of the users whose reviews express this statement. The three types of scores are all in the range $[0, 1]$. From the formulas above, it is obvious that the three types of scores are defined in an intertwined relationship. The measurement of trust is propagated along the structural connections. For example, a user's honesty score is dependent on the trustworthiness of the statements in his reviews, thus the trust is propagated from his statements to the user himself, and further propagates to his reviews and back to his statements. Each type of score gets feedbacks from the other two, which allows reinforcement based on the connections among the nodes. The scores of nodes are computed in an iterative

---

Algorithm 1. Iterative framework to compute trust-related scores

---
Input:
  Collections of users $\mathcal{U}$, reviews $\mathcal{R}$, and statements $\mathcal{S}$;
  Initial sentiment polarities for all statements in $\mathcal{S}$;
  Interpolation parameters α, μ;
Output:
  honesty scores $h(u)$ for all users in $\mathcal{U}$, faithfulness scores $f(r)$ for all reviews in $\mathcal{R}$,
  and truthfulness scores $t(s)$ for all statements in $\mathcal{S}$;
  repeat
    Compute the honesty scores for all users using (3)
    Compute the truthfulness scores for all statements using (5)
    Compute the faithfulness scores for all reviews using (4)
    Normalize each type of score so that the largest is 1
  until converged

---

computational framework, as shown in Algorithm 1. After the model converges, it outputs the final result.

5   Experiments

5.1   Dataset

We used two datasets in the experiments. We first used the SemEval dataset [30], which contains 3,041 sentences from restaurant reviews, to train our classifier. In SemEval, each sentence is labeled with one or multiple aspect categories (i.e., food, service, price, ambience, and anecdotes/miscellaneous) and the corresponding sentiment polarities (i.e., positive, neutral, negative, and conflict). As discussed in Sect. 3, the "conflict" sentiment category is not considered in our model. We then split this dataset in 4:1 ratio with a training dataset and a testing dataset of 2,432 and 609 labeled sentences, respectively.

We tested our content-aware trust propagation scheme on a second dataset with restaurant reviews from Yelp.com, which is a subset of the dataset that we crawled from Yelp.com in 2013. The entire dataset contains 9,314,945 reviews about 125,815 restaurants in 12 U.S. cities from 1,246,453 reviewers between 2004 and 2013. In this experiment, we extracted a dataset for the city of Palo Alto, California. It contains 128,361 reviews about 1,144 restaurants from 45,180 reviewers. Although our dataset contains rich information about the reviewers, such as the total number of reviews, average ratings, social relationships, etc., we only used review content in this study.

5.2   Aspect Category and Sentiment Polarity Classifications

Aspect Category Classification. We use Support Vector Machine (SVM) in the Python machine learning library scikit-learn [29] as the classifier for opinion extraction. For feature extractions, we used the bag-of-words model and extracted the tf-idf weights as features. The classifiers for aspect categories and sentiment polarities are trained separately at sentence level.

Since SVM is a binary classifier, a trained SVM classifier can only classify whether a sentence contains a category or not. However, a single sentence may contain multiple aspect categories, which cannot be classified with a single SVM classifier. Therefore, we trained five separate binary one-vs-all SVM classifiers independently, one for each aspect category. For example, if a sentence contains opinions about "food", it is classified into the "food" category by the "food" classifier. If it contains opinions about both "food" and "price", the "food" classifier identifies the sentence as "food", and the "price" classifier also identities it as "price" at the same time. The results of aspect category classification are shown in Table 1.

Table 1. Classification performance of aspect category classifiers

Label	Precision	Recall	F1-score	Support	Accuracy
food	0.81	0.78	0.80	238	0.844
not_food	0.86	0.88	0.87	371	
avg/total	0.84	0.84	0.84	609	
price	0.91	0.62	0.73	65	0.952
not_price	0.96	0.99	0.97	544	
avg/total	0.95	0.95	0.95	609	
service	0.82	0.69	0.75	122	0.906
not_service	0.92	0.96	0.94	487	
avg/total	0.90	0.91	0.90	609	
ambience	0.83	0.52	0.64	84	0.920
not_ambience	0.93	0.98	0.95	525	
avg/total	0.91	0.92	0.91	609	
anecdotes/miscellaneous	0.77	0.70	0.73	243	0.796
not_anecdotes/miscellaneous	0.81	0.86	0.84	366	
avg/total	0.79	0.80	0.79	609	

For "avg/total" in the table, "avg" means the average of precision, recall, and f1-score, respectively, and "total" denotes the total support of each category. Among the five categories, the category anecdotes/miscellaneous has the worst performance (with the lowest precision of 0.77). This category contains the aspects that do not belong to any one of the other four categories. Since it is always easier to determine if a sentence does not belong to the anecdotes/miscellaneous category than it does, the precision, recall, and f1-score of the "not_anecdotes/miscellaneous" category are higher than the "anecdotes/miscellaneous" category.

Interestingly, we find that the food category, which is a most popular aspect category for restaurant reviews, has the second-worst performance among the five categories. This may because many different terms and aspects representing the food category. Using tf-idf weights as the features, it is difficult to have a unified representation of the category. Therefore, it is relatively more difficult to train an effective classifier for the food category than for others such as price or service.

Sentiment Polarity Classification. After we obtain the classified results of aspect categories, we apply the sentiment classifier in each category to compute the category-based sentiment polarities. One review may contain multiple opinions about multiple categories. For example, after sentiment polarity classification, we can extract opinions as "food-positive", "price-neutral", and "service-negative" from a single review. We show the results of sentiment polarity classification in Table 2.

Table 2. Classification performance of category-based sentiment polarities

Label	Precision	Recall	F1-score	Support	Accuracy
food, negative	0.39	0.36	0.38	33	0.740
food, neutral	0.50	0.04	0.07	25	
food, positive	0.80	0.92	0.85	169	
avg/total	0.71	0.74	0.70	227	
price, negative	0.55	0.44	0.49	25	0.635
price, neutral	0.00	0.00	0.00	2	
price, positive	0.67	0.81	0.73	36	
avg/total	0.60	0.63	0.61	63	
service, negative	0.66	0.69	0.67	48	0.698
service, neutral	0.00	0.00	0.00	7	
service, positive	0.73	0.79	0.76	61	
avg/total	0.66	0.70	0.68	116	
ambience, negative	0.64	0.30	0.41	23	0.675
ambience, neutral	0.00	0.00	0.00	5	
ambience, positive	0.68	0.92	0.78	49	
avg/total	0.62	0.68	0.62	77	
anecdotes/miscellaneous, negative	0.11	0.10	0.10	31	0.547
anecdotes/miscellaneous, neutral	0.60	0.49	0.54	96	
anecdotes/miscellaneous, positive	0.60	0.73	0.66	107	
avg/total	0.54	0.55	0.54	234	

From the two tables, we can see that the performance of aspect category classification is much better than category-based sentiment polarity classification. This is because using bag-of-words model, it is easier to find representative features for categories than for sentiment polarities. Sometimes, the sentiment polarities are implicit and context-dependent. In addition, the category-based sentiment analysis takes the classification results as aspect categories. Any inaccuracy from previous classification results affects the overall performance. It is worth noting that the classification performance of sentiment polarity in our scheme is comparable to the baseline (e.g., some submissions in the SemEval 14 contest [30]).

The classifications of aspect categories and category-based sentiment polarities are used as input to the trust propagation model. SVM does not yield the best results, but the current classification results do provide a good set of inputs to the trust propagation computation. Other supervise- and unsupervised-classification models may yield higher precision and thus improve the performance of our model.

5.3   Trustworthiness Scores Computation

We construct the proposed three-layer trust propagation model using the structural relationships among reviewers, reviews, and statements, where the statements are aspect-specific opinions about the restaurants. There are in total three different sentiment polarities, but for each restaurant there exists at most one statement for a specific aspect category. Since the score of statement nodes depends on the feedback from the other two types of nodes as well as the support value on the link, the sentiment of statement can be set to any arbitrary polarity.

We conduct two sets of experiments which initialize the statement sentiments with two different settings. In the first set of experiments, we initialize the statement sentiments based on the aggregated opinions, and in the second set of experiments, we set all the statement sentiments as positive. In the experiments, we only use four aspect categories, i.e., *food, price, service,* and *ambience*. The category of miscellaneous is ignored since it is not an informative category related to trustworthiness analysis. Finally, in the experiments, we set the values of α and μ to 0.5.

Results. The average truthfulness scores of four aspect categories under two statement sentiment initialization settings are shown in Table 3. In both settings, food-related statements receive the highest scores.

Table 3. Average truthfulness scores of the statements of different categories under different initialization settings

Category	Initialized with majority opinions	Initialized to be all positive
Food	0.744	0.620
Price	0.554	0.196
Service	0.471	0.276
Ambience	0.676	0.102

Compared with scores using the first setting, the results from the second setting all decreased. Truthfulness scores about categories except food have obvious deduction, especially for the category ambience. This means that statements on categories like price and ambience are more controversial and subjective, since many "positive" statements about these categories are considered of low trustfulness.

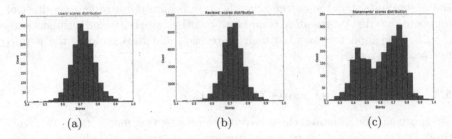

Fig. 2. Distribution of scores when sentiments of statements are set based on majority opinions. (a) Distribution of honesty scores for users. (b) Distribution of faithfulness scores for reviews. (c) Distribution of truthfulness scores for statements

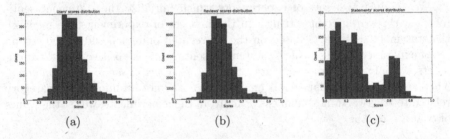

Fig. 3. Distribution of the scores when sentiments of statements are all set to be positive. (a) Distribution of honesty scores for users. (b) Distribution of faithfulness scores for reviews. (c) Distribution of truthfulness scores for statements

For score distributions, we first presents the results of the experiment where we set the sentiments based on majority opinions, as it is the most intuitive setting. The distributions of the honesty, faithfulness, and truthfulness scores are shown in Fig. 2. The results show that for users and reviews, the scores roughly follows normal distributions with mean both around 0.75, which indicate that most of the users and reviews are somehow with some biased opinions but still honest. The distribution of truthfulness scores are somehow skewed and pushed to the right side. The results indicate that most claims are of high truthfulness since they are initialized based on majority opinions.

We did the experiments under another setting to make sure our model works as we expected. In the second setting, we make initial sentiment polarities of the statements set to be positive. Under this setting, what can be expected is that some statements would be false since in reality, they do not have that kind of positive feedback, and thus these statements will receive much lower truthfulness scores. However, the scores of users and reviews will not be affected too much since our model will always award the users who express opinions that are consistent with the majorities and penalized those who do not. In the second set of experiments, when setting sentiment polarities of all statements to be positive, the distributions of scores are shown in Fig. 3. The most obvious change is that the distribution of truthfulness of statements is divided into two parts, which is as expected from the results in Table 3. A part of statements receive scores lower than 0.4, indicating these are the statements that becomes false because of the arbitrary initialization of positive sentiment. Another part of statements still have relatively high scores as they are still true under this setting. Note that the changes in the distribution of statements scores do not mean that our model is sensitive to initializations. The drop of truthfulness scores of some statements is caused by the unreliable initial values of statements (arbitrarily set to positive). During the trust propagation in our model, some of truthfulness of the statements are penalized since the majority do not agree that it should be positive. In fact, the changes of statements scores reflect how our model treats unreliable statements and it is exactly what we expect to see. For honesty and faithfulness, the distributions left shift a bit as some users are affected by the false statement. As expected, the honesty scores and faithfulness scores did not change much.

Evaluations. In this work, we did two kinds of evaluations, add synthetic data and use human evaluators. The purpose of using synthetic data is to test whether our model works in the way as we expected. To achieve this goal, we modified the data of 20 users in the dataset and changed them to extreme cases. 10 users are changed to fully support all the statements their reviews expressed. The rest 10 users are changed to reject all the statements their reviews expressed. With the modified data, we conducted experiments under the setting that sentiments of statements are set based on majority opinions. The distributions of the scores are shown in Fig. 4. The scores of the users with synthetic data are shown in Table 4.

Table 4. Average honesty scores of users with synthetic data

Synthetic type	Min	Average	Median	Max
Support	0.784	0.865	0.863	0.942
Reject	6.842e−12	0.002	6.842e−12	0.013

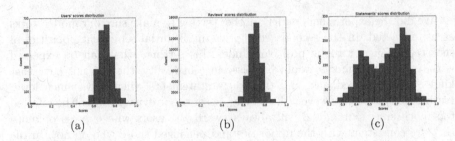

Fig. 4. Distribution of the scores with synthetic data. (a) Distribution of honesty scores for users. (b) Distribution of faithfulness scores for reviews. (c) Distribution of truthfulness scores for statements

The results of the synthetic data show that the model works the way as expected, to award users who agree with majority opinions and penalize users who do not. As mentioned before, we argue that the majority opinions reflect the truth about the qualities of items in the review system, thus the honesty and faithfulness scores we defined reflect the trustworthiness of users and reviews respectively.

As for human evaluation, three human evaluators were involved. We randomly selected twenty users from our dataset as the subject of the evaluation. For each user, eight reviews were randomly picked. Since every two users could form a pair, there would be 190 pairs in total. We randomly picked 20 pairs to compare the users' extent of honesty by asking the three evaluators to read their reviews. For every pair of users, the evaluators were instructed to make a judgment of which user was more honest. For example, for two users u_1 and u_2, the judgment of honesty is either $u_1 > u_2$ or $u_1 < u_2$. We conducted two steps of evaluations. In the first step, the only information about a user we provided for the human evaluator was the eight randomly selected reviews. In the second step, along with the reviews, we provided the ratings about the reviewed restaurants on Yelp.com as facts of qualities.

Table 5. Agreement in first evaluation

	Our model	Evaluator 1	Evaluator 2	Evaluator 3
Our model		13	5	7
Evaluator 1	13		10	10
Evaluator 2	5	10		12
Evaluator 3	7	10	12	

After we got the judgment from the evaluators, we compared the judgments with the results of users' honesty scores from our proposed model. In the model, for each user, his/her honesty score can be computed. For each pair of users

Table 6. Agreement in second evaluation

	Our model	Evaluator 1	Evaluator 2	Evaluator 3
Our model		13	12	13
Evaluator 1	13		7	12
Evaluator 2	12	7.		13
Evaluator 3	13	12	13	

u_1 and u_2, we can make a judgment according to the honesty scores $h(u_1)$ and $h(u_2)$. The agreements among human evaluators and our model of first level and second level evaluations are shown in Tables 5 and 6 respectively. Here the agreement means that the judgment of whether a user is more honest than the other is consistent between two results. For example, in not meaningful. In the first evaluation, human evaluators just read the randomly selected reviews and have no reference for quality judgment. By barely reading reviews, the evaluators tend to make arbitrary judgments and the agreements between human and our model are relatively low. In the second evaluation, evaluators have ratings of the reviewed targets from Yelp.com as the group truth for qualities. Comparing the results of the two tables, we find that the agreements between human evaluators and our model in the second evaluation are higher than in the first one, which means that by providing the actual ratings of the restaurants as facts, the human evaluators were able to make more reasonable judgments and achieved better consistency with our model. Also, comparing to the intra-human agreements, the agreements between human evaluators and our model are pretty acceptable.

To further analyze the agreements between evaluators and our model, out of the agreements between each pair of evaluators, we computed the ratio of overlapping agreements of model and the pair of evaluators (e.g., the ratio of agreements of model, evaluator 1, and evaluator 2 over the agreements between evaluator 1 and evaluator 2). The computed ratios for the two evaluations are shown in Table 7. The increased ratio of overlapping agreements in the second evaluation indicates that with proper reference of quality, evaluators tend to make similar judgments with our model. The judgments that our model disagree with the evaluators, the evaluators themselves are also unlikely to agree with each other. The evaluations show that our model is able to achieve higher consistency with the human evaluators when they have fair reference of qualities. Thus, our model is able to evaluate the extent of honesty of users.

Table 7. Ratio of overlapping agreements between model and each pair of evaluators in two evaluations

	Evaluator 1 & 2	Evaluator 1 & 3	Evaluator 2 & 3
First evaluation	0.400	0.500	0.167
Second evaluation	0.857	0.750	0.692

6 Conclusion

In this work, we study the problem of inferring trustworthiness from the content of online reviews. We first apply opinion-mining techniques using supervised learning algorithms to extract opinions that are expressed in the reviews. Then, we integrate the opinions to obtain opinion vectors for individual reviews and statements. Finally, we develop an iterative content-based computational model to compute honesty scores for users, reviews, and statements. According to the results, there exist differences of statement truthfulness across different categories. Our model shows that the trustworthiness of a user is closely related to the content of her reviews. The review dataset we used was collected in 2013. The structures and content in the dataset are static and there is no dynamic changes considered in our model. However, the reviews and qualities of restaurants tend to change with time. In order to take the dynamic changes into account, we plan to add a temporal dimension in our model in the future. For the opinion mining task, we applied a supervised learning model and used a labeled dataset. However, manually labeling dataset is usually both labor-intensive and time consuming. In our next step, we will apply unsupervised learning methods such as word2vec to group the aspect categories, and thus to automate the opinion mining process.

References

1. Akoglu, L., Chandy, R., Faloutsos, C.: Opinion fraud detection in online reviews by network effects. In: ICWSM (2013)
2. Baccianella, S., Esuli, A., Sebastiani, F.: Sentiwordnet 3.0: an enhanced lexical resource for sentiment analysis and opinion mining. In: LREC, vol. 10, pp. 2200–2204 (2010)
3. Brody, S., Elhadad, N.: An unsupervised aspect-sentiment model for online reviews. In: Human Language Technologies: The 2010 Annual Conference of the North American Chapter of the Association for Computational Linguistics, pp. 804–812. Association for Computational Linguistics (2010)
4. Chen, Z., Mukherjee, A., Liu, B.: Aspect extraction with automated prior knowledge learning. In: ACL, vol. 1, pp. 347–358 (2014)
5. Fahrni, A., Klenner, M.: Old wine or warm beer: target-specific sentiment analysis of adjectives. In: Proceedings of the Symposium on Affective Language in Human and Machine, AISB, pp. 60–63 (2008)
6. Fei, G., Mukherjee, A., Liu, B., Hsu, M., Castellanos, M., Ghosh, R.: Exploiting burstiness in reviews for review spammer detection. In: ICWSM (2013)
7. Ganu, G., Elhadad, N., Marian, A.: Beyond the stars: improving rating predictions using review text content. WebDB **9**, 1–6 (2009)
8. Guha, R., Kumar, R., Raghavan, P., Tomkins, A.: Propagation of trust and distrust. In: Proceedings of the 13th International Conference on World Wide Web, pp. 403–412. ACM (2004)
9. Hai, Z., Chang, K., Kim, J.J., Yang, C.C.: Identifying features in opinion mining via intrinsic and extrinsic domain relevance. IEEE Trans. Knowl. Data Eng. **26**(3), 623–634 (2014)

10. Hu, M., Liu, B.: Mining and summarizing customer reviews. In: Proceedings of the tenth ACM SIGKDD International Conference on Knowledge Discovery and Data Mining, pp. 168–177. ACM (2004)
11. Jijkoun, V., Hofmann, K.: Generating a non-English subjectivity lexicon: relations that matter. In: Proceedings of the 12th Conference of the European Chapter of the Association for Computational Linguistics, pp. 398–405. Association for Computational Linguistics (2009)
12. Jindal, N., Liu, B.: Opinion spam and analysis. In: Proceedings of the 2008 International Conference on Web Search and Data Mining, WSDM 2008, pp. 219–230. ACM, New York (2008)
13. Jo, Y., Oh, A.H.: Aspect and sentiment unification model for online review analysis. In: Proceedings of the Fourth ACM International Conference on Web Search and Data Mining, pp. 815–824. ACM (2011)
14. Jøsang, A., Marsh, S., Pope, S.: Exploring different types of trust propagation. In: Stølen, K., Winsborough, W.H., Martinelli, F., Massacci, F. (eds.) iTrust 2006. LNCS, vol. 3986, pp. 179–192. Springer, Heidelberg (2006). doi:10.1007/11755593_14
15. Kennedy, A., Inkpen, D.: Sentiment classification of movie reviews using contextual valence shifters. Comput. Intell. 22(2), 110–125 (2006)
16. Kleinberg, J.M.: Authoritative sources in a hyperlinked environment. J. ACM (JACM) 46(5), 604–632 (1999)
17. Konishi, T., Tezuka, T., Kimura, F., Maeda, A.: Estimating aspects in online reviews using topic model with 2-level learning. In: Proceedings of the International MultiConference of Engineers and Computer Scientists, vol. 1, pp. 120–126 (2012)
18. Lim, E.P., Nguyen, V.A., Jindal, N., Liu, B., Lauw, H.W.: Detecting product review spammers using rating behaviors. In: Proceedings of the 19th ACM International Conference on Information and Knowledge Management, CIKM 2010, pp. 939–948. ACM, New York (2010)
19. Liu, Q., Gao, Z., Liu, B., Zhang, Y.: A logic programming approach to aspect extraction in opinion mining. In: 2013 IEEE/WIC/ACM International Joint Conferences on Web Intelligence (WI) and Intelligent Agent Technologies (IAT), vol. 1, pp. 276–283. IEEE (2013)
20. Luca, M., Zervas, G.: Fake it till you make it: reputation, competition, and yelp review fraud. Manag. Sci. 62(12), 3412–3427 (2016)
21. Massa, P., Avesani, P.: Trust-aware recommender systems. In: Proceedings of the 2007 ACM conference on Recommender systems, pp. 17–24. ACM (2007)
22. Miller, G.A.: Wordnet: a lexical database for english. Commun. ACM 38(11), 39–41 (1995)
23. Mukherjee, A., Liu, B., Wang, J., Glance, N., Jindal, N.: Detecting group review spam. In: Proceedings of the 20th International Conference Companion on World Wide Web, WWW 2011, pp. 93–94 (2011)
24. Mukherjee, S., Dutta, S., Weikum, G.:
25. Ott, M., Choi, Y., Cardie, C., Hancock, J.T.: Finding deceptive opinion spam by any stretch of the imagination. In: Proceedings of the 49th Annual Meeting of the Association for Computational Linguistics: Human Language Technologies, vol. 1 (2011)
26. Page, L., Brin, S., Motwani, R., Winograd, T.: The PageRank citation ranking: bringing order to the web (1999)

27. Pang, B., Lee, L.: A sentimental education: sentiment analysis using subjectivity summarization based on minimum cuts. In: Proceedings of the 42nd Annual Meeting on Association for Computational Linguistics, p. 271. Association for Computational Linguistics (2004)

28. Pang, B., Lee, L., Vaithyanathan, S.: Thumbs up?: sentiment classification using machine learning techniques. In: Proceedings of the ACL 2002 Conference on Empirical Methods in Natural Language Processing, vol. 10, pp. 79–86. Association for Computational Linguistics (2002)

29. Pedregosa, F., Varoquaux, G., Gramfort, A., Michel, V., Thirion, B., Grisel, O., Blondel, M., Prettenhofer, P., Weiss, R., Dubourg, V., Vanderplas, J., Passos, A., Cournapeau, D., Brucher, M., Perrot, M., Duchesnay, E.: Scikit-learn: machine learning in Python. J. Mach. Learn. Res. **12**, 2825–2830 (2011)

30. Pontiki, M., Galanis, D., Pavlopoulos, J., Papageorgiou, H., Androutsopoulos, I., Manandhar, S.: Semeval-2014 task 4: aspect based sentiment analysis. In: Proceedings of the 8th International Workshop on Semantic Evaluation (SemEval 2014), pp. 27–35 (2014)

31. Popescu, A.M., Etzioni, O.: Extracting product features and opinions from reviews. In: Kao, A., Poteet, S.R. (eds.) Natural Language Processing and Text Mining, pp. 9–28. Springer, London (2007)

32. Poria, S., Cambria, E., Ku, L.W., Gui, C., Gelbukh, A.: A rule-based approach to aspect extraction from product reviews. In: Proceedings of the Second Workshop on Natural Language Processing for Social Media (SocialNLP), pp. 28–37 (2014)

33. Qiu, G., Liu, B., Bu, J., Chen, C.: Opinion word expansion and target extraction through double propagation. Comput. Linguist. **37**(1), 9–27 (2011)

34. Salvetti, F., Lewis, S., Reichenbach, C.: Automatic opinion polarity classification of movie. Colo. Res. Linguist. **17**(1), 2 (2004)

35. Than, C., Han, S.: Improving recommender systems by incorporating similarity, trust and reputation. J. Internet Serv. Inf. Secur. (JISIS) **4**(1), 64–76 (2014)

36. Titov, I., McDonald, R.: Modeling online reviews with multi-grain topic models. In: Proceedings of the 17th International Conference on World Wide Web, pp. 111–120. ACM (2008)

37. Turney, P.D.: Thumbs up or thumbs down?: semantic orientation applied to unsupervised classification of reviews. In: Proceedings of the 40th Annual Meeting on Association for Computational Linguistics, pp. 417–424. Association for Computational Linguistics (2002)

38. Vydiswaran, V., Zhai, C., Roth, D.: Content-driven trust propagation framework. In: Proceedings of the 17th ACM SIGKDD International Conference on Knowledge Discovery and Data Mining, pp. 974–982. ACM (2011)

39. Wang, G., Xie, S., Liu, B., Yu, P.S.: Identify online store review spammers via social review graph. ACM Trans. Intell. Syst. Technol. **3**(4), 61:1–61:21 (2012)

40. Wilson, T., Wiebe, J., Hoffmann, P.: Recognizing contextual polarity: an exploration of features for phrase-level sentiment analysis. Comput. Linguist. **35**(3), 399–433 (2009)

41. Xie, S., Wang, G., Lin, S., Yu, P.S.: Review spam detection via temporal pattern discovery. In: Proceedings of the 18th ACM SIGKDD International Conference on Knowledge Discovery and Data Mining, KDD 2012, pp. 823–831. ACM, New York (2012)

42. Xue, H., Li, F., Seo, H., Pluretti, R.: Trust-aware review spam detection. In: 2015 IEEE Trustcom/BigDataSE/ISPA, vol. 1, pp. 726–733. IEEE (2015)

43. Yin, X., Han, J., Philip, S.Y.: Truth discovery with multiple conflicting information providers on the web. IEEE Trans. Knowl. Data Eng. **20**(6), 796–808 (2008)

Securing Networks Against Unpatchable and Unknown Vulnerabilities Using Heterogeneous Hardening Options

Daniel Borbor[1]([✉]), Lingyu Wang[1], Sushil Jajodia[2], and Anoop Singhal[3]

[1] Concordia Institute for Information Systems Engineering,
Concordia University, Montreal, Canada
{d_borbor,wang}@ciise.concordia.ca

[2] Center for Secure Information Systems, George Mason University, Fairfax, USA
jajodia@gmu.edu

[3] Computer Security Division,
National Institute of Standards and Technology, Gaithersburg, USA
anoop.singhal@nist.gov

Abstract. The administrators of a mission critical network usually have to worry about non-traditional threats, e.g., how to live with known, but unpatchable vulnerabilities, and how to improve the network's resilience against potentially unknown vulnerabilities. To this end, network hardening is a well-knowfn preventive security solution that aims to improve network security by taking proactive actions, namely, hardening options. However, most existing network hardening approaches rely on a single hardening option, such as disabling unnecessary services, which becomes less effective when it comes to dealing with unknown and unpatchable vulnerabilities. There lacks a heterogeneous approach that can combine different hardening options in an optimal way to deal with both unknown and unpatchable vulnerabilities. In this paper, we propose such an approach by unifying multiple hardening options, such as firewall rule modification, disabling services, service diversification, and access control, under the same model. We then apply security metrics designed for evaluating network resilience against unknown and unpatchable vulnerabilities, and consequently derive optimal hardening solutions that maximize security under given cost constraints.

1 Introduction

Today's computing networks are playing the role of nerve systems in many mission critical infrastructures, such as cloud data centers and smart grids. However, the scale and severity of security breaches in such networks have continued to grow at an ever-increasing pace, which is evidenced by many high-profile security incidents, such as the recent large scale DDoS attacks caused by the Mirai Botnet on the Dyn DNS, and the cyber-physical attack on the Ukrainian power grid in 2015. The so-called zero-day attacks, which exploit either previously unknown or

G. Livraga and S. Zhu (Eds.): DBSec 2017, LNCS 10359, pp. 509–528, 2017.
DOI: 10.1007/978-3-319-61176-1_28

known, but unpatched vulnerabilities, are usually behind such security incidents, e.g., Stuxnet employs four different zero day vulnerabilities to target SCADA. Therefore, administrators of a mission critical network usually need to worry about not only patching known vulnerabilities and deploying traditional defense mechanisms (e.g., firewalls, IDSs, and IPSs), but also non-traditional security threats, e.g., how to live with known, but unpatchable vulnerabilities, and how to improve the network's resilience against potentially unknown vulnerabilities.

In fact, it is well known that both cybercriminals and governmental agencies stockpile vulnerabilities that are not publicly known (e.g., the NSA reportedly spent more than 25 million a year to acquire software vulnerabilities, and private vendors are providing at least 85 zero-day exploits on any given day [16]). On the other hand, even for known vulnerabilities, patching is not always a viable option. For example, a patch may not be readily available at the time of the attack, or the system may have reached their end-of-support with no more patch available; patching a vulnerability may cause unacceptable service disruptions on a regular basis (e.g., Windows updates); even worse, patching a vulnerability may sometimes reintroduce other security vulnerabilities that have previously been fixed (e.g., Apache MINA SSHD 2.0.14 introduces an SSL regression previously fixed in 2.0.13 [20]).

Consequently, security professionals need to block the exploitation of such vulnerabilities through other means, such as modifying firewall rules, service diversification, or access control. A critical question is *How to optimally combine such options in order to both improve the security and lower the cost?* To this end, network hardening is a well-known preventive security solution that aims to improve network security by taking proactive actions, namely, hardening options. However, most existing network hardening approaches rely on a single hardening option, such as disabling unnecessary services [9,21] or service diversification [6] (a detailed review of related work will be given later in Sect. 5). Such a solution becomes less effective when it comes to dealing with unknown and unpatchable vulnerabilities. There lacks a heterogeneous approach that can combine different hardening options in an optimal way to deal with such vulnerabilities.

Running Example. We first consider a concrete example to demonstrate why deriving an optimal hardening solution with heterogeneous hardening options would demand a systematic and automated approach. Figure 1 shows a hypothetical network roughly based on Cisco's cloud data center concept [5] as well as the OpenStack architecture [11]. Despite its relatively small scale, it mimics a typical cloud network, e.g., the client layer connects the cloud network to the internet through the CRS 7600; a firewall (ASA v1000) separates the outside network from the inner one. There is a security/authentication layer (authentication server, Neutron server, etc.) as well as a VM and Application layer (Web and application servers). Finally, a storage layer is separated and protected by another firewall (ASA 5500) and an MDS 9000.

We make the following assumptions about the network. We assume the two firewalls and other host-based security mechanisms (e.g., personal firewalls or iptables) together enforce the connectivity described inside the connectivity table

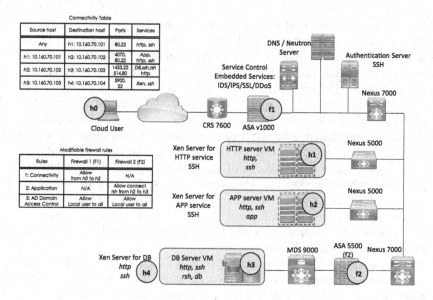

Fig. 1. An example cloud network.

shown in the figure. External users (including attackers) are represented with host $h0$, and the most critical asset is assumed to be the Xen database server ($h4$), which may be accessed through the three-tier architecture involving hosts $h1$, $h2$, and $h3$. We assume the network is free of any known vulnerabilities, except for an unpatchable vulnerability on the application server running SecurityCenter 5.5 (which cannot be changed due to functionality requirements), and another one on the database server running MySQL 5.7 which may be changed to MSQL 2012 or PostgreSQL 9. For simplicity, we exclude exploits and conditions that involve firewalls in this example.

To measure the network's resilience against zero-day attacks, we apply the k-*zero-day safety metric* ($k0d$) [25]. This metric basically counts how many distinct services must be compromised using unknown vulnerabilities before an attacker may compromise the critical asset (i.e., the number of distinct services along the shortest path). In addition, we refine the metric by taking into consideration the potentially uneven distribution of distinct services along the shortest path [29,32] (e.g., a path consisting of three *http* and one *Xen* would be considered slightly "shorter", or less secure, than a path consisting of two *http* and two *Xen*, although both paths have the same number of resource instances and resource types).

For hardening options, we consider changes of both the firewall rules and service types. First, we assume the administrator may enable or disable firewall rules on both the firewall ASA v1000 ($f1$) and on the firewall ASA 5500 ($f2$). On $f1$ he has a rule that allows the connection from the cloud user ($h0$) to the *app* VM ($h2$); he also has the option to allow local user access to $h1$ and $h2$. The firewall ASA 5500 ($f2$) has a rule where he allows the *rsh* connection on

$h3$ from $h2$, as well as local user access to $h3$ and $h4$. Second, we assume the administrator has the option of replacing the Apache Mina 2.0.14 *ssh* servers with either Copssh 5.8, OpenSSH 7.4, or Attachmate 8.0; the Web servers with either Apache 2.4, IIS 8.5, NGINX 1.9 or a Litespeed 5.0.14 Web server; the *rsh* service only uses MVRSHD 2.2.

Clearly, even with such a small scale network, the administrator now faces a number of hardening options, including disabling service instances, diversifying service types, and changing firewall rules, each of which may incur certain installation/maintenance costs (we will discuss the cost model in more details later in Sect. 2). To maximize the resilience of the network against both unknown and unpatchable vulnerabilities, the administrator must decide what would be the optimal combination of such hardening options in order to maximize the aforementioned security metric, while respecting given cost constraints. Such a task would obviously be tedious and error-prone, if done manually, and demands a systematic and automated approach.

In this paper, we develop such an approach to optimally combine heterogeneous hardening options in order to increase a network's resilience again both unknown and unpatchable vulnerabilities under various cost constraints. Specifically, we first devise our model of different hardening options, costs, and the security metric. We then develop optimization and heuristic algorithms to derive optimal hardening solutions under given cost constraints. We evaluate our approach through simulations in order to study the effect of optimization parameters on accuracy and running time, and the effectiveness of optimization for different types of networks. In summary, the main contribution of this paper is the following.

- To the best of our knowledge, this is the first effort on network hardening using heterogeneous hardening options.
- In contrast to previous works, we provide a refined security metric and an improved cost model that takes into account real world variables in calculating hardening costs.
- Our method is practically relevant to the defense of mission critical networks in which unknown and unpatchable vulnerabilities are realistic security concerns.

The remainder of this paper is organized as follows: In Sect. 2, we present the model and formulate the optimization problem, and in Sect. 3 we discuss the methodology and show case studies. Section 4 shows simulation results. Section 5 reviews related work and Sect. 6 concludes the paper.

2   The Model

We first introduce the extended resource graph model to capture network services and their relationships, then we present the heterogeneous hardening control and cost model, followed by the problem formulation.

2.1   Extended Resource Graph

To model network services and their relationships, we revise the *Extended Resource Graph* concept introduced in our previous work [6] in order to model both unpatchable and unknown vulnerabilities, as well as heterogeneous hardening options. The extended resource graph of the running example is shown in Fig. 2 and detailed below.

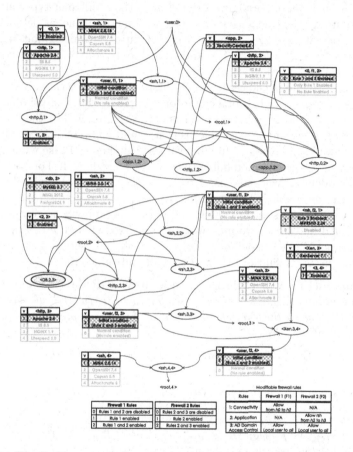

Fig. 2. The extended resource graph of our running example.

In Fig. 2, each pair shown in a rectangle is a security-related condition. If the condition is a privilege, it is represented as ⟨*privilege, host*⟩; if it is connectivity, it is represented as ⟨*source, destination*⟩. If a firewall affects a security-related condition, it is represented as ⟨*privilege, firewall, host*⟩ or as ⟨*source, firewall, destination*⟩. Each one of the rows below the rectangle indicate different hardening options available for that condition. The option currently in use is indicated by the highlighted integer (e.g., 0 means disabled; in the case of service diversification, 1 means Apache, and 2 means IIS) and other potential instances are in a lighter text. For the conditions modifiable by a firewall rule, the rows below the rectangle indicate the firewall rules that affect it.

Each exploit node (oval) is a tuple that consists of a service running on a destination host, the source host, and the destination host (e.g., the tuple $\langle http, 1, 2 \rangle$ indicates a potential zero-day vulnerability in the $http$ service on host 2, which is exploitable from host 1). If the exploit is unpatchable, but diversifiable, it is represented by a double oval; if it is neither patchable nor diversifiable, it is represented as a colored oval (those different types of exploits will contribute to the calculation of the security metric value, as detailed later). The self-explanatory edges point from preconditions to an exploit (e.g., from $\langle 0, 1 \rangle$ and $\langle http, 1 \rangle$ to $\langle http, 0, 1 \rangle$), and from the exploit to its post-conditions (e.g., from $\langle http, 0, 1 \rangle$ to $\langle user, 1 \rangle$).

We make two design choices here. The first is to associate the service instance concept as a property (label) of a condition (e.g., $\langle http, 1 \rangle$), instead of an exploit (as in our previous work [6]). This label can then be inherited by the corresponding exploits. The second design choice is that, while some conditions indicate the involved firewall rules, the actual label values that they will take will depend on the number of predefined modifiable rules in the firewall itself. For each firewall, instead of modeling service instances, we model the number of modifiable firewall rules that can be enabled. This would help to avoid the need for introducing new conditions and exploits into the extended resource graph when firewall rules are to be disabled and hence we may work with a fixed structure of the extended resource graph. While the definitions of service pool and service instance remain the same as in [6], Definitions 1 and 2 formally introduce the revised concepts.

Definition 1 (Firewall Rule Pool and Firewall Rule). *Denote F the set of all firewalls and Z the set of integers, for each firewall $f \in F$, the function $r(.) : F \to Z$ gives the firewall rule pool of f which represent all modifiable firewall rules of that firewall.*

Definition 2 (Extended Resource Graph). *Given a network composed of*

- *a set of hosts H,*
- *a set of services S, with the service mapping $serv(.) : H \to 2^S$,*
- *the collection of service pools $SP = \{sp(s) \mid s \in S\}$,*
- *the collection of firewall rules $FR = \{r(f) \mid f \in F\}$,*
- *a set of firewalls F, with the rule mapping $r(.) : F \to | FR |$,*
- *and the labeling function $v(.) = v_f(.) \cup v_c(.)$ where $v_f(.) : f \to F$ and $v_c(.) : C \to SP$.*

Let E be the set of zero-day exploits $\{\langle s, h_s, h_d \rangle \mid h_s \in H, h_d \in H, s \in serv(h_d)\}$, and $R_r \subseteq C \times E$ and $R_i \subseteq E \times C$ be the collection of pre and post-conditions in C. We call the labeled directed graph, $\langle G(E \cup C, R_r \cup R_i), v \rangle$ the extended resource graph.

2.2 Heterogeneous Hardening Control and Cost Model

We introduce the notion of *heterogeneous hardening control* as a model to account for all hardening options in a network where we represent each initial condition as an optimization variable. We formulate the heterogeneous hardening control vectors using those variables as follows. We note that the number of optimization variables present in a network will depend on the number of initial conditions that are affected by one or more hardening options. Since we only consider remotely accessible services in the extended resource graph model, we would expect in practice the number of optimization variables to grow linearly in the size of the network (i.e., the number of hosts). We will further evaluate and discuss the scalability of our solution in Sect. 4.

Definition 3 (Optimization Variable and Heterogeneous Hardening Control). *Given an extended resource graph* $\langle G, v \rangle$, $\forall c \in C$ *and* $\forall f \in F$, $v(c)$ *and* $v(f)$ *are optimization variables. A hardening control vector is the integer valued vector* $\boldsymbol{V} = (v(c_1), v(c_2), ..., v(c_{|C|}) \cup (v(f_1), v(f_2), ..., v(f_{|F|})$

Changing the value of an optimization variable has an associated *hardening cost* and the collection of such costs are given in a *hardening cost matrix* in a self-explanatory manner. We make use of Gartner's 2003 *Total Cost of Ownership* (TCO) analysis report [19] to establish a realistic cost estimation of the cost of different hardening options. Table 1 provides a reference as to which criteria is applicable to different hardening options costs.

Table 1. Criteria to be used when calculating hardening costs for different hardening options based on Gartner's TCO [19]

Hardening option cost selection criteria				
Gartner's TCO criteria	Firewall connectivity	Firewall layer 3	Firewall access control	Diversity
Downtime costs				x
Operational costs	x	x	x	x
Support costs	x		x	x
Changes in upgrade costs	x	x	x	x
Monitoring costs			x	x
Production costs				x
Security management and failure control costs	x	x	x	x

Definition 4 (Hardening Cost). *Given* $s \in S$ *and* $sp(s)$, *and given* $f \in F$ *and* $r(f)$, *the cost to change from one specific hardening option to another is defined as the hardening cost.*

Definition 5 (Hardening Cost Matrix). *The collection of all hardening costs for all hardening options are given as a hardening cost matrix HCM. For* *the different hardening options, the element at i^{th} row and j^{th} column indicates the hardening cost of changing the ith hardening option to be the jth hardening option.*

Definition 6 (Total Hardening Cost). *Let $v_s(c_i)$ be the service associated with the optimization variable $v(c_i)$ and $\boldsymbol{V}_{c0}$ the initial service instance values for each of the conditions in the network. Let $v_f(f_i)$ be the firewall associated with the optimization variable $v(f_i)$ and $\boldsymbol{V}_{f0}$ the initial firewall rule set values for each of the firewalls in the network. The total hardening cost, Q_d, given by the heterogeneous hardening vector $\boldsymbol{V}$ is obtained by*

$$Q_d = \sum_{i=1}^{|C|} CM_{v_s(c_i)}(\boldsymbol{V}_{c0}(i), \boldsymbol{V_c}(i)) + \sum_{i=1}^{|F|} CM_{v_f(f_i)}(\boldsymbol{V}_{f0}(i), \boldsymbol{V_f}(i))$$

We note that the above definition of hardening cost between each pair of service instances has some advantages. For example, in practice we can easily imagine cases where the cost is not symmetric, i.e., changing one service instance to another (e.g., from Apache to IIS) carries a cost that is not necessarily the same as the cost of changing it back (from IIS to Apache). Our approach of using a collection of two-dimensional matrices allows us to account for cases like this. Additionally, by considering instance 0, it provides us the advantage to model disabling a service as a special case of service diversification if the hardening option allows it. Nonetheless, our cost model can certainly be further improved, as discussed in Sect. 6.

2.3   Problem Formulation

As mentioned in Sect. 1, the security metric that we will be using, denoted as d, is based on the minimum number of distinct resources, excluding those with unpatchable vulnerabilities, on the shortest attack path in the resource graph [25], with the extension for considering the uneven distribution of services along that path [29,32], as formally defined below.

Definition 7 (d-Safety Metric). *Given an extended resource graph $\langle G(E \cup C, R_r \cup R_i), v \rangle$, and a goal condition $c_g \in C$; let $t = \sum_{i=1}^{n} 2^{-n} \mid serv(h_i))$ | (total number of service instances), and let $p_j = \frac{|h_i:s_j \in serv(h_i))|}{t}$ $(1 \leq i \leq n, 1 \leq j \leq n)$ (relative frequency of each resource). For each $c \in C$ and $q \in seq(c)$ (attack path), denote $R(q)$ for $s : s \in R$, r appears in q, r is not unpatchable, we define the network's d-safety metric (where min(.) returns the minimum value in a set) $d = min_{q \in seq(c_g)} r(R(q))$; where $r(R(q))$ is the attack path's effective richness of the services, defined as $r(G) = \frac{1}{\prod_1^n p_i^{p_i}}$ [29].*

With the aforementioned models, the network hardening problem is to maximize the d value by changing the hardening options while respecting the available budget in terms of given cost constraints. In the following, we formally formulate this as an optimization problem.

Problem 1 (d-Optimization Problem).   Given an extended resource graph $\langle G, v \rangle$, find a heterogeneous hardening control vector V which maximizes $min(d(\langle G(V), v \rangle))$ subject to the constraint $Q \leq B$, where B is the available budget and Q is the total hardening cost as given in Definition 6.

Since our problem formulation is based on an extended version of the resource graph, which is syntactically equivalent to attack graphs, many existing tools developed for the latter (e.g., the tool in [15] has seen many real applications to enterprise networks) may be easily extended to generate extended resource graphs which we need as inputs. Additionally, our problem formulation assumes a very general model of budget B and cost Q, which allows us to account for different types of budgets and cost constraints that an administrator might encounter in practice, as will be demonstrated in the following section.

3   The Methodology

This section details our optimization and heuristic algorithms used for solving the formulated heterogeneous hardening problem. We also illustrate the optimization process through a few case studies.

3.1   Optimization Algorithm

Our first task is to select an optimization algorithm that would fit our hardening problem. First, it is well known that most gradient-based methods require to satisfy mathematical properties like convexity or differentiability, which are not applicable to our problem. Second, the problem we want to solve includes different if-then-else constructs to account for the different hardening technique used, and thus, an algorithm that allows to insert this construct is necessary. Additionally, since our optimization problem uses variables that are defined as discrete (discrete variable space), a simple and robust search method and optimization technique is needed. We find that metaheuristic algorithms provide these advantages. Specifically, the Genetic Algorithm (GA) provides a simple and clever way to encode candidate solutions to the problem [8]. One of the main advantages is that we do not have to worry about explicit mathematical definitions. For our automated optimization approach, we chose GA because it requires little information to search effectively in a large search space in contrast to other optimization methods (e.g., the mixed integer programming [4]).

The extended resource graph is the input to our automated optimization algorithm where the fitness function is d. One important point to consider when optimizing the d function on the extended resource graph is that, for each generation of the GA, the graph's labels selected will dynamically change. This in turn will change the value of d, since the shortest path may have changed with each successive generation of GA and the change in the hardening options will enable or disable certain paths. Our optimization tool takes this into consideration. Additionally, if there are more than one shortest path that provides the optimized d, our optimization tool gives priority to the paths by considering the uneven distribution and relative frequency of resources in that path, thus addressing one of the limitations that was present in [6] where no priority was provided.

The constraints are defined as a set of inequalities in the form of $q \leq b$, where q represents one or more constraint conditions and b represents one or more budgets. These constraint conditions can be overall constraints (e.g., the total hardening cost Q_d) or specific constraints to address certain requirements or priorities while implementing the heterogeneous hardening options. The number of independent variables used by the GA (genes) are the optimization variables given by the extended resource graph. For our particular network hardening problem, the GA will be dealing with integer variables representing the selection of a hardening option. Because $v(.)$ is defined as an integer, the optimization variables need to be given a minimum value and a maximum value. This range is determined by the number of instances provided in the service pool of each service and firewall rule pool of each firewall. The initial service instance for each of the services and the initial set of firewall rules are given by the extended resource graph while the final heterogeneous hardening control vector V is obtained after running the GA.

3.2   Case Studies

In the following, we demonstrate different use cases of our method with varying cost constraints and hardening options. For these test cases, the population size defined for our tool is set to be at least the value of optimization variables (more details will be provided in the coming section). This way we ensure the individuals in each population span the search space. We ensure the population diversity by testing with different settings in genetic operations (like crossover and mutation). For all the test cases, we have used the following algorithm parameters: population size = 100, number of generations = 150, crossover probability = 0.8, and mutation probability = 0.2 (Fig. 3).

Test case A: $Q_d \leq 500$ units with firewall rule constraints. We start with the simple case of one overall budget constraint ($Q_d \leq 500$). There are 11 different services-based optimization variables and 2 firewall-based optimization variables. If no firewall rules are changed, the solution provided by the GA yields $d = 2.7529$. In this case, because of the firewall rules that are enabled, the metric cannot be increased any further.

On the other hand, if we allow the firewall rules to be modified, while maintaining the overall budget $Q_d \leq 500$, the optimization results will be quite different. The solution provided by the GA is a d metric of 3.3895. This total hardening cost satisfies both the overall budget constraints. We can see that the hardening options enforced by the firewall rules in our optimization tool can affect the optimization. Nevertheless, additional budget constraints might not allow achieving the maximum d possible.

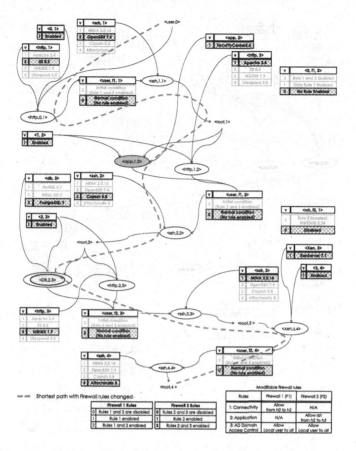

Fig. 3. Test case A: Effect of modifiable hardening options and budget constraints.

Test case B: $Q_d \leq 500$ units with a critical service with an unpatched vulnerability. While test case A shows how enabling or disabling predefined firewall rules can affect the d metric optimization, when considering the effects of unpatchable vulnerabilities the d metric value will change. This test case models such a scenario by assigning a restriction for the *ssh* services not to be diversified or disabled.

In the graph, we can see that the *ssh* service is highlighted to represent the fact that it cannot be patched. The solution provided by the GA is $d = 2.8284$. While the increase is less than when the *ssh* service can be diversified, we can still have an increase in the d metric even with unpatchable vulnerabilities on the network (Fig. 4).

As seen from the above test cases, our model and problem formulation makes it relatively straightforward to apply any standard optimization techniques, such as the GA, to optimize the d metric through combining different network hardening options while dealing with unpatchable and unknown vulnerabilities and respecting given cost constraints.

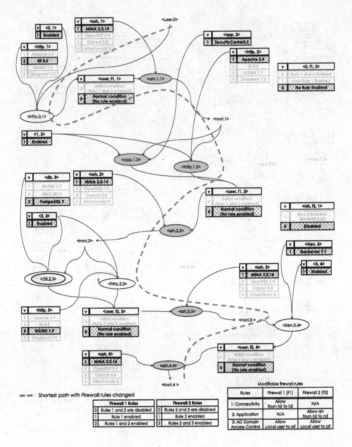

Fig. 4. Test case B: Effect of having an unpatchable vulnerability in the network.

3.3 Heuristic Algorithm

All the test cases described above rely on the assumption that all the attack paths are readily available. However, this is not always the case in practice. Due to the well-known complexity that resource graphs have inherited from attack graphs due to their common syntax [29,32], it is usually computationally infeasible to enumerate all the available attack paths in a resource graph for large networks. Therefore, we present a modified version of the heuristic algorithm [6] to reduce the search complexity when calculating and optimizing the d metric by only storing the m-shortest paths at each step. The following briefly describes the modified algorithm.

The algorithm starts by finding the initial conditions that are affected by the modifiable firewall rules and stores them on a list γ. After that, it topologically sorts the graph and proceeds to go through each one of the nodes on the resource graph. If an exploit is a post-condition of one of the conditions in γ, it is not included in the set of exploits $\sigma()$. The main loop cycles through each unprocessed node. If a node is an initial conditions, the algorithm assumes that the node itself is the only path to it and it marks it as processed. For each exploit e, all of its preconditions are placed in a set. The collection of attack paths $\alpha(e)$ is constructed from the attack paths of those preconditions. In a similar way, $\sigma'(ov(e))$ is constructed with the function $ov()$ which, aside from using the exploits, includes value of elements of the hardening control vector that supervises that exploit.

If there are more than m paths to that node, the algorithm will first look for the relative frequency of each unique combination of exploit and service instance in $\alpha'(ov(e))$. Then, the algorithm creates a dictionary structure where the key is a path from $\alpha(e)$ and the value is the effective richness of service/service instance combinations given by each one of the respective paths in $\alpha'(ov(e))$. A function $ShortestM()$ selects the top m keys whose values are the smallest and returns the m paths with the smallest effective richness value. If there are less than m paths, it will return all of the paths. After this, it marks the node as processed. The process is similar when going through each one of the intermediate conditions. Finally, the algorithm returns the collection of m paths that can reach the goal condition c_g. It is worth noting that by considering the effective richness of each path, the algorithm provides a path a priority based on the relative frequency of the combination of unique service with service instance.

4   Simulations

In this section, we show simulation results. All simulations are performed using a computer equipped with a 3.0 GHz CPU and 8GB RAM in the Python 2.7.10 environment under Ubuntu 12.04 LTS and MATLAB 2015a's GA toolbox. To generate a large number of resource graphs for simulations, we first construct a small number of seed graphs based on realistic cloud networks and then generate larger graphs from those seed graphs by injecting new hosts and assigning resources in a random but realistic fashion (e.g., the number of pre-conditions of each exploit is varied within a small range since real world exploits usually have a constant number of pre-conditions).

For the different hardening options that are implemented through firewall rules, we randomly select 10% of the initial conditions. Additionally, to analyze the effect of unpatchable vulnerabilities, our graphs include randomly assigned unpatchable services. The resource graphs are used as the input for the optimization toolbox where the objective function is to maximize the minimum d value subject to budget constraints. In all the simulations, we employ the heuristic algorithm described in Sect. 3.3.

To determine the genetic operators, we used the hill climbing algorithm. Our simulations showed that (detailed simulation results are omitted here due to page limitations), using the GA with a crossover probability of 80%, a mutation rate of 20%, and setting the number of generations to 70 will be sufficient to obtain good results. Additionally, our experiences also show that, because our largest resource graph had a heterogeneous hardening control vector of fewer

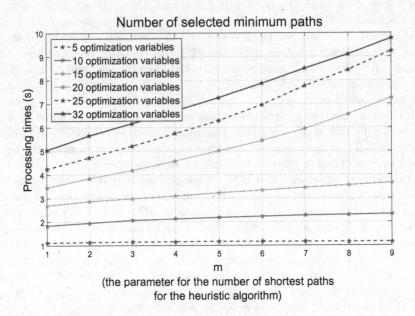

Fig. 5. The processing time.

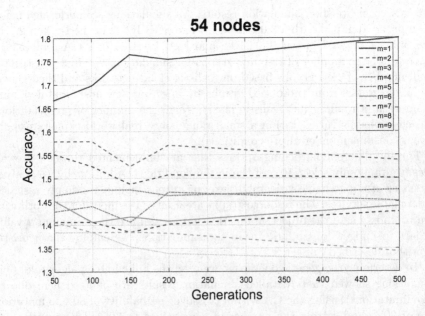

Fig. 6. The accuracy vs. m (the parameter of the heuristic algorithm).

than 100 variables, we could set the population size equal to 200; nevertheless, we believe that when dealing with a bigger number of optimization variables, the population size should be at least twice the number of variables.

The complexity of our proposed solution will depend on the objective function, the population size, and the length of hardening control vector. We note that the optimization problem here is NP-hard since the sub-problem of finding the shortest paths (within the objective function) in resource graphs is already intractable by the well know results in attack graphs [29,32] and the common syntax between resource graphs and attack graphs. We will therefore rely on the heuristic algorithm presented in Sect. 3.3. Figure 5 shows that the processing time increases almost linearly as we increase the number of optimization variables or the parameter m of the heuristic algorithm. The results show that the algorithm is relatively scalable with a linear processing time.

The accuracy of the results presented in Fig. 5 is also an important issue to be considered. This is address through the simulations depicted in Fig. 6. Here the accuracy refers to the approximation ratio between the result obtained for the d metric using our heuristic algorithm and that of simply enumerating and searching all the paths while assuming all services and service instances are different ($\frac{d_{Heuristic}}{d_{BruteForce}}$). The heterogeneous hardening control vector provided by the GA is used to calculate the accuracy. A ration close to 1 indicates that our algorithm can provide a solution that is closer to the one provided by enumerating all paths (brute force). From the results, we can see that when m is greater or equal to 4 the approximation ratio reaches an acceptable level. For the following simulations, we have settled with an m value of 9.

We also consider the ratio between the difference in the d metric before and after optimization, ($\frac{d_{Optimized} - d_{NotOptimized}}{d_{NotOptimized}}$), which will be called the *gain* of the d metric (or simply the gain). The gain provides us with an idea on how much room there is to improve the security with respect to given cost constraints using our method. Figure 7 shows that the gain will increase linearly as we increase the number of firewall-based hardening options. These results confirm that firewall-based hardening options can positively affect our effort to provide better resilience for cloud networks against zero-day attacks. Additionally, the figure shows that the number of unpatchable vulnerabilities that are present in the network will significantly reduce the gain that can be achieved through other hardening techniques. Since it is not probable to find a large number of unpatchable vulnerabilities all at the same time within a network, we only consider up to three unpatchable vulnerabilities.

In Fig. 8, we analyze the average gain in the optimized results for different sizes of graphs. In this figure, we can see that we have a good enough gain for graphs with a relatively high number of nodes. As expected, as we increase the number of unpatchable vulnerabilities, the gain will decrease. However, we can also see this decrease is linear. In the case where no unpatchable vulnerabilities are present, we can see that the gain stops to increase after reaching a certain size of the graph, which can be explained as that the number of available service

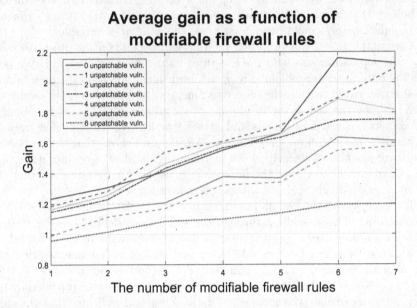

Fig. 7. The average gain based on the number of modifiable firewall rules.

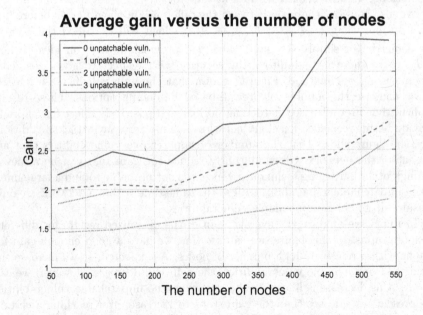

Fig. 8. The average gain vs. the number of nodes.

instances is not large enough (in constrast to the increasing size of the graph) to allow to optimize the d metric any further.

5   Related Work

In general, the security of networks may be qualitatively modeled using attack trees [9,10,22] or attack graphs [2,23]. A majority of existing quantitative models of network security focus on known attacks [1,28], while few works have tackled zero day attacks [25,26,29,32] which are usually considered unmeasurable due to the uncertainties involved [17]. In terms of security metrics, most of the current works deal with assigning numeric scores to rank known vulnerabilities (mostly based on the CVSS) [18] to be able to model the impact that they have on a network. This ranking is based on how likely and easily exploitable the known vulnerabilities are. This, however, is not the case for unknown vulnerabilities.

Early works on network hardening typically rely on qualitative models while improving the security of a network [23,24,27]. Those works secure a network by breaking all the attack paths that an attacker can follow to compromise an asset, either in the middle of the paths or at the beginning (disabling initial conditions). Also, those works do not consider the implications when dealing with budget constraints nor include cost assignments, and tend to leave that as a separate task for the network administrators. While more recent works [1,31] generally provide a cost model to deal with budget constraints, one of the first attempts to systematically address this issue is by Gupta et al. [14]. The authors employed genetic algorithms to solve the problem of choosing the best set of security hardening options while reducing costs.

Dewri et al. [9] build on top of Gupta's work to address the network hardening problem using a more systematic approach. They start by analyzing the problem as a single objective optimization problem and then consider multiple objectives at the same time. Their work consider the damage of compromising any node in the cost model in order to determine the most cost-effective hardening solution. Later on, in [10] and in [30], the authors extrapolate the network hardening optimization problem as vulnerability analysis with the cost/benefit assessment, and risk assessment respectively. In [21] Poolsappasit et al. extend Dewri's model to also take into account dynamic conditions (conditions that may change or emerge while the model is running) by using Bayesian attack graphs in order to consider the likelihood of an attack. Unlike our work, most existing work is limited to known vulnerabilities and focus on disabling existing services.

There exist a rich literature on employing diversity for security purposes. The idea of using design diversity for tolerating faults has been investigated for a long time, such as the N-version programming approach [3], and similar ideas have been employed for preventing security attacks, such as the N-Variant system [7], and the behavioral distance approach [12]. In addition to design diversity and generated diversity, recent work employ opportunistic diversity which already exists among different software systems. For example, the practicality of employing OS diversity for intrusion tolerance is evaluated in [13]. More recently, the authors in [29,32] adapted biodiversity metrics to networks and lift the diversity metrics to the network level. While those works on diversity provide motivation and useful models, they do not directly provide a systematic solution for improving diversity. So far, the work done by [6], is one of the first work that has tried to provide a solution for this problem; their limitation, however, is that their metric is too simplistic and does not consider additional hardening options.

6   Conclusions

In this paper, we have provided a heterogeneous approach to network hardening to increase the resilience of a network against both unknown and unpatchable vulnerabilities. By unifying different hardening options within the same model, we derived a more general method than most existing efforts that rely on a single hardening option. Our automated approach employed a heuristic algorithm that helped to manage the complexity of evaluating the security metric as well as limiting the time for optimization to an acceptable level. We have addressed one limitation of our previous work by considering the uneven distribution of services along an attack path. We have devised a more realistic cost model. We have tested the efficiency and accuracy of the proposed algorithms through simulation results, and we have also discussed how the gain in the d value will be affected by the number of available modifiable firewall rules, unpatchable vulnerabilities, and the different sizes and shapes of the resource graphs.

The following lists several future direction of our approach.

- While this paper has proven that we can integrate different network hardening options (e.g., firewalls and diversity) under the same model, some hardening options may not easily fit into this model (e.g., service relocation).
- The security metric we applied relies on the number of unknown vulnerabilities, which may be refined by further considering known and patchable vulnerabilities (even though those would carry less weight).
- This study relies on a static network configuration. A future research direction would be to consider a dynamic network model in which both attackers and defenders may cause incremental changes in the network.
- We note that, although we assume that the costs are linearly additive, there could be cases where the exact costs may depend on the actual combination of controls (which would make the problem significantly more complex). We believe this could be explored in a future work.
- We will evaluate other optimization algorithms in addition to GA to find the most efficient solution for our problem.

Acknowledgements. The authors thank the anonymous reviewers for their valuable comments. Authors with Concordia University were partially supported by the Natural Sciences and Engineering Research Council of Canada under Discovery Grant N01035. Sushil Jajodia was supported in part by the National Science Foundation under grant IIP-1266147; by the Army Research Office under grants W911NF-13-1-0421 and W911NF-13-1-0317; and by the Office of Naval Research under grants N00014-15-1-2007 and N00014-13-1-0703.

References

1. Albanese, M., Jajodia, S., Noel, S.: Time-efficient and cost-effective network hardening using attack graphs. In: 2012 42nd Annual IEEE/IFIP International Conference on Dependable Systems and Networks (DSN), pp. 1–12. IEEE (2012)
2. Ammann, P., Wijesekera, D., Kaushik, S.: Scalable, graph-based network vulnerability analysis. In: Proceedings of the 9th ACM Conference on Computer and Communications Security, pp. 217–224. ACM (2002)
3. Avizienis, A., Chen, L.: On the implementation of n-version programming for software fault tolerance during execution. In: Proceedings of IEEE COMPSAC, vol. 77, pp. 149–155 (1977)
4. Md Azamathulla, H., Wu, F.C., Ab Ghani, A., Narulkar, S.M., Zakaria, N.A., Chang, C.K.: Comparison between genetic algorithm and linear programming approach for real time operation. J. Hydro Environ. Res. **2**(3), 172–181 (2008)
5. Bakshi, K.: CISCO cloud computing-data center strategy, architecture, and solutions. CISCO White Paper (2009). Accessed 13 Oct 2010
6. Borbor, D., Wang, L., Jajodia, S., Singhal, A.: Diversifying network services under cost constraints for better resilience against unknown attacks. In: Ranise, S., Swarup, V. (eds.) DBSec 2016. LNCS, vol. 9766, pp. 295–312. Springer, Cham (2016). doi:10.1007/978-3-319-41483-6_21
7. Cox, B., Evans, D., Filipi, A., Rowanhill, J., Wei, H., Davidson, J., Knight, J., Nguyen-Tuong, A., Hiser, J.: N-variant systems: a secretless framework for security through diversity. In: USENIX Security, vol. 6, pp. 105–120 (2006)
8. Deb, K.: An efficient constraint handling method for genetic algorithms. Comput. Methods Appl. Mech. Eng. **186**(2), 311–338 (2000)
9. Dewri, R., Poolsappasit, N., Ray, I., Whitley, D.: Optimal security hardening using multi-objective optimization on attack tree models of networks. In: Proceedings of the 14th ACM Conference on Computer and Communications Security, pp. 204–213. ACM (2007)
10. Dewri, R., Ray, I., Poolsappasit, N., Whitley, D.: Optimal security hardening on attack tree models of networks: a cost-benefit analysis. Int. J. Inf. Secur. **11**(3), 167–188 (2012)
11. Fifield, T., Fleming, D., Gentle, A., Hochstein, L., Proulx, J., Toews, E. and Topjian, J.: OpenStack Operations Guide. O'Reilly Media, Inc. (2014)
12. Gao, D., Reiter, M.K., Song, D.: Behavioral distance measurement using hidden Markov models. In: Zamboni, D., Kruegel, C. (eds.) RAID 2006. LNCS, vol. 4219, pp. 19–40. Springer, Heidelberg (2006). doi:10.1007/11856214_2
13. Garcia, M., Bessani, A., Gashi, I., Neves, N., Obelheiro, R.: OS diversity for intrusion tolerance: myth or reality? In: 2011 IEEE/IFIP 41st International Conference on Dependable Systems and Networks (DSN), pp. 383–394. IEEE (2011)
14. Gupta, M., Rees, J., Chaturvedi, A., Chi, J.: Matching information security vulnerabilities to organizational security profiles: a genetic algorithm approach. Decis. Support Syst. **41**(3), 592–603 (2006)
15. Jajodia, S., Noel, S., O'Berry, B.: Topological analysis of network attack vulnerability. In: Kumar, V., Srivastava, J., Lazarevic, A. (eds.) Managing Cyber Threats: Issues, Approaches and Challenges. Kluwer Academic Publisher, New York (2003)
16. Krebs, B.: How many zero-days hit you today? (2013). http://krebsonsecurity.com/2013/12/how-many-zero-days-hit-you-today/
17. McHugh, J.: Quality of protection: measuring the unmeasurable? In Proceedings of the 2nd ACM workshop on Quality of protection, pp. 1–2. ACM (2006)

18. Mell, P., Scarfone, K., Romanosky, S.: Common vulnerability scoring system. IEEE Secur. Priv. **4**(6), 85–89 (2006)
19. Mieritz, L., Kirwin, B.: Defining gartner total cost of ownership (2005)
20. Apache mina project, October 2016. https://mina.apache.org/mina-project/
21. Poolsappasit, N., Dewri, R., Ray, I.: Dynamic security risk management using Bayesian attack graphs. IEEE Trans. Dependable Secur. Comput. **9**(1), 61–74 (2012)
22. Ray, I., Poolsapassit, N.: Using attack trees to identify malicious attacks from authorized insiders. In: Vimercati, S.C., Syverson, P., Gollmann, D. (eds.) ESORICS 2005. LNCS, vol. 3679, pp. 231–246. Springer, Heidelberg (2005). doi:10.1007/11555827_14
23. Sheyner, O., Haines, J., Jha, S., Lippmann, R., Wing, J.M.: Automated generation and analysis of attack graphs. In: Proceedings of the 2002 IEEE Symposium on Security and privacy, pp. 273–284. IEEE (2002)
24. Wang, L., Albanese, M., Jajodia, S.: Network Hardening: An Automated Approach to Improving Network Security. Springer, Heidelberg (2014)
25. Wang, L., Jajodia, S., Singhal, A., Cheng, P., Noel, S.: k-zero day safety: a network security metric for measuring the risk of unknown vulnerabilities. IEEE Trans. Dependable Secur. Comput. **11**(1), 30–44 (2014)
26. Wang, L., Jajodia, S., Singhal, A., Noel, S.: k-zero day safety: measuring the security risk of networks against unknown attacks. In: Gritzalis, D., Preneel, B., Theoharidou, M. (eds.) ESORICS 2010. LNCS, vol. 6345, pp. 573–587. Springer, Heidelberg (2010). doi:10.1007/978-3-642-15497-3_35
27. Wang, L., Noel, S., Jajodia, S.: Minimum-cost network hardening using attack graphs. Compu. Commun. **29**(18), 3812–3824 (2006)
28. Wang, L., Singhal, A., Jajodia, S.: Measuring the overall security of network configurations using attack graphs. In: Barker, S., Ahn, G.-J. (eds.) DBSec 2007. LNCS, vol. 4602, pp. 98–112. Springer, Heidelberg (2007). doi:10.1007/978-3-540-73538-0_9
29. Wang, L., Zhang, M., Jajodia, S., Singhal, A., Albanese, M.: Modeling network diversity for evaluating the robustness of networks against zero-day attacks. In: Kutyłowski, M., Vaidya, J. (eds.) ESORICS 2014. LNCS, vol. 8713, pp. 494–511. Springer, Cham (2014). doi:10.1007/978-3-319-11212-1_28
30. Wang, S., Zhang, Z., Kadobayashi, Y.: Exploring attack graph for cost-benefit security hardening: a probabilistic approach. Comput. Secur. **32**, 158–169 (2013)
31. Yigit, B., Gur, G., Alagoz, F.: Cost-aware network hardening with limited budget using compact attack graphs. In: 2014 IEEE Military Communications Conference (MILCOM), pp. 152–157. IEEE (2014)
32. Zhang, M., Wang, L., Jajodia, S., Singhal, A., Albanese, M.: Network diversity: a security metric for evaluating the resilience of networks against zero-day attacks. IEEE Trans. Inf. Forensics Secur. (TIFS) **11**(5), 1071–1086 (2016)

A Distributed Mechanism to Protect Against DDoS Attacks

Negar Mosharraf[1]([⊠]), Anura P. Jayasumana[2], and Indrakshi Ray[3]

[1] Forcepoint Security Labs, Forcepoint LLC, San Diego, CA, USA
nmosharraf@forcepoint.com
[2] Department of Electrical and Computer Engineering,
Colorado State University, Fort Collins, CO, USA
anura.jayasumana@colostate.edu
[3] Department of Computer Science, Colorado State University,
Fort Collins, CO, USA
indrakshi.ray@colostate.edu

Abstract. Distributed Denial of Service (DDoS) attacks remain one of the most serious threats on the Internet. Combating such attacks to protect the victim and network infrastructure requires a distributed real-time defense mechanism. We propose Responsive Point Identification using Hop distance and Attack estimation rate (RPI-HA) that when deployed is able to filter out attack traffic and allow legitimate traffic in the event of an attack. It dynamically activates detection and blocks attack traffic while allowing legitimate traffic, as close to the source nodes as possible so that network resources are not wasted in propagating the attack. RPI-HA identifies the most effective points in the network where the filter can be placed to minimize attack traffic in the network and maximize legitimate traffic for the victim during the attack period. Extensive OPNET© based simulations with a real network topology and CAIDA attack data set shows that the method is able to place all filtering routers within three routers of the attacker nodes and stop 95% of attack traffic while allowing 77% of legitimate traffic to reach victim node.

1 Introduction

Denial of Service (DoS) attacks, that make network service unavailable to legitimate users, have been known since early 1980s. Distributed Denial of Service (DDoS) attacks originate and propagate in a distributed manner making them harder to mitigate. DDoS attacks against commercial websites like Yahoo, Ebay and E*Trade have provided evidence of how DDoS attacks block legitimate users and cause financial loss [9]. Moreover, emergency and essential

This work was partially supported by NSF I/UCRC Award Number 1650573 and funding from CableLabs. The views and conclusions contained in this document are those of the authors and should not be automatically interpreted as representing the official policies, either expressed or implied of NSF and CableLabs.

services rely on the network infrastructure, and thus DDoS attacks may have severe consequences, such as loss of life. Consequently, techniques for preventing, detecting, and surviving such attacks [2] are needed. Despite significant research into countermeasures, DDoS attacks still remain a major threat [3]. DDoS attack can appear like a flash crowd, i.e., a large number of legitimate users connecting to a server/site simultaneously [10]. A comprehensive defense mechanism should include preventing, detecting, and responding techniques to counter DDoS attacks since there is no one-size-fit-all solution to the DDoS problems [15]. Prevention mechanisms aim to stop the occurrences of attacks [9,14], while detection mechanisms aim to identify attack traffic [5,7]. The response mechanisms attempt to identify the sources of attack and react to those [6]. Our work belongs to this last category and tries to identify the attack source and prevent the propagation of attack traffic.

Responsive techniques identify the source and mitigate DDoS attacks by filtering or limiting attack packets [1,3,4]. Such schemes comprise two parts: attack detection and packet filtering. The characteristics of attack packets, such as source IP address or marked IP header values [11,14], are often used to detect and identify attack traffic and packet filtering. Note that packet filtering can be applied either close to the attack node [5,7] or close to the victim node [14] where all the attack aggregate. However, applying filtering close to at the victim has two drawbacks. First, the victim may crash while dealing with an overwhelming volume of attack traffic. Second, the high volume of attack traffic may still overwhelm upstream Internet resources. At these traffic intensities, the infrastructure upstream from the intended victim becomes severely affected necessitating attack traffic be filtered as close as possible to the attack sources. However, it is difficult to anticipate and identify such nodes as the attack may originate at widely distributed nodes and spread through various routes [15]. Our approach aims to solve this problem.

This paper proposes a novel distributed DDoS defense mechanism for achieving Responsive Point Identification algorithm using Hop distance and Attack estimation rate (RPI-HA), which does not consume any router resources in the process of identifying routers for upstream filtering. The approach tries to minimize the modifications required to the routers and the current protocols to combat DDoS attacks and such modifications have a low complexity and are scalable. The mechanism aims to maximize the arrival rate for legitimate traffic and minimize the attack flow during the attack. The approach consists of four parts. The first part develops rules to create a history-based profile of high confidence legitimate IP addresses that serve to differentiate the good traffic from the malicious [8]. The second part represents the IP address history in the form of a Bloom filter for efficient transfer. The third and fourth parts, the main contribution of this paper, identify how and where this history is used to prevent the attacks. Placing filters in upstream routers incur storage and performance costs since the filter must be applied to multiple routers. Placing the filter closer to the victim causes the link capacity to become saturated and wastes network resources. Our scheme introduces an algorithm that identifies the routers where the filters can be placed. To the best of our knowledge, this is the first work that considers the optimal placement of the filters to mitigate DDoS attacks. Section 2 describes our responsive defense mechanism. Section 3 presents our simulation results. Section 4 concludes the paper.

2   Distributed Responsive Defense Approach

This section presents our scheme to identify upstream routers, and block the DDoS attacks at these routers to minimize the impact both to the victim and to the upstream network during the attack time. The DDoS mitigation mechanism consists of the following components: (1) identification model to discriminate attack traffic from legitimate traffic based on a history-based profile, (2) capture the history-based profile in the form of a Bloom filter for efficient transfer, (3) identify the responsive points (router/switch) which carry the attack traffic, and (4) activate packet filtering at selected points. We use the mechanisms detailed in [8] for the first two steps. The main contributions of this paper are the last two components, and the resulting overall architecture.

2.1   Identification Model

History based profiles the specific attack features as well as normal traffic characteristics for history based profiles to discriminate between the attack traffic from legitimate traffic are investigated in [8]. A key observation is that the DDOS attacks tend to use randomly spoofed IP addresses [8] and other packet features, such as port number and size of packet are randomly distributed as well. Our experiments [8] with the CAIDA 2007 dataset [12] indicated that such as filtering model can protect the victim node from 95% of attack traffic while allowing 70% of legitimate traffic.

2.2   Bloom Filter Mechanism

The filtering mechanism must be applied at upstream routers which must process all packets targeted towards the victim node. Since the network bandwidth may already be saturated during an attack, transferring the entire history and looking it up in the upstream routers is rather expensive. A Bloom filter is thus proposed for representing the contents of the IP based history [8]. Such a filter helps reduce the communication and computation costs and also the storage requirements at upstream routers that check for malicious traffic. There are three fundamental performance metrics for Bloom filters where the size of the Bloom filter is an adjustable parameter based on the accepted false positive rate as well as number of hash functions.

2.3 Responsive Points' Identification

The third step constitutes the main contribution of this paper and it is how and where to use this filter to minimize the impact of the attack. Proposed solution addresses this problem by using a recently developed technology, typically implemented as Small Formfactor Probes (SFP) using Field Programmable Gate Arrays (FPGAs). Our proposed approach monitors traffic by using SFPs to efficiently identify router/switch which carry high volume of attack traffic and then applies packet filtering at selected routers as the responsive point of defense mechanism. An example of such hardware is JDSU SFProbes and Packet Portal [16]. SFProbes can plug into any SFP compatible elements such as switches/routers in such a way that it taps into the normal fiber without interfering with the traffic flow. It can be programmed over the network using the same fiber to do tasks such as counting the number of packets with certain values in header fields and forward information about link traffic to a remote base station. Our approach uses these probes at a subset of ports in the network to identify the upstream links, and thus nodes, which carry attack traffic. A main advantage of using SFProbes is that they plug into the routers/switches and do not require modifying the router's operation and software to apply our scheme. This feature is used to send the history based profiles to identify the paths with high intensity of attack traffic. Moreover, the portal base station (PBS) has knowledge about SFProbes attached to the routers throughout the network and can collect data from SFProbes and perform the computation needed for our scheme, obviating the need for routers performing such computations.

Upon detection of an attack, the proposed approach starts to protect a victim node as illustrated in Fig. 1. At this point, the victim network sends the Bloom filters that it had created to PBS. The PBS sends the Bloom filter to those SFProbes that are plugged into different routers as shown in Fig. 1. SFProbes start monitoring the intensity of traffic directed toward the victim node.

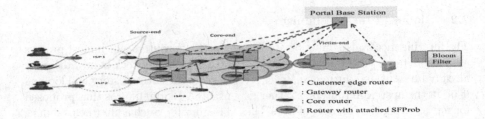

Fig. 1. Responsive defense mechanism

Let $t_1, t_2, ...t_m$ be discrete time slots and $X(t_m, i)$ be the number of packets received by a router during time slot m at SFProbe i destined to the victim node. Equation (1) is defines the historical estimate of the average number of packets received by a router, where α is a weighted value between 0 and 1.

$$\bar{X}(t_m, i) = (1 - \alpha)\bar{X}(t_{m-1}, i) + \alpha X(t_m, i) \tag{1}$$

Let $A_j(t_m, i)$ represent a Boolean variable which equals 1 if the packet P_j is received at router i at time slot t_m and matches the corresponding Bloom filter $B(v)$ for point v, and is 0 otherwise, i.e.,

$$A_j(t_m, i) = \begin{cases} 1 & \text{if} \quad P_j(t_m, i) \in B(v) \\ 0 & \text{otherwise} \end{cases} \tag{2}$$

Let $\bar{W}(t_m, i)$ be the historical estimate of the average number of packets received by the router at SFProbe i during time slot m that match the Bloom filter. Thus, $\bar{W}(t_m, i)$ Eq. (3) shows the average number of packets that is considered as the legitimate traffic by the Bloom filter directed towards the victim where, n is the total number of packets flowing toward the victim node in time slot t_m:

$$\bar{W}(t_m, i) = (1 - \alpha)\bar{W}(t_{m-1}, i) + (\alpha)\sum_{j=1}^{n} \frac{A_j(t_m, i)}{n} \tag{3}$$

During the attack time, if the number of IP addresses that do not match Bloom filter is higher than a specific threshold β, then the router is likely to be carrying significant attack traffic. Such routers are candidates for filter placement. We define Eq. (4) Attack Estimation Rate (ASR) $R(t_m, i)$ to determine the average number of packets that do not match Bloom filter.

$$R(t_m, i) = \frac{\bar{X}(t_m, i) - \bar{W}(t_m, i)}{\bar{X}(t_m, i)} \tag{4}$$

Upon detection of an attack, the SFProbes start monitoring traffic going towards the victim nodes and send $R(t_m, i)$ estimate to PBS. Next we decide the points at which the filters are to be placed. Save network resources, the best routers to apply the filtering mechanism must be as far away as possible from the victim node. The volume of potential attack traffic passing through a router has to be considered as well. Thus, we use two factors to determine the best routers to place the SFProbes - the Attack Estimation Rate of Eq. (4) and the hop distance $H_i(v)$ that shows how far the SFProbe i is from the victim node v. For each SFProbe i computer the weighted attack estimation rate $H_i(v)$ is given by:

$$S(t_m, i) = \frac{\bar{X}(t_m, i) - \bar{W}(t_m, i)}{\bar{X}(t_m, i)} * H_i(v) \tag{5}$$

The routers with higher value of $S(t_m, i)$ are selected routers to apply filtering mechanism. We call this approach as the Responsive Point's Identification algorithm using Hop distance and Attack estimation rate (RPI-HA). In addition to the hop distance and the volume of attack traffic, other issues must also be considered while placing the filters. One such additional factor is the number of routers on which filters are placed based on the distribution of the attack traffic. We present a new formula that adjusts the number of routers according to the attack traffic distribution that we refer to as the Responsive Point's Identification algorithm using Hop distance, Transmission rate and Attack estimation rate (RPI-HTA). In this scheme, the transmission rate of traffic directed towards the victim node as well as hop distance is considered to determine the best filtering points. SFProbes collect the information of attack estimation rate $R(t_m, i)$ and traffic transmission rate $T(t_m, i)$ during m time slots and send this information to the PBS. To select the filtering points the attack estimation rate $D(t_m, i)$ is computed as follows:

$$D(t_m, i) = \frac{\bar{X}(t_m, i) - \bar{W}(t_m, i)}{\bar{X}(t_m, i)} * \frac{\bar{T}(t_m, i)}{C(v)} * H_i(v) \tag{6}$$

$$\bar{D}(t_m, i) = (1 - \alpha)\bar{D}(t_m, i) + (\alpha)\sum_{i=1}^{n} \frac{D(t_m, i)}{n} \tag{7}$$

In Eq. (6) we consider the distribution of traffic towards the victim node as an important factor that helps to determine how the attack has followed, where traffic transmission rate $T(t_m, i)$ and capacity of the victim node $C(v)$ are considered. $D(t_m, i)$ determines if the attack is distributed or centralized. If the attack is highly distributed, we need to consider more filtering points to stop the attack whereas if the attack is more centralized we apply filters on few of the routers. The historical average attack estimation rate is given by $\bar{D}(t_m, i)$ in Eq. (7) where n indicates total number of participating SFProbes that collects the data. We select only those routers $\bar{D}(t_m, i)$ higher than average attack estimation rate as filtering points. Thus, the number of filters in the RPI-HTA depends on traffic transmission rate, hop distance, as well as attack estimation rate and it will be vary according to $\bar{D}(t_m, i)$ and $D(t_m, i)$.

2.4   Packet Filtering

The last step of the proposed approach is activating packet filtering at selected points. According to the previous section, PBS selects those routers which carry the attack traffic for applying Bloom filtering. So PBS sends created Bloom filters to these selected routers and the routers start to filter incoming traffic directed towards the victim node. This process continues during the attack.

3   Evaluation

Performance of our approach is evaluated next using a real network topology from Oregon route-views between March 31 and May 26 2001 [18] and set it up on OPNET©. We test the effectiveness of the responsive defense mechanism using the DARPA 1998 intrusion detection dataset [17] which contains 7 weeks of training datasets that we use to establish an IP address history and 2 weeks of testing dataset to evaluate our techniques. The first step is creating the IP address history from the DARPA training dataset and then create corresponding Bloom filter, the details of which was presented in [8]. The next step, evaluates the responsive defense approach based on the created IP address history in OPNET©. We also have validated our model by real network traffic collected at University of Auckland [13] and CAIDA attack dataset [12] in Sect. 3.5

3.1 Metrics

Responsive defense mechanism is evaluated, (i) *Attack Traffic Detection Rate* (ii) *Normal Traffic Detection Rate*, and (iii) *Link Utilization Rate*. Attack Detection Rate is the percentage of attack dataset that is correctly detected as the attack and cannot pass through the Bloom filter to reach the victim node. Normal Traffic Detection Rate is defined as the percentage of normal traffic that can correctly pass through the Bloom filter during the attack period. False Negative Rate is defined as the percentage of attack traffic that is incorrectly marked as normal traffic and therefore can pass through the Bloom filter. Link Utilization Rate is defined as the percentage of the network's bandwidth that is currently being consumed by the network traffic.

3.2 Percentage of Collaborative SFProbes

The effectiveness of the responsive defense mechanism relies on the collaboration of SFProbes through the network. Increasing the number of SFProbes attached to the routers enables more close monitoring and more effective filtering. We look at four different scenarios to validate this. We assume a different percentage of routers (80% to 25%) have SFProbes attached to them to monitor network traffic to address the case where only a fraction of routers implement the mitigation technique. The number of filtering routers is either fixed according to the RPI-HA or variable based on RPI-HTA algorithm. In the RPI-HA algorithm, the number of filtering routers considered 8, 5 and 3 as 20%, 12.5% and 7.5% of the total routers through the network. We also looked at a fifth scenario where the filters are placed in random locations throughout the network without applying any algorithm to stop the attack traffic. Figure 2 shows the average attack detection rate over 5 runs. We use a time slot of 60 s. The results demonstrate the effectiveness of using the RPI-HTA algorithm, which considers all the three features (hop distance, transformation rate and attack estimation rate) together. The RPI-HTA algorithm produces an attack traffic detection rate of 91% when 80% of routers use SFProbes. Note that, by applying the random selection to determine placement of filters the attack traffic detection rate reduces by more than 14% and up to 23%. These results show how the location of filters plays an important role in protecting the victim node. Recall that the number of filtering points for RPI-HTA algorithm is variable and depends on attack distribution, transmission rate, and hop distance. 7 filters for networks having 80% and 60% probes and 6 filters for those having 40% and 25% probes are selected according to the RPI-HTA algorithm. As shown in Fig. 3 the attack traffic detection rate for RPI-HTA algorithm for 80% and 60% probes is equal or higher than when

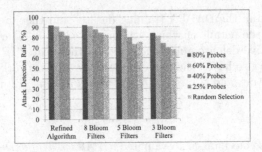

Fig. 2. Attack traffic detection rate

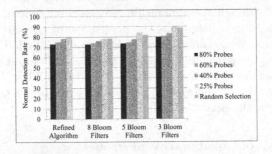

Fig. 3. Normal detection rate

RPI-HA algorithm is applied by 8 filtering routers through the network. This means that the RPI-HTA algorithm can provide comparable or better filtering mechanism by using lower number of filtering routers by placing the filters in the appropriate locations. The other important parameter to evaluate is how many normal packets can reach the victim node. As shown in Fig. 3, normal or legitimate traffic detection rate is around 72% for RPI-HTA algorithm with 80% probes and it increases to around 80% if fewer filters are used to stop attack traffic. Thus, there is a trade off between accuracy of the attack traffic detection rate and the normal detection rate.

3.3 Efficiency of Distributed Approach

Figure 4 depicts the fraction of attack traffic dropped at different hop distances from the victim, and thus the source. It shows that 60% of the attacks in total are detected and blocked in the first two routers from attacker, with 23% in the first and 37% in the second. Thus, it shows that RPI-HTA algorithm can effectively select routers further from the victim node and close to attacker. Furthermore, it shows that having more probes is more effective and we can select farther routers as well. For instance, with 80% probes all the filtering points are selected at least 3 hop distances away from victim node, while with 25% probes the filtering points must be within 1 and 2 hop distances from victim node. In this experiment, the 25*th*, 50*th* (median), 75*th* percentiles, minimum and

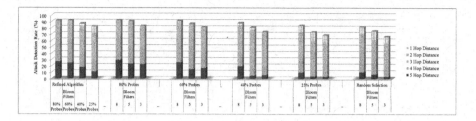

Fig. 4. Attack detection rate and location of selected routes

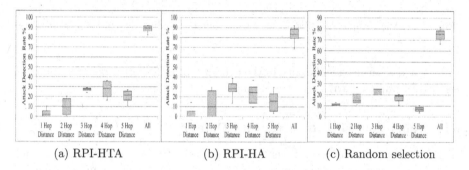

(a) RPI-HTA	(b) RPI-HA	(c) Random selection

Fig. 5. Result of attack detection rate

maximum value of attack traffic detection rate for all 3 scenarios are computed as well. As shown in Fig. 5(a), the first layer of router from attacker (fifth hop distance from victim) has relatively good attack traffic detection rate close to 20% for 50% of simulations in the RPI-HTA algorithm, whereas, this detection rate reduces to less than 10% for random selection scenario. This proves that RPI-HTA algorithm can accurately select desirable filtering points during the attack period. Thus, it can be observed that the core of detection and prevention mechanism is located at the router with hop distance 4, 3, and 5 from victim node in that order.

3.4   End User's Utilization

The other part of our evaluation is specifying utilization of the victim node and other end-users before and after applying filtering mechanism. In Fig. 6(a), the last link utilization of victim node without applying filtering approach shows that it is fully utilized during the attack time. However, the link utilization reduces to around 60% after deploying Bloom filter through the network based on RPI-HTA algorithm. The result shows that 80% probes through the network give 50% link utilization rate for the victim node - this is the least link utilization rate that we get in our experiments. Figure 6(b) shows the last link's utilization for other end-users increases with applying the filtering routers. It means the other end-users can receive normal traffic during the attack time. This is good

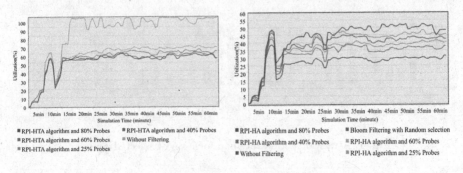

Fig. 6. (a) Victim node's utilization. (b) Average of last link utilization of end-users

as it provides service availability in the presence of DDoS attacks and minimizes the attack impact during an attack time.

3.5 Validation with Real Network Dataset

In this experiment, the effectiveness of responsive defense approach is tested using real network trace from University of Auckland in New Zealand. The packet trace contains 6.5 weeks IP header trace taken with 155 Mbps Internet links [13]. We use The CAIDA attack dataset 2007 [12] in the experiments as attack traffic. The dataset is run with same topology that was used in previous part by distributing the dataset traffic over the network. History-based profile of normal traffic going to the victim node is created using the trace collected from the University of Auckland. The corresponding Bloom filter will be created based on the scheme [8] in the second step and then represented approach is applied during the attack time. Figure 7 shows the attack traffic detection rate against the CAIDA attack traffic. Attack traffic detection rate is around 95% with 80% probes in the RPI-HTA algorithm where it was around 91% for DARPA dataset. Overall the attack traffic detection rate increases slightly compared with DARPA dataset, where the Bloom filter accuracy played a role in this situation. In Fig. 8, the 25th, 50th (median), 75th percentiles, minimum and maximum value of attack traffic detection rate of RPI-HTA algorithm is shown for CAIDA dataset traffic; this can be compared with Fig. 5(a) for DARPA dataset as well. As shown for 75% of simulations, the most portion of attack is detected and blocked at the first layer of routers from the attacker (hop distance 5 from victim node) followed by the one in the second layer of routers from attacker. The router with fifth hop distance from victim has a better attack traffic detection rate close to 25% for 50% of simulations in the RPI-HTA algorithm, whereas this detection rate was 20% for DARPA dataset. Moreover, it shows that router with hop distance 5, 4 and 3 in that order were the core of attack detection. This proves that RPI-HTA algorithm accurately selected desirable filtering points during the attack period for CAIDA attack traffic as well.

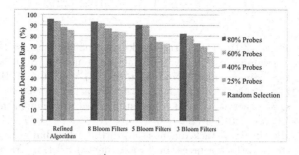

Fig. 7. Attack traffic detection rate

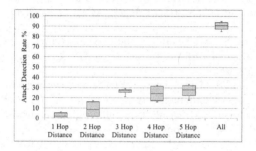

Fig. 8. Attack traffic detection rate of RPI-HTA for CAIDA attack traffic

4 Conclusion

A responsive defense approach to defend against DDoS attacks was presented. A key contribution is the distributed mechanism that identifies in real-time the best response points where filters are to be activated so as to minimize attack traffic and maximize legitimate traffic during the attack. The technique has been validated with two real-world data sets. Results for CAIDA attack set, e.g., indicate that the responsive mechanism protects the victim nodes from 95% of attack traffic close to the source of attack, while allowing 77% of legitimate traffic. The method is light in terms of computational and communication overheads. Results also demonstrate the effectiveness of the mechanism in preserving valuable network resources and link utilizations for other end-users during the attack time, thus preserving the service availability and minimizing the attack impact. Our future work includes validating the scheme with very recent real-world network dataset. A part of our future work also includes extending our identification scheme for IPv6 addresses.

Acknowledgment. The authors gratefully thank Forcepoint LLC for their funding support.

References

1. Aghaei Foroushani, Z.H.: TDFA: traceback-based defense against DDoS flooding attacks. In: Proceedings of 28th International Conference on Advanced Information Networking and Applications (AINA), Victoria, BC, pp. 710–715. IEEE (2014)
2. Cabrera, J.B.D., Lewis, L., Qin, X.Z., et al.: Proactive intrusion detection and distributed denial of service attacks-a case study in security management. J. Netw. Syst. Manag. **10**(2), 225–254 (2002)
3. Chen, C., Park, J.-M.: Attack diagnosis: throttling distributed denial-of-service attacks close to the attack sources. In: Proceedings of IEEE International Conference on Computer Communications and Networks, pp. 275–280 (2005)
4. Francois, J., Aib, I., et al.: FireCol: a collaborative protection network for the detection of flooding DDoS attacks. IEEE/ACM Trans. Netw. **20**(6), 1828–1841 (2012)
5. Gil, T.M., Poletto, T.: MULTOPS: a data-structure for bandwidth attack detection. In: Proceedings of 10th Conference on USENIX Security Symposium, Washington, D.C., USA (2001)
6. John, A., Sivakumar, T.: DDoS: survey of traceback methods. Int. J. Recent Trends Eng. ACEEE (Assoc. Comput. Electron. Electr. Eng.) 1(2) (2009)
7. Mahajan, R., Bellovin, S.M., et al.: Controlling high bandwidth aggregates in the network. ACM SIGCOMM Comput. Commun. Rev. **32**(3), 62–73 (2002)
8. Mosharraf, N., Jayasumana, A.P., Ray, I.: A responsive defense mechanism against DDoS attacks. In: Cuppens, F., Garcia-Alfaro, J., Zincir Heywood, N., Fong, P.W.L. (eds.) FPS 2014. LNCS, vol. 8930, pp. 347–355. Springer, Cham (2015). doi:10.1007/978-3-319-17040-4_23
9. Peng, T., Leckie, C., Ramamohanarao, K.: Proactively detecting distributed denial of service attacks using source IP address monitoring. In: Mitrou, N., Kontovasilis, K., Rouskas, G.N., Iliadis, I., Merakos, L. (eds.) NETWORKING 2004. LNCS, vol. 3042, pp. 771–782. Springer, Heidelberg (2004). doi:10.1007/978-3-540-24693-0_63
10. Peng, T., Leckie, C., Ramamohanarao, K.: Survey of network-based defense mechanisms countering the DoS and DDoS problems. ACM Comput. Surv. **39**(1), 1–42 (2007)
11. RioRey, Inc. 2009-2012: RioRey taxonomy of DDoS attacks, RioReyTaxonomyRev2.32012,2012. http://www.riorey.com/xresources/2012/RioRe
12. The CAIDA DDoS Attack 2007 Dataset. http://www.Caida.org/data/passive/ddos-20070804dataset.xml
13. W.A.N.D.R. Group. http://wand.cs.waikato.ac.nz/wand/wits/auck
14. Yaar, Y., Perrig, A., Song, D.: Pi: a path identification mechanism to defend against DDoS attacks. In: Proceedings of the 2003 IEEE Symposium on Security and Privacy, Pittsburgh, PA (2003)
15. Zargar, S., Joshi, J., Tipper, D.: A survey of defense mechanisms against distributed denial of service (DDoS) flooding attacks. IEEE Commun. Surv. Tutorials **15**(4), 2046–2069 (2013)
16. http://www.jdsu.com/en-us/Test-and-Measurement/Products/a-z-productlist/Pages/packetportal.aspx
17. http://www.ll.mit.edu/mission/communications/cyber/CSTcorpora/ideval/data1998data.html
18. https://snap.stanford.edu/data/oregon1.html,   http://www.darkreading.com/attacks-and-breaches/ddos-attack-hits-400-gbit-s-breaks-record/d/d-id/1113787

Securing Web Applications with Predicate Access Control

Zhaomo Yang[1][✉] and Kirill Levchenko[2]

[1] University of California, San Diego, USA
zhy001@cs.ucsd.edu
[2] University of California, San Diego, USA
klevchenko@cs.ucsd.edu

Abstract. Web application security is an increasingly important concern as we entrust these applications to handle sensitive user data. Security vulnerabilities in these applications are quite common, however, allowing malicious users to steal other application users' data. A more reliable mechanism for enforcing application security policies is needed. Most applications rely on a database to store user data, making it a natural point to introduce additional access controls. Unfortunately, existing database access control mechanisms are too coarse-grained to express an application security policy. In this paper we propose and implement a fine-grained access control mechanism for controlling access to user data. Application access control policy is expressed using row-level access predicates, which allow an application's access control policy to be extended to the database. These predicates are expressed using the SQL syntax familiar to developers, minimizing the developer effort necessary to take advantage of this mechanism. We implement our predicate access control system in the PostgreSQL 9.2 DBMS and evaluate our system by developing an access control policy for the Drupal 7 and Spree Commerce. Our mechanism protected Drupal and Spree against five known security vulnerabilities.

1   Introduction

We depend on Web applications to handle more and more of our private and sensitive data. However, unauthorized data accesses are still very common today. A modern Web application consists of three distinct parts: client-side code running in the browser, server-side code running directly or in an application server, and a database often running on a separate server. Since the client side is completely under the user's control, in a typical application code that checks data accesses will be intermingled with code implementing the server-side functionality, split across multiple components of the application's server-side code. All checking code combined together consists of the application's security policy. Because there is not a centralized policy, developers may forget it when adding a new file or editing an existing file, thus introduce data access vulnerabilities.

© IFIP International Federation for Information Processing 2017
Published by Springer International Publishing AG 2017. All Rights Reserved
G. Livraga and S. Zhu (Eds.): DBSec 2017, LNCS 10359, pp. 541–554, 2017.
DOI: 10.1007/978-3-319-61176-1_30

Most Web applications rely on a database to store and query user data. As the custodian of this data, the database management system (DBMS) seems a natural place to centralize those portions of security mechanism and policy that control access to user data. Unfortunately, the SQL access control mechanism is too coarse, providing only column-level access control. Moreover, a DBMS has no notion of application users, and so cannot protect one application user's data from another's. Such Web applications, therefore, do not use the SQL access control mechanism; instead, the database becomes a convenient data structure which the application uses to manage its data. From the DBMS point of view, there is only one database user, the application, which has complete access to all tables in the database.

In this work, we set out to design the most developer-friendly application user access control mechanism possible. We believe that it is simple, intuitive, and compatible with most applications' user protection boundaries. Our mechanism is implemented in the DBMS, where it can guarantee that the access control policy is enforced *even if the attacker gains direct access to the database*. The application developer specifies the access control policy to be enforced by the DBMS row-level predicates, which are checked on every query. These predicates are expressed using the familiar SQL syntax of WHERE clauses, attached to SQL GRANT statements.

In addition to specifying the desired policy, the security mechanism needs a way to tell the DBMS, at run time, on which application user's behalf the application is currently acting, in order for the DBMS to apply the policy correctly. We do this by allowing the developer to specify an authentication function that provides a means for the DBMS to authenticate a user.

We implemented our system as an extension to the PostgreSQL 9.2 database management system. The application policy, consisting of authentication functions and GRANT–WHERE clauses, is compiled by a separate tool into SQL statements accepted by the modified DBMS. Besides, we created security policies for several modules of the Drupal content management system and the Spree e-commerce platform. We appeal to the reader's own judgement and experience as a programmer and security practitioner to judge whether our means of expressing security policy is more clear and direct than the state of the art. The security policies provide protection against at least five known vulnerabilities in Drupal and Spree server-side code (Table 2).

The rest of the paper is organized as follows. Section 2, next, provides the necessary technical background for the rest of the paper. Section 3 provides a toy application example we use to illustrate our mechanism. Section 4 describes the application interface to the mechanism, that is, how the developer specifies the desired policy. Section 5 describes our implementation. Section 6 evaluates our system using Drupal and Spree. Section 7 describes related work. Section 8 concludes the paper.

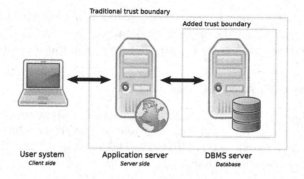

Fig. 1. Structure of a modern Web application.

2   Background

2.1   Modern Web Application Structure

A typical modern Web application consists of three parts, illustrated in Fig. 1: client-side code, server-side code, and a database.

Client-side. The client side of a Web application forms its user interface. It interacts with the server side via HTTP requests. This part of the application is in the difficult position of being trusted neither by user nor by the rest of the application itself.

Server-side. The server side of an application contains the major logic of the application. Each client interacts with a distinct instance of the server, and this interaction between a client and an instance of the server is termed a *session*. Applications may associate a set of access privileges with a session, which delineate what a client is allowed to do. The server side also interacts with a central data store, typically an SQL database. User and application data in the database is subject to access controls, and these are often implemented in the server side of the application as well.

Database. Many Web applications rely on an SQL database to manage application data. Web applications interact with the database by issuing SQL queries, either directly or via a framework (e.g., Ruby on Rails). Each server side instance, corresponding to a session, opens a connection to the database. In performing its function, the application server side may issue queries to the database or modify data in the database. In this sense, the server side *mediates* user access to the database.

Neither the operating system nor the DBMS has any notion of application users, and so can offer the application no help in implementing its access control policy; the server side of the application must implement all necessary access controls. A database has its own notion of users, managed by a database administrator, distinct from OS users and application users. An application connects to the database as *a single database user*. Because of a single database user representing the application, the application database user must have the union of all access permissions necessary to carry out every application function.

2.2 Current SQL Access Controls

Our mechanism extends existing SQL access controls. SQL access controls work at the granularity of columns. A database user may be granted any combination of permissions to issue SELECT, UPDATE, DELETE, and INSERT statements affecting a set of columns. Syntactically, this is accomplished using the GRANT statement. For example, to grant user "Alice" SELECT and UPDATE privileges on table Table 1, the table owner would issue:

```
GRANT SELECT, UPDATE ON Table1 TO Alice;
```

After this statement is executed, the user "Alice" will be able to SELECT (read) and UPDATE (modify) values in table Table 1.

2.3 Threat Model

In this work, we assume that an attacker can co-opt the *server side* into letting him send arbitrary queries to the DBMS. Based on this threat model, the trusted computing base consists of the DBMS server and its operating system, the DBMS itself, and any user-defined functions installed in the application database when the database was first created. This model is assumed in Roichman *et al.* [9] and Oracle's VPD, and is stronger than the model of Nemesis [5], GuardRails [3], Scuta [12] and Diesel [6].

3 Toy Application: Gradebook

To make things concrete, we use a toy example application to illustrate both the traditional application security mechanism and our fine-grained mechanism. The Gradebook application is a simple Web application for students and instructors in a course. It allows students to check their own grades and allows instructors to view all grades, enter new grades and update old grades. Figure 2 shows the tables used by the application. The Gradebook application consists of two tables, Users and Grades, shown in Fig. 2. The Users table stores user names, a password salt, and a password hash. In addition, the instr column stores a boolean flag indicating if the user is an instructor or a student. The Grades table stores student grades for each assignment.

To authenticate a user, the Gradebook application issues the following query, which implements a salted hash:

```
SELECT user_id, instr
  FROM Users
  WHERE user_name = ?
    AND pass_hash = SHA1(pass_salt || ?);
```

Users	
user_id	INTEGER
instr	BOOLEAN
user_name	TEXT
pass_salt	TEXT
pass_hash	TEXT

Grades	
user_id	INTEGER
assignment	TEXT
score	INTEGER

Fig. 2. SQL database tables for the Gradebook toy application example.

Here the "?" stands for the user identifier saved by the application when it authenticated the user on login. (We show all queries as prepared statements, although our access control mechanism does not depend on their use.) To retrieve a student's grade, for example, the server side issues the query:

```
SELECT assignment, grade
  FROM Grades
  WHERE user_id = ?;
```

4   Application Interface

In this section we describe our access control mechanism. Our design was guided by two requirements:

- **Keep it simple.** Expressing a policy should be as simple and intuitive as possible.
- **Assume the worst.** The application should not be trusted; assume it is completely compromised.

First we introduce a straightforward concept called user-defined authentication function, which encapsulates the application's authentication mechanism. Then we extend the familiar SQL GRANT statement with a WHERE clause, as proposed by Chaudhuri *et al.* [4].

4.1   User Authentication Function

The first element of our mechanism is *authentication function*. This function is just like a normal user-defined function that returns table rows, but the results of the last evaluation of this function within a session are remembered in a per-session *authentication table* with the same name as the authentication function.

To illustrate, recall that the Gradebook application authenticates a user by querying a user table with the user name and password supplied by the user. If the user name and password match, the query returns the user_id and a flag indicating if the user is an instructor or not. (This information is used throughout the rest of the session.) To use our mechanism, the developer wraps this query in an authentication function:

```
CREATE AUTHENTICATION FUNCTION Auth(TEXT, TEXT)
RETURNS TABLE(user_id INTEGER, instr BOOLEAN)
AS $$
   SELECT user_id, instr
     FROM Users
     WHERE user_name = $1
       AND pass_hash = SHA1(pass_salt || $2);
$$ LANGUAGE SQL;
```

The authentication function Auth takes two text arguments, the user name and password of the user, represented by "$1" and "$2" in the function body. It returns a table row containing the user_id and instr flag of the row corresponding to the user, or no rows if the user name or password do not match. Note the similarity of the enclosed query to the authentication query given in Sect. 3. To authenticate a user, the application then issues the query:

```
SELECT user_id, instr FROM Auth($,$);
```

The result of the query is the same as in the corresponding query in Sect. 3, however because the query was issued through an authentication function, the result is now saved in an authentication table named Auth. This special table is available for use in access control predicates, as we will describe in the next section.

Note that some applications have more than one way to authenticate users. For example, applications which have a notion of session can authenticate a user using a session token stored in a cookie. To handle other authentication mechanisms, multiple instances of the authentication function taking different parameters can be defined.

4.2 Predicate Access Control

To protect individual rows in a table, the developer specifies a predicate defining which rows an application user may access. Syntactically, the predicate is expressed using a GRANT statement with a USING clause and WHERE clause. The USING clause lists the tables used in the predicate. The WHERE clause then specifies the predicate itself.

For example, in the Gradebook application, to express the policy that a student user may only see his own grades, the application developer would use the statement:

```
GRANT SELECT ON Grades TO Gradebook
USING Auth
WHERE Auth.user_id = Grades.user_id
   OR Auth.instr;
```

This statement grants a database user Gradebook permission to SELECT from the table Grades only those rows of the table where the user_id column is equal to the user_id of the currently authenticated user. The second term of the WHERE clause allows users with the instr flag to access all rows. All SELECT statements executed by the Gradebook database user will only see those rows of the Grades table that satisfy the predicate. The "SELECT ... FROM Grades" query from Sect. 3 is now, in effect:

```
SELECT assignment, grade
  FROM Grades, Auth
  WHERE user_id = ?
    AND (Auth.user_id = Grades.user_id
         OR Auth.instr);
```

The result of inner join above will return only those rows of the **Grades** table allowed by the policy.

The USING clause specifies the authentication table to be used for this access control check. Ordinary tables may also be specified in the USING clause of the GRANT statement, which will result in them being added to the join. DELETE, UPDATE, and INSERT privileges may be granted in a similar way.

4.3   Composition

Grants are a monotonic operation: a GRANT statement can only increase the access privileges of the grantee. Multiple GRANT statements for the same privilege (e.g., SELECT) on the same table will result in the grantee having access to all rows satisfying at least one of the WHERE clauses of the grant. In other words, the WHERE clauses of the two GRANT statements are OR'ed together. Thus, the "**GRANT SELECT**" statement in Sect. 4.2 has the same effect as two separate grants of the SELECT privilege with predicates "**Auth.user_id = Grades.user_id**" and "**Auth.instr**." Traditional (unpredicated) GRANT statements can be mixed with predicated GRANT statements. An unpredicated grant is equivalent to predicated GRANT with WHERE clause **WHERE TRUE**.

4.4   Revocation, Ownership and De-authentication

The REVOKE statement, which revokes access privileges from a user, is unconditional. That is, the result of executing a revoke statement is that the named user will not have the specified access privilege to the named tables. Predicates are not preserved when privileges are revoked. Granting the revoked privileges to the same user again will not restore the predicate in place before the REVOKE.

Access privileges on a table or view can only be granted by its owner. If the application database user (**Gradebook** in the examples above) is to be treated as untrusted, then the protected tables must have a different owner. Otherwise, an attacker connected to the database as the application database user can restore to himself all privileges. All application tables and views should be owned by a user other than the database user used by the application to service application user requests.

To de-authenticate, the application can re-authenticate as another user or call the authentication function in a way that is guaranteed to return no rows. For the **Auth** function used in the Gradebook application, calling it with either user name or password NULL will result in no rows being returned. This clears the authentication table associated with the authentication function.

5   Implementation

We implemented the predicate access access control mechanism described above by extending the PostgreSQL DBMS. Our implementation consists of two parts: a minimally-modified PostgreSQL 9.2 DBMS and a separate tool to compile security policies into lower-level constructs available in PostgreSQL.

5.1   Architecture

We chose to prototype most of the mechanism as a separate policy compiler, rather than in the DBMS itself, for convenience. In the intended ultimate implementation, the GRANT–WHERE predicate access control mechanism would be part of the DBMS.

In our current implementation, all schema declarations, including CREATE TABLE, CREATE VIEW, and CREATE FUNCTION statements are sent directly to the database. All security policy statements, namely GRANT (with and without a WHERE clause), REVOKE, and CREATE AUTHENTICATION FUNCTION, are sent to our tool for compilation into rewriting rules, triggers, and function definitions understood by PostgreSQL. The compiled statements are then sent to the database.

Because the compiler composes predicates as described in Sect. 4.3 and then packages them to be installed at the start of each session, the security policy must be presented all at once to the compiler.

5.2   PostgreSQL

The PostgreSQL DMBS supports query rewriting via its rule system, and we use this facility to implement access control predicates. To avoid large, unwieldy rules covering every database user, we modified PostgreSQL to support session-local rules, allowing the appropriate set of rules to be installed for each database user at the start of their session.[1] We call such session-local rules *temporary rules*, by analogy to temporary tables which exist for the duration of a session and are visible only inside the session in which they are created.

Temporary rules are defined using the CREATE TEMPORARY RULE statement, which, except for the addition of the TEMPORARY keyword, is syntactically identical to an ordinary CREATE RULE statement. A temporary rule is visible and effective only inside the session in which it is created.

Temporary rules differ from ordinary rules in one other way. PostgreSQL applies ordinary rewriting rules recursively to each table in a query until no further rules apply. This means, in particular, that rewriting rules cannot be recursive (or the rewriting step would not terminate). However, temporary rules derived from access control predicates will result in rewritten queries referring

---

[1] Temporary rules are also necessary because the current app. user's information is cached in a temporary table and the rules which referring to this table must also be temporary. Thus, modifying PostgreSQL cannot be avoided completely.

to the same table. For this reason, temporary rules are applied once and before permanent rules are applied. This allows temporary rules to rewrite queries using the same table.

5.3 On CONNECT Trigger

Predicate access controls are enforced using temporary rules specific to the database user of each session. These rules must be put into place at the start of a session, before the user issues any queries. To implement this, we added a special kind of trigger function to PostgreSQL which is executed when a user connects. This trigger is specified using a CREATE ON CONNECT TRIGGER statement. To prevent a database administrator from accidentally locking herself out of the database, ON CONNECT triggers are disabled for the database superuser. In our implementation, only the database superuser can create an ON CONNECT trigger.

5.4 Policy Compiler

We implemented the security mechanism described in Sect. 4 as a standalone security policy compiler that compiles GRANT, REVOKE, and CREATE AUTHENTICATION FUNCTION statements into temporary rewriting rules that are put into place at the start of a session by an ON CONNECT trigger.

6 Evaluation

To evaluate our predicate access control mechanism, we developed security policies for two modules of Drupal 7.1, a popular open-source content management system written in PHP, and Spree 1.3.1, a popular open-source eCommerce application using Ruby on Rails. Policies are based on what we inferred to be the intended access control policies of these applications. A set of general security policies for these two different popular applications allowed us to exercise all parts of our prototype and evaluate its performance on two complex applications with large user bases. Due to the space limits, the security policies are not shown in the paper.

6.1 Expressiveness

The ability to express an application's intended access control policy is our main criterion for evaluation. Moreover, our access control mechanism allows a policy to be expressed in a declarative manner and in one place. Both Drupal and Spree expressed access control policy in their implementation programming language (Drupal in PHP and Spree in Ruby) at the point in the code program where the access control check took place. The resulting policy and mechanism were spread across multiple locations in code.

Drupal. Drupal uses role-based access control, storing the assigned permissions in the database. Both login and session authentication functions cache the current application user's permissions and user ID. Drupal manages sessions by storing session-to-user mapping in the database. We protected 14 key tables using 51 GRANT statements, each of which expresses a separate access condition.[2] Drupal's existing access control checks occur at 15 separate locations in the source code.

Spree. Spree access control is also role based. Our authentication functions cache the current application user's roles along with his user ID. By default, Spree manages session information using the Ruby on Rails `CookieStore` mechanism, which stores session information in a cryptographically-signed cookie. While it would be possible to implement cookie signature checking in a database user-defined function, we configured Spree to use the `ActiveRecordStore` mechanism, which stores session information on the server in the same way as Drupal. We protected 53 tables using 29 GRANT statements, each of which expresses a separate access control condition. Spree's existing access control checks are spread across 11 separate locations in the source code.

6.2   Security

Our security policies mitigated three known Drupal vulnerabilities and two known Spree vulnerabilities (see Table 2). By their nature, GRANT-based polices define what data a user is allowed to access rather than defining disallowed behavior; such polices are much more likely to be effective against future vulnerabilities than policies covering specific vulnerabilities.

6.3   Performance

To assess the overhead imposed by our prototype, we measured the performance of Drupal and Spree using Apache JMeter, a popular server performance testing tool. All the experiments use a Intel Core i7-3632QM 2.20 GHz laptop with 4 GB of RAM and all of the client side, the server side and database were running on the same machine. Table 1 shows the time to perform each benchmarked task with and without our security policy. We report the mean of 50 runs.

Drupal. For the Drupal tests, we populated the database with 2,000 users and 5,000 nodes (articles). Each view article task results in 114 queries issued to the database. Each edit article task generates in 257 queries and data modification statements.

Spree. For the Spree tests, we populated the database with 50 users, 100 items and 200 orders. The benchmark task simulates the entire process of a user adding an item to the cart and checking out. The task results in 2,987 queries and data modification statements issued to the database.

---

[2] Alternatively, we could combine several access control into one statement in a disjunction.

Table 1. Drupal and Spree performance with and without security policies (*Before* and *After* columns, respectively).

Task	Before	After
Drupal view article	0.44 s	0.54 s
Drupal edit article	1.29 s	1.61 s
Spree buy item	18.40 s	24.12 s

Spree uses persistent connections more aggressively than Drupal, so Drupal creates database connections more frequently, which requires installing all the rules at the start of each session in the ON CONNECT trigger, which icnreases the performance penalty. Spree also issues more queries against tables that do not require row-level protection (e.g. `spree_countries`, which stores country information). On the other hand, Spree policies are more detailed, incurring a higher overhead. For example, the access control predicate for INSERT operations on the `spree_orders` table checks if the new row's total cost, order state and payment state are valid. In other words, our security policy encoded additional application constraints beyond plain access control.[3]

Although the performance overhead is not negligible, the absolute increase in query times is likely acceptable in many applications. Furthermore, these performance measurements are for an unoptimized prototype; an implementation integrated into the DBMS would likely perform better.

7   Related Work

Fine-grained and predicate access control has a long history, and it is not our intention to reinvent it. The aim of this work is to synthesize the most developer-friendly access control mechanism from the vast body of existing work, without sacrificing security guarantees.

7.1   Database Mechanisms

Rizvi *et al.* [8] present a predicate access control model using *authorization views*. Queries issued to the database must satisfy using authorization views granted to a user. Authorization views are defined using special authorization parameters, however how such parameters are communicated to the database is considered out of scope. Roichman *et al.* [9] also describe a similar solution using views parametrized by an authentication token. The application needs to be modified to store this token and transmit it back with each query.

---

[3] In effect, we are using the predicate access control to implement sophisticated foreign key and value constraints. A lower overhead, pure access control policy is also possible.

Table 2. Known vulnerabilities in Drupal 7.1 and Spree 1.3.1 which are mitigated by our policies.

Appl.	Vulnerability	Description
Drupal	CVE-2012-1590	User permissions are not checked properly for unpublished forum nodes, which allows remote authenticated users to obtain sensitive information such as the post title via the forum overview page
Drupal	CVE-2012-2153	Tag `node_access` is not added to queries thus queries are not rewritten properly and restrictions of content access module are ignored
Drupal	CVE-2011-2687	Proper tables are not joined when node access queries are rewritten thus access restrictions of content access module are ignored
Spree	*(No CVE ID)*[a]	By passing a crafted string to the API as the API token, a user may authenticate as a random user, potentially an administrator
Spree	CVE-2013-2506	Mass assignment is not performed safely, which allows remote authenticated users to assign any roles to themselves

[a] https://spreecommerce.com/blog/exploits-found-within-core-and-api

The Oracle DBMS introduced the Virtual Private Database (VPD) feature in release 8.1.5 to provide fine-grained access control [2]. Compared to our approach, the VPD mechanism is considerably more complex to use. The developer must supply a database function that returns a string containing a dynamically-generated WHERE clause fragment. User authentication information must be communicated via application contexts (key-value stores); the authentication function must explicitly store values in the context to be available by the dynamically-generated queries.

Row-level security introduced to PostgreSQL since version 9.5 allows the DB administrator to restrict which rows can be accessed or modified on a per-database user basis [1]. Although it provides a finer-grained access control compared to the standard privilege system, the DBMS is still not aware of application users, thus the per-application user access control cannot be achieved.

The GRANT-WHERE syntax we use was previously proposed by Chaudhuri et al. [4]. Like Rizvi et al. [8], they do not address the problem of authentication, assuming the availability of a function called `userId` conveyed securely by the database driver (e.g. via ODBC, JDBC, etc.), which supplies a user identifier. We are inspired by their clean and familiar syntax for expressing access control policy, which we pair with an intuitive way to authenticate a user to the application. There is no implementation of their system.

7.2   Non-DBMS Mechanisms

Several proposals from the computer security community solve the problem by mediating the communication between application and database. The Nemesis [5] system relies on taint tracking in a specialized PHP interpreter to automatically infer when a user has authenticated to a database and then applies appropriate access controls. Taint tracking is also used by GuardRails, a source-to-source translator for Ruby on Rails programs, which together with developer annotations allows a security policy to be applied to program objects [3]. Both Nemesis and GuardRails cannot protect data in the database if the application is tricked into sending arbitrary queries to DB. The CLAMP [7] system solves the authentication and access control problem by extracting from the application the authentication mechanism and access control logic to create a "User Authenticator"(UA) and a "Query Restrictor"(QR). These two components are isolated from the application itself, and so, should the attacker gain control of the application, would not be corrupted. Building a UA and a QR consisting of authentication logic and data access logic respectively, however, may be too complex for developers.

The fine-grained access control we propose can be applied at the level of users, the number of which is not known *a priori*. Two systems, Scuta [12] and Diesel [6] provide protection at the level of application modules or components. This provides isolation between untrusted modules of an application. Scuta works by placing components into different access control rings, which are mapped to multiple database users with the necessary table-level access controls. Diesel, on the other hand, intercepts and rewrites queries sent to the database. The level of access is indicated to the rewriting proxy at the start of a session, in effect allowing an application to drop privileges before issuing queries on behalf of an untrusted module. Besides, Son *et al.* [10,11] attempt to find and repair vulnerabilities in applications using static analysis. This is a very promising approach, but is likely still too complex for developers to use.

8   Conclusion

In this paper we described a row-level access control mechanism implemented in the database aimed at Web applications. Our mechanism allows the application developer to express her application's security policy using familiar SQL syntax.

We implemented our system as an extension to PostgreSQL and developed a security policy for two core Drupal modules and the Spree e-commerce platform. Our prototype has acceptable performance overhead, yet provides protection against five known and future unknown vulnerabilities. Moreover, our mechanism allows the security policy for these systems to be expressed in a centralized manner.

References

1. PostgreSQL's row-level security. https://www.postgresql.org/docs/9.5/static/ddl-rowsecurity.html
2. Virtual Private Database. http://www.oracle.com/technetwork/database/security/index-088277.html
3. Burket, J., Mutchler, P., Weaver, M., Zaveri, M., Evans, D.: GuardRails: a data-centric web application security framework. In: Proceedings of the 2nd USENIX Conference on Web Application Development (2011)
4. Chaudhuri, S., Dutta, T., Sudarashan, S.: Fine grained authorization through predicated grants. In: Proceedings of the 23rd IEEE International Conference on Data Engineering (2007)
5. Dalton, M., Kozyrakis, C., Zeldovich, N.: Nemesis: preventing authentication and access control vulnerabilities in web applications. In: Proceedings of the 18th USENIX Security Symposium (2009)
6. Felt, A.P., Finifter, M., Weinberger, J., Wagner, D. : Diesel: applying privilege separation to database access. In: Proceedings of 6th ACM Symposium on Information, Computer and Communication Security (2011)
7. Parno, B., McCune, J., Wendlandt, D., Andersen, D., Perrig, A.: CLAMP: practical prevention of large-scale data leaks. In: Proceedings of the 30th IEEE Symposium on Security and Privacy (2009)
8. Rizvi, S., Mendelzon, A., Sudarshan, S., Roy, P.: Extending query rewriting techniques for fine-grained access control. In: Proceedings of the 2004 ACM SIGMOD international conference on Management of Data (2004)
9. Roichman, A., Gudes, E.: Fine-grained access control to web databases. In: Proceedings of the 12th ACM Symposium on Access Control Models and Technologies (2007)
10. Son, S., McKinley, K.S., Shmatikov, V.: RoleCast: finding missing security checks when you do not know what checks are. In: Proceedings of the 2011 ACM Conference on Object Oriented Programing Systems Languages and Applications (2011)
11. Son, S., McKinley, K.S., Shmatikov, V., Up, F.M.: Repairing Access-Control Bugs in Web Applications. In Proceedings of the 10th USENIX Symposium on Networked Systems Design and Implementation (2013)
12. Tan, X., Du, W., Luo, T., Soundararaj, K.D.: SCUTA: a server-side access control system for web applications. In: Proceedings of the 17th ACM Symposium on Access Control Models and Technologies (2012)

Author Index

Printed in the United States
By Bookmasters